NONLINEAR CONTROL
OF
DYNAMIC NETWORKS

AUTOMATION AND CONTROL ENGINEERING
A Series of Reference Books and Textbooks

Series Editors

FRANK L. LEWIS, Ph.D.,
Fellow IEEE, Fellow IFAC
Professor
The Univeristy of Texas Research Institute
The University of Texas at Arlington

SHUZHI SAM GE, Ph.D.,
Fellow IEEE
Professor
Interactive Digital Media Institute
The National University of Singapore

PUBLISHED TITLES

Nonlinear Control of Dynamic Networks,
Tengfei Liu; Zhong-Ping Jiang; David J. Hill

Modeling and Control for Micro/Nano Devices and Systems,
Ning Xi; Mingjun Zhang; Guangyong Li

Linear Control System Analysis and Design with MATLAB®, Sixth Edition,
Constantine H. Houpis; Stuart N. Sheldon

Real-Time Rendering: Computer Graphics with Control Engineering,
Gabriyel Wong; Jianliang Wang

Anti-Disturbance Control for Systems with Multiple Disturbances,
Lei Guo; Songyin Cao

Tensor Product Model Transformation in Polytopic Model-Based Control,
Péter Baranyi; Yeung Yam; Péter Várlaki

Fundamentals in Modeling and Control of Mobile Manipulators, *Zhijun Li; Shuzhi Sam Ge*

Optimal and Robust Scheduling for Networked Control Systems, *Stefano Longo; Tingli Su; Guido Herrmann; Phil Barber*

Advances in Missile Guidance, Control, and Estimation, *S.N. Balakrishna; Antonios Tsourdos; B.A. White*

End to End Adaptive Congestion Control in TCP/IP Networks,
Christos N. Houmkozlis; George A Rovithakis

Robot Manipulator Control: Theory and Practice, *Frank L. Lewis; Darren M Dawson; Chaouki T. Abdallah*

Quantitative Process Control Theory, *Weidong Zhang*

Classical Feedback Control: With MATLAB® and Simulink®, Second Edition,
Boris Lurie; Paul Enright

Intelligent Diagnosis and Prognosis of Industrial Networked Systems,
Chee Khiang Pang; Frank L. Lewis; Tong Heng Lee; Zhao Yang Dong

Synchronization and Control of Multiagent Systems, *Dong Sun*

Subspace Learning of Neural Networks, *Jian Cheng; Zhang Yi; Jiliu Zhou*

Reliable Control and Filtering of Linear Systems with Adaptive Mechanisms,
Guang-Hong Yang; Dan Ye

Reinforcement Learning and Dynamic Programming Using Function Approximators, *Lucian Busoniu; Robert Babuska; Bart De Schutter; Damien Ernst*

Modeling and Control of Vibration in Mechanical Systems, *Chunling Du;*
Lihua Xie

Analysis and Synthesis of Fuzzy Control Systems: A Model-Based Approach,
Gang Feng

Lyapunov-Based Control of Robotic Systems, *Aman Behal; Warren Dixon;*
Darren M. Dawson; Bin Xian

System Modeling and Control with Resource-Oriented Petri Nets,
MengChu Zhou; Naiqi Wu

Sliding Mode Control in Electro-Mechanical Systems, Second Edition,
Vadim Utkin; Juergen Guldner; Jingxin Shi

Autonomous Mobile Robots: Sensing, Control, Decision Making and
Applications, *Shuzhi Sam Ge; Frank L. Lewis*

Linear Control Theory: Structure, Robustness, and Optimization,
Shankar P. Bhattacharyya; Aniruddha Datta; Lee H.Keel

Optimal Control: Weakly Coupled Systems and Applications, *Zoran Gajic*

Deterministic Learning Theory for Identification, Recognition, and Control,
Cong Wang; David J. Hill

Intelligent Systems: Modeling, Optimization, and Control, *Yung C. Shin;*
Myo-Taeg Lim; Dobrila Skataric; Wu-Chung Su; Vojislav Kecman

FORTHCOMING TITLES

Modeling and Control Dynamic Sensor Network,
Silvia Ferrari; Rafael Fierro; Thomas A. Wettergren

Cooperative Control of Multi-agent Systems: A Consensus Region Approach,
Zhongkui Li; Zhisheng Duan

Optimal Networked Control Systems, *Jagannathan Sarangapani; Hao Xu*

NONLINEAR CONTROL
OF
DYNAMIC NETWORKS

Tengfei Liu
Polytechnic Institute of New York University
Department of Electrical & Computer Engineering

Zhong-Ping Jiang
Polytechnic Institute of New York University
Department of Electrical & Computer Engineering

David J. Hill
University of Hong Kong
Department of Electrical & Electronic Engineering

CRC Press
Taylor & Francis Group
Boca Raton London New York

CRC Press is an imprint of the
Taylor & Francis Group, an **informa** business

CRC Press
Taylor & Francis Group
6000 Broken Sound Parkway NW, Suite 300
Boca Raton, FL 33487-2742

First issued in paperback 2017

© 2014 by Taylor & Francis Group, LLC
CRC Press is an imprint of Taylor & Francis Group, an Informa business

No claim to original U.S. Government works

Version Date: 20140224

ISBN 13: 978-1-138-07661-7 (pbk)
ISBN 13: 978-1-4665-8459-4 (hbk)

Visit the Taylor & Francis Web site at
http://www.taylorandfrancis.com

and the CRC Press Web site at
http://www.crcpress.com

Dedication

This work is dedicated to
Lina and Debbie (TFL)
Xiaoming, Jenny, and Jack (ZPJ)
Gloria (DJH)

Contents

List of Figures

Preface

The rapid development of computing, communications, and sensing technologies has been enabling new potential applications of advanced control of complex systems like smart power grids, biological processes, distributed computing networks, transportation systems, and robotic networks. Significant problems are to integrally deal with the fundamental system characteristics such as nonlinearity, dimensionality, uncertainty, and information constraints, and diverse kinds of networked behaviors, which may arise from quantization, data sampling, and impulsive events.

Physical systems are inherently nonlinear and interconnected in nature. Significant progress has been made on nonlinear control systems in the past three decades. However, new system analysis and design tools that are capable of addressing more communication and networking issues are still highly desired to handle the emerging theoretical challenges underlying the new engineering problems. As an example, small quantization errors may cause the performance of a "well-designed" nonlinear control system to deteriorate. The need for new tools motivates this book, the purpose of which is to present a set of novel analysis and design tools to address the newly arising theoretical problems from the viewpoint of dynamic networks. The results are intended to help solve real-world nonlinear control problems, including quantized control and distributed control aspects.

In this book, dynamic networks are regarded as systems composed of structurally interconnected subsystems. Such systems often display complex dynamic behaviors. The control problem of such a complex system could be simplified with the notion of a dynamic network if the subsystems have some common characteristic which, together with the structural feature of the dynamic network, can guarantee the achievement of the control objective. For the research in this book, one such characteristic is Sontag's input-to-state stability (ISS), based on which, refined small-gain theorems are extremely useful in solving control problems of complex systems by taking advantage of the structural feature.

By bridging the gap between the stability concepts defined in the input–output and the state–space contexts, the notion of ISS has proved to be extremely useful in analysis and control design of nonlinear systems with the influence of external inputs represented by nonlinear gains. Its essential relationship with robust stability provides an effective approach to robust control by means of input-to-state stabilization. For a dynamic network of ISS subsystems, the small-gain theorem is capable of testing the overall ISS by directly checking compositions of the ISS gains of the subsystems. Based on the ISS small-gain theorem, complex systems can be input-to-state stabilized by appropriately designing the subsystems.

This book is based on the authors' recent research results on nonlinear control of dynamic networks. In particular, it contains refined small-gain results for dynamic networks and their applications in solving the control problems of nonlinear uncertain systems subject to sensor noise, quantization error, and information exchange constraints. The widely known Lyapunov functions approach is mainly used for proofs and discussions. The relationship between the new tools and the existing nonlinear control methods is highlighted. In this way, not only control researchers but also students interested in related topics may understand and use the tools for control designs.

The organization of the book is as follows. To make the book self-contained, Chapter 1 provides some prerequisite knowledge on useful characteristics of Lyapunov stability and ISS. Chapter 2 presents ISS small-gain results for interconnected systems composed of two subsystems. Both trajectory-based and Lyapunov-based formulations of the ISS small-gain theorem are reviewed with proofs. For dynamic networks that may contain more than two subsystems, Chapter 3 introduces more readily usable cyclic-small-gain methods to reduce the complexity of analysis and control design problems for more general dynamic networks. Detailed proofs of some of the background theorems in Chapters 1–3, which need a higher level of mathematical sophistication and are available in the literature, are not provided. However, the basic ideas are highlighted.

The applications of the cyclic-small-gain theorem to nonlinear control designs are studied in Chapters 4–6. Specifically, Chapter 4 investigates the important measurement feedback control problem for uncertain nonlinear systems with disturbed measurements. In Chapter 5, the quantized nonlinear control problem is studied. Chapter 6 discusses the distributed control problem for coordination of groups of nonlinear systems under information exchange constraints. The control problems are transformed into input-to-state stabilization problems of dynamic networks, and the influence of the uncertain sources, i.e., sensor noise, quantization, and information exchange constraints, are explicitly evaluated and attenuated by new cyclic-small-gain designs.

Certainly, most of the results presented in this book can be extended for more general systems. Some of the easier extensions mentioned in the book are not thoroughly discussed and may be used as exercises for interested readers. Several future challenges in this research direction are outlined in Chapter 7. The Appendix gives supplementary materials on graph theory and discontinuous systems, and the proofs of the technical lemmas which seem too mathematical to be placed in the main chapters. Finally, historical discussion will be confined to brief notes at appropriate points in the text.

TFL wishes to express his sincere gratitude to his coauthors, Professor Zhong-Ping Jiang and Professor David Hill, who are also TFL's postdoctoral adviser and PhD supervisor, respectively. They introduced TFL to nonlinear control, drew his attention to ISS and small-gain, and offered him precious opportunities for working on the frontier research subjects in the field. Their persistent support, expert guidance, and willingness to share wisdom have been invaluable for TFL's academic career. TFL would also like to give thanks to his previous and current labmates in Canberra and New York. He has benefited a lot from the discussions/debates with them. TFL would need more than a lifetime to thank his wife, Lina Zhang, for her understanding and patience and their daughter, Debbie Liu, for a lot of happiness.

ZPJ would like to thank, from the bottom of his heart, all his coauthors and friends for sharing their passion for nonlinear control. The ideas and methods presented in this book truly reflect their wisdom and vision for the nonlinear control of dynamic networks. Special thanks go to Iven Mareels, Laurent Praly (his former PhD adviser), Andy Teel, and Yuan Wang for collaborations on the very first, nonlinear ISS small-gain theorems, and to Hiroshi Ito, Iasson Karafyllis, Pierdomenico Pepe, and again Yuan Wang for recent joint work on various extensions of the small-gain theorem for dynamic networks. Finally, it is only under the strong and constant support and love of his family that ZPJ can discover the beauty of nonlinear feedback and control, while having fun doing research.

DJH firstly thanks his coauthors for their hard work and collaboration throughout the research leading to this book. It has been a pleasure to see the ideas of state-space-based small-gain theorems progress through all our PhD theses, as well as work with colleagues—with special mention of Iven Mareels, and now into this book. (As a memorial note, his thesis was over 30 years ago, following the seminal work on dissipative systems by Jan Willems, who sadly passed away during our writing.) Personally, DJH would like to thank his wife, Gloria Sunnie Wright, whose positive supportive approach and excitement for life are a perfect match for an academic who (as she often hears) has "got to run" for deadlines.

The authors are grateful to the series editors, Frank Lewis and Sam Ge, for the opportunity to publish the book. The authors would also like to thank the editorial staff, in particular, Nora Konopka, Michele Smith, Amber Donley, Michael Davidson, John Gandour, and Shashi Kumar, of Taylor & Francis for their efforts in publishing the book.

The research presented in this book was supported partly by the NYU-Poly Faculty Fellowship provided to the first author during his visit at the Polytechnic Institute of New York University, partly by the U.S. National Science Foundation and by the Australian Research Council.

Tengfei Liu   New York, USA
Zhong-Ping Jiang   New York, USA
David J. Hill   Hong Kong, China

Author Biographies

Dr. Tengfei Liu received a B.E. degree in automation and a M.E. degree in control theory and control engineering from South China University of Technology, in 2005 and 2007, respectively. He received a Ph.D. in engineering from the Australian National University in 2011. Tengfei Liu is a visiting assistant professor at the Polytechnic Institute of New York University. His research interests include stability theory, robust nonlinear control, quantized control, distributed control, and their applications in mechanical systems, power systems, and transportation systems.

Dr. Liu, with Z. P. Jiang and D. J. Hill, received the Guan Zhao-Zhi Best Paper Award at the 2011 Chinese Control Conference.

Professor Zhong-Ping Jiang received a B.Sc. degree in mathematics from the University of Wuhan, Wuhan, China, in 1988, a M.Sc. degree in statistics from the University of Paris XI, France, in 1989, and a Ph.D. degree in automatic control and mathematics from the Ecole des Mines de Paris, France, in 1993.

Currently, he is a professor of electrical and computer engineering at the Polytechnic School of Engineering of New York University. His main research interests include stability theory, robust and adaptive nonlinear control, adaptive dynamic programming, and their applications to underactuated mechanical systems, communication networks, multi-agent systems, smart grid, and systems neuroscience. He is coauthor of the book *Stability and Stabilization of Nonlinear Systems* (with Dr. I. Karafyllis, Springer 2011).

A Fellow of both the IEEE and IFAC, Dr. Jiang is an editor for the *International Journal of Robust and Nonlinear Control* and has served as an associate editor for several journals including *Mathematics of Control, Signals and Systems* (MCSS), *Systems and Control Letters, IEEE Transactions on Automatic Control, European Journal of Control*, and *Science China: Information Sciences*. Dr. Jiang is a recipient of the prestigious Queen Elizabeth II Fellowship Award from the Australian Research Council, the CAREER Award from the U.S. National Science Foundation, and the Young Investigator Award from the NSF of China. He received the Best Theory Paper Award (with Y. Wang) at the 2008 WCICA, and with T. Liu and D.J. Hill, the Guan Zhao Zhi Best Paper Award at the 2011 CCC. The paper with his PhD student Y. Jiang entitled "Robust Adaptive Dynamic Programming for Optimal Nonlinear Control Design" received the Shimemura Young Author Prize at the 2013

Asian Control Conference in Istanbul, Turkey.

Professor David J. Hill received BE (electrical engineering) and BSc (mathematics) degrees from the University of Queensland, Australia, in 1972 and 1974, respectively. He received a PhD degree in electrical engineering from the University of Newcastle, Australia, in 1977.

He holds the Chair of Electrical Engineering in the Department of Electrical and Electronic Engineering at the University of Hong Kong. He is also a part-time professor in the Centre for Future Energy Networks at the University of Sydney, Australia. During 2005–2010, he was an Australian Research Council Federation Fellow at the Australian National University and, from 2006, also a chief investigator and theme leader (complex networks) in the ARC Centre of Excellence for Mathematics and Statistics of Complex Systems. Since 1994, he has held various positions at the University of Sydney, Australia, including the Chair of Electrical Engineering until 2002 and again during 2010–2013 along with an ARC Professorial Fellowship. He has also held academic and substantial visiting positions at the universities of Melbourne, California (Berkeley), Newcastle (Australia), Lund (Sweden), Munich, and Hong Kong (City and Polytechnic). During 1996–1999 and 2001–2004, he served as head of the respective departments in Sydney and Hong Kong. He currently holds honorary professorships at City University of Hong Kong, South China University of Technology, Wuhan University, and Northeastern University, China.

His general research interests are in control systems, complex networks, power systems, and stability analysis. His work is now mainly on control and planning of future energy networks and basic stability questions for dynamic networks.

Professor Hill is a Fellow of the Institute of Electrical and Electronics Engineers, USA, the Society for Industrial and Applied Mathematics, USA, the Australian Academy of Science, and the Australian Academy of Technological Sciences and Engineering. He is also a foreign member of the Royal Swedish Academy of Engineering Sciences.

Notations

$\mathbb{C}$	The set of complex numbers
$\mathbb{R}$	The set of real numbers
$\mathbb{R}_+$	The set of nonnegative real numbers
$\mathbb{R}^n$	The n-dimensional Euclidean space
$\mathbb{Z}$	The set of integers
$\mathbb{Z}_+$	The set of nonnegative integers
$\mathbb{N}$	The set of natural numbers
x^T	The transpose of vector x
$|x|$	Euclidean norm of vector x
$|A|$	Induced Euclidean norm of matrix A
$\mathrm{sgn}(x)$	The sign of $x \in \mathbb{R}$: $\mathrm{sgn}(x) = 1$ if $x > 0$; $\mathrm{sgn}(x) = 0$ if $x = 0$; $\mathrm{sgn}(x) = -1$ if $x < 0$
$a \bmod b$	Remainder of the Euclidean division of a by b for $a \in \mathbb{R}, b \in \mathbb{R}\backslash\{0\}$
$\|u\|_\Delta$	$\mathrm{ess\,sup}_{t\in\Delta}\,|u(t)|$ with $\Delta \subseteq \mathbb{R}_+$ for $u : \mathbb{R}_+ \to \mathbb{R}^n$
$\|u\|_\infty$	$\|u\|_\Delta$ with $\Delta = [0, \infty)$
$:=$ or $\overset{\mathrm{def}}{=}$	Equal by definition
$\equiv$	Identically equal
$f \circ g$	Composition of functions f and g
$\lambda_{\max}\ (\lambda_{\min})$	Largest (smallest) eigenvalue
$t^+\ (t^-)$	Time right after (right before) t
∂	Partial derivative
$\nabla V(x)$	Gradient vector of function V at x
Id	The identity function
$\mathcal{B}^n$	The unit ball centered at the origin in $\mathbb{R}^n$
$\mathrm{cl}(\mathcal{S})$	The closure of set $\mathcal{S}$
$\mathrm{int}(\mathcal{S})$	The interior of set $\mathcal{S}$
$\mathrm{co}(\mathcal{S})$	The convex hull of set $\mathcal{S}$
$\overline{\mathrm{co}}(\mathcal{S})$	The closed convex hull of set $\mathcal{S}$
$\mathrm{dom}(F)$	The domain of map F
$\mathrm{graph}(F)$	The graph of map F
$\mathrm{range}(F)$	The range of map F

ABBREVIATIONS

AG	Asymptotic Gain
AS	Asymptotic Stability
GAS	Global Asymptotic Stability
GS	Global Stability
IOpS	Input-to-Output Practical Stability

IOS	Input-to-Output Stability
ISpS	Input-to-State Practical Stability
ISS	Input-to-State Stability
OAG	Output Asymptotic Gain
RS	Robust Stability
UBIBS	Uniform Bounded-Input Bounded-State Stability
UO	Unboundedness Observability
WRS	Weakly Robust Stability

1 Introduction

1.1 CONTROL PROBLEMS WITH DYNAMIC NETWORKS

The basic idea for control of dynamic networks is to consider complex systems as structural interconnections of subsystems with specific properties, and solve their control problems using the subsystem and structural features. Such ideas can be traced back to the original development of circuit theory. The rapid development of computing, communication, and sensing technology has enabled new potential applications of advanced control to complex systems. Significant problems are to integrally deal with the fundamental system characteristics, such as nonlinearity, dimensionality, uncertainty and information constraints, and diverse kinds of networked behaviors like quantization, data sampling, and impulsive events. With the development of new tools, this book studies the analysis and control problems of complex systems from the viewpoint of dynamic networks.

Even the single-loop control system may be considered as a dynamic network if detailed behaviors of the sensor and the actuator are taken into account. In a typical single-loop state-feedback control system, as shown in Figure 1.1, the state of the plant is measured by the sensor and sent to the controller, which computes the needed control actions. These are implemented by the actuator for a desired behavior of the plant. A key issue with control systems is stability. By designing an asymptotically stable control system, the error between the actual state signal and a desired signal is expected to converge to zero ultimately.

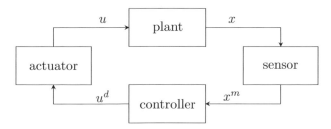

FIGURE 1.1 State-feedback control system: x is the state of the plant, u is the control input, x^m is the measurement of x, and u^d is the desired control input computed by the controller.

Practical control systems are inevitably subject to uncertainties, which may be caused by the sensing and actuation components, and the unmodeled dynamics of the plant. By considering a control system as an interconnection of the perturbation-free nominal system and the perturbation terms, the basic

1

idea of robust control is to design the nominal system to be robust to the perturbations.

Based on this idea, a linear state-feedback control system

$$\dot{x} = Ax + Bu \tag{1.1}$$

$$u^d = -Kx^m \tag{1.2}$$

can be rewritten as the closed-loop nominal system with the perturbation terms:

$$\dot{x} = Ax + B(-K(x + \tilde{x}) - \tilde{u})$$
$$= (A - BK)x - BK\tilde{x} - B\tilde{u}, \tag{1.3}$$

where $\tilde{x} = x^m - x$ and $\tilde{u} = u^d - u$. Suppose that the control objective is to make the system practically stable at the origin, i.e., to steer the state x to within a specific bounded neighborhood of the origin. If $\tilde{x}, \tilde{u}$ are bounded, then such an objective can be achieved if $(A - BK)$ is Hurwitz, i.e., all the eigenvalues of $(A - BK)$ are on the open left-half of the complex plane. For such a linear system, we can study the influence of $\tilde{u}$ and $\tilde{x}$ separately, due to the well-known Superposition Principle. If the eigenvalues of $(A - BK)$ can be arbitrarily assigned by an appropriate choice of K (with complex eigenvalues occurring in conjugate pairs), then the influence of $\tilde{u}$ can be attenuated to within an arbitrarily small level. But this may not be the case for the perturbation term $BK\tilde{x}$ (because it depends on K).

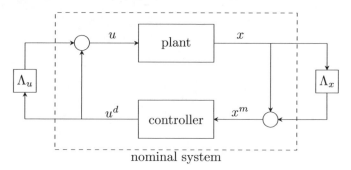

$$\text{nominal system}$$

FIGURE 1.2 Robust control configuration, where Λ_u, Λ_x represent perturbation terms.

The problem can still be handled even if the perturbation terms may not be bounded. For example, they can satisfy the following properties:

$$|\tilde{x}| \leq \bar{\delta}_x |x| + \bar{c}_x \tag{1.4}$$

$$|\tilde{u}| \leq \bar{\delta}_u |u^d| + \bar{c}_u, \tag{1.5}$$

where $\bar{\delta}_x, \bar{c}_x, \bar{\delta}_u, \bar{c}_u$ are nonnegative constants. Such perturbations are said to have the sector bound property.

One may denote $\tilde{x} = \delta_x(t)x + c_x(t)$ and $\tilde{u} = \delta_u(t)K(1 + \delta_x(t))x + \delta_u(t)Kc_x(t) + c_u(t)$ with $|\delta_x(t)| \leq \bar{\delta}_x$, $|\delta_u(t)| \leq \bar{\delta}_u$, $|c_x(t)| \leq \bar{c}_x$, $|c_u(t)| \leq \bar{c}_u$ for $t \geq 0$. Then, system (1.3) can be represented by $\dot{x} = (A - BK)x + w$ with

$$w(t) = -BK(\delta_x(t) + \delta_u(t)(1 + \delta_x(t)))x(t)$$
$$- B(Kc_x(t) + K\delta_u(t)c_x(t) + c_u(t))$$
$$:= \phi(x(t), c_x(t), c_u(t), t). \tag{1.6}$$

It can be directly checked that $|\phi(x, c_x, c_u, t)| \leq a_1|x| + a_2|c_x| + a_3|c_u|$ for all $t \geq 0$ with constants $a_1, a_2, a_3 \geq 0$.

As shown in Figure 1.3, the system is transformed into the interconnection of the nominal system and the perturbation term. There have been standard methods to solve this kind of problem in robust linear control theory [288]. One of them is the classical small-gain theorem, due to Sandberg and Zames. Interested readers may consult [48, Chapter 5] and [54, Chapter 4] for the details. See also [207, Section V] for a small-gain result of large-scale systems.

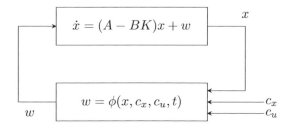

FIGURE 1.3 An interconnected system.

For genuinely nonlinear control systems, the problems discussed above will typically be more complicated. Consider the popular strict-feedback nonlinear system:

$$\dot{x}_i = \Delta_i(\bar{x}_i, w) + x_{i+1}, \quad i = 1, \dots, n-1 \tag{1.7}$$
$$\dot{x}_n = \Delta_n(\bar{x}_n, w) + u, \tag{1.8}$$

where $[x_1, \dots, x_n]^T := x \in \mathbb{R}^n$ is the state, $\bar{x}_i = [x_1, \dots, x_i]^T$, $u \in \mathbb{R}$ is the control input, $w \in \mathbb{R}^{n_w}$ represents the external disturbances, and $\Delta_i : \mathbb{R}^i \to \mathbb{R}$ for $i = 1, \dots, n$ are locally Lipschitz functions. For this system, we consider x_1 as the output. Recursive designs have proved to be useful for the control of such system; see e.g., [153, 235, 151]. By representing the system as a dynamic network composed of x_i-subsystems for $i = 1, \dots, n$, the basic idea is to recursively design control laws for the $\bar{x}_i$-subsystems by considering x_{i+1} as the control inputs until the true control input u occurs. For such system, the influence of the disturbance w might be amplified through the numerous interconnections between the subsystems as shown in Figure 1.4. The problem would be more complicated if the system is subject to sensor noise. As shown in Example 4.1, even for a first-order nonlinear system, small sensor noise may drive the system state to infinity, although the state of the noise-free system asymptotically converges to the origin. Quantized control provides another interesting example for robust control of nonlinear systems with measurement errors satisfying the sector bound property; see Section 5.1 for details.

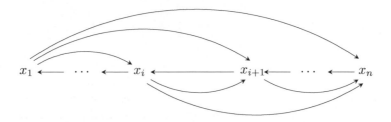

FIGURE 1.4 A high-order nonlinear system as a dynamic network.

Dynamic networks also occur in distributed control of interconnected systems, for which each i-th subsystem takes the following form:

$$\dot{x}_i = f_i(x_i, u_i) \tag{1.9}$$

$$y_i = h_i(x_i), \tag{1.10}$$

where $y_i \in \mathbb{R}^{p_i}$ is the output, $x_i \in \mathbb{R}^{n_i}$ is the state, $u_i \in \mathbb{R}^{m_i}$ is the control input, and $f_i : \mathbb{R}^{m_i+n_i} \to \mathbb{R}^{n_i}, h_i : \mathbb{R}^{n_i} \to \mathbb{R}^{p_i}$ are properly defined functions. In a distributed control structure, each subsystem may be equipped with a controller. Through information exchange, the controllers for the subsystems coordinate with each other, and the outputs of the subsystems achieve some desired group behavior, e.g., $\lim_{t\to\infty}(y_i(t) - y_j(t)) = 0$. In this case, the interconnections in the dynamic network are formed by the information exchange between the controllers, and in some cases, e.g., power systems and telephone networks, by direct physical interconnections.

This book develops new design tools for nonlinear control of dynamic networks, which are applicable to measurement feedback control, quantized control, and distributed control. With the new tools, the related control problems can be transformed into solvable stability problems of dynamic networks composed of subsystems admitting the input-to-state stability (ISS) property or the more general input-to-output stability (IOS) property. To introduce these basic notions, we begin with Lyapunov stability for systems without external inputs.

1.2 LYAPUNOV STABILITY

The stabilization problem is one of the most important problems in control theory. In general terms, a control system is stabilizable if one can find a control law which makes the closed-loop system stable at an equilibrium point. This section reviews some basic concepts of Lyapunov stability [195, 78, 144] for systems with no external inputs.

The comparison functions defined below are used to characterize Lyapunov stability and the notions of ISS and IOS.

Definition 1.1 *A function $\alpha : \mathbb{R}_+ \to \mathbb{R}_+$ is said to be positive definite if $\alpha(0) = 0$ and $\alpha(s) > 0$ for $s > 0$.*

Definition 1.2 *A continuous function $\alpha : \mathbb{R}_+ \to \mathbb{R}_+$ is said to be a class $\mathcal{K}$ function, denoted by $\alpha \in \mathcal{K}$, if it is strictly increasing and $\alpha(0) = 0$; it is said to be a class $\mathcal{K}_\infty$ function, denoted by $\alpha \in \mathcal{K}_\infty$, if it is a class $\mathcal{K}$ function and satisfies $\alpha(s) \to \infty$ as $s \to \infty$.*

Definition 1.3 *A continuous function $\beta : \mathbb{R}_+ \times \mathbb{R}_+ \to \mathbb{R}_+$ is said to be a class $\mathcal{KL}$ function, denoted by $\beta \in \mathcal{KL}$, if, for each fixed $t \in \mathbb{R}_+$, function $\beta(\cdot, t)$ is a class $\mathcal{K}$ function and, for each fixed $s \in \mathbb{R}_+$, function $\beta(s, \cdot)$ is decreasing and satisfies $\lim_{t \to \infty} \beta(s, t) = 0$.*

For convenience of the further discussions, we also give the following definitions on Lipschitz continuity.

Definition 1.4 *A function $h : \mathcal{X} \to \mathcal{Y}$ with $\mathcal{X} \subseteq \mathbb{R}^n$ and $\mathcal{Y} \subseteq \mathbb{R}^m$ is said to be Lipschitz continuous, or simply Lipschitz, on $\mathcal{X}$, if there exists a constant $L_h \geq 0$, such that for any $x_1, x_2 \in \mathcal{X}$,*

$$|h(x_1) - h(x_2)| \leq L_h |x_1 - x_2|. \tag{1.11}$$

Definition 1.5 *A function $h : \mathcal{X} \to \mathcal{Y}$ with $\mathcal{X} \subseteq \mathbb{R}^n$ being open and connected, and $\mathcal{Y} \subseteq \mathbb{R}^m$ is said to be locally Lipschitz on $\mathcal{X}$, if each $x \in \mathcal{X}$ has a neighborhood $\mathcal{X}_0 \subseteq \mathcal{X}$ such that h is Lipschitz on $\mathcal{X}_0$.*

Definition 1.6 *A function $h : \mathcal{X} \to \mathcal{Y}$ with $\mathcal{X} \subseteq \mathbb{R}^n$ and $\mathcal{Y} \subseteq \mathbb{R}^m$ is said to be Lipschitz on compact sets, if h is Lipschitz on every compact set $\mathcal{D} \subseteq \mathcal{X}$.*

Consider the system

$$\dot{x} = f(x), \tag{1.12}$$

where $f : \mathbb{R}^n \to \mathbb{R}^n$ is a locally Lipschitz function. Assume that the origin is an equilibrium of the nonlinear system, i.e., $f(0) = 0$. Note that if an equilibrium other than the origin, say x^e, is of interest, one may use a coordinate transformation $x' = x - x^e$ to move the equilibrium to the origin. Therefore, the assumption of the equilibrium at the origin is with no loss of generality. Denote $x(t, x_0)$ or simply $x(t)$ as the solution of system (1.12) with initial condition $x(0) = x_0$, and let $[0, T_{\max})$ with $0 < T_{\max} \leq \infty$ be the right maximal interval for the definition of $x(t, x_0)$.

The standard definition of Lyapunov stability is usually given by using "ϵ-δ" terms, which can be found in the standard textbooks on nonlinear systems; see, e.g., [78] and [144, Chapter 4]. Definition 1.7 employs the comparison functions $\alpha \in \mathcal{K}$ and $\beta \in \mathcal{KL}$ for convenience of the comparison between Lyapunov stability and ISS. A proof of the equivalence between the standard definition and Definition 1.7 can be found in [144, Appendix C.6]. See also the discussions in [78, Definitions 2.9 and 24.2].

Definition 1.7 *System* (1.12) *is*

- *stable at the origin if there exist an $\alpha \in \mathcal{K}$ and a constant $c > 0$ such that for any $|x_0| \le c$,*

$$|x(t, x_0)| \le \alpha(|x_0|) \tag{1.13}$$

  for all $t \ge 0$;
- *globally stable (GS) at the origin if property (1.13) holds for all initial states $x_0 \in \mathbb{R}^n$;*
- *asymptotically stable (AS) at the origin if there exist a $\beta \in \mathcal{KL}$ and a constant $c > 0$ such that for any $|x_0| \le c$,*

$$|x(t, x_0)| \le \beta(|x_0|, t) \tag{1.14}$$

  for all $t \ge 0$;
- *globally asymptotically stable (GAS) at the origin if condition (1.14) holds for any initial state $x_0 \in \mathbb{R}^n$.*

With the standard definition, GAS at the origin can be defined based on GS by adding the global convergence property at the origin: $\lim_{t \to \infty} x(t, x_0) = 0$ for all $x_0 \in \mathbb{R}^n$; see [144, Definition 4.1]. It can be observed that GAS is more than global convergence.

Theorem 1.1, which is known as Lyapunov's Second Theorem (or the Lyapunov Direct Method), gives sufficient conditions for stability and AS.

Theorem 1.1 *Let the origin be an equilibrium of system* (1.12) *and* $\Omega \subset \mathbb{R}^n$ *be a domain containing the origin. Let* $V : \Omega \to \mathbb{R}_+$ *be a continuously differentiable function such that*

$$V(0) = 0, \tag{1.15}$$
$$V(x) > 0 \text{ for } x \in \Omega \backslash \{0\}, \tag{1.16}$$
$$\nabla V(x) f(x) \le 0 \text{ for } x \in \Omega. \tag{1.17}$$

Then, system (1.12) *is stable at the origin. Moreover, if*

$$\nabla V(x) f(x) < 0 \text{ for } x \in \Omega \backslash \{0\}, \tag{1.18}$$

then system (1.12) *is AS at the origin.*

A function V that satisfies (1.15)–(1.17) is called a Lyapunov function. If moreover, V satisfies (1.18), then it is called a strict Lyapunov function [16].

It is natural to ask whether the condition for AS in Theorem 1.1 can guarantee GAS by directly replacing the Ω with $\mathbb{R}^n$. Example 1.1, which was given in [78, p. 109], answers this question.

Example 1.1 *Consider system*

$$\dot{x}_1 = \frac{-6x_1}{(1+x_1^2)^2} + 2x_2, \tag{1.19}$$

$$\dot{x}_2 = \frac{-2(x_1+x_2)}{(1+x_1^2)^2}. \tag{1.20}$$

Let

$$V(x) = \frac{x_1^2}{1+x_1^2} + x_2^2. \tag{1.21}$$

It can be directly verified that V satisfies all the conditions for AS at the origin given by Theorem 1.1 with $n = 2$ and $\Omega = \mathbb{R}^2$. By testing the vector field on the boundary of hyperbola $x_2 = 2/(x_1 - \sqrt{2})$, the trajectories to the right of the branch in the first quadrant cannot cross that branch. This means that the system is not GAS at the origin.

Theorem 1.2 gives extra conditions on the Lyapunov function V for GAS.

Theorem 1.2 *Let the origin be an equilibrium of system (1.12). Let $V : \mathbb{R}^n \to \mathbb{R}_+$ be a continuously differentiable function such that*

$$V(0) = 0, \tag{1.22}$$
$$V(x) > 0 \text{ for } x \in \mathbb{R}^n \backslash \{0\}, \tag{1.23}$$
$$|x| \to \infty \Rightarrow V(x) \to \infty, \tag{1.24}$$
$$\nabla V(x)f(x) < 0 \text{ for } x \in \mathbb{R}^n \backslash \{0\}. \tag{1.25}$$

Then, system (1.12) is globally asymptotically stable at the origin.

According to Theorem 1.2, it is not sufficient to guarantee GAS by simply replacing the Ω in the condition for AS in Theorem 1.1 with $\mathbb{R}^n$. Condition (1.24) is also needed for GAS.

Condition (1.22)–(1.24) is equivalent to the statement that V is positive definite and radially unbounded, which can be represented with comparison functions $\underline{\alpha}, \overline{\alpha} \in \mathcal{K}_\infty$ as

$$\underline{\alpha}(|x|) \leq V(x) \leq \overline{\alpha}(|x|) \tag{1.26}$$

for all $x \in \mathbb{R}^n$. Moreover, condition (1.25) is equivalent to the existence of a continuous and positive definite function α such that

$$\nabla V(x) f(x) \leq -\alpha(V(x)) \tag{1.27}$$

holds for all $x \in \mathbb{R}^n$. See [144, Lemma 4.3] for the details.

Theorems 1.1 and 1.2 give sufficient conditions for stability, AS and GAS. A proof of the converse Lyapunov theorem for the necessity of the conditions can be found in [144].

1.3 INPUT-TO-STATE STABILITY

For systems with external inputs, the notion of input-to-state stability (ISS), invented by Sontag, has proved to be powerful for evaluating the influence of the external inputs.

1.3.1 DEFINITION

Consider the system

$$\dot{x} = f(x, u), \tag{1.28}$$

where $x \in \mathbb{R}^n$ is the state, $u \in \mathbb{R}^m$ represents the input, and $f : \mathbb{R}^n \times \mathbb{R}^m \to \mathbb{R}^n$ is a locally Lipschitz function and satisfies $f(0,0) = 0$. By considering the input u as a function of time, assume that u is measurable and locally essentially bounded. Recall that u is locally essentially bounded if for any $t \geq 0$, $\|u\|_{[0,t]}$ exists. Denote $x(t, x_0, u)$, or simply $x(t)$, as the solution of system (1.12) with initial condition $x(0) = x_0$ and input u.

In [241], the original definition of ISS is given in the "plus" form; see (1.31). For convenience of discussions, we mainly use the definition in the equivalent "max" form. The equivalence is discussed later.

Definition 1.8 *System* (1.28) *is said to be input-to-state stable (ISS) if there exist* $\beta \in \mathcal{KL}$ *and* $\gamma \in \mathcal{K}$ *such that for any initial state* $x(0) = x_0$ *and any measurable and locally essentially bounded input* u, *the solution* $x(t)$ *satisfies*

$$|x(t)| \leq \max\{\beta(|x_0|, t), \gamma(\|u\|_\infty)\} \tag{1.29}$$

for all $t \geq 0$.

Here, γ is called the ISS gain of the system. Notice that, if $u \equiv 0$, then Definition 1.8 is reduced to Definition 1.7 for GAS at the origin. Due to causality, $x(t)$ depends on x_0 and the past inputs $\{u(\tau) : 0 \leq \tau \leq t\}$, and thus, the $\|u\|_\infty$ in (1.29) can be replaced with $\|u\|_{[0,t]}$.

Since

$$\max\{a, b\} \leq a + b \leq \max\{(1 + 1/\delta)a, (1 + \delta)b\} \tag{1.30}$$

for any $a, b \geq 0$ and any $\delta > 0$, property (1.29) in the "max" form is equivalent to

$$|x(t)| \leq \beta'(|x_0|, t) + \gamma'(\|u\|_\infty), \tag{1.31}$$

where $\beta' \in \mathcal{KL}$ and $\gamma' \in \mathcal{K}$. It should be noted that although the transformation from (1.29) to (1.31) can be done by directly replacing the "max" operation with the "+" operation without changing functions β and γ, the transformation from (1.31) to (1.29) may result in a pair of β and γ different from the pair of β' and γ'. To get a γ very close to γ', one may choose a very small δ for the transformation, but this could result in a very large β.

With property (1.31), $x(t)$ asymptotically converges to within the region defined by $|x| \leq \gamma'(\|u\|_\infty)$, i.e.,

$$\overline{\lim_{t \to \infty}} |x(t)| \leq \gamma'(\|u\|_\infty). \tag{1.32}$$

As shown in Figure 1.5, γ' describes the "steady-state" performance of the system, and is usually called the asymptotic gain (AG), while the "transient performance" is described by β'.

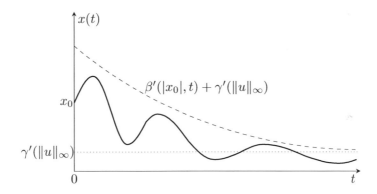

FIGURE 1.5 Asymptotic gain property.

Intuitively, since only large values of t determine the value $\overline{\lim}_{t \to \infty} |x(t)|$, one may replace the $\gamma'(\|u\|_\infty)$ in (1.32) with $\gamma'(\overline{\lim}_{t \to \infty} |u(t)|)$ or $\overline{\lim}_{t \to \infty} \gamma'(|u(t)|)$. See [250, 247] for more detailed discussions.

When system (1.28) is reduced to a linear system, a necessary and sufficient condition for the ISS property can be derived.

Theorem 1.3 *A linear time-invariant system*

$$\dot{x} = Ax + Bu \tag{1.33}$$

is ISS if and only if A is Hurwitz.

Proof. With initial condition $x(0) = x_0$ and input u, the solution of system (1.33) is

$$x(t) = e^{At}x_0 + \int_0^t e^{A(t-\tau)}Bu(\tau)d\tau, \tag{1.34}$$

which implies

$$|x(t)| \le |e^{At}||x_0| + \left(\int_0^\infty |e^{A\tau}|d\tau\right)|B|\|u\|_\infty. \tag{1.35}$$

If A is Hurwitz, i.e., every eigenvalue of A has negative real part, then $\int_0^\infty |e^{As}|ds < \infty$. Define $\beta'(s,t) = |e^{At}|s$ and $\gamma'(s) = \left(\int_0^\infty |e^{A\tau}|d\tau\right)|B|s$ for $s, t \in \mathbb{R}_+$. Clearly, $\beta' \in \mathcal{KL}$ and $\gamma' \in \mathcal{K}_\infty$. Then, the linear system is ISS in the sense of (1.31). The sufficiency part is proved.

For the necessity, one may consider the case of $u \equiv 0$. In this case, the ISS of system (1.33) implies GAS of system

$$\dot{x} = Ax \tag{1.36}$$

at the origin. According to linear systems theory [28], system (1.36) is GAS at the origin if and only if A is Hurwitz. $\diamondsuit$

Based on the proof of Theorem 1.3, one may consider the ISS property (1.31) as a nonlinear modification of property (1.35) of linear systems. Lemma 1.1 shows that any $\mathcal{KL}$ function $\beta(s,t)$ can be considered as a nonlinear modification of function se^{-t}.

Lemma 1.1 *For any $\beta \in \mathcal{KL}$, there exist $\alpha_1, \alpha_2 \in \mathcal{K}_\infty$ such that*

$$\beta(s,t) \le \alpha_2(\alpha_1(s)e^{-t}) \tag{1.37}$$

for all $s, t \ge 0$.

See [243, Proposition 7] and its proof therein.

According to Lemma 1.1, if property (1.31) holds, then there exist $\alpha_1', \alpha_2' \in \mathcal{K}_\infty$ such that

$$|x(t)| \le \alpha_2'(\alpha_1'(|x_0|)e^{-t}) + \gamma'(\|u\|_\infty), \tag{1.38}$$

which shows a close analogy of ISS to the solution property (1.35) of linear system (1.33) with A being Hurwitz.

Also, with Lemma 1.1, property (1.29) implies

$$|x(t)| \le \max\{\alpha_2(\alpha_1(|x_0|)e^{-t}), \gamma(\|u\|_\infty)\}, \tag{1.39}$$

where α_1, α_2 are appropriate class $\mathcal{K}_\infty$ functions. This means, for any x_0 and $\|u\|_\infty$ satisfying $\alpha_2 \circ \alpha_1(|x_0|) > \gamma(\|u\|_\infty)$, there exists a finite time $t^* = \log(\alpha_1(|x_0|)) - \log(\alpha_2^{-1} \circ \gamma(\|u\|_\infty))$, after which solution $x(t)$ is within the range defined by $|x| \leq \gamma(\|u\|_\infty)$. This shows the difference between the ISS gain γ defined in (1.29) and the asymptotic gain γ' defined in (1.31).

Theorem 1.3 means that a linear system is ISS if the corresponding input-free system is GAS at the origin. But this may not be true for nonlinear systems. Consider Example 1.2 given by [243].

Example 1.2 *Consider the nonlinear system*

$$\dot{x} = -x + ux \tag{1.40}$$

with $x, u \in \mathbb{R}$. If $u \equiv 0$, then the resulting system $\dot{x} = -x$ is GAS at the origin. But system (1.40) is not ISS. Just consider the class of constant inputs $u > 1$.

However, it has been proved that AS at the origin of system (1.28) with $u \equiv 0$ is equivalent to a local ISS property of system (1.28) [250]. The definition of local ISS is given by Definition 1.9.

Definition 1.9 *System (1.28) is said to be locally input-to-state stable if there exist $\beta \in \mathcal{KL}$, $\gamma \in \mathcal{K}$, and constants $\rho^x, \rho^u > 0$ such that for any initial state $x(0) = x_0$ satisfying $|x_0| \leq \rho^x$ and any measurable and locally essentially bounded input u satisfying $\|u\|_\infty \leq \rho^u$, the solution $x(t)$ satisfies*

$$|x(t)| \leq \max\{\beta(|x_0|, t), \gamma(\|u\|_\infty)\} \tag{1.41}$$

for all $t \geq 0$.

Theorem 1.4 presents the equivalence between AS and local ISS.

Theorem 1.4 *System (1.28) is locally ISS if and only if the zero-input system*

$$\dot{x} = f(x, 0) \tag{1.42}$$

is AS at the origin.

Proof. The proof of Theorem 1.4 is motivated by the proof of [78, Theorems 56.3 and 56.4] on the equivalence between total stability and AS at the origin, and the proof of [250, Lemma I.2] on the sufficiency of GAS for local ISS.

The necessity part is obvious. We prove the sufficiency part. By using the converse Lyapunov theorem (see e.g., [144]), the AS of system (1.42) at the origin implies the existence of a Lyapunov function $V : \Omega \to \mathbb{R}_+$ with $\Omega \subseteq \mathbb{R}^n$ being a domain containing the origin such that properties (1.15)–(1.18) hold. For such V, one can find an $\Omega' \subseteq \Omega$ still containing the origin such that for all $x \in \Omega'$,

$$\underline{\alpha}(|x|) \leq V(x) \leq \overline{\alpha}(|x|) \tag{1.43}$$
$$\nabla V(x) f(x, 0) \leq -\alpha(V(x)), \tag{1.44}$$

where $\underline{\alpha}, \overline{\alpha} \in \mathcal{K}_\infty$ and α is a continuous and positive definite function.

By using the continuity of ∇V and f, for any $x \in \Omega' \backslash \{0\}$, one can find a $\delta > 0$ such that

$$|\nabla V(x)f(x,\epsilon) - \nabla V(x)f(x,0)| \leq \frac{1}{2}\alpha(V(x)) \qquad (1.45)$$

for all $|\epsilon| \leq \delta$. Thus, there is a positive definite function χ_0 such that for any $x \in \Omega'$, property (1.45) holds for all $|\epsilon| \leq \chi_0(|x|)$.

Then, we choose Ω_0 as a compact set containing the origin and belonging to Ω', and choose $\chi \in \mathcal{K}$ such that

$$\chi(s) \leq \chi_0(s) \qquad (1.46)$$

for all $0 \leq s \leq \max\{|x| : x \in \Omega_0\}$. It can be directly proved that if $x \in \Omega_0$ and $\chi(|u|) \leq |x|$, then

$$\nabla V(x)f(x,u) \leq -\frac{1}{2}\alpha(V(x)). \qquad (1.47)$$

Thus, with (1.43), property (1.47) holds if

$$V(x) \leq \max\left\{\underline{\alpha}(|x|) : x \in \Omega_0\right\}, \qquad (1.48)$$
$$V(x) \geq \overline{\alpha} \circ \chi(\|u\|_\infty) := \gamma(\|u\|_\infty). \qquad (1.49)$$

Then, the sufficiency part can be proved following the same line as (1.54)–(1.56) given later for ISS-Lyapunov functions. The interested reader may also consult the proof of [250, Lemma I.2]. $\qquad \diamond$

From Definition 1.8, an ISS system is always forward complete, i.e., for any initial state $x(0) = x_0$ and any measurable and locally essentially bound input u, the solution $x(t)$ is defined for all $t \geq 0$. Moreover, it has the uniformly bounded-input bounded-state (UBIBS) property.

Definition 1.10 *System (1.28) is said to have the UBIBS property if there exists $\sigma_1, \sigma_2 \in \mathcal{K}$ such that for any initial state $x(0) = x_0$ and any measurable and locally essentially bounded input u,*

$$|x(t)| \leq \max\{\sigma_1(|x_0|), \sigma_2(\|u\|_\infty)\} \qquad (1.50)$$

for all $t \geq 0$.

Recall Definition 1.3 for class $\mathcal{KL}$ functions. If system (1.28) is ISS satisfying (1.29), then it admits property (1.50) by defining $\sigma_1(s) = \beta(s,0)$ and $\sigma_2(s) = \gamma(s)$ for $s \in \mathbb{R}_+$.

More importantly, ISS is equivalent to the conjunction of UBIBS and AG [250]. This result can be used for the proof of the ISS small-gain theorem for interconnected nonlinear systems; see detailed discussions in Chapter 2.

Theorem 1.5 *System (1.28) is ISS if and only if it has the properties of UBIBS and AG in the sense of (1.50) and (1.32), respectively.*

1.3.2 ISS-LYAPUNOV FUNCTION

Similar to stability, we can employ ISS-Lyapunov functions to formulate the notion of ISS. For system (1.28), the equivalence between ISS and the existence of ISS-Lyapunov functions is originally presented in [249].

Theorem 1.6 *System* (1.28) *is ISS if and only if it admits a continuously differentiable function* $V : \mathbb{R}^n \to \mathbb{R}_+$, *for which*

1. *there exist* $\underline{\alpha}, \overline{\alpha} \in \mathcal{K}_\infty$ *such that*

$$\underline{\alpha}(|x|) \leq V(x) \leq \overline{\alpha}(|x|), \quad \forall x, \qquad (1.51)$$

2. *there exist a* $\gamma \in \mathcal{K}$ *and a continuous, positive definite* α *such that*

$$V(x) \geq \gamma(|u|) \Rightarrow \nabla V(x) f(x,u) \leq -\alpha(V(x)), \quad \forall x, \ u. \qquad (1.52)$$

A function V satisfying (1.51) and (1.52) is called an ISS-Lyapunov function and γ is called the Lyapunov-based ISS gain. ISS-Lyapunov functions defined with (1.52) are said to be in the gain margin form. It can be observed that, under condition (1.52), the state x ultimately converges to within the region such that $V(x) \leq \gamma(\|u\|_\infty)$. If input $u \equiv 0$, then the sufficiency part of Theorem 1.6 is reduced to Theorem 1.2 for GAS.

An equivalent formulation to (1.52) is in the dissipation form:

$$\nabla V(x) f(x,u) \leq -\alpha'(V(x)) + \gamma'(|u|), \qquad (1.53)$$

where $\alpha' \in \mathcal{K}_\infty$ and $\gamma' \in \mathcal{K}$.

The proof of the sufficiency part of Theorem 1.6 can be found in the original ISS paper [241], while the necessity part is proved for the first time in [249]. Here, according to [241, 249], we give a sketch of the proof, which could be helpful in understanding ISS-Lyapunov functions.

With property (1.52), it can be proved that there exists a $\beta \in \mathcal{KL}$ satisfying $\beta(s, 0) = s$ for all $s \in \mathbb{R}_+$ such that

$$V(x(t)) \leq \beta(V(x(0)), t), \qquad (1.54)$$

as long as $V(x(t)) \geq \gamma(\|u\|_\infty)$. This means

$$V(x(t)) \leq \max\{\beta(V(x(0)), t), \gamma(\|u\|_\infty)\} \qquad (1.55)$$

for all $t \geq 0$. Define $\bar{\beta}(s, t) = \underline{\alpha}^{-1}(\beta(\overline{\alpha}(s), t))$ and $\bar{\gamma}(s) = \underline{\alpha}^{-1} \circ \gamma(s)$ for $s, t \in \mathbb{R}_+$. Then, $\bar{\beta} \in \mathcal{KL}$, $\bar{\gamma} \in \mathcal{K}$, and

$$|x(t)| \leq \max\{\bar{\beta}(|x(0)|, t), \bar{\gamma}(\|u\|_\infty)\} \qquad (1.56)$$

holds for all $t \geq 0$. The sufficiency part of Theorem 1.6 is proved.

It should be noted that an ISS-Lyapunov function is not necessarily continuously differentiable. Sometimes, it is more convenient to construct locally Lipschitz ISS-Lyapunov functions, which are still sufficient for ISS. According to Rademacher's theorem [59, p. 216], a locally Lipschitz function is continuously differentiable almost everywhere. For a locally Lipschitz ISS-Lyapunov function V, condition (1.52) holds for almost all x. In this case, the arguments used in the original ISS paper [241] are still valid to show that the existence of such a V implies ISS.

The necessity part of Theorem 1.6 can be proved by constructing ISS-Lyapunov functions. The proof given in [249] employs the notion of weakly robust stability (WRS), and the basic idea is shown in Figure 1.6.

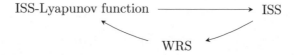

FIGURE 1.6 Equivalence between ISS and the existence of an ISS-Lyapunov function.

The WRS property describes the capability of a system to handle state-dependent perturbations. System (1.28) is said to be WRS if it admits a stability margin $\rho \in \mathcal{K}_\infty$ such that system

$$\dot{x} = f(x, d(t)\rho(|x|)) \tag{1.57}$$

is GAS at the origin uniformly with respect to time, for all possible $d : \mathbb{R}_+ \to \mathcal{B}^m$. Recall that $\mathcal{B}^m$ represents the unit ball with center at the origin in $\mathbb{R}^m$. It is proved in [249] that ISS implies WRS for system (1.28), and the Lyapunov function of system (1.57) can be used as the ISS-Lyapunov function for system (1.28).

The proof of the existence of a Lyapunov function for a WRS system is related to the converse Lyapunov theorem. Reference [170] presents a result on the construction of smooth Lyapunov functions for weakly robustly stable systems. In [137, Chapter 3], converse Lyapunov results for general nonlinear systems were proved by using the relationship between exponential convergence and $\mathcal{KL}$ convergence. According to the proof given in [249], a Lyapunov function for system (1.57) can be used as an ISS-Lyapunov function for system (1.28). The interested reader may consult [249] for a detailed proof.

A property stronger than WRS is the robust stability (RS) property [249], which considers systems with state-dependent perturbations in the more general form:

$$\dot{x} = f(x, \Delta(t, x)), \tag{1.58}$$

where the perturbation term $\Delta(t, x)$ might be caused by uncertainty of the system dynamics. System (1.28) is said to be RS with a gain margin $\rho \in \mathcal{K}$

if the perturbed system (1.58) is uniformly GAS at the origin as long as $|\Delta(t,x)| \leq \rho(|x|)$. The equivalence between RS and ISS has also been proved in [249]. In Chapter 2, we consider the RS property of ISS as a special case of the ISS small-gain theorem.

In the context of dissipativity, the α' and γ' functions in (1.53) are known as the supply functions. Reference [248] presents a result on the freedom of modifying the supply functions and its application to stability analysis and the construction of Lyapunov functions for cascade systems. Connections between ISS and dissipativity [277, 89, 92] were made in [248, 249], and have persisted throughout the ISS approach, but generally there are many differences between the dissipative systems and ISS approaches which would be interesting to explore as an extension of [248, 243, 53, 7, 108] on variants of both. Further, more effort is needed for systems with supply functions in more general forms. As seen in the basic ideas [277, 87], the supply functions can take very general forms of which quadratics and monotone gains are just examples. Also, some generalization of passivity concepts away from the quadratic form can be found in [235, 62, 63].

1.4   INPUT-TO-OUTPUT STABILITY

Consider the system

$$\dot{x} = f(x, u) \tag{1.59}$$
$$y = h(x), \tag{1.60}$$

where $x \in \mathbb{R}^n$ is the state, $u \in \mathbb{R}^m$ is the input, $y \in \mathbb{R}^l$ is the output, and $f : \mathbb{R}^n \times \mathbb{R}^m \to \mathbb{R}^n$ and $h : \mathbb{R}^n \to \mathbb{R}^l$ are locally Lipschitz functions. It is assumed that $f(0,0) = 0$ and $h(0) = 0$.

The notion of input-to-output stability (IOS) can be derived by directly replacing the state x on the left-hand side of (1.29) with the output y.

Definition 1.11 *System* (1.59)–(1.60) *is said to be IOS if there exist a* $\beta \in \mathcal{KL}$ *and a* $\gamma \in \mathcal{K}$ *such that for any initial state* $x(0) = x_0$, *any measurable and locally essentially bounded* u, *and any* t *where* $x(t)$ *is defined, it holds that*

$$|y(t)| \leq \max\{\beta(|x_0|, t), \gamma(\|u\|_\infty)\}. \tag{1.61}$$

Here, γ is called the IOS gain of system (1.59)–(1.60).

As for ISS, property (1.61) can be equivalently represented by

$$|y(t)| \leq \beta'(|x_0|, t) + \gamma'(\|u\|_\infty) \tag{1.62}$$

with $\beta' \in \mathcal{KL}$ and $\gamma' \in \mathcal{K}$.

Corresponding to the AG property of ISS, an IOS system has the output asymptotic gain (OAG) property: for any initial state $x(0) = x_0$, any measurable and locally essentially bounded input u, and any t where $x(t)$ is defined,

$$\varlimsup_{t \to \infty} |y(t)| \leq \gamma'(\|u\|_\infty), \tag{1.63}$$

where γ' is called the output asymptotic gain.

The IOS studied in this book has connections with the classical input-output stability, but the two concepts are not entirely equivalent. To avoid confusion, we use I/O stability as the abbreviation of input-output stability. The study of the stability problem for the systems with input-output (I/O) representations in the operator setting goes back to functional analysis and other approaches developed in the 1960s; see [233, 283, 48]. The work of Hill [88] and Mareels and Hill [199] proposed a generalized notion of I/O stability by introducing monotone gains. This is a nonlinear extension of the classical finite-gain stability. References [92, 93, 87] made the role of initial condition in I/O stability explicit to find connections to Lyapunov stability/instability, but did not assume asymptotic stability *a priori*. By introducing the $\mathcal{KL}$ function (see the β function in (1.61)), IOS provides an explicit description of the converging effect of the initial condition and a link to partial stability of internal states [269].

For systems with outputs, observability describes the capability to estimate the internal state by using the input and output data. In the literature, several observability notions have been used for guaranteeing asymptotic stability in nonlinear systems; see e.g., [276, 91, 128, 130, 85] and the references therein. Now, we recall the notion of unboundedness observability (UO) from [130] which, together with IOS, will be used in the following chapters.

Definition 1.12 *System* (1.59)–(1.60) *is said to be unboundedness observable if there exist* $\alpha^O \in \mathcal{K}_\infty$ *and constant* $D^O \geq 0$ *such that for each measurable and locally essentially bounded input* u *and for any initial condition* $x(0) = x_0$, *the solution* $x(t)$ *of the system satisfies*

$$|x(t)| \leq \alpha^O(|x(0)| + \|u\|_{[0,t]} + \|y\|_{[0,t]}) + D^O \tag{1.64}$$

for all t *where* $x(t)$ *is defined.*

If system (1.59)–(1.60) has the UO property in the form of (1.64) with $D^O = 0$, then it is said to be UO with zero offset.

This section only briefly reviews the notions that are used in the following chapters. The interested reader may consult the original papers [104, 130, 131, 141, 142, 223, 241, 242, 243, 247, 249, 250, 251, 252] for more characterizations and properties of ISS and IOS. Specifically, discussions related to ISS and robust stability of systems describing difference equations or difference inclusions can be found in [131, 142, 143, 22].

1.5 INPUT-TO-STATE STABILIZATION AND AN OVERVIEW OF THE BOOK

By characterizing model uncertainty as well as external inputs as a perturbation of a nominal model, the goal of robust control is to design feedback control laws such that the closed-loop system is robust with respect to a certain level of perturbations. Due to its relationship to robust stability, ISS is a natural tool for robust control. One approach is input-to-state stabilization, i.e., designing ISS systems via feedback, with the influence of the perturbations represented by ISS gains. A one-to-one correspondence between input-to-state stability and input-to-state stabilization has been discussed in [137]. Notice that another line of research to cope with uncertainty leads to adaptive control designs, for which ISS has also been used as a powerful tool [115].

The above-mentioned systems with uncertainties and disturbances can be considered as an interconnected system. Other interconnected systems occur in emerging control applications ranging from conventional and smart electric grids, robotic networks and transportation networks to communication and biological networks. The ISS small-gain theorem is capable of testing the ISS of an interconnected system by directly checking the composition of ISS gains for the subsystems. It may significantly reduce the complexity of analyzing and designing interconnected systems. For interconnected systems, with the ISS small-gain theorem, the input-to-state stabilization problem can be solved by designing control laws to make the subsystems ISS with appropriate gains. The basic ideas of the ISS small-gain and the more general IOS small-gain theorems together with an introduction to their use in control designs are reviewed in Chapter 2.

Inspired by ISS small-gain methods, the basic idea of this book is to transform several robust nonlinear control problems into input-to-state stabilization problems of large-scale dynamic networks, which may contain more than two subsystems. For this purpose, a cyclic-small-gain theorem is developed in Chapter 3 for large-scale dynamic networks composed of ISS and the more general IOS subsystems. With the cyclic-small-gain theorem, the problem of testing the ISS property of a dynamic network is reduced to checking the specific compositions of the ISS gains along the simple cycles in the network. Moreover, as shown in the following chapters, this technique is extremely efficient for analyzing the influence of perturbations through the numerous links and loops in a complex dynamic network.

The robust control problem for nonlinear uncertain systems subject to sensor noise is challenging, yet important. Chapter 4 contributes new cyclic-small-gain design methods to solve the problems caused by sensor noise. Measurement feedback control issues in the settings of static state-feedback, dynamic state-feedback and output-feedback, and the applications to event-based control, synchronization, and robust adaptive control are thoroughly studied. With small-gain control designs, the closed-loop systems can be transformed into large-scale dynamic networks of ISS subsystems, for which the influence of sensor noise can be explicitly described by ISS gains and attenuated to the level of sensor noise.

In modern automatic control systems, signals are usually quantized before transmission via communication channels. A quantizer can be mathematically modeled as a discontinuous map from a continuous region to a discrete set of numbers, which leads to a special class of system uncertainties. In Chapter 5, quantized control of nonlinear systems with static quantization is first solved through a cyclic-small-gain design. Due to the finite word-length of digital devices, a practical quantizer has a finite number of quantization levels. In Chapter 5, dynamic quantization is developed for high-order nonlinear systems such that the quantization levels can be dynamically scaled during the control procedure for semiglobal quantized stabilization. Quantized output-feedback control is also studied.

The trend of controlling complex systems composed of spatially distributed subsystems motivates the idea of distributed control. In a distributed control system, the subsystems are controlled by local controllers with information exchange between these controllers for coordination purposes. Formation control of mobile robots is an example. Chapter 6 develops cyclic-small-gain methods for distributed control of nonlinear systems. By representing the information exchange in distributed control systems with directed graphs (digraphs), a cyclic-small-gain result in digraphs is first proposed. Then, we use this result to solve the problems with distributed output-feedback control and distributed formation control.

Based on the results in this book, nontrivial efforts are required for further development of nonlinear control of more general dynamic networks. Some challenging problems in this research direction are listed in Chapter 7.

2 Interconnected Nonlinear Systems

The small-gain theorem is an extremely useful tool for the analysis and control design of interconnected systems. This chapter introduces the first, fundamentally nonlinear variant of the small-gain theorem, known as the ISS small-gain theorem, as well as the related methods. With the ISS small-gain theorem, the ISS property of an interconnected system composed of two ISS subsystems can be tested by checking the composition of the ISS gains.

We start with a simple case in which the interconnected system is composed of one dynamic subsystem and one static subsystem and the interconnected system does not have any external input. The block diagram of the system is shown in Figure 2.1. Suppose that subsystem Ξ_1 is in the form of (1.28) and subsystem Ξ_2 is defined as

$$u = \Delta(x, t). \tag{2.1}$$

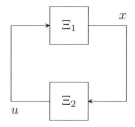

FIGURE 2.1 An interconnected system with no external input.

With the robust stability property of ISS, the interconnected system is GAS at the origin if subsystem Ξ_1 is ISS in the sense of (1.29) and there exists an appropriate $\rho \in \mathcal{K}_\infty$ such that subsystem Ξ_2 satisfies

$$|\Delta(x, t)| \le \rho(|x|) \tag{2.2}$$

for all x and all $t \ge 0$. A sufficient condition for ρ to guarantee GAS is

$$\rho(\gamma(s)) < s \tag{2.3}$$

for all $s > 0$, i.e.,

$$\rho \circ \gamma < \mathrm{Id}. \tag{2.4}$$

As shown later, the robust stability property of ISS can be considered as a special case of the ISS small-gain theorem. Here, we give a proof of the sufficiency of condition (2.4) for GAS, which is a reduced version of the original proof of the ISS small-gain theorem given by Jiang, Teel, and Praly [130]. The proof is carried out in two steps. We first show the GS of the interconnected system at the origin, and then prove the GAS with the help of Lemma D.1. Lemma D.1 is a slight modification of [130, Lemma A.1].

Suppose that the solution $x(t)$ of the interconnected system is right maximally defined on $[0, T)$ with $0 < T \leq \infty$. By applying $|u| = |\Delta(x, t)| \leq \rho(|x|)$ to (1.29), we have

$$
\begin{aligned}
|x(t)| &\leq \max\{\beta(|x(0)|, 0), \gamma(\|u\|_{[0,T)})\} \\
&\leq \max\{\beta(|x(0)|, 0), \gamma \circ \rho(\|x\|_{[0,T)})\}
\end{aligned} \tag{2.5}
$$

for all $0 \leq t < T$. Then, taking the supremum of $|x|$ over $[0, T)$ implies

$$
\|x\|_{[0,T)} \leq \max\{\beta(|x(0)|, 0), \gamma \circ \rho(\|x\|_{[0,T)})\}. \tag{2.6}
$$

With condition (2.4) satisfied, $\gamma \circ \rho(\|x\|_{[0,T)}) < \|x\|_{[0,T)}$, and thus

$$
\|x\|_{[0,T)} \leq \beta(|x(0)|, 0) := \sigma(|x(0)|). \tag{2.7}
$$

It can be directly verified that if (2.7) is not true, then (2.6) cannot be satisfied.

This means that $x(t)$ is defined on $[0, \infty)$, and the T in (2.7) can be directly replaced by ∞, which implies

$$
|x(t)| \leq \sigma(|x(0)|) \tag{2.8}
$$

for all $t \geq 0$. GS at the origin, in the sense of Definition 1.7, is proved.

For system Ξ_1, with the time-invariance property, (1.29) implies

$$
|x(t)| \leq \max\{\beta(|x(t_0)|, t - t_0), \gamma(\|u\|_{[t_0,\infty)})\} \tag{2.9}
$$

for any $0 \leq t_0 \leq t$. Then, by choosing $t_0 = t/2$ and using $|u| \leq \rho(|x|)$, we have

$$
\begin{aligned}
|x(t)| &\leq \max\left\{\beta\left(\left|x\left(\frac{t}{2}\right)\right|, \frac{t}{2}\right), \gamma\left(\|u\|_{[t/2,\infty)}\right)\right\} \\
&\leq \max\left\{\beta\left(\left|x\left(\frac{t}{2}\right)\right|, \frac{t}{2}\right), \gamma \circ \rho\left(\|x\|_{[t/2,\infty)}\right)\right\}.
\end{aligned} \tag{2.10}
$$

Note that property (2.8) implies $|x(t/2)| \leq \sigma(|x(0)|)$ for all $t \geq 0$. Thus,

$$
\begin{aligned}
|x(t)| &\leq \max\left\{\beta\left(\sigma(|x(0)|), \frac{t}{2}\right), \gamma \circ \rho\left(\|x\|_{[t/2,\infty)}\right)\right\} \\
&:= \max\left\{\bar{\beta}\left(|x(0)|, t\right), \gamma \circ \rho\left(\|x\|_{[t/2,\infty)}\right)\right\}.
\end{aligned} \tag{2.11}
$$

With Lemma D.1, there exists a $\hat{\beta} \in \mathcal{KL}$ such that

$$|x(t)| \leq \hat{\beta}(|x(0)|, t). \tag{2.12}$$

The GAS property is thus proved.

Recall that GAS is equivalent to the conjunction of GS and global convergence; see the discussions below Definition 1.7. An alternative proof that does not use Lemma D.1 can be carried out by proving the global convergence property. This is not provided in this book, but one may consider it as a special case of the proof of Theorem 2.1 for interconnected ISS systems by using the AG arguments in Appendix D.2. A drawback of this approach is that one may not get an explicit $\hat{\beta} \in \mathcal{KL}$ for the GAS property (2.12).

Property (2.2) of subsystem Ξ_2 implies that

$$|u(t)| \leq \rho(\|x\|_\infty) \tag{2.13}$$

for all $t \geq 0$, which can be considered as a special case of (1.29) with ρ considered as the gain. Then, condition (2.4) means that the composition of the gains of the two subsystems is less than the identity function. Such a condition is called the nonlinear small-gain condition.

This chapter reviews the small-gain results for interconnected systems composed of ISS or more general IOS subsystems. Specifically, Section 2.1 extends the above robustness analysis to the trajectory-based small-gain results developed in [130]. Due to the importance of Lyapunov functions, the Lyapunov-based ISS small-gain theorem originally developed in [126] is reviewed in Section 2.2. Section 2.3 introduces the basic idea of the important gain assignment technique [130, 223, 123, 125] for small-gain designs and provides a case study of applying the small-gain theorem to robust control of nonlinear uncertain systems. Note that small-gain results for more general systems, e.g., systems modeled by retarded functional differential equations, have also been developed in the literature; see e.g., [137]. Some related topics on large-scale dynamic networks are discussed in Chapter 3 in the cyclic-small-gain framework.

2.1 TRAJECTORY-BASED SMALL-GAIN THEOREM

Consider an interconnected system composed of two subsystems

$$\dot{x}_1 = f_1(x, u_1) \tag{2.14}$$

$$\dot{x}_2 = f_2(x, u_2), \tag{2.15}$$

where $x = [x_1^T, x_2^T]^T$ with $x_1 \in \mathbb{R}^{n_1}$ and $x_2 \in \mathbb{R}^{n_2}$ is the state, $u_1 \in \mathbb{R}^{m_1}$ and $u_2 \in \mathbb{R}^{m_2}$ are external inputs, and $f_1 : \mathbb{R}^{n_1+n_2} \times \mathbb{R}^{m_1} \to \mathbb{R}^{n_1}$ and $f_2 : \mathbb{R}^{n_1+n_2} \times \mathbb{R}^{m_2} \to \mathbb{R}^{n_2}$ are locally Lipschitz functions satisfying $f_1(0,0) = 0$ and $f_2(0,0) = 0$. For convenience of notation, define $u = [u_1^T, u_2^T]^T$. By considering

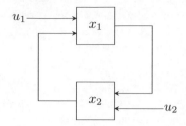

FIGURE 2.2 An interconnected system with external inputs.

u as a function of time, assume that it is measurable and locally essentially bounded.

For $i = 1, 2$, assume that each x_i-subsystem is ISS with x_{3-i} and u_i as the inputs. Specifically, for each $i = 1, 2$, there exist $\beta_i \in \mathcal{KL}$ and $\gamma_{i(3-i)}, \gamma_i^u \in \mathcal{K}$ such that for any initial state $x_i(0) = x_{i0}$ and any measurable and locally essentially bounded inputs x_{3-i}, u_i, it holds that

$$|x_i(t)| \leq \max\{\beta_i(|x_{i0}|, t), \gamma_{i(3-i)}(\|x_{3-i}\|_\infty), \gamma_i^u(\|u_i\|_\infty)\} \qquad (2.16)$$

for all $t \geq 0$. Here, the ISS property of the subsystems are in the "max" form, which is equivalent to the "plus" form used in [130].

With the discussions below, we show that the interconnected system is ISS with u as the input if

$$\gamma_{12} \circ \gamma_{21} < \mathrm{Id}. \qquad (2.17)$$

It should be noted that for any $\gamma_{12}, \gamma_{21} \in \mathcal{K}$,

$$\gamma_{12} \circ \gamma_{21} < \mathrm{Id} \Leftrightarrow \gamma_{21} \circ \gamma_{12} < \mathrm{Id}. \qquad (2.18)$$

Indeed, for the implication "$\Rightarrow$", assume that $\gamma_{21} \circ \gamma_{12} < \mathrm{Id}$ does not hold. That is, there exists a positive s such that $\gamma_{21}(\gamma_{12}(s)) \geq s$. Then, $\gamma_{12} \circ \gamma_{21}(\gamma_{12}(s)) \geq \gamma_{12}(s)$, which leads to a contradiction with $\gamma_{12} \circ \gamma_{21} < \mathrm{Id}$. By symmetry, the other implication "$\Leftarrow$" holds.

Theorem 2.1 presents a trajectory-based ISS small-gain result.

Theorem 2.1 *Consider the interconnected system composed of two subsystems in the form of (2.14)–(2.15) satisfying (2.16). The interconnected system is ISS with u as the input if the small-gain condition (2.17) is satisfied.*

Proof. The proof is basically a reduced version of the proof for the IOS small-gain theorem given in [130]. We only make slight modifications to handle the difference between the two forms of the ISS property. Pick any specific initial state $x(0)$ and any measurable and locally essentially bounded input u.

Step 1–UBIBS: Suppose that the solution $x(t)$ of the interconnected system is defined on $[0, T)$ with $T > 0$. Define $\sigma_i(s) = \beta_i(s, 0)$ for $s \in \mathbb{R}_+$. For $i = 1, 2$, by using the ISS property (2.16), one has

$$|x_i(t)| \leq \max\{\sigma_i(|x_i(0)|), \gamma_{i(3-i)}(\|x_{3-i}\|_{[0,T)}), \gamma_i^u(\|u_i\|_\infty)\} \qquad (2.19)$$

for $0 \leq t < T$, and thus, by taking the supremum of $|x_i(t)|$ over $[0, T)$, we have

$$\|x_i\|_{[0,T)} \leq \max\{\sigma_i(|x_i(0)|), \gamma_{i(3-i)}(\|x_{3-i}\|_{[0,T)}), \gamma_i^u(\|u_i\|_\infty)\}. \qquad (2.20)$$

By substituting (2.20) with i replaced by $3 - i$ in the right-hand side of (2.19), it is achieved that

$$
\begin{aligned}
|x_i(t)| \leq \max\{ & \sigma_i(|x_i(0)|), \gamma_{i(3-i)} \circ \sigma_{3-i}(|x_{3-i}(0)|), \\
& \gamma_{i(3-i)} \circ \gamma_{(3-i)i}(\|x_i\|_{[0,T)}), \\
& \gamma_{i(3-i)} \circ \gamma_{3-i}^u(\|u_{3-i}\|_\infty), \gamma_i^u(\|u_i\|_\infty)\} \\
\leq \max\{ & \sigma_i(|x_i(0)|), \gamma_{i(3-i)} \circ \sigma_{3-i}(|x_{3-i}(0)|), \\
& \gamma_{i(3-i)} \circ \gamma_{(3-i)i}(\|x_i\|_{[0,T)}), \\
& \gamma_{i(3-i)} \circ \gamma_{3-i}^u(\|u_{3-i}\|_\infty), \gamma_i^u(\|u_i\|_\infty)\}. \qquad (2.21)
\end{aligned}
$$

Define

$$\bar{\sigma}_{i1}(s) = \max\{\sigma_i(s), \gamma_{i(3-i)} \circ \sigma_{3-i}(s)\}, \qquad (2.22)$$

$$\bar{\sigma}_{i2}(s) = \max\{\gamma_i^u(s), \gamma_{i(3-i)} \circ \gamma_{3-i}^u(s)\} \qquad (2.23)$$

for $s \in \mathbb{R}_+$. By taking the supremum of $x_i(t)$ over $[0, T)$ and using (2.21), one has

$$
\begin{aligned}
\|x_i\|_{[0,T)} & \leq \max\{\bar{\sigma}_{i1}(|x(0)|), \bar{\sigma}_{i2}(\|u\|_\infty), \gamma_{i(3-i)} \circ \gamma_{(3-i)i}(\|x_i\|_{[0,T)})\} \\
& \leq \max\{\bar{\sigma}_{i1}(|x(0)|), \bar{\sigma}_{i2}(\|u\|_\infty)\}, \qquad (2.24)
\end{aligned}
$$

where the small-gain condition (2.17) is used for the last inequality. This means that $|x_i(t)|$ is defined on $[0, \infty)$. With the T in (2.24) replaced by ∞, it is achieved that

$$|x_i(t)| \leq \max\{\bar{\sigma}_{i1}(|x(0)|), \bar{\sigma}_{i2}(\|u\|_\infty)\} \qquad (2.25)$$

for all $t \geq 0$. UBIBS property is proved as property (2.25) holds for any initial state $x(0)$ and any measurable and locally essentially bounded input u.

Step 2–ISS: Denote $x_i^* = \max\{\bar{\sigma}_{i1}(|x(0)|), \bar{\sigma}_{i2}(\|u\|_\infty)\}$ for $i = 1, 2$.

By means of time invariance and causality, (2.16) implies

$$|x_i(t)| \leq \max\left\{\beta_i(|x_i(t_0)|, t - t_0), \gamma_{i(3-i)}(\|x_{3-i}\|_{[t_0,t]}), \gamma_i^u(\|u_i\|_\infty)\right\} \qquad (2.26)$$

for any $0 \leq t_0 \leq t$, and thus, by choosing $t_0 = t/2$, we have

$$|x_i(t)| \leq \max\left\{\beta_i\left(\left|x_i\left(\frac{t}{2}\right)\right|, \frac{t}{2}\right), \gamma_{i(3-i)}\left(\|x_{3-i}\|_{[t/2,t]}\right), \gamma_i^u(\|u_i\|_\infty)\right\}$$

$$\leq \max\left\{\beta_i\left(x_i^*, \frac{t}{2}\right), \gamma_{i(3-i)}\left(\|x_{3-i}\|_{[t/2,t]}\right), \gamma_i^u(\|u_i\|_\infty)\right\} \quad (2.27)$$

for $i = 1, 2$.

By taking the maximum of $x_i(t)$ over $[t/2, t]$, it is achieved that

$$\|x_i\|_{[t/2,t]}$$

$$\leq \max_{t/2 \leq \tau \leq t}\left\{\beta_i\left(\left|x_i\left(\frac{\tau}{2}\right)\right|, \frac{\tau}{2}\right), \gamma_{i(3-i)}\left(\|x_{3-i}\|_{[\tau/2,\tau]}\right), \gamma_i^u(\|u_i\|_\infty)\right\}$$

$$\leq \max\left\{\beta_i\left(x_i^*, \frac{t}{4}\right), \gamma_{i(3-i)}\left(\|x_{3-i}\|_{[t/4,t]}\right), \gamma_i^u(\|u_i\|_\infty)\right\} \quad (2.28)$$

for $i = 1, 2$.

Then, by substituting (2.28) with i replaced by $3 - i$ into (2.27), one has

$$|x_i(t)| \leq \max\left\{\beta_i\left(x_i^*, \frac{t}{2}\right), \gamma_{i(3-i)} \circ \beta_{3-i}\left(x_{3-i}^*, \frac{t}{4}\right),\right.$$

$$\gamma_{i(3-i)} \circ \gamma_{(3-i)i}\left(\|x_i\|_{[t/4,t]}\right), \gamma_{i(3-i)} \circ \gamma_{3-i}^u(\|u_{3-i}\|_\infty),$$

$$\left.\gamma_i^u(\|u_i\|_\infty)\right\}. \quad (2.29)$$

Recall the $x_i^* = \max\{\bar{\sigma}_{i1}(|x(0)|), \bar{\sigma}_{i2}(\|u\|_\infty)\}$ for $i = 1, 2$. Property (2.29) implies that

$$|x_i(t)| \leq \max\left\{\bar{\beta}_i(|x(0)|, t), \gamma_{i(3-i)} \circ \gamma_{(3-i)i}\left(\|x_i\|_{[t/4,t]}\right), \bar{\gamma}_i^u(\|u\|_\infty)\right\} \quad (2.30)$$

for all $t \geq 0$, where

$$\bar{\beta}_i(s, t) = \max\left\{\beta_i\left(\bar{\sigma}_{i1}(s), \frac{t}{2}\right), \gamma_{i(3-i)} \circ \beta_{3-i}\left(\bar{\sigma}_{(3-i)1}(s), \frac{t}{4}\right)\right\}, \quad (2.31)$$

$$\bar{\gamma}_i^u(s) = \max\left\{\gamma_i^u(s), \beta_i\left(\bar{\sigma}_{i2}(s), 0\right), \gamma_{i(3-i)} \circ \gamma_{3-i}^u(s),\right.$$

$$\left.\gamma_{i(3-i)} \circ \beta_{3-i}\left(\bar{\sigma}_{(3-i)2}(s), 0\right)\right\}. \quad (2.32)$$

Clearly, $\bar{\beta}_i \in \mathcal{KL}$, $\bar{\gamma}_i^u \in \mathcal{K}$.

Then, by using Lemma D.1, there exists a $\hat{\beta}_i \in \mathcal{KL}$ such that

$$|x_i(t)| \leq \max\left\{\hat{\beta}_i(|x(0)|, t), \bar{\gamma}_i^u(\|u\|_\infty)\right\} \quad (2.33)$$

for all $t \geq 0$, for $i = 1, 2$. Note that property (2.33) holds for any initial state $x(0)$ and any measurable and locally essentially bounded u. The ISS of the interconnected system is proved. $\diamond$

It should be noted that by using Theorem 1.5, Theorem 2.1 can also be proved by showing the UBIBS and AG properties of the interconnected system. The proof of AG is provided in Appendix D.2 for interested readers.

If γ_{12} or γ_{21} is zero, then the interconnected system is reduced to a cascade system, for which the small-gain condition is satisfied automatically. If moreover, $u_1 = u_2 = 0$, then Theorem 2.1 is reduced to [144, Lemma 4.7] for GAS. Also note that if $\gamma_1^u = \gamma_2^u = 0$ and one of the subsystems, say the x_2-subsystem, has $\beta_2 = 0$, then the result of Theorem 2.1 is reduced to the robust stability result given at the beginning of this chapter.

The small-gain result developed in [130] can cover the more general case in which the subsystems are interconnected with each other by outputs instead of states. Consider the following interconnected system:

$$\dot{x}_i = f_i(x_i, y_{3-i}, u_i) \tag{2.34}$$
$$y_i = h_i(x_i) \tag{2.35}$$

where, for $i = 1, 2$, $x_i \in \mathbb{R}^{n_i}$ is the state, $u_i \in \mathbb{R}^{m_i}$ is the input, $y_i \in \mathbb{R}^{l_i}$ is the output, and f_i, h_i are locally Lipschitz functions satisfying $f_i(0, 0, 0) = 0$ and $h_i(0) = 0$.

Assume that each i-th subsystem is UO with zero offset and IOS with y_{3-i}, u_i as the inputs and y_i as the output. Specifically, there exist $\alpha_i^O \in \mathcal{K}_\infty$, $\beta_i \in \mathcal{KL}$, $\gamma_{i(3-i)} \in \mathcal{K}$, and $\gamma_i^u \in \mathcal{K}$ such that

$$|x_i(t)| \leq \alpha_i^O \left(|x_i(0)| + \|y_{3-i}\|_{[0,t]} + \|u_i\|_{[0,t]} \right) \tag{2.36}$$
$$|y_i(t)| \leq \max\{\beta_i(|x_i(0)|, t), \gamma_{i(3-i)}(\|y_{3-i}\|_{[0,t]}), \gamma_i^u(\|u_i\|_{[0,t]})\} \tag{2.37}$$

for all $t \in [0, T_{\max})$, where $[0, T_{\max})$ with $0 < T_{\max} \leq \infty$ is the right maximal interval for the definition of $(x_1(t), x_2(t))$.

Theorem 2.2 gives a small-gain result for the interconnected IOS system.

Theorem 2.2 *Consider the interconnected system* (2.34)–(2.35) *satisfying* (2.36) *and* (2.37) *for* $i = 1, 2$. *Then the interconnected system is UO and IOS if*

$$\gamma_{12} \circ \gamma_{21} < \mathrm{Id}. \tag{2.38}$$

Theorem 2.2 does not assume the forward completeness of the subsystems. Following the discussions in [130], IOS together with UO implies the forward completeness of the subsystems. If the small-gain condition is satisfied, then the forward completeness of the interconnected system is guaranteed by the IOS and UO properties of the subsystems. In [134], Theorem 2.2 is generalized for large-scale dynamic networks composed of more than two subsystems. This result is reviewed in Chapter 3.

Reference [130] also takes into account the practical convergence issues by introducing the notion of input-to-output practical stability (IOpS) property, and the small-gain theorem therein is more general than Theorem 2.2. Further extensions of [130] can be found in [133]. References [135, 167, 169] as well as the book [137] show the extensions of the small-gain theorem to more general complex systems such as hybrid systems and systems modeled by retarded functional differential equations.

2.2 LYAPUNOV-BASED SMALL-GAIN THEOREM

Lyapunov functions play an irreplaceable role in the analysis and control of nonlinear systems. With the Lyapunov-based formulation of ISS, the ISS property of nonlinear systems is often tested by constructing ISS-Lyapunov functions. This section reviews the Lyapunov-based ISS small-gain theorem developed in [126] for feedback systems. In particular, it is shown that if an interconnected system satisfies the Lyapunov-based ISS small-gain condition, then ISS-Lyapunov functions can be constructed for the system by using the ISS-Lyapunov functions of the subsystems.

For the interconnected system (2.14)–(2.15), assume that each x_i-subsystem for $i = 1, 2$ admits a continuously differentiable ISS-Lyapunov function $V_i : \mathbb{R}^{n_i} \to \mathbb{R}_+$ satisfying the following:

1. there exist $\underline{\alpha}_i, \overline{\alpha}_i \in \mathcal{K}_\infty$ such that

$$\underline{\alpha}_i(|x_i|) \leq V_i(x_i) \leq \overline{\alpha}_i(|x_i|), \quad \forall x_i; \tag{2.39}$$

2. there exist $\chi_{i(3-i)}, \chi_i^u \in \mathcal{K}$ and a continuous, positive definite α_i such that

$$V_i(x_i) \geq \max\{\chi_{i(3-i)}(V_{3-i}(x_{3-i})), \chi_i^u(|u_i|)\}$$
$$\Rightarrow \nabla V_i(x_i) f_i(x, u_i) \leq -\alpha_i(V_i(x_i)), \quad \forall x, \ u_i. \tag{2.40}$$

Theorem 2.3 gives a Lyapunov formulation of the ISS small-gain theorem.

Theorem 2.3 *Interconnected system* (2.14)–(2.15) *with each x_i-subsystem admitting an ISS-Lyapunov function V_i satisfying* (2.39)–(2.40) *is ISS if the following small-gain condition is satisfied:*

$$\chi_{12} \circ \chi_{21} < \text{Id}. \tag{2.41}$$

Proof. Theorem 2.3 is proved by constructing an ISS-Lyapunov function V for the interconnected system.

For $\chi_{12}, \chi_{21} \in \mathcal{K}$ satisfying the small-gain condition (2.41), we find a $\sigma \in \mathcal{K}_\infty$ such that it is continuously differentiable on $(0, \infty)$ and satisfies

$$\sigma > \chi_{21}, \quad \sigma^{-1} > \chi_{12}. \tag{2.42}$$

This can be achieved because for $\chi_{12}, \chi_{21} \in \mathcal{K}$ satisfying condition (2.41), there exists a $\hat{\chi}_{12} \in \mathcal{K}_\infty$ such that $\hat{\chi}_{12} > \chi_{12}$ and $\hat{\chi}_{12} \circ \chi_{21} < \mathrm{Id}$. One can always find a $\sigma \in \mathcal{K}_\infty$ such that it is continuously differentiable on $(0, \infty)$ and satisfies

$$\chi_{21} < \sigma < \hat{\chi}_{12}^{-1}, \tag{2.43}$$

which guarantees the satisfaction of (2.42). See [126] for the detailed proof of the existence of such σ.

An ISS-Lyapunov function candidate for the interconnected system is defined as

$$V(x) = \max \left\{ \sigma(V_1(x_1)), V_2(x_2) \right\}. \tag{2.44}$$

Clearly, V is positive definite and radially unbounded. Also, V is continuously differentiable almost everywhere.

Let $f(x, u) = [f_1^T(x, u_1), f_2^T(x, u_2)]^T$. In the following procedure, we prove that there exist a $\chi \in \mathcal{K}$ and a continuous, positive definite α such that

$$V(x) \geq \chi(|u|) \Rightarrow \nabla V(x) f(x, u) \leq -\alpha(V(x)) \tag{2.45}$$

for almost all x and all u.

For this purpose, define the following sets, as shown in Figure 2.3:

$$A = \{(x_1, x_2) : V_2(x_2) < \sigma(V_1(x_1))\}, \tag{2.46}$$
$$B = \{(x_1, x_2) : V_2(x_2) > \sigma(V_1(x_1))\}, \tag{2.47}$$
$$O = \{(x_1, x_2) : V_2(x_2) = \sigma(V_1(x_1))\}. \tag{2.48}$$

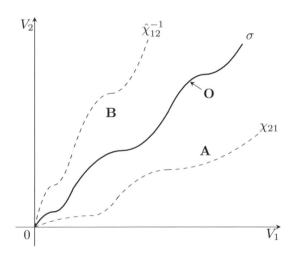

FIGURE 2.3 Definitions of sets A, B and O.

For any fixed point $p = (p_1, p_2) \neq (0, 0)$ and a control value $v = (v_1, v_2)$, consider the following three cases.

Case 1: $p \in A$

In this case, $V(x) = \sigma(V_1(x_1))$ in a neighborhood of p, and consequently

$$\nabla V(p) f(p, v) = \frac{\partial \sigma(V_1(p_1))}{\partial V_1(p_1)} \nabla V_1(p_1) f_1(p, v_1). \qquad (2.49)$$

For $p \in A$, it holds that $V_2(p_2) < \sigma(V_1(p_1))$, and based on the definition of σ, $V_1(p_1) > \chi_{12}(V_2(p_2))$. With (2.40), this implies

$$\nabla V_1(p_1) f_1(p, v_1) \le -\alpha_1(v_1(p_1)) \qquad (2.50)$$

whenever $V_1(p_1) \ge \sigma \circ \chi_1^u(|v_1|)$. It follows that, for $p \in A$,

$$\nabla V(p) f(p, v) \le -\hat{\alpha}_1(V(p)) \qquad (2.51)$$

whenever $V(p) \ge \hat{\chi}_1^u(|v_1|)$, where

$$\hat{\chi}_1^u(s) = \sigma \circ \chi_1^u(s) \qquad (2.52)$$

for $s \in \mathbb{R}_+$, and $\hat{\alpha}_1$ is a continuous and positive definite function such that

$$\hat{\alpha}_1(s) \le \sigma^d(\sigma^{-1}(s)) \alpha_1(\sigma^{-1}(s)) \qquad (2.53)$$

for $s > 0$, with $\sigma^d(s) = d\sigma(s)/ds$.

Case 2: $p \in B$

In this case, by using similar arguments as in Case 1, it can be proved that

$$\nabla V(p) f(p, v) \le -\hat{\alpha}_2(V(p)) \qquad (2.54)$$

whenever $V(p) \ge \hat{\chi}_2^u(|v_2|)$, where $\hat{\alpha}_2 = \alpha_2$ and $\hat{\chi}_2^u = \chi_2^u$.

Case 3: $p \in O$

First note that it holds for the locally Lipschitz function V that

$$\nabla V(p) f(p, v) = \left. \frac{d}{dt} \right|_{t=0} V(\varphi(t)) \qquad (2.55)$$

for almost all p and all v, with $\varphi(t) = [\varphi_1^T(t), \varphi_2^T(t)]^T$ being the solution of the initial-value problem

$$\dot{\varphi}(t) = f(\varphi(t), v), \quad \varphi(0) = p. \qquad (2.56)$$

In this case, assume $p = (p_1, p_2) \ne (0, 0)$ and

$$V_1(p_1) \ge \chi_1^u(|v_1|), \qquad (2.57)$$
$$V_2(p_2) \ge \chi_2^u(|v_2|). \qquad (2.58)$$

Then, by using similar arguments as for Cases 1 and 2, one has

$$\nabla\sigma(V_1(p_1))f_1(p,v_1) \leq -\hat{\alpha}_1(V(p)) \tag{2.59}$$

$$\nabla V_2(p_2)f_2(p,v_2) \leq -\hat{\alpha}_2(V(p)) \tag{2.60}$$

where $\hat{\alpha}_1$ and $\hat{\alpha}_2$ are continuous and positive definite functions.

Note that in this case $p_1 \neq 0$ and $p_2 \neq 0$. Then, because of the continuous differentiability of σ, V_1, and V_2, and the continuity of f, there exist neighborhoods $\mathcal{X}_1$ of p_1 and $\mathcal{X}_2$ of p_2 such that

$$\nabla\sigma(V_1(x_1))f_1(x,v_1) \leq -\hat{\alpha}_1(V(p)), \tag{2.61}$$

$$\nabla V_2(x_2)f_2(x,v_2) \leq -\hat{\alpha}_2(V(p)) \tag{2.62}$$

for all $x \in \mathcal{X}_1 \times \mathcal{X}_2$. Note also that there exists $\delta > 0$ such that $\varphi(t) \in \mathcal{X}_1 \times \mathcal{X}_2$ for all $0 \leq t < \delta$.

Now pick $\Delta t \in (0,\delta)$. If $\varphi(\Delta t) \in A \cup O$, then

$$V(\varphi(\Delta t)) - V(p) = \sigma(V_1(\varphi_1(\Delta t))) - \sigma(V_1(p_1))$$

$$\leq -\frac{1}{2}\hat{\alpha}_1(V(p))\Delta t. \tag{2.63}$$

Similarly, if $\varphi(\Delta t) \in B \cup O$, then

$$V(\varphi(\Delta t)) - V(p) = V_2(\varphi_2(\Delta t)) - V_2(p_2)$$

$$\leq -\frac{1}{2}\hat{\alpha}_2(V(p))\Delta t. \tag{2.64}$$

Hence, if V is differentiable at p, then

$$\nabla V(p)f(p,v) \leq -\alpha(V(p)) \tag{2.65}$$

where $\alpha(s) = \min\{\hat{\alpha}_1(s)/2, \hat{\alpha}_1(s)/2\}$ for $s \in \mathbb{R}_+$. Note that conditions (2.57) and (2.58) can be guaranteed by

$$V(p) \geq \max\{\hat{\chi}_1^u(|v_1|), \hat{\chi}_2^u(|v_2|)\}. \tag{2.66}$$

By combining the three cases, it can be concluded that

$$V(p) \geq \max\{\hat{\chi}_1^u(|v_1|), \hat{\chi}_2^u(|v_2|)\} \Rightarrow \nabla V(p)f(p,v) \leq -\alpha(V(p)). \tag{2.67}$$

Since V is continuously differentiable almost everywhere, (2.67) holds for almost all p and all v. Property (2.45) is then proved by defining $\chi(s) = \max\{\hat{\chi}_1^u(s), \hat{\chi}_2^u(s)\}$ for $s \in \mathbb{R}_+$. Thus, V is an ISS-Lyapunov function of the interconnected system. Theorem 2.3 is proved. $\diamond$

Actually, the selected σ is not necessarily continuously differentiable on $(0, \infty)$, as the constructed V is not required to be continuously differentiable. A similar proof can be carried out with a $\sigma \in \mathcal{K}_\infty$ which is locally Lipschitz on $(0, \infty)$. More generally, the small-gain condition and the construction of the ISS-Lyapunov function are still valid even if V_1 and V_2 are locally Lipschitz on $\mathbb{R}^{n_1}\backslash\{0\}$ and $\mathbb{R}^{n_2}\backslash\{0\}$, respectively. With the techniques presented in [170], a smooth ISS-Lyapunov function can be constructed based on the V defined above.

2.3   SMALL-GAIN CONTROL DESIGN

If a system can be transformed into an interconnection of ISS subsystems through control design, then one may employ the ISS small-gain theorem to analyze the stability property of the closed-loop system. This section introduces a small-gain control design method for nonlinear uncertain systems based on the gain assignment technique [130, 223, 123, 125].

2.3.1   GAIN ASSIGNMENT

Assigning an appropriate ISS gain to a system by means of feedback is a key step in applying the ISS small-gain theorem to nonlinear control design. This subsection introduces the basic idea of the gain assignment technique, whereby a system is transformed by feedback into one with a given Lyapunov function and specific ISS gains. We illustrate with the following first-order system:

$$\dot{\eta} = \phi(\eta, w_1, w_2) + \bar{\kappa} \tag{2.68}$$

where $\eta \in \mathbb{R}$ is the state, $\bar{\kappa} \in \mathbb{R}$ is the control input, $w_1 \in \mathbb{R}^{m_1}, w_2 \in \mathbb{R}^{m_2}$ represent external disturbance inputs, and the nonlinear function $\phi : \mathbb{R}^{m_1+m_2+1} \to \mathbb{R}$ satisfies

$$|\phi(\eta, w_1, w_2)| \le \psi_\phi^\eta(|\eta|) + \sum_{i=1,2} \psi_\phi^{w_i}(|w_i|), \quad \forall \eta, \ w_1, \ w_2, \tag{2.69}$$

with $\psi_\phi^\eta, \psi_\phi^{w_1}, \psi_\phi^{w_2} \in \mathcal{K}_\infty$. Define

$$V(\eta) = \alpha_V(|\eta|) \tag{2.70}$$

with $\alpha_V(s) = s^2/2$ for $s \in \mathbb{R}_+$. We look for a feedback control law in the form of

$$\bar{\kappa} = \kappa(\eta) \tag{2.71}$$

such that the closed-loop system composed of (2.68) and (2.71) with w_1, w_2 as the external inputs is ISS with V defined in (2.70) as an ISS-Lyapunov function. Moreover, the closed-loop system will be designed to have specific ISS gains $\chi_\eta^{w_1}, \chi_\eta^{w_2} \in \mathcal{K}_\infty$ corresponding to the external inputs. To further realize small-gain-based recursive control design, the control law κ is expected to be continuously differentiable.

For any constants $\epsilon, \ell > 0$, we find a $\nu : \mathbb{R}_+ \to \mathbb{R}_+$ which is positive, nondecreasing, and continuously differentiable on $(0, \infty)$ such that

$$\nu(s)s \ge \psi_\phi^\eta(s) + \sum_{i=1,2} \psi_\phi^{w_i} \circ \left(\chi_\eta^{w_i}\right)^{-1} \circ \alpha_V(s) + \frac{\ell}{2}s \tag{2.72}$$

for all $s \geq \sqrt{2}\epsilon$. This is achievable by using Lemma C.8 because the right-hand side of (2.72) is a class $\mathcal{K}_\infty$ function of s. Inequality (2.72) enables construction of a feedback law to ensure the specified ISS properties for the closed loop.

Define

$$\kappa(r) = -\nu(|r|)r \tag{2.73}$$

for $r \in \mathbb{R}$. Clearly, κ is odd, strictly decreasing, radially unbounded, and continuously differentiable on $(-\infty, 0) \cup (0, \infty)$. Direct calculation yields:

$$\lim_{r \to 0^+} \frac{d\kappa(r)}{dr} = \lim_{r \to 0^+} \left(-\nu(r) - \frac{d\nu(r)}{dr} r \right), \tag{2.74}$$

$$\lim_{r \to 0^-} \frac{d\kappa(r)}{dr} = \lim_{r \to 0^-} \left(-\nu(-r) + \frac{d\nu(-r)}{d(-r)} r \right)$$

$$= \lim_{r' \to 0^+} \left(-\nu(r') - \frac{d\nu(r')}{d(r')} r' \right), \tag{2.75}$$

which implies $\lim_{r \to 0^+} \frac{d\kappa(r)}{dr} = \lim_{r \to 0^-} \frac{d\kappa(r)}{dr}$. Thus, κ is continuously differentiable on $\mathbb{R}$.

Recall the definition of V in (2.70). Consider the case of

$$V(\eta) \geq \max_{i=1,2} \{\chi_\eta^{w_i}(|w_i|), \epsilon\}. \tag{2.76}$$

In this case, $|w_i| \leq \left(\chi_\eta^{w_i}\right)^{-1} \circ \alpha_V(|\eta|)$ for $i = 1, 2$, and $|\eta| \geq \sqrt{2}\epsilon$. Direct calculation yields:

$$\begin{aligned}
\nabla V(\eta)(\phi(\eta, w_1, w_2) + \kappa(\eta)) &= \eta(\phi(\eta, w_1, w_2) + \bar{\kappa}) \\
&= \eta(\phi(\eta, w_1, w_2) - \nu(|\eta|)\eta) \\
&\leq |\eta||\phi(\eta, w_1, w_2)| - \nu(|\eta|)|\eta|^2 \\
&\leq |\eta| \left(\psi_\phi^\eta(|\eta|) + \sum_{i=1,2} \psi_\phi^{w_i}(|w_i|) - \nu(|\eta|)|\eta| \right) \\
&\leq -\frac{\ell}{2}|\eta|^2 = -\ell V(\eta).
\end{aligned} \tag{2.77}$$

As a result, for any specific $\chi_\eta^{w_1}, \chi_\eta^{w_2} \in \mathcal{K}_\infty$ and constants $\epsilon, \ell > 0$, one can find a continuously differentiable, odd, strictly decreasing, and radially unbounded κ in the form of (2.73), such that V satisfies

$$V(\eta) \geq \max_{i=1,2} \{\chi_\eta^{w_i}(|w_i|), \epsilon\}$$

$$\Rightarrow \nabla V(\eta)(\phi(\eta, w_1, w_2) + \kappa(\eta)) \leq -\ell V(\eta), \quad \forall \eta, \ w_1, \ w_2. \tag{2.78}$$

It should be noted that even if $w_1 = w_2 = 0$, property (2.78) can only guarantee that $V(\eta(t))$ ultimately converges to within the region $V(\eta) \leq \epsilon$. This means practical convergence. If we also consider ϵ as an external input, then the system is ISS with V as an ISS-Lyapunov function. More precisely, the system is said to be input-to-state practically stable (ISpS); see [130] for the definition of ISpS. With the gain assignment technique, the Lyapunov-based ISS gains $\chi_\eta^{w_1}, \chi_\eta^{w_2}$ can be chosen to be any class $\mathcal{K}_\infty$ functions.

By using Lemma C.8, if ψ_η^η and $\psi_\phi^{w_i} \circ \left(\chi_\phi^{w_i}\right)^{-1} \circ \alpha_V$ for $i = 1, 2$ are Lipschitz on compact sets, then we can choose $\epsilon = 0$ for (2.72). If ψ_ϕ^η and $\psi_\phi^{w_i}$ are Lipschitz on compact sets, then we may choose $\chi_\eta^{w_i} = \alpha_V \circ \left(\vartheta_\eta^{w_i}\right)^{-1}$ or $\chi_\eta^{w_i} = \left(\vartheta_\eta^{w_i}\right)^{-1} \circ \alpha_V$ with $\vartheta_\eta^{w_i} \in \mathcal{K}_\infty$ being Lipschitz on compact sets.

In the following chapters, nontrivial modifications of the gain assignment technique will be made to solve the specific problems.

2.3.2 SMALL-GAIN CONTROL DESIGN: A CASE STUDY

Reference [123] successfully applied the IOS small-gain theorem to recursive control design of general cascade nonlinear systems with dynamic uncertainties. In this subsection, we consider a much simpler case to show the basic approach.

Consider the following nonlinear system in the strict-feedback form [153]:

$$\dot{x}_i = x_{i+1} + \Delta_i(\bar{x}_i), \quad i = 1, \ldots, n-1 \tag{2.79}$$

$$\dot{x}_n = u + \Delta_n(\bar{x}_n) \tag{2.80}$$

where $[x_1, \ldots, x_n]^T \in \mathbb{R}^n$ is the state, $\bar{x}_i = [x_1, \ldots, x_i]^T$ and $u \in \mathbb{R}$ is the control input. It is assumed that, for each $i = 1, \ldots, n$, there exists a $\psi_{\Delta_i} \in \mathcal{K}_\infty$ such that

$$|\Delta_i(\bar{x}_i)| \leq \psi_{\Delta_i}(|\bar{x}_i|). \tag{2.81}$$

Through small-gain design, the $[x_1, \ldots, x_n]^T$-system is recursively transformed into a new $[e_1, \ldots, e_n]^T$-system with ISS e_i-subsystems by defining coordinate transformation

$$\begin{bmatrix} e_1 \\ e_2 \\ \vdots \\ e_n \end{bmatrix} = \begin{bmatrix} x_1 \\ x_2 - \kappa_1(e_1) \\ \vdots \\ x_n - \kappa_{n-1}(e_{n-1}) \end{bmatrix} \tag{2.82}$$

and control law

$$u = \kappa_n(e_n) \tag{2.83}$$

where $\kappa_1, \ldots, \kappa_n : \mathbb{R} \to \mathbb{R}$ are appropriately chosen functions.

Define an ISS-Lyapunov function candidate for each e_i-subsystem as

$$V_{e_i}(e_i) = \alpha_V(|e_i|) \tag{2.84}$$

with $\alpha_V(s) = s^2/2$ for $s \in \mathbb{R}_+$. For convenience of notation, denote $\bar{e}_i = [e_1, \ldots, e_i]^T$ and $e_{n+1} = x_{n+1} - \kappa_n(e_n)$ with $x_{n+1} = u$. The e_i-subsystems for $i = 1, \ldots, n$ are designed to be ISS one-by-one.

Initial Step: The e_1-subsystem

The e_1-subsystem is in the following form:

$$\dot{e}_1 = x_2 + \Delta_1(e_1). \tag{2.85}$$

Recall that $e_2 = x_2 - \kappa_1(e_1)$. Then, the e_1-subsystem can be rewritten as

$$\begin{aligned} \dot{e}_1 &= \kappa_1(e_1) + e_2 + \Delta_1(e_1) \\ &:= \kappa_1(e_1) + \Delta_1^*(\bar{e}_2). \end{aligned} \tag{2.86}$$

With (2.81) satisfied, from the definition of Δ_1^*, we can find $\psi_{\Delta_i^*}^{e_1}, \psi_{\Delta_i^*}^{e_2} \in \mathcal{K}_\infty$ such that $|\Delta_1^*(\bar{e}_2)| \leq \psi_{\Delta_i^*}^{e_1}(|e_1|) + \psi_{\Delta_i^*}^{e_2}(|e_2|)$. With the gain assignment technique introduced in Subsection 2.3.1, for any specified constants $\epsilon_{e_1}, \ell_{e_1} > 0$ and $\gamma_{e_1}^{e_2} \circ \alpha_V \in \mathcal{K}_\infty$, we can find a continuously differentiable, odd, strictly decreasing, and radially unbounded function κ_1 such that

$$\begin{aligned} V_{e_1}(e_1) &\geq \max\{\gamma_{e_1}^{e_2} \circ \alpha_V(|e_2|), \epsilon_{e_1}\} \\ &\Rightarrow \nabla V_{e_1}(e_1)(\kappa_1(e_1) + \Delta_1^*(\bar{e}_2)) \leq -\ell_{e_1} V_{e_1}(e_1), \end{aligned} \tag{2.87}$$

and thus,

$$\begin{aligned} V_{e_1}(e_1) &\geq \max\{\gamma_{e_1}^{e_2}(V_{e_2}(e_2)), \epsilon_{e_1}\} \\ &\Rightarrow \nabla V_{e_1}(e_1)(\kappa_1(e_1) + \Delta_1^*(\bar{e}_2)) \leq -\ell_{e_1} V_{e_1}(e_1). \end{aligned} \tag{2.88}$$

Recursive Step: The e_i-subsystem ($i = 2, \ldots, n$)

Suppose that the $\bar{e}_{i-1}$-subsystem has been designed to be in the following form:

$$\dot{e}_1 = \kappa_1(e_1) + \Delta_1^*(\bar{e}_2) \tag{2.89}$$

$$\vdots$$

$$\dot{e}_{i-1} = \kappa_{i-1}(e_{i-1}) + \Delta_{i-1}^*(\bar{e}_i) \tag{2.90}$$

where $\kappa_1, \ldots, \kappa_{i-1}$ are appropriately chosen continuously differentiable, odd, strictly decreasing, and radially unbounded functions.

For convenience of notation, denote $\dot{\bar{e}}_{i-1} = F_{i-1}(\bar{e}_i)$.

Also suppose that the $\bar{e}_{i-1}$-subsystem is ISS with an ISS-Lyapunov function $V_{\bar{e}_{i-1}}$ satisfying

$$\underline{\alpha}_{\bar{e}_{i-1}}(|\bar{e}_{i-1}|) \leq V_{\bar{e}_{i-1}}(\bar{e}_{i-1}) \leq \overline{\alpha}_{\bar{e}_{i-1}}(|\bar{e}_{i-1}|), \tag{2.91}$$

and

$$V_{\bar{e}_{i-1}}(\bar{e}_{i-1}) \geq \max\{\gamma^{e_i}_{\bar{e}_{i-1}}(V_{e_i}(e_i)), \epsilon_{\bar{e}_{i-1}}\}$$
$$\Rightarrow \nabla V_{\bar{e}_{i-1}}(\bar{e}_{i-1})F_{i-1}(\bar{e}_i) \leq -\alpha_{\bar{e}_{i-1}}(V_{\bar{e}_{i-1}}(\bar{e}_{i-1})) \quad \text{a.e.}, \qquad (2.92)$$

where $\underline{\alpha}_{\bar{e}_{i-1}}, \overline{\alpha}_{\bar{e}_{i-1}}, \gamma^{e_i}_{\bar{e}_{i-1}} \in \mathcal{K}_\infty$, $\epsilon_{\bar{e}_{i-1}} > 0$ is a constant and $\alpha_{\bar{e}_{i-1}}$ is a continuous and positive definite function.

By taking the derivative of e_i, we have

$$\dot{e}_i = \dot{x}_i - \frac{\partial \kappa_{i-1}(e_{i-1})}{\partial e_{i-1}}\dot{e}_{i-1}$$

$$= x_{i+1} + \Delta_i(\bar{x}_i) - \frac{\partial \kappa_{i-1}(e_{i-1})}{\partial e_{i-1}}(\kappa_{i-1}(e_{i-1}) + \Delta^*_{i-1}(\bar{e}_i)). \qquad (2.93)$$

With the recursive definition (2.82), we can represent $\bar{x}_i$ with $\bar{e}_i$. Also note that $e_{i+1} = x_{i+1} - \kappa_i(e_i)$. Then, the e_i-subsystem can be rewritten as

$$\dot{e}_i = \kappa_i(e_i) + e_{i+1} + \Delta_i(\bar{x}_i) - \frac{\partial \kappa_{i-1}(e_{i-1})}{\partial e_{i-1}}(\kappa_{i-1}(e_{i-1}) + \Delta^*_{i-1}(\bar{e}_i))$$

$$:= \kappa_i(e_i) + \Delta^*_i(\bar{e}_{i+1}). \qquad (2.94)$$

It can be proved that there exist $\psi^{\bar{e}_{i-1}}_{\Delta^*_i}, \psi^{e_i}_{\Delta^*_i}, \psi^{e_{i+1}}_{\Delta^*_i} \in \mathcal{K}_\infty$ such that $|\Delta^*_i(\bar{e}_{i+1})| \leq \psi^{\bar{e}_{i-1}}_{\Delta^*_i}(|\bar{e}_{i-1}|) + \psi^{e_i}_{\Delta^*_i}(|e_i|) + \psi^{e_{i+1}}_{\Delta^*_i}(|e_{i+1}|)$. With the gain assignment technique, for any specified constants $\epsilon_{e_i}, \ell_{e_i} > 0$ and $\gamma^{\bar{e}_{i-1}}_{e_i} \circ \overline{\alpha}_{\bar{e}_{i-1}}, \gamma^{e_{i+1}}_{e_i} \circ \alpha_V \in \mathcal{K}_\infty$, we can find a continuously differentiable, odd, strictly decreasing, and radially unbounded function κ_i such that

$$V_{e_i}(e_i) \geq \max\{\gamma^{\bar{e}_{i-1}}_{e_i} \circ \overline{\alpha}_{\bar{e}_{i-1}}(|\bar{e}_{i-1}|), \gamma^{e_{i+1}}_{e_i} \circ \alpha_V(|e_{i+1}|), \epsilon_{e_i}\}$$
$$\Rightarrow \nabla V_{e_i}(e_i)(\kappa_i(e_i) + \Delta^*_i(\bar{e}_{i+1})) \leq -\ell_{e_i}V_i(e_i), \qquad (2.95)$$

and thus,

$$V_{e_i}(e_i) \geq \max\{\gamma^{\bar{e}_{i-1}}_{e_i}(V_{\bar{e}_{i-1}}(\bar{e}_{i-1})), \gamma^{e_{i+1}}_{e_i}(V_{e_{i+1}}(e_{i+1})), \epsilon_{e_i}\}$$
$$\Rightarrow \nabla V_{e_i}(e_i)(\kappa_i(e_i) + \Delta^*_i(\bar{e}_{i+1})) \leq -\ell_{e_i}V_i(e_i). \qquad (2.96)$$

The $\bar{e}_i$-subsystem is an interconnection of the $\bar{e}_{i-1}$-subsystem and the e_i-subsystem, as shown in Figure 2.4. For convenience of notation, denote $\dot{\bar{e}}_i = F_i(\bar{e}_{i+1})$. With the Lyapunov-based ISS small-gain theorem given in Section 2.2, the $\bar{e}_i$-subsystem is ISS if we choose $\gamma^{\bar{e}_{i-1}}_{e_i}$ such that

$$\gamma^{\bar{e}_{i-1}}_{e_i} \circ \gamma^{e_i}_{\bar{e}_{i-1}} < \text{Id}. \qquad (2.97)$$

Moreover, we can construct an ISS-Lyapunov function $V_{\bar{e}_i}$ for the $\bar{e}_i$-subsystem as

$$V_{\bar{e}_i}(\bar{e}_i) = \max\{\sigma_{\bar{e}_{i-1}}(V_{\bar{e}_{i-1}}(\bar{e}_{i-1})), V_{e_i}(e_i)\} \qquad (2.98)$$

where $\sigma_{\bar{e}_{i-1}} \in \mathcal{K}_\infty$ is continuously differentiable on $(0, \infty)$ and satisfies $\sigma_{\bar{e}_{i-1}} > \gamma_{e_i}^{\bar{e}_{i-1}}$ and $\sigma_{\bar{e}_{i-1}} \circ \gamma_{\bar{e}_{i-1}}^{e_i} < \text{Id}$.

Then, there exist $\underline{\alpha}_{\bar{e}_i}, \overline{\alpha}_{\bar{e}_i}, \gamma_{\bar{e}_i}^{e_{i+1}} \in \mathcal{K}_\infty$, constants $\epsilon_{\bar{e}_i} > 0$ and $\alpha_{\bar{e}_i}$ that are continuous and positive definite such that

$$\underline{\alpha}_{\bar{e}_i}(|\bar{e}_i|) \leq V_{\bar{e}_i}(\bar{e}_i) \leq \overline{\alpha}_{\bar{e}_i}(|\bar{e}_i|), \tag{2.99}$$

$$V_{\bar{e}_i}(\bar{e}_i) \geq \max\{\gamma_{\bar{e}_i}^{e_{i+1}}(V_{e_{i+1}}(e_{i+1})), \epsilon_{\bar{e}_i}\}$$
$$\Rightarrow \nabla V_{\bar{e}_i}(\bar{e}_i) F_i(\bar{e}_{i+1}) \leq -\alpha_{\bar{e}_i}(V_{\bar{e}_i}(\bar{e}_i)) \quad \text{a.e.} \tag{2.100}$$

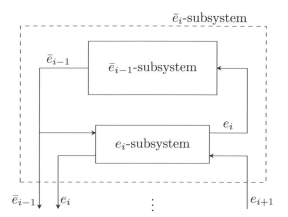

FIGURE 2.4 Small-gain-based recursive control design.

In the case of $i = n$, the control input $u = x_{n+1}$ occurs and $e_{n+1} = 0$. We can construct an ISS-Lyapunov function $V_{\bar{e}_n}$ for the $\bar{e}_n$-system, i.e., the e-system, which satisfies (2.100) with $i = n$ and $e_{i+1} = 0$. In this way, the closed-loop system is designed to be ISS with the ϵ_{e_i}'s as the inputs, or more precisely, practically stable [78, 156]. With the gain assignment technique, the ϵ_{e_i}'s can be chosen arbitrarily small, and the closed-loop signals e_i's ultimately converge to within an arbitrarily small neighborhood of the origin.

Actually, for systems in the form of (2.79)–(2.80), if each Δ_i satisfies (2.81) with $\psi_{\Delta_i} \in \mathcal{K}_\infty$ being Lipschitz on compact sets, then asymptotic stabilization can be achieved with $\epsilon_{\bar{e}_n} = 0$. For this purpose, one may choose the $\gamma_{(\cdot)}^{(\cdot)}$'s and the $\sigma_{(\cdot)}$'s such that their inverse functions are Lipschitz on compact sets. Then, each ISS-Lyapunov function $V_{\bar{e}_i}$, recursively constructed with $V_{e_1}, \ldots, V_{e_i}$, satisfies property (2.99) with $\overline{\alpha}_{\bar{e}_i}$ in the form of

$$\overline{\alpha}_{\bar{e}_i} = (\vartheta_{\bar{e}_i})^{-1} \circ \alpha_V \tag{2.101}$$

where $\vartheta_{\bar{e}_i} \in \mathcal{K}_\infty$ is Lipschitz on compact sets.

Then, the $\gamma_{e_i}^{\bar{e}_{i-1}} \circ \overline{\alpha}_{\bar{e}_{i-1}}$ used for the design of each e_i-subsystem is in the form of $\left(\vartheta'_{\bar{e}_{i-1}}\right)^{-1} \circ \alpha_V$ with $\vartheta'_{\bar{e}_{i-1}} = \vartheta_{\bar{e}_{i-1}} \circ \left(\gamma_{e_i}^{\bar{e}_{i-1}}\right)^{-1}$ being Lipschitz on compact sets. Note that $\gamma_{e_i}^{\bar{e}_{i-1}} \circ \overline{\alpha}_{\bar{e}_{i-1}}$ and $\gamma_{e_i}^{e_{i+1}} \circ \alpha_V$ correspond to the $\chi_\phi^{w_1}$ and $\chi_\phi^{w_2}$ in (2.72) for gain assignment. This fulfills the condition given at the end of Subsection 2.3.1 for zero ϵ.

2.4 NOTES

The small-gain condition, i.e., a loop-gain of less than unity, is one way to ensure stability of interconnected systems. In the past twenty years, tremendous efforts have been made in stability analysis and control design of interconnected nonlinear systems. The idea of the small-gain theorem was originally studied with the gain property taking a linear or affine form; see, e.g., [48, 283] for input-output feedback systems, as well as the recent works [20, 76]. The small-gain theorem for nonlinear feedback systems with non-affine gains was presented in [88, 199] within the input-output context.

Taking explicit advantage of Sontag's seminal work on ISS [249, 241, 242], the first generalized, nonlinear ISS small-gain theorem was proposed in [130]. The IOS counterpart of the small-gain theorem is also available in [130]. As a fundamental difference with respect to the earlier small-gain theorems, in the ISS or IOS framework, the role of the initial conditions is made explicit to ensure asymptotic stability in the Lyapunov sense as well as bounded-input bounded-output stability. A new small-gain design tool was presented for the first time in [123, 130] for robust global stabilization of nonlinear systems with dynamic uncertainties. In parallel, Teel presents a small-gain tool for the analysis and synthesis of control systems with saturation in [260]. A Lyapunov reformulation of the ISS small-gain theorem can be found in [126]. Necessary and sufficient small-gain conditions for interconnected integral input-to-state stable (iISS) systems can be found in [107, 108, 110]. Further extensions of the small-gain theorem have also been made for general nonlinear systems possibly with time delays using the concept of vector Lyapunov functions [138]. As a powerful tool, the ISS small-gain theorem has been included in standard textbooks on nonlinear systems; see, e.g., [106, 144]. See also the book [137] and the references cited therein for other more recent developments along the line of ISS small-gain.

This chapter mainly focuses on continuous-time interconnected systems described by differential equations, while the counterparts of the results for discrete-time systems [131, 154] and hybrid systems [208, 167, 211, 169] have also been developed based on the corresponding extensions of ISS. The interconnected hybrid systems studied in [169] may involve both stable and unstable dynamics.

There have also been numerous successful applications of the small-gain theorem to nonlinear control designs. The applications of the small-gain theorem to output regulation and global stabilization of nonlinear feedforward systems can be found in [100, 31, 32, 30, 99]. References [208, 25, 169] employ the small-gain theorem for networked and quantized control designs. In [220], the authors employ a modified small-gain theorem to solve the stability problem arising from observer-based control designs. Another interesting application of the small-gain theorem lies in robust adaptive dynamic programming; see e.g., [120].

Chapter 3 introduces an extension of the small-gain theorem to large-scale dynamic networks which contain more than two subsystems. More related discussions are given in Chapter 3.

3 Large-Scale Dynamic Networks

The small-gain theorem introduced in Chapter 2 has found wide application in stability analysis, stabilization, robust adaptive control, observer design, and output regulation for interconnected nonlinear systems. Although one may use the small-gain theorem recursively for interconnected systems involving more than one cycle, refined small-gain criteria are highly desired to handle large-scale dynamic networks more efficiently.

Example 3.1 shows a control system which is transformable into an interconnected system of three ISS subsystems and contains more than one cycle in the system structure.

Example 3.1 *Consider the single-mode approximation of the PDE model of an axial compressor introduced in [206]:*

$$\dot{R} = \sigma R(-2\phi - \phi^2 - R), \quad R(t) \geq 0 \tag{3.1}$$

$$\dot{\phi} = -\psi - \frac{3}{2}\phi + \frac{1}{2} - \frac{1}{2}(\phi+1)^3 - 3(\phi+1)R \tag{3.2}$$

$$\dot{\psi} = \frac{1}{\beta^2}(\phi + 1 - v), \tag{3.3}$$

where ϕ and ψ are the deviations of the mass flow and the pressure rise from their set points, R is the nonnegative magnitude of the first stall mode, the control input v is the flow through the throttle, and σ, β are positive constants. For this system, ψ and R are not measurable. The control objective is to stabilize the system and make ϕ asymptotically converge to the origin.

In [153, Section 2.4], a state-feedback control law is designed for global asymptotic stabilization. Reference [152] improves the design by only using ϕ and ψ for feedback. By only using the measurement of ϕ, [196] realize semi-global stabilization with a high-gain observer. Global asymptotic stabilization is achieved in [10] through an ISS-induced output-feedback design.

For convenience of discussions, we denote $z = R$, $x_1 = \phi$, $x_2 = -\psi$, $y = x_1$, and $u = v/\beta^2$, and rewrite the system as

$$\dot{z} = g(z, x_1) \tag{3.4}$$

$$\dot{x}_1 = f_1(x_1, z) + x_2 \tag{3.5}$$

$$\dot{x}_2 = f_2(x_1) + u \tag{3.6}$$

$$y = x_1. \tag{3.7}$$

In [10], by using y and u, a reduced-order observer with state $\hat{x}_2$ is designed to estimate the unmeasurable x_2 such that the estimation error system with state $\hat{x}_2 - x_2$ is ISS with z as the input. A control law $u = u(y, \hat{x}_2)$ is designed such that the (x_1, x_2)-subsystem is ISS with both $\hat{x}_2 - x_2$ and z as the inputs. The z-subsystem is also proved to be ISS with x_1 as the input. Thus, the closed-loop system is a dynamic network composed of three ISS subsystems. The system structure is shown in Figure 3.1.

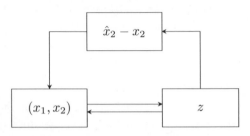

FIGURE 3.1 The block diagram of the closed-loop system in Example 3.1.

The axial compressor model considered in this example is in the widely recognized output-feedback form. With the tools developed in this book, we can solve more complicated control problems, e.g., quantized control and distributed control, with this class of nonlinear systems.

To illustrate the need of small-gain results for network analysis in more general terms than the example, consider a nonlinear dynamic network composed of three subsystems:

$$\dot{x}_i = f_i(x), \quad i = 1, 2, 3, \tag{3.8}$$

where $x_i \in \mathbb{R}^{n_i}$ is the state of the i-th subsystem, $x = [x_1^T, x_2^T, x_3^T]^T$, and $f_i : \mathbb{R}^{n_1 + n_2 + n_3} \to \mathbb{R}^{n_i}$ is a locally Lipschitz function satisfying $f_i(0) = 0$.

Suppose that each x_i-subsystem has an ISS-Lyapunov function V_i, which is positive definite and radially unbounded, and satisfies

$$V_i(x_i) \geq \max_{j \neq i} \{\gamma_{ij}(V_j(x_j))\} \Rightarrow \nabla V_i(x_i) f_i(x) \leq -\alpha_i(V_i(x_i)), \quad \forall x, \tag{3.9}$$

where $\gamma_{ij} \in \mathcal{K} \cup \{0\}$ represents the ISS gains and α_i is a continuous and positive definite function. We consider the case in which only $\gamma_{12}, \gamma_{13}, \gamma_{21}, \gamma_{32}, \gamma_{31}$ are nonzero ISS gains.

The gain interconnection structure of the dynamic network can be represented with a digraph, called the gain digraph, by considering the subsystems as vertices and the nonzero gain interconnections as directed links. Since the gain digraph describes the relation between the Lyapunov functions, each x_i-subsystem is represented with its ISS-Lyapunov function V_i. The gain digraph of the dynamic network defined above is shown in Figure 3.2.

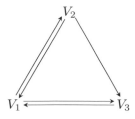

FIGURE 3.2 The gain digraph of the dynamic network (3.8).

We consider using the small-gain theorem introduced in Chapter 2 twice to analyze the stability property of the dynamic network. First, we divide the dynamic network into two parts: the (x_1, x_2)-subsystem with x_3 as the input and the x_3-subsystem with (x_1, x_2) as the input.

The (x_1, x_2)-subsystem is ISS because the small-gain condition is satisfied:

$$\gamma_{12} \circ \gamma_{21} < \text{Id}. \tag{3.10}$$

We construct an ISS-Lyapunov function for the (x_1, x_2)-subsystem as

$$V_{(1,2)}(x_1, x_2) = \max\{V_1(x_1), \sigma(V_2(x_2))\}, \tag{3.11}$$

where σ is a class $\mathcal{K}_\infty$ function that is continuously differentiable on $(0, \infty)$ and satisfies

$$\sigma > \gamma_{12}, \quad \sigma^{-1} > \gamma_{21}. \tag{3.12}$$

Then, it holds that

$$\begin{aligned}
&V_{(1,2)}(x_1, x_2) \geq \gamma_{13}(V_3(x_3)) \\
&\Rightarrow \nabla V_{(1,2)}(x_1, x_2) f_{(1,2)}(x) \leq -\alpha_{(1,2)}(V_{(1,2)}(x_1, x_2)) \quad \text{a.e.,}
\end{aligned} \tag{3.13}$$

where $f_{(1,2)}(x) := [f_1^T(x), f_2^T(x)]^T$ and $\alpha_{(1,2)}$ is a continuous and positive definite function.

From (3.9), the influence of $V_{(1,2)}(x_1, x_2)$ to $V_3(x_3)$ can be represented by

$$\begin{aligned}
&V_3(x_3) \geq \gamma_{3(1,2)}(V_{(1,2)}(x_1, x_2)) \\
&\Rightarrow \nabla V_3(x_3) f_3(x) \leq -\alpha_3(V_3(x_3)),
\end{aligned} \tag{3.14}$$

where $\gamma_{3(1,2)}(s) := \max\{\gamma_{31}(s), \gamma_{32} \circ \sigma^{-1}(s)\}$ for $s \geq 0$.

Then, we consider the interconnection of the (x_1, x_2)-subsystem and the x_3-subsystem. The dynamic network is asymptotically stable at the origin if it satisfies the small-gain condition $\gamma_{13} \circ \gamma_{3(1,2)} < \text{Id}$, or equivalently,

$$\gamma_{13} \circ \gamma_{31} < \text{Id}, \tag{3.15}$$

$$\gamma_{13} \circ \gamma_{32} \circ \sigma^{-1} < \text{Id}. \tag{3.16}$$

The satisfaction of condition (3.16) depends on the choice of σ, which is subject to constraints $\sigma > \gamma_{12}$ and $\sigma^{-1} > \gamma_{21}$. Note that (3.10) guarantees the existence of σ to satisfy the constraints. By choosing σ such that $\sigma^{-1} > \gamma_{21}$ and σ^{-1} is very close to γ_{21}, (3.16) can be guaranteed by

$$\gamma_{13} \circ \gamma_{32} \circ \gamma_{21} < \text{Id}. \tag{3.17}$$

Thus, the dynamic network is asymptotically stable at the origin if (3.10), (3.15), and (3.17) are satisfied. This means that the composition of the ISS gains along every simple cycle in the gain digraph should be less than Id. This condition is referred to as the cyclic-small-gain condition.

Considering the wide interest in studying large-scale dynamic networks, it is natural to ask:

1. Is the cyclic-small-gain condition valid for general large-scale dynamic networks composed of ISS subsystems?
2. How do we construct an ISS-Lyapunov function for a dynamic network if it satisfies the cyclic-small-gain condition?

In this chapter, we develop cyclic-small-gain results to solve the problems for continuous-time, discrete-time, and more general hybrid dynamic networks. To make the results more accessible, we mainly consider ISS systems in this chapter, while some extensions to IOS systems are also provided.

The first three sections of this chapter study continuous-time, discrete-time, and hybrid dynamic networks, respectively. The fourth section discusses the development of the relevant literature.

3.1 CONTINUOUS-TIME DYNAMIC NETWORKS

Consider the following large-scale dynamic network containing N subsystems:

$$\dot{x}_i = f_i(x, u_i), \quad i = 1, \ldots, N, \tag{3.18}$$

where $x = \begin{bmatrix} x_1^T, \ldots, x_N^T \end{bmatrix}^T$ with $x_i \in \mathbb{R}^{n_i}$ is the state, $u_i \in \mathbb{R}^{m_i}$ represents the external inputs, and each $f_i : \mathbb{R}^{n+m_i} \to \mathbb{R}^{n_i}$ with $n = \sum_{j=1}^{N} n_j$ is a locally Lipschitz function satisfying $f_i(0, 0) = 0$. The external input $u = \begin{bmatrix} u_1^T, \ldots, u_N^T \end{bmatrix}^T$ is a measurable and locally essentially bounded function from $\mathbb{R}_+$ to $\mathbb{R}^m$ with $m = \sum_{i=1}^{N} m_i$. Denote $f(x, u) = [f_1^T(x, u_1), \ldots, f_N^T(x, u_N)]^T$.

Assume that for $i = 1, \ldots, N$, each x_i-subsystem admits a continuously differentiable ISS-Lyapunov function $V_i : \mathbb{R}^{n_i} \to \mathbb{R}_+$ satisfying the following

1. there exist $\underline{\alpha}_i, \overline{\alpha}_i \in \mathcal{K}_\infty$ such that

$$\underline{\alpha}_i(|x_i|) \leq V_i(x_i) \leq \overline{\alpha}_i(|x_i|), \quad \forall x_i; \tag{3.19}$$

2. there exist $\gamma_{ij} \in \mathcal{K} \cup \{0\}$ $(j = 1, \ldots, N, j \neq i)$ and $\gamma_{ui} \in \mathcal{K} \cup \{0\}$ such that

$$V_i(x_i) \geq \max_{j \neq i} \{\gamma_{ij}(V_j(x_j)), \gamma_{ui}(|u_i|)\}$$
$$\Rightarrow \nabla V_i(x_i) f_i(x, u_i) \leq -\alpha_i(V_i(x_i)), \quad \forall x, \ u_i, \qquad (3.20)$$

where α_i is a continuous and positive definite function.

For systems that are formulated in the dissipation form, property 2 above should be replaced by

$2'$. there exist $\alpha_i' \in \mathcal{K}_\infty$, $\sigma_{ij}' \in \mathcal{K} \cup \{0\}$ $(j = 1, \ldots, N, j \neq i)$ and $\sigma_{ui}' \in \mathcal{K} \cup \{0\}$ such that

$$\nabla V_i(x_i) f_i(x, u_i) \leq -\alpha_i'(V_i(x_i)) + \max\{\sigma_{ij}'(V_j(x_j)), \sigma_{ui}'(|u_i|)\}. \qquad (3.21)$$

Due to the equivalence of the two forms for continuous-time systems, we only consider the gain margin form in the following discussions.

By considering the subsystems as vertices and the nonzero gain interconnections as directed links, the gain interconnection structure of the dynamic network can be represented by a digraph, called the gain digraph. Then, concepts from graph theory, such as path, reachability, and simple cycle, can be used to describe the gain interconnections in the dynamic network. Since the gains are defined with Lyapunov functions, for $i = 1, \ldots, N$, each x_i-subsystem is represented by its Lyapunov function V_i. Appendix A gives the definitions of the related notions in graph theory.

Theorem 3.1 answers the question on the validity of the cyclic-small-gain condition for continuous-time large-scale dynamic networks with subsystems admitting ISS-Lyapunov functions.

Theorem 3.1 *Consider the continuous-time dynamic network* (3.18) *with each x_i-subsystem admitting a continuously differentiable ISS-Lyapunov function V_i satisfying* (3.19)–(3.20). *Then, it is ISS with x as the state and u as the input if for every simple cycle $(V_{i_1}, V_{i_2}, \ldots, V_{i_r}, V_{i_1})$ in the gain digraph,*

$$\gamma_{i_1 i_2} \circ \gamma_{i_2 i_3} \circ \cdots \circ \gamma_{i_r i_1} < \mathrm{Id}, \qquad (3.22)$$

where $r = 2, \ldots, N$ and $1 \leq i_j \leq N$, $i_j \neq i_{j'}$ if $j \neq j'$.

Condition (3.22) means that the composition of the ISS gains along every simple cycle in the gain digraph is less than the identity function Id. We prove Theorem 3.1 by constructing ISS-Lyapunov functions. In this way, the problem of constructing ISS-Lyapunov functions for the large-scale dynamic networks is solved at the same time.

3.1.1   BASIC IDEA OF CONSTRUCTING ISS-LYAPUNOV FUNCTIONS

The small-gain theorem introduced in Chapter 2 considers the case in which dynamic network (3.18) contains two subsystems, i.e., $N = 2$. In this case, if $\gamma_{12} \circ \gamma_{21} < \text{Id}$, then the dynamic network is ISS and an ISS-Lyapunov function can be constructed as:

$$V(x) = \max\{V_1(x_1), \sigma(V_2(x_2))\}, \qquad (3.23)$$

where $\sigma \in \mathcal{K}_\infty$ is continuously differentiable on $(0, \infty)$ and satisfies

$$\sigma > \gamma_{12}, \quad \sigma^{-1} > \gamma_{21}. \qquad (3.24)$$

Recall the fact that $\gamma_{12} \circ \gamma_{21} < \text{Id} \Leftrightarrow \gamma_{21} \circ \gamma_{12} < \text{Id}$. By using Lemma C.1 twice, there exist $\hat{\gamma}_{12}, \hat{\gamma}_{21} \in \mathcal{K}_\infty$ which are continuously differentiable on $(0, \infty)$ and satisfy $\hat{\gamma}_{12} > \gamma_{12}, \hat{\gamma}_{21} > \gamma_{21}$ and $\hat{\gamma}_{12} \circ \hat{\gamma}_{21} < \text{Id}$. Thus, with γ_{12}, γ_{21} replaced by $\hat{\gamma}_{12}, \hat{\gamma}_{21}$ (as shown Figure 3.3), the small-gain condition is still satisfied.

FIGURE 3.3   The replacement of the ISS gains.

If we choose $\sigma = \hat{\gamma}_{12}$, then condition (3.24) is satisfied and the resulting ISS-Lyapunov function is

$$V(x) = \max\{V_1(x_1), \hat{\gamma}_{12}(V_2(x_2))\}. \qquad (3.25)$$

Since $\hat{\gamma}_{12}$ is a modification of the ISS gain γ_{12}, the term $\hat{\gamma}_{12}(V_2(x_2))$ can be considered as the "potential influence" of V_2 acting on V_1 with modified gain $\hat{\gamma}_{12}$.

3.1.2   A CLASS OF ISS-LYAPUNOV FUNCTIONS FOR DYNAMIC NETWORKS

Based on the idea of potential influence, a class of ISS-Lyapunov functions are constructed for large-scale dynamic networks satisfying the cyclic-small-gain condition.

Recall the fact that for any $\chi_1, \chi_2 \in \mathcal{K} \cup \{0\}$, $\chi_1 \circ \chi_2 < \text{Id} \Leftrightarrow \chi_2 \circ \chi_1 < \text{Id}$. Consider a dynamic network in the form of (3.18) with the cyclic-small-gain condition (3.22) satisfied. For each $i^* = 1, \ldots, N$, it holds that

$$\gamma_{i^* i_2} \circ \gamma_{i_2 i_3} \circ \cdots \circ \gamma_{i_r i^*} < \text{Id} \qquad (3.26)$$

for $r = 2, \ldots, N$, $1 \le i_j \le N$, $i_j \ne i^*$, $i_j \ne i_{j'}$ if $j \ne j'$. With Lemma C.1, if $\gamma_{i^* i_2} \ne 0$, then one can find a $\hat{\gamma}_{i^* i_2} \in \mathcal{K}_\infty$ which is continuously differentiable on $(0, \infty)$ and satisfies $\hat{\gamma}_{i^* i_2} > \gamma_{i^* i_2}$ such that (3.26) still holds with $\gamma_{i^* i_2}$ replaced by $\hat{\gamma}_{i^* i_2}$.

By repeating this procedure for all the $\gamma_{i^* i_2}$ with $i^* = 1, \ldots, N$, $i_2 \ne i^*$, there exist $\hat{\gamma}_{(\cdot)}$'s such that

1. $\hat{\gamma}_{(\cdot)} \in \mathcal{K}_\infty$ and $\hat{\gamma}_{(\cdot)} > \gamma_{(\cdot)}$ if $\gamma_{(\cdot)} \in \mathcal{K}$; $\hat{\gamma}_{(\cdot)} = 0$ if $\gamma_{(\cdot)} = 0$.
2. $\hat{\gamma}_{(\cdot)}$'s are continuously differentiable on $(0, \infty)$.
3. for each $r = 2, \ldots, N$,

$$\hat{\gamma}_{i_1 i_2} \circ \cdots \circ \hat{\gamma}_{i_r i_1} < \mathrm{Id} \tag{3.27}$$

holds for all $1 \le i_j \le N$ and $i_j \ne i_{j'}$ if $j \ne j'$.

Through the approach above, all the nonzero gains in the dynamic network are replaced by the $\hat{\gamma}_{(\cdot)}$'s, which are of class $\mathcal{K}_\infty$ and continuously differentiable on $(0, \infty)$ such that the cyclic-small-gain condition is still satisfied. Note that the replacement of the nonzero gains does not influence the gain digraph.

In the large-scale dynamic network, the potential influence acting on the p-th subsystem from all the subsystems can be described as

$$\mathbb{V}^{[p]} = \bigcup_{j=1, \ldots, N} \mathbb{V}_j^{[p]}(x) \tag{3.28}$$

with

$$\mathbb{V}_j^{[p]}(x) = \left\{ \hat{\gamma}_{i_1^{[p]} i_2^{[p]}} \circ \cdots \circ \hat{\gamma}_{i_{j-1}^{[p]} i_j^{[p]}} \left(V_{i_j^{[p]}} \left(x_{i_j^{[p]}} \right) \right) \right\},$$

where $i_1^{[p]} = p$, $i_k^{[p]} \in \{1, \ldots, N\}$, $k \in \{1, \ldots, j\}$, $i_k^{[p]} \ne i_{k'}^{[p]}$ if $k \ne k'$, for $j = 1, \ldots, N$. Clearly, each element in $\mathbb{V}_j^{[p]}(x)$ corresponds to a simple path ending at V_p in the gain digraph.

Note that $\hat{\gamma}_{(\cdot)} \in \mathcal{K}_\infty \cup \{0\}$. It is easy to verify that $\max \mathbb{V}^{[p]}$ is positive definite and radially unbounded with respect to the Lyapunov functions of the subsystems with indices belonging to $\mathcal{RS}(p)$.

Correspondingly, the potential influence of the external input $u = [u_1^T, \ldots, u_N^T]^T$ acting on the p-th subsystem can be described as

$$\mathbb{U}^{[p]} = \bigcup_{j=1, \ldots, N} \mathbb{U}_j^{[p]} \tag{3.29}$$

with

$$\mathbb{U}_j^{[p]} = \left\{ \hat{\gamma}_{i_1^{[p]} i_2^{[p]}} \circ \cdots \circ \hat{\gamma}_{i_{j-1}^{[p]} i_j^{[p]}} \circ \gamma_{u i_j^{[p]}} \left(|u_{i_{j-1}^{[p]}}| \right) \right\} \tag{3.30}$$

for $j = 1, \ldots, N$.

Define

$$V_\Pi(x) = \max \mathbb{V}_\Pi(x) = \max \left(\bigcup_{p \in \Pi} \mathbb{V}^{[p]}(x) \right) \qquad (3.31)$$

where set $\Pi \subseteq \{1, \ldots, N\}$ satisfies $\bigcup_{p \in \Pi}(\mathcal{RS}(p)) = \{1, \ldots, N\}$.

It can be directly verified that $\max \left(\bigcup_{p \in \Pi} \mathbb{V}^{[p]} \right)$ is positive definite and radially unbounded with respect to $\max \{V_1, \ldots, V_N\}$ and thus with respect to x, i.e., there exist $\underline{\alpha}, \overline{\alpha} \in \mathcal{K}_\infty$ such that $\underline{\alpha}(|x|) \leq V_\Pi(x) \leq \overline{\alpha}(|x|)$ for all x. It can also be observed that V_Π is locally Lipschitz on $\mathbb{R}^n \setminus \{0\}$. Thanks to Rademacher's theorem (see, e.g., [59, p. 216]), V_Π is differentiable almost everywhere.

Correspondingly, denote

$$u_\Pi = \max \mathbb{U}_\Pi = \max \left(\bigcup_{p \in \Pi} \mathbb{U}^{[p]} \right). \qquad (3.32)$$

It can be verified that there exists a $\gamma^u \in \mathcal{K}_\infty$ such that $u_\Pi \leq \gamma^u(|u|)$ for all u.

In Subsection 3.1.3, we show that $V_\Pi(x)$ is an ISS-Lyapunov function (not necessarily continuously differentiable) of the dynamic network with u_Π as the new input; see (3.54). Then, it is directly proved that the dynamic network is ISS with u as the input; see (3.55).

3.1.3 PROOF OF THE CYCLIC-SMALL-GAIN THEOREM FOR CONTINUOUS-TIME DYNAMIC NETWORKS

Throughout the proof, we consider the case where $V_\Pi(x) \geq u_\Pi$ and $x \neq 0$. Intuitively, if V_Π is strictly decreasing on the timeline, then it is an ISS-Lyapunov function of the dynamic network. The decreasing property of $V_\Pi = \max \mathbb{V}_\Pi$ is determined by the decrease in all the elements in $\mathbb{V}_\Pi$ that take the value of V_Π.

Note that the elements in $\mathbb{V}_\Pi$ are defined with the compositions of ISS gains along specific simple paths and each element corresponds to the ISS-Lyapunov function of one subsystem. For convenience of notation, we use a to mark the simple path corresponding to an element taking the value of V_Π and use A to denote the set of all such a's. Consider a specific simple path $m^a := (V_{i_j^a}, \ldots, V_{i_2^a}, V_{i_1^a})$ in the gain digraph. The simple path is highlighted by the thick arrows in Figure 3.4. According to the definition, for a fixed x, it holds that

$$\hat{\gamma}_{i_1^a i_2^a} \circ \cdots \circ \hat{\gamma}_{i_{k-1}^a i_k^a} \circ \cdots \circ \hat{\gamma}_{i_{j-1}^a i_j^a} \left(V_{i_j^a} \left(x_{i_j^a} \right) \right) = V_\Pi(x), \qquad (3.33)$$

where $i_k^a \in \{1, \ldots, N\}$, $k \in \{1, \ldots, j\}$, and $i_k^a \neq i_{k'}^a$ if $k \neq k'$, for all $a \in A$.

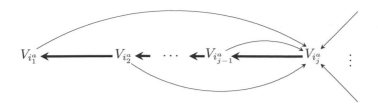

FIGURE 3.4 The j subsystems on a specified simple path m^a ending at $V_{i_1^a}$.

With the cyclic-small-gain condition (3.22) satisfied, by using property (3.33), we study the relation between $V_{i_j^a}$ and the Lyapunov functions of the other subsystems.

1. *Relation between $V_{i_j^a}$ and V_l with $l \in \{i_1^a, \ldots, i_{j-1}^a\}$*
   If $j \geq 2$, for all $k = 1, \ldots, j-1$, property (3.33) implies

$$\hat{\gamma}_{i_1^a i_2^a} \circ \cdots \circ \hat{\gamma}_{i_{k-1}^a i_k^a}\left(V_{i_k^a}(x_{i_k^a})\right)$$
$$\leq \hat{\gamma}_{i_1^a i_2^a} \circ \cdots \circ \hat{\gamma}_{i_{k-1}^a i_k^a} \circ \cdots \circ \hat{\gamma}_{i_{j-1}^a i_j^a}\left(V_{i_j^a}(x_{i_j^a})\right), \tag{3.34}$$

and thus,

$$V_{i_k^a}(x_{i_k^a}) \leq \hat{\gamma}_{i_k^a i_{k+1}^a} \circ \cdots \circ \hat{\gamma}_{i_{j-1}^a i_j^a}\left(V_{i_j^a}(x_{i_j^a})\right) \tag{3.35}$$

by canceling out $\hat{\gamma}_{i_1^a i_2^a} \circ \cdots \circ \hat{\gamma}_{i_{k-1}^a i_k^a}$.
Note that $V_{i_k^a}$ with $k = 1, \ldots, j-1$ represents the Lyapunov functions of all the subsystems on simple path m^a. Property (3.35) describes the relation between $V_{i_k^a}$ and $V_{i_j^a}$. If $j = 1$, then there is only the subsystem corresponding to $V_{i_1^a}$ on simple path m^a.
Condition (3.27) means that the composition of the modified gains $\hat{\gamma}_{(\cdot)}$'s along every simple cycle in the gain digraph is less than Id. Specifically, for the subsystems along simple path m^a, it holds that

$$\hat{\gamma}_{i_j^a i_k^a} \circ \hat{\gamma}_{i_k^a i_{k+1}^a} \circ \cdots \circ \hat{\gamma}_{i_{j-1}^a i_j^a} < \text{Id}. \tag{3.36}$$

Then, property (3.35) can be simplified as:

$$\hat{\gamma}_{i_j^a i_k^a}\left(V_{i_k^a}(x_{i_k^a})\right) \leq \hat{\gamma}_{i_j^a i_k^a} \circ \hat{\gamma}_{i_k^a i_{k+1}^a} \circ \cdots \circ \hat{\gamma}_{i_{j-1}^a i_j^a}\left(V_{i_j^a}(x_{i_j^a})\right)$$
$$< V_{i_j^a}(x_{i_j^a}) \tag{3.37}$$

for all $k = 1 \ldots, j-1$. Equivalently,

$$V_{i_j^a}(x_{i_j^a}) > \hat{\gamma}_{i_j^a l}\left(V_l(x_l)\right) \tag{3.38}$$

holds for all $l \in \{i_1^a, \ldots, i_{j-1}^a\}$.

2. *Relation between $V_{i_j^a}$ and V_l with $l \in \{1, \ldots, N\}\backslash\{i_1^a, \ldots, i_j^a\}$*

We first consider the case of $j \leq N-1$. For each $l \in \{1, \ldots, N\}\backslash\{i_1^a, \ldots, i_j^a\}$, if $\hat{\gamma}_{i_{j-1}^a l} \neq 0$, then $\hat{\gamma}_{i_1^a i_2^a} \circ \cdots \circ \hat{\gamma}_{i_{j-2}^a i_{j-1}^a} \circ \hat{\gamma}_{i_{j-1}^a l}(V_l(x_l))$ belongs to $\mathbb{V}_\Pi(x)$; otherwise $\hat{\gamma}_{i_1^a i_2^a} \circ \cdots \circ \hat{\gamma}_{i_{j-2}^a i_{j-1}^a} \circ \hat{\gamma}_{i_{j-1}^a l}(V_l(x_l)) = 0$. Thus, if $j \leq N-1$, then

$$\hat{\gamma}_{i_1^a i_2^a} \circ \cdots \circ \hat{\gamma}_{i_{j-2}^a i_{j-1}^a} \left(V_{i_{j-1}^a}(x_{i_{j-1}^a}) \right)$$
$$\geq \hat{\gamma}_{i_1^a i_2^a} \circ \cdots \circ \hat{\gamma}_{i_{j-2}^a i_{j-1}^a} \circ \hat{\gamma}_{i_{j-1}^a l}(V_l(x_l)), \tag{3.39}$$

holds for all $l \in \{1, \ldots, N\}\backslash\{i_1^a, \ldots, i_j^a\}$, which can be simplified by canceling out the common terms on both sides as

$$V_{i_j^a}(x_{i_j^a}) \geq \hat{\gamma}_{i_j^a l}(V_l(x_l)). \tag{3.40}$$

If $j = N$, then all the subsystems of the dynamic network are on simple path m^a.

Properties (3.38) and (3.40) together imply

$$V_{i_j^a}(x_{i_j^a}) \geq \max_{l \in \{1, \ldots, N\}\backslash\{i_j^a\}} \left\{ \hat{\gamma}_{i_j^a l}(V_l(x_l)) \right\}$$
$$\geq \max_{l \in \{1, \ldots, N\}\backslash\{i_j^a\}} \left\{ \gamma_{i_j^a l}(V_l(x_l)) \right\}. \tag{3.41}$$

From the definition of the new input u_Π, it can also be guaranteed that

$$V_\Pi(x) \geq u_\Pi \Rightarrow V_{i_j^a}(x) \geq \hat{\gamma}_{u i_j^a}(|u_{i_j^a}|). \tag{3.42}$$

For each $a \in A$, if conditions (3.41) and (3.42) hold, then the property of the ISS-Lyapunov functions, given in (3.20), yields:

$$\nabla V_{i_j^a}(x_{i_j^a}) f_{i_j^a}(x, u_{i_j^a}) \leq -\alpha_{i_j^a} \left(V_{i_j^a}(x_{i_j^a}) \right) \leq -\hat{\alpha}_{i_j^a}(V_\Pi(x)) \tag{3.43}$$

where $\hat{\alpha}_{i_j^a} := \alpha_{i_j^a} \circ \hat{\gamma}_{i_{j-1}^a i_j^a}^{-1} \circ \cdots \circ \hat{\gamma}_{i_1^a i_2^a}^{-1}$ and the second inequality holds due to (3.33).

Define

$$\hat{\gamma}_{m^a} = \hat{\gamma}_{i_1^a i_2^a} \circ \cdots \circ \hat{\gamma}_{i_{j-1}^a i_j^a}, \tag{3.44}$$
$$V_{m^a}(x_{i_j^a}) = \hat{\gamma}_{m^a} \left(V_{i_j^a}(x_{i_j^a}) \right). \tag{3.45}$$

For each $a \in A$, since $\hat{\gamma}_{i_1^a i_2^a} \circ \cdots \circ \hat{\gamma}_{i_{j-1}^a i_j^a} \in \mathcal{K}_\infty$ and the $\hat{\gamma}_{(.)}$'s are continuously differentiable on $(0, \infty)$, with (3.43), there exists a continuous and positive definite function $\hat{\alpha}_{m^a}$ such that

$$\nabla V_{m^a}(x_{m^a}) f_{i_j^a}(x, u_{i_j^a}) = \hat{\gamma}'_{m^a} \left(V_{i_j^a}(x_{i_j^a}) \right) \nabla V_{i_j^a}(x_{i_j^a}) f_{i_j^a}(x, u_{i_j^a})$$
$$\leq - \left(\hat{\gamma}'_{m^a} \circ \hat{\gamma}_{m^a}^{-1}(V_\Pi(x)) \right) \hat{\alpha}_{i_j^a}(V_\Pi(x))$$
$$\leq -\hat{\alpha}_{m^a}(V_\Pi(x)) \tag{3.46}$$

if $V_\Pi(x) \neq 0$ and $V_\Pi(x) \geq u_\Pi$. Property (3.46) means that all the elements that take the value of $V_\Pi(x)$ are decreasing in the case of $V_\Pi(x) \geq u_\Pi$. In the following procedure, we study the decreasing property of $V_\Pi(x)$.

Denote the size of A as N_A. We consider two cases: $N_A = 1$ and $N_A \geq 2$.

Case 1: $N_A = 1$. The decreasing property of V_Π is determined by V_{m^a} (and thus $V_{i_j^a}$) in a neighborhood of x, which can be guaranteed by (3.46) as

$$\nabla V_\Pi(x) f(x, u) = \nabla V_{m^a}(x_{m^a}) f_{i_j^a}(x, u_{i_{j-1}^a}) \leq -\hat{\alpha}_{m^a}(V_\Pi(x)) \qquad (3.47)$$

whenever $V_\Pi(x) \geq u_\Pi$.

Case 2: $N_A \geq 2$. Recall that $x \neq 0$. Then, for all $a \in A$, $x_{i_j^a} \neq 0$. Using the continuous differentiability of $\hat{\gamma}_{(.)}$'s and $V_{i_j^a}$ and the continuity of $f_{i_j^a}$, one sees that $\nabla V_{m^a}(x_{m^a}) f_{i_j^a}(x, u_{i_j^a})$ is continuous with respect to x for a specified $u_{i_j^a}$, and there exists a neighborhood $\mathcal{X} = \mathcal{X}_1 \times \cdots \times \mathcal{X}_N$ of x such that

$$\nabla V_{m^a}(x_{m^a}) f_{i_j^a}(\xi, u_{i_j^a}) \leq -\frac{1}{2}\hat{\alpha}_{m^a}(V_\Pi(x)) \qquad (3.48)$$

holds for all $\xi \in \mathcal{X}$ and all $a \in A$.

For the locally Lipschitz function V_Π, it holds for almost all pairs of (x, u) that

$$\nabla V_\Pi(x) f(x, u) = \left.\frac{d}{dt}\right|_{t=0} V_\Pi(\phi(t)), \qquad (3.49)$$

where $\phi(t) = [\phi_1^T(t), \ldots, \phi_N^T(t)]^T$ is the solution of the initial-value problem

$$\dot{\phi}(t) = f(\phi(t), u), \quad \phi(0) = x. \qquad (3.50)$$

Because of the continuity of $\phi(t)$ with respect to t, there exists a $\bar{\delta} > 0$ such that $\phi(t) \in \mathcal{X}$ and any element in V_Π corresponding to a simple path marked by b with $b \notin A$ satisfies $V_{m^b}(\phi_{i_j^b}(t)) < \max\left\{V_{m^a}(\phi_{i_j^a}(t)) : a \in A\right\}$ for $t \in [0, \bar{\delta})$.

For any $t \in (0, \bar{\delta})$, irrespective of which element in $\left\{V_{m^a}(\phi_{i_j^a}(t)) : a \in A\right\}$, $V_\Pi(\phi(t))$ takes the value of, there exists a continuous and positive definite function $\hat{\alpha}_A$ such that

$$\frac{V_\Pi(\phi(t)) - V_\Pi(x)}{t} \leq -\hat{\alpha}_A(V_\Pi(x)). \qquad (3.51)$$

For instance, we can take $\hat{\alpha}_A(s) = \min_{a \in A}\{\hat{\alpha}_{m^a}(s)/3\}$ for all $s \geq 0$.

Hence, if V_Π is differentiable at x, then

$$\nabla V_\Pi(x) f(x, u) \leq -\hat{\alpha}_A(V_\Pi(x)). \qquad (3.52)$$

By combining (3.47) and (3.52), it follows that if V_Π is differentiable at x, then

$$V_\Pi(x) \geq u_\Pi \Rightarrow \nabla V_\Pi(x) f(x, u) \leq -\hat{\alpha}_A(V_\Pi(x)). \qquad (3.53)$$

Note that for different x, different elements may take the value of $V_\Pi(x)$ and set A may be different. Define $\alpha_\Pi(s)$ as the minimum of all the possible $\hat{\alpha}_A(s)$'s for $s \geq 0$. Then, α_Π is a continuous and positive definite function. For any x, if V_Π is differentiable at x, then

$$V_\Pi(x) \geq u_\Pi \Rightarrow \nabla V_\Pi(x) f(x,u) \leq -\alpha_\Pi\left(V_\Pi(x)\right). \qquad (3.54)$$

Since V_Π is differentiable almost everywhere, (3.54) holds almost everywhere. Recall that the definition of u_Π in (3.32) implies $u_\Pi \leq \gamma^u(|u|)$ for all u with $\gamma^u \in \mathcal{K}_\infty$. Thus, as a direct consequence of (3.54), we have

$$V_\Pi(x) \geq \gamma^u(|u|) \Rightarrow \nabla V_\Pi(x) f(x,u) \leq -\alpha_\Pi\left(V_\Pi(x)\right). \qquad (3.55)$$

This proves that V_Π is an ISS-Lyapunov function of the dynamic network with u as the input, and at the same time, proves the cyclic-small-gain theorem for continuous-time dynamic networks.

The ISS-Lyapunov function V_Π proposed above is not continuously differentiable. With the technique given in [170], one can further construct smooth ISS-Lyapunov functions based on V_Π.

For simplicity of discussions, the ISS-Lyapunov functions of the subsystems are assumed to be continuously differentiable. The construction method is still valid for systems with Lyapunov functions that are continuously differentiable almost everywhere and satisfy (3.20) almost everywhere. In that case, it can still be proved that the constructed V_Π satisfies (3.54) almost everywhere. Moreover, the modified gains $\hat{\gamma}_{(\cdot)}$'s are not required to be continuously differentiable on $(0, \infty)$ either. The ISS-Lyapunov function V_Π can still be constructed by using $\hat{\gamma}_{(\cdot)}$'s which are continuously differentiable almost everywhere.

Example 3.2 *Consider the $N = 3$ dynamic network (3.8). With no external input, it satisfies the cyclic-small-gain condition and is asymptotically stable at the origin according to Theorem 3.1.*

For the gain digraph shown in Figure 3.2, $\mathcal{RS}(i) = \{1, 2, 3\}$ for $i = 1, 2, 3$. Different ISS-Lyapunov functions V_Π's can be constructed by choosing different Π's. For example,

$$V_{\{1\}}(x) = \max\left\{V_1(x_1), \hat{\gamma}_{12}(V_2(x_2)), \hat{\gamma}_{13} \circ \hat{\gamma}_{32}(V_2(x_2)), \hat{\gamma}_{13}(V_3(x_3))\right\}, \quad (3.56)$$
$$V_{\{2\}}(x) = \max\left\{V_2(x_2), \hat{\gamma}_{21}(V_1(x_1)), \hat{\gamma}_{21} \circ \hat{\gamma}_{13}(V_3(x_3))\right\}. \qquad (3.57)$$

There are two terms depending on $V_2(x_2)$ in the definition of $V_{\{1\}}(x)$, because there are two simple paths leading from V_2 to V_1 in the gain digraph.

Example 3.3 *If the gain digraph of a dynamic network is disconnected, then it is impossible to find one single subsystem which is reachable from all the other subsystems, and the Π should contain more than one element to construct a positive definite and radially unbounded V_Π. Consider a dynamic network with the gain digraph shown in Figure 3.5.*

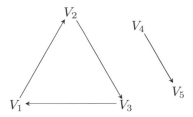

FIGURE 3.5   A disconnected gain digraph.

Since $\mathcal{RS}(1) \cup \mathcal{RS}(5) = \{1, 2, 3, 4, 5\}$, *we choose* $\Pi = \{1, 5\}$, *with which an ISS-Lyapunov function is constructed as*

$$V_\Pi(x) = \max\left\{V_1(x_1), \hat{\gamma}_{13}(V_3(x_3)), \hat{\gamma}_{13} \circ \hat{\gamma}_{32}(V_2(x_2)), V_5(x_5), \hat{\gamma}_{54}(V_4(x_4))\right\} \tag{3.58}$$

where $\hat{\gamma}_{(\cdot)}$*'s are the appropriately modified ISS gains. It can be observed that* $\max\{V_1(x_1), \hat{\gamma}_{13}(V_3(x_3)), \hat{\gamma}_{13} \circ \hat{\gamma}_{32}(V_2(x_2))\}$ *and* $\max\{V_5(x_5), \hat{\gamma}_{54}(V_4(x_4))\}$ *are the Lyapunov functions of the* (x_1, x_2, x_3)*-subsystem (the part on the left-hand side in Figure 3.5) and the* (x_4, x_5)*-subsystem (the part on the right-hand side), respectively. In fact, the Lyapunov function for a disconnected dynamic network can be directly defined as the maximum of the Lyapunov functions of all the disconnected parts.*
  With $\Pi = \{1, 5\}$, *define*

$$u_\Pi = \max\left\{u_1, \hat{\gamma}_{13} \circ \gamma_{u3}(u_3), \hat{\gamma}_{13} \circ \hat{\gamma}_{32} \circ \gamma_{u2}(u_2), u_5, \hat{\gamma}_{54} \circ \gamma_{u4}(u_4)\right\}. \tag{3.59}$$

Then, there exists a continuous and positive definite function α_Π *such that*

$$V_\Pi(x) \geq u_\Pi \Rightarrow \nabla V_\Pi(x) f(x, u) \leq -\alpha_\Pi(V_\Pi(x)), \quad a.e. \tag{3.60}$$

  To analyze the influence of the external inputs to each subsystem, one may first transform the Lyapunov-based ISS property into the trajectory-based ISS property: for any initial state $x(0) = x_0$,

$$V_\Pi(x(t)) \leq \max\left\{\beta(V_\Pi(x_0), t), \|u_\Pi\|_{[0,t]}\right\} \tag{3.61}$$

where $\beta \in \mathcal{KL}$. *Consider the* x_2*-subsystem, for example. From the definition of* V_Π, $V_2(x_2) \leq \hat{\gamma}_{32}^{-1} \circ \hat{\gamma}_{13}^{-1}(V_\Pi(x))$ *for all* x, *and the influence of the external inputs on the* x_2*-subsystem can be easily estimated through the following IOS property by considering* $V_2(x_2)$ *as the output:*

$$V_2(x_2(t)) \leq \max\left\{\hat{\gamma}_{32}^{-1} \circ \hat{\gamma}_{13}^{-1}\left(\beta(V_\Pi(x_0), t)\right), \hat{\gamma}_{32}^{-1} \circ \hat{\gamma}_{13}^{-1}\left(\|u_\Pi\|_{[0,t]}\right)\right\} \tag{3.62}$$

for any initial state $x(0) = x_0$. *According to the definition of* u_Π *in (3.59), property (3.62) implies that the* x_2*-subsystem is influenced by* u_4 *and* u_5. *However, due to the disconnected system structure,* u_4 *and* u_5 *do not influence the*

x_2-subsystem. For a more accurate estimation, one may just use the Lyapunov function of the (x_1, x_2, x_3)-subsystem to estimate the influence of u_1, u_2, and u_3 on the x_2-subsystem.

3.1.4 DISCONTINUOUS DYNAMIC NETWORKS

Reference [84] extends the concepts of ISS and the ISS-Lyapunov function to discontinuous systems and also proposes an extended Filippov solution for interconnected discontinuous systems by using differential inclusions. Appendix B provides some related concepts on discontinuous systems. Based on the concept of extended Filippov solution, an ISS small-gain theorem has been developed for discontinuous systems. Based on the results in [84], we develop a cyclic-small-gain theorem for discontinuous dynamic networks with the subsystems represented by differential inclusions:

$$\dot{x}_i \in F_i(x, u_i), \quad i = 1, \dots, N, \tag{3.63}$$

where $F_i : \mathbb{R}^{n+m_i} \rightsquigarrow \mathbb{R}^{n_i}$ is a convex, compact, and upper semi-continuous set-valued map satisfying $0 \in F_i(0,0)$, and the variables are defined in the same way as for (3.18).

Assume that each x_i-subsystem in (3.63) admits an ISS-Lyapunov function V_i satisfying (3.19) and

$$V_i(x_i) \geq \max_{j=1,\dots,N; j \neq i} \{\gamma_{ij}(V_j(x_j)), \gamma_i^u(|u_i|)\}$$
$$\Rightarrow \max_{f_i \in F_i(x,u_i)} \nabla V_i(x_i) f_i \leq -\alpha_i(V_i(x_i)) \tag{3.64}$$

wherever ∇V_i exists. Clearly, (3.64) is a direct modification of (3.20).

We have such a cyclic-small-gain result for discontinuous dynamic networks: if the cyclic-small-gain condition (3.22) is satisfied, then the discontinuous dynamic network is ISS and an ISS-Lyapunov function V_Π can be constructed as in (3.31) such that

$$V_\Pi(x) \geq u_\Pi \Rightarrow \max_{f \in F(x,u)} \nabla V_\Pi(x) f \leq -\alpha_\Pi(V_\Pi(x)) \tag{3.65}$$

wherever ∇V exists, with $F(x, u) = [F_1^T(x, u_1), \dots, F_N^T(x, u_N)]^T$. Note that property (3.65) is an extension of property (3.54).

3.1.5 DYNAMIC NETWORKS OF IOS SUBSYSTEMS

Corresponding to Theorem 2.2, this subsection presents cyclic-small-gain results for large-scale dynamic networks composed of IOS subsystems. Time-delay issues are also discussed.

Consider a large-scale dynamic network in the form of

$$\dot{x}_1 = f_1(x_1, y_2, y_3, \ldots, y_n, u_1) \tag{3.66}$$
$$\dot{x}_2 = f_2(x_2, y_1, y_3, \ldots, y_n, u_2) \tag{3.67}$$
$$\vdots$$
$$\dot{x}_n = f_n(x_n, y_1, y_2, \ldots, y_{n-1}, u_n) \tag{3.68}$$

with output maps

$$y_i = h_i(x_i), \quad i = 1, \ldots, n. \tag{3.69}$$

For each i-th subsystem, $x_i \in \mathbb{R}^{n_i}$ is the state, $u_i \in \mathbb{R}^{m_i}$ is the input, $y_i \in \mathbb{R}^{l_i}$ is the output, and f_i, h_i are locally Lipschitz functions. Denote $x = [x_1^T, \ldots, x_n^T]^T$, $y = [y_1^T, \ldots, y_n^T]^T$, and $u = [u_1^T, \ldots, u_n^T]^T$. By considering u as a function of time, assume that u is measurable and locally essentially bounded.

Suppose that each i-th subsystem is UO with zero offset and IOS with y_j for $j \neq i$ and u_i as the inputs and y_i as the output. Specifically, there exist $\alpha_i^O \in \mathcal{K}_\infty$, $\beta_i \in \mathcal{KL}$, $\gamma_{ij} \in \mathcal{K}$ and $\gamma_i^u \in \mathcal{K}$ such that

$$|x_i(t)| \leq \alpha_i^O \left(|x_i(0)| + \sum_{j \neq i} \|y_j\|_{[0,t]} + \|u_i\|_{[0,t]} \right) \tag{3.70}$$

$$|y_i(t)| \leq \max_{j \neq i} \{ \beta_i(|x_i(0)|, t), \gamma_{ij}(\|y_j\|_{[0,t]}), \gamma_i^u(\|u_i\|_\infty) \} \tag{3.71}$$

for all $t \in [0, T_{\max})$, where $[0, T_{\max})$ with $0 < T_{\max} \leq \infty$ is the right maximal interval for the definition of $(x_1(t), \ldots, x_n(t))$.

A cyclic-small-gain theorem for large-scale dynamic networks composed of IOS subsystems is given in Theorem 3.2.

Theorem 3.2 *Consider dynamic network (3.66)–(3.69) satisfying (3.70)–(3.71) for $i = 1, \ldots, n$. Then the dynamic network is UO and IOS if the cyclic-small-gain condition (3.22) is satisfied.*

Reference [134] presents a cyclic-small-gain theorem for large-scale dynamic networks composed of output-Lagrange input-to-output stable (OLIOS) subsystems with an induction-based proof. For the OLIOS systems, the cyclic-small-gain theorem can be proved by using the equivalence between OLIOS and the conjunction of UBIBS and the output asymptotic gain property. But this method seems not directly applicable to the systems with only UO and IOS properties. For Theorem 3.2, Appendix D.3 presents a sketch of a proof, which can be considered as a combination of the methods in [130] and [134].

The cyclic-small-gain condition is also valid for the large-scale dynamic networks with interconnection time delays. This topic has been studied in

[137, 265]. Consider a dynamic network in the following form:

$$\dot{x}_1(t) = f_1(x_1(t), y_2(t - \tau_{12}), y_3(t - \tau_{13}), \ldots, y_n(t - \tau_{1n}), u_1(t)) \qquad (3.72)$$

$$\dot{x}_2(t) = f_2(x_2(t), y_1(t - \tau_{21}), y_3(t - \tau_{23}), \ldots, y_n(t - \tau_{2n}), u_2(t)) \qquad (3.73)$$

$$\vdots$$

$$\dot{x}_n(t) = f_n(x_n(t), y_1(t - \tau_{n1}), y_2(t - \tau_{n2}), \ldots, y_{n-1}(t - \tau_{n(n-1)}), u_n(t)) \qquad (3.74)$$

with output maps defined in (3.69), where $\tau_{ij} : \mathbb{R}_+ \to [0, \theta]$ for $i \neq j$ represents the time delay of the interconnection from the j-th subsystem to the i-th subsystem with constant $\theta \geq 0$ being the largest time delay. The analogous definitions of UO and IOS for systems with delays can be found in [265].

Intuitively, (but maybe not mathematically rigorously), since

$$|y_i(t - \tau_{ji})| \leq \|y_i\|_{[-\theta, \infty)}, \qquad (3.75)$$

one may consider the time-delay components (shown in Figure 3.6) as subsystems with the identity gain, so they should not cause violation of the cyclic-small-gain condition for a system when introduced.

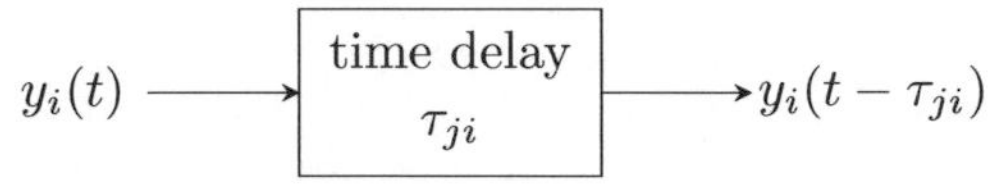

$$y_i(t) \longrightarrow \boxed{\begin{array}{c} \text{time delay} \\ \tau_{ji} \end{array}} \longrightarrow y_i(t - \tau_{ji})$$

FIGURE 3.6 A time-delay component.

Theorem 3.3 gives a cyclic-small-gain result for large-scale dynamic networks with time-delays.

Theorem 3.3 *Consider dynamic network (3.72)–(3.74) with output maps defined by (3.69). Suppose that if the time delays do not exist, i.e., $\theta = 0$, each i-th subsystem with $i = 1, \ldots, n$ satisfies (3.70)–(3.71). Then the dynamic network with $\theta \geq 0$ is UO and IOS if the cyclic-small-gain condition (3.22) is satisfied.*

3.2 DISCRETE-TIME DYNAMIC NETWORKS

In view of the critical importance of discrete-time system theory in computer-aided control engineering applications, in this section, we generalize the ISS cyclic-small-gain theorem introduced in Section 3.1 to discrete-time dynamic networks. Due to phenomena which are particular to discrete-time systems (see Example 3.4 below), such a generalization is nontrivial.

Analogous to the continuous-time dynamic network studied in Section 3.1, the discrete-time dynamic network addressed in this section is composed of

N discrete-time subsystems in the following form:

$$x_i(T+1) = f_i(x(T), u_i(T)), \quad i = 1, \ldots, N, \tag{3.76}$$

where $x = [x_1^T, \ldots, x_N^T]^T$ with $x_i \in \mathbb{R}^{n_i}$ is the state, $u_i \in \mathbb{R}^{n_{ui}}$ represent the external inputs, and $f_i : \mathbb{R}^{n+n_{ui}} \to \mathbb{R}^{n_i}$ with $n := \sum_{i=1}^{N} n_i$ is continuous. T takes values in $\mathbb{Z}_+$. It is assumed that $f_i(0,0) = 0$, and the external input $u = [u_1^T, \ldots, u_N^T]^T$ is bounded. Denote $f(x, u) = [f_1^T(x, u_1), \ldots, f_N^T(x, u_N)]^T$.

There are two kinds of Lyapunov formulations for discrete-time ISS systems: the dissipation form and the gain margin form. We first give the dissipation form.

For $i = 1, \ldots, N$, each x_i-subsystem admits a continuous ISS-Lyapunov function $V_i : \mathbb{R}^{n_i} \to \mathbb{R}_+$ satisfying the following:

1. there exist $\underline{\alpha}_i, \overline{\alpha}_i \in \mathcal{K}_\infty$ such that

$$\underline{\alpha}_i(|x_i|) \le V_i(x_i) \le \overline{\alpha}_i(|x_i|), \quad \forall x_i; \tag{3.77}$$

2. there exist $\alpha_i \in \mathcal{K}_\infty$, $\sigma_{ij} \in \mathcal{K} \cup \{0\}$ and $\sigma_{ui} \in \mathcal{K} \cup \{0\}$ such that

$$\begin{aligned} &V_i(f_i(x, u_i)) - V_i(x_i) \\ &\le -\alpha_i(V_i(x_i)) + \max_{j \ne i} \{\sigma_{ij}(V_j(x_j)), \sigma_{ui}(|u_i|)\}, \quad \forall x, \ u_i. \end{aligned} \tag{3.78}$$

Without loss of generality, we assume $(\mathrm{Id} - \alpha_i) \in \mathcal{K}$. Note that if $(\mathrm{Id} - \alpha_i) \notin \mathcal{K}$, one can always find an $\alpha_i' < \alpha_i$ such that $(\mathrm{Id} - \alpha_i') \in \mathcal{K}$ and property (3.78) holds with α_i replaced by α_i'. Then, we consider

$$\hat{\gamma}_{ij} = \alpha_i^{-1} \circ (\mathrm{Id} - \rho_i)^{-1} \circ \sigma_{ij} \tag{3.79}$$

as the ISS gain from V_j to V_i, with ρ_i being a continuous and positive definite function and satisfying $(\mathrm{Id} - \rho_i) \in \mathcal{K}_\infty$.

Correspondingly, the ISS gain from the external input u_i to V_i is defined as

$$\hat{\gamma}_{ui} = \alpha_i^{-1} \circ (\mathrm{Id} - \rho_i)^{-1} \circ \sigma_{ui}. \tag{3.80}$$

The gain margin formulation of the ISS-Lyapunov functions can be obtained by modifying property 2 as: there exist a continuous and positive definite function α_i' and $\gamma_{ij}', \gamma_{ui}' \in \mathcal{K} \cup \{0\}$ such that

$$\begin{aligned} &V_i(x_i) \ge \max_{j \ne i} \{\gamma_{ij}'(V_j(x_j)), \gamma_{ui}'(|u_i|)\} \\ &\Rightarrow V_i(f_i(x, u_i)) - V_i(x_i) \le -\alpha_i'(V_i(x_i)), \quad \forall x, \ u_i. \end{aligned} \tag{3.81}$$

In contrast to the continuous-time systems discussed in Section 3.1, the trajectories of a discrete-time system may "jump out" of the region determined by the gain margin in (3.81), which means that the γ_{ij}' and the γ_{ui}' in (3.81) may not be the true ISS gains. Consider Example 3.4.

Example 3.4 *Consider a discrete-time system*

$$z(T+1) = g(z(T), |w(T)|), \tag{3.82}$$

where $z \in \mathbb{R}$ is the state, $w \in \mathbb{R}^m$ is the external input, and $g : \mathbb{R}^{m+1} \to \mathbb{R}$ defined in Figure 3.7 is continuous. Define $V_z(z) = |z|$. Then, one can find a small $\delta > 0$ such that

$$V_z(z) \geq (1+\delta)|w| \Rightarrow V_z(g(z, |w|)) - V_z(z) \leq -\alpha_z(V_z(z)), \tag{3.83}$$

where α_z is a continuous and positive definite function. Then, $(1+\delta)$ is the "ISS gain" defined by the gain margin formulation (3.81).

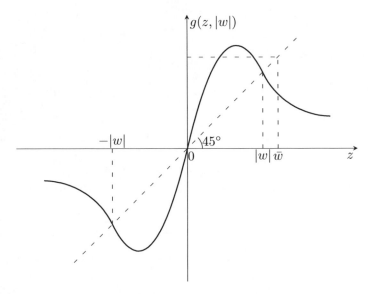

FIGURE 3.7 An example of the gain margin property of discrete-time systems, where $\bar{w} = (1+\delta)|w|$.

However, from Figure 3.7, it is possible that $V_z(g(z, |w|)) > (1+\delta)|w|$, even if $V_z(z) \leq (1+\delta)|w|$, which means that the state of the discrete-time nonlinear system may "jump out" of the region defined by the gain margin. To solve this problem, one may find an $\alpha_g \in \mathcal{K}$ such that $|g(z, |w|)| \leq \alpha_g(|w|)$ whenever $|z| \leq \alpha_g(|w|)$, and define $\gamma_w(s) = \max\{(1+\delta)s, \alpha_g(s)\}$ for $s \geq 0$. Then, the phenomenon of "jump out" can be avoided as

$$V_z(z) \geq \gamma_w(|w|) \Rightarrow V_z(g(z, |w|)) - V_z(z) \leq -\alpha_z(V_z(z)) \tag{3.84}$$

$$V_z(z) \leq \gamma_w(|w|) \Rightarrow V_z(g(z, |w|)) \leq \gamma_w(|w|), \tag{3.85}$$

and γ_w can be used as the ISS gain for the discrete-time system.

Following the idea in Example 3.4, to take into account the "jump out" issue, the gain margin formulation (3.81) is consolidated with

$$V_i(x_i) \leq \max_{j \neq i} \left\{ \gamma'_{ij}(V_j(x_j)), \gamma'_{ui}(|u_i|) \right\}$$
$$\Rightarrow V_i(f_i(x, u_i)) \leq (\mathrm{Id} - \delta'_i) \left(\max \left\{ \gamma'_{ij}(V_j(x_j)), \gamma'_{ui}(|u_i|) \right\} \right), \tag{3.86}$$

where δ'_i is a continuous and positive definite function satisfying $(\mathrm{Id} - \delta'_i) \in \mathcal{K}_\infty$.

By combining (3.81) and (3.86), the refined gain margin formulation is described with property 1 above and

2′. there exist $\hat{\gamma}_{ij} \in \mathcal{K} \cup \{0\}$ and $\hat{\gamma}_{ui} \in \mathcal{K} \cup \{0\}$ such that

$$V_i(f_i(x, u_i)) \leq (\mathrm{Id} - \delta_i) \left(\max_{j \neq i} \left\{ \hat{\gamma}_{ij}(V_j(x_j)), V_i(x_i), \hat{\gamma}_{ui}(|u_i|) \right\} \right), \quad \forall x, \ u_i,$$
$$\tag{3.87}$$

where δ_i is a continuous and positive definite function satisfying $(\mathrm{Id} - \delta_i) \in \mathcal{K}_\infty$.

We present the cyclic-small-gain results for discrete-time dynamic networks formulated in the dissipation form and the gain margin form in Theorems 3.4 and 3.5, respectively.

Theorem 3.4 *Consider the discrete-time dynamic network (3.76) with each x_i-subsystem having a continuous ISS-Lyapunov function V_i satisfying (3.77) and (3.78). Then, it is ISS with x as the state and u as the input, if there exist continuous and positive definite functions ρ_i satisfying $(\mathrm{Id} - \rho_i) \in \mathcal{K}_\infty$ such that for every simple cycle $(V_{i_1}, V_{i_2}, \ldots, V_{i_r}, V_{i_1})$ in the gain digraph,*

$$\hat{\gamma}_{i_1 i_2} \circ \hat{\gamma}_{i_2 i_3} \circ \ldots \circ \hat{\gamma}_{i_r i_1} < \mathrm{Id}, \tag{3.88}$$

where $r = 2, \ldots, N$ and $1 \leq i_j \leq N$, $i_j \neq i_{j'}$ if $j \neq j'$.

Theorem 3.5 *Consider the discrete-time dynamic network (3.76) with each x_i-subsystem having a continuous ISS-Lyapunov function V_i satisfying (3.77) and (3.87). Then, the dynamic network is ISS with x as the state and u as the input if for every simple cycle $(V_{i_1}, V_{i_2}, \ldots, V_{i_r}, V_{i_1})$ in the gain digraph,*

$$\hat{\gamma}_{i_1 i_2} \circ \hat{\gamma}_{i_2 i_3} \circ \ldots \circ \hat{\gamma}_{i_r i_1} < \mathrm{Id}, \tag{3.89}$$

where $r = 2, \ldots, N$, $1 \leq i_j \leq N$, $i_j \neq i_{j'}$ if $j \neq j'$.

As for continuous-time dynamic networks, Theorems 3.4 and 3.5 are proved by constructing ISS-Lyapunov functions. The ISS-Lyapunov function candidates for discrete-time dynamic networks are constructed like continuous-time dynamic networks in Subsection 3.1.2. One difference is that the ISS-Lyapunov functions constructed for discrete-time systems are only required to be continuous.

The proofs of Theorems 3.4 and 3.5 are given in Subsections 3.2.1 and 3.2.2, respectively.

3.2.1 PROOF OF THE CYCLIC-SMALL-GAIN THEOREM FOR DISCRETE-TIME DYNAMIC NETWORKS IN DISSIPATION FORM

In the proof for continuous-time dynamic networks, we only consider the behavior of the largest elements in $\mathbb{V}_\Pi$. However, for discrete-time dynamic networks, we should study the motion of all the elements in $\mathbb{V}_\Pi$. Denote the largest element of $\mathbb{V}_\Pi$ at time T as $V_\Pi^*(x^*(T))$. Then, $V_\Pi(x(T+1)) - V_\Pi(x(T))$ is determined by all the elements in $\mathbb{V}_\Pi$, not only by the largest elements. This leads to another difference between discrete-time systems and continuous-time systems.

According to the definition of $\mathbb{V}_\Pi$, each element of $\mathbb{V}_\Pi(x)$ corresponds to a simple path in the gain interconnection digraph. Consider any specific element in $\mathbb{V}_\Pi(x)$ that corresponds to a simple path $n^a = (V_{i_j^a}, \ldots, V_{i_2^a}, V_{i_1^a})$, as shown in Figure 3.4. The definitions of V_Π and u_Π imply

$$\hat{\gamma}_{i_1^a i_2^a} \circ \cdots \circ \hat{\gamma}_{i_{k-1}^a i_k^a} \circ \cdots \circ \hat{\gamma}_{i_{j-1}^a i_j^a}(V_{i_j^a}(x_{i_j^a})) \leq V_\Pi(x), \tag{3.90}$$

$$\hat{\gamma}_{i_1^a i_2^a} \circ \cdots \circ \hat{\gamma}_{i_{j-1}^a i_j^a} \circ \hat{\gamma}_{u i_j^a}(|u_{i_j^a}|) \leq u_\Pi. \tag{3.91}$$

We first study the relation between V_Π and the ISS-Lyapunov functions of the subsystems. We consider the following two cases.

1. *Relation between V_Π and V_l with $l \in \{i_1^a, i_2^a, \ldots, i_{j-1}^a\}$*
   If $j = 1$, then the simple path n^a contains only the i_1^a-th subsystem. If $j \geq 2$, then for all $k \in \{1, \ldots, j-1\}$, we have

$$\hat{\gamma}_{i_1^a i_2^a} \circ \cdots \circ \hat{\gamma}_{i_{k-1}^a i_k^a}(V_{i_k^a}(x_{i_k^a})) \leq V_\Pi(x). \tag{3.92}$$

With the satisfaction of the cyclic-small-gain condition, it holds that

$$\hat{\gamma}_{i_k^a i_{k+1}^a} \circ \cdots \circ \hat{\gamma}_{i_{j-1}^a i_j^a} \circ \hat{\gamma}_{i_j^a i_k^a} < \mathrm{Id} \tag{3.93}$$

for all $k \in \{1, \ldots, j-1\}$.
Then, (3.92) and (3.93) together imply

$$\hat{\gamma}_{i_1^a i_2^a} \circ \cdots \circ \hat{\gamma}_{i_{k-1}^a i_k^a} \circ \hat{\gamma}_{i_k^a i_{k+1}^a} \circ \cdots \circ \hat{\gamma}_{i_{j-1}^a i_j^a} \circ \hat{\gamma}_{i_j^a i_k^a}(V_{i_k^a}(x_{i_k^a})) \leq V_\Pi(x) \tag{3.94}$$

for all $k \in \{1, \ldots, j-1\}$, and equivalently,

$$\hat{\gamma}_{i_1^a i_2^a} \circ \cdots \circ \hat{\gamma}_{i_{j-1}^a i_j^a} \circ \hat{\gamma}_{i_j^a l}(V_l(x_l)) \leq V_\Pi(x) \tag{3.95}$$

for all $l \in \{i_1^a, i_2^a, \ldots, i_{j-1}^a\}$.

2. *Relation between V_Π and V_l with $l \in \{1, \ldots, N\} \setminus \{i_1^a, i_2^a, \ldots, i_{j-1}^a\}$*
   If $j = N$, then the simple path n^a contains all the subsystems in the dynamic network. If $j \leq N - 1$, then for all $l \in \{1, \ldots, N\} \setminus \{i_1^a, i_2^a, \ldots, i_j^a\}$, we directly have

$$\hat{\gamma}_{i_1^a i_2^a} \circ \cdots \circ \hat{\gamma}_{i_{j-1}^a i_j^a} \circ \hat{\gamma}_{i_j^a l}(V_l(x_l)) \leq V_\Pi(x), \tag{3.96}$$

because $\hat{\gamma}_{i_1^a i_2^a} \circ \cdots \circ \hat{\gamma}_{i_{j-1}^a i_j^a} \circ \hat{\gamma}_{i_j^a l}(V_l(x_l))$ is an element of $\mathbb{V}_\Pi$.

By using the definition of $\hat{\gamma}_{i_j^a l}$, properties (3.95) and (3.96) can be equivalently represented by

$$(\mathrm{Id} - \rho_{i_j^a}) \circ \alpha_{i_j^a} \circ \hat{\gamma}_{i_{j-1}^a i_j^a}^{-1} \circ \cdots \circ \hat{\gamma}_{i_1^a i_2^a}^{-1}(V_\Pi(x)) \geq \max_{l \neq i_j^a} \left\{ \sigma_{i_j^a l}(V_l(x_l)) \right\}. \tag{3.97}$$

Based on the relation between V_Π and the Lyapunov functions of the subsystems given in (3.97), we prove that V_Π is an ISS-Lyapunov function of the discrete-time dynamic network and satisfies the refined gain margin property defined by (3.87).

1. *Case 1: $V_\Pi(x) \geq u_\Pi$.*
   Using (3.91), we have

$$\hat{\gamma}_{u i_j^a}(|u_{i_j^a}|) \leq \hat{\gamma}_{i_{j-1}^a i_j^a}^{-1} \circ \cdots \circ \hat{\gamma}_{i_1^a i_2^a}^{-1}(V_\Pi(x)) \tag{3.98}$$

i.e.,

$$\sigma_{u i_j^a}(|u_{i_j^a}|) \leq (\mathrm{Id} - \rho_{i_j^a}) \circ \alpha_{i_j^a} \circ \hat{\gamma}_{i_{j-1}^a i_j^a}^{-1} \circ \cdots \circ \hat{\gamma}_{i_1^a i_2^a}^{-1}(V_\Pi(x)). \tag{3.99}$$

By combining (3.78), (3.90), (3.97), and (3.99), we have

$$\begin{aligned}
V_{i_j^a}(f_{i_j^a}(x, u_{i_j^a})) &\leq (\mathrm{Id} - \alpha_{i_j^a})(V_{i_j^a}(x_{i_j^a})) + \max_{l \neq i_j^a} \left\{ \sigma_{i_j^a l}(V_l(x_l)), \sigma_{u i_j^a}(|u_{i_j^a}|) \right\} \\
&\leq (\mathrm{Id} - \alpha_{i_j^a}) \circ \hat{\gamma}_{i_{j-1}^a i_j^a}^{-1} \circ \cdots \circ \hat{\gamma}_{i_1^a i_2^a}^{-1}(V_\Pi(x)) \\
&\quad + (\mathrm{Id} - \rho_{i_j^a}) \circ \alpha_{i_j^a} \circ \hat{\gamma}_{i_{j-1}^a i_j^a}^{-1} \circ \cdots \circ \hat{\gamma}_{i_1^a i_2^a}^{-1}(V_\Pi(x)) \\
&= \hat{\gamma}_{i_{j-1}^a i_j^a}^{-1} \circ \cdots \circ \hat{\gamma}_{i_1^a i_2^a}^{-1}(V_\Pi(x)) \\
&\quad - \rho_{i_j^a} \circ \alpha_{i_j^a} \circ \hat{\gamma}_{i_{j-1}^a i_j^a}^{-1} \circ \cdots \circ \hat{\gamma}_{i_1^a i_2^a}^{-1}(V_\Pi(x)).
\end{aligned} \tag{3.100}$$

With Lemma C.3, by considering $V_{i_j^a}(x_{i_j^a})$ as s, $\hat{\gamma}_{i_{j-1}^a i_j^a}^{-1} \circ \cdots \circ \hat{\gamma}_{i_1^a i_2^a}^{-1}(V_\Pi(x))$ as s', $\rho_{i_j^a} \circ \alpha_{i_j^a}$ as α, and $\hat{\gamma}_{i_1^a i_2^a} \circ \cdots \circ \hat{\gamma}_{i_{j-1}^a i_j^a}$ as χ, there exists a continuous and positive definite function $\tilde{\alpha}_{n^a}$ such that

$$\begin{aligned}
\hat{\gamma}_{i_1^a i_2^a} \circ \cdots \circ \hat{\gamma}_{i_{j-1}^a i_j^a}(V_{i_j^a}(f_{i_j^a}(x, u_{i_j^a}))) &- V_\Pi(x) \\
&\leq - \tilde{\alpha}_{n^a} \circ \hat{\gamma}_{i_{j-1}^a i_j^a}^{-1} \circ \cdots \circ \hat{\gamma}_{i_1^a i_2^a}^{-1}(V_\Pi(x)) \\
&\leq - \bar{\alpha}_{n^a}(V_\Pi(x)),
\end{aligned} \tag{3.101}$$

i.e.,

$$\hat{\gamma}_{i_1^a i_2^a} \circ \cdots \circ \hat{\gamma}_{i_{j-1}^a i_j^a}(V_{i_j^a}(f_{i_j^a}(x, u_{i_j^a}))) \le (\mathrm{Id} - \bar{\alpha}_{n^a})(V_\Pi(x)), \qquad (3.102)$$

where $\bar{\alpha}_{n^a}$ is positive definite and satisfies $(\mathrm{Id} - \bar{\alpha}_{n^a}) \in \mathcal{K}_\infty$.

2. *Case 2:* $V_\Pi(x) < u_\Pi$.

Property (3.91) can be rewritten as

$$\sigma_{u_{i_j^a}}(|u_{i_j^a}|) \le (\mathrm{Id} - \rho_{i_j^a}) \circ \alpha_{i_j^a} \circ \hat{\gamma}_{i_{j-1}^a i_j^a}^{-1} \circ \cdots \circ \hat{\gamma}_{i_1^a i_2^a}^{-1}(u_\Pi). \qquad (3.103)$$

From property (3.97), one can observe

$$(\mathrm{Id} - \rho_{i_j^a}) \circ \alpha_{i_j^a} \circ \hat{\gamma}_{i_{j-1}^a i_j^a}^{-1} \circ \cdots \circ \hat{\gamma}_{i_1^a i_2^a}^{-1}(u_\Pi) \ge \max_{l \ne i_j^a}\left\{\sigma_{i_j^a l}(V_l(x_l))\right\}. \qquad (3.104)$$

Combining (3.78), (3.90), (3.103), and (3.104), we obtain

$$\begin{aligned}
V_{i_j^a}(f_{i_j^a}(x, u_{i_j^a})) &\le (\mathrm{Id} - \alpha_{i_j^a})(V_{i_j^a}(x_{i_j^a})) \\
&\quad + \max_{l \ne i_j^a}\left\{\sigma_{i_j^a l}(V_l(x_l)), \sigma_{u_{i_j^a}}(|u_{i_j^a}|)\right\} \\
&\le (\mathrm{Id} - \alpha_{i_j^a}) \circ \hat{\gamma}_{i_{j-1}^a i_j^a}^{-1} \circ \cdots \circ \hat{\gamma}_{i_1^a i_2^a}^{-1}(u_\Pi) \\
&\quad + (\mathrm{Id} - \rho_{i_j^a}) \circ \alpha_{i_j^a} \circ \hat{\gamma}_{i_{j-1}^a i_j^a}^{-1} \circ \cdots \circ \hat{\gamma}_{i_1^a i_2^a}^{-1}(u_\Pi) \\
&= \hat{\gamma}_{i_{j-1}^a i_j^a}^{-1} \circ \cdots \circ \hat{\gamma}_{i_1^a i_2^a}^{-1}(u_\Pi) \\
&\quad - \rho_{i_j^a} \circ \alpha_{i_j^a} \circ \hat{\gamma}_{i_{j-1}^a i_j^a}^{-1} \circ \cdots \circ \hat{\gamma}_{i_1^a i_2^a}^{-1}(u_\Pi). \qquad (3.105)
\end{aligned}$$

With Lemma C.3, as in Case 1, one can achieve

$$\hat{\gamma}_{i_1^a i_2^a} \circ \cdots \circ \hat{\gamma}_{i_{j-1}^a i_j^a}(V_{i_j^a}(f_{i_j^a}(x, u_{i_j^a}))) - u_\Pi \le -\bar{\alpha}_{n^a}(u_\Pi), \qquad (3.106)$$

i.e.,

$$\hat{\gamma}_{i_1^a i_2^a} \circ \cdots \circ \hat{\gamma}_{i_{j-1}^a i_j^a}(V_{i_j^a}(V_{i_j^a}(f_{i_j^a}(x, u_{i_j^a})))) \le (\mathrm{Id} - \bar{\alpha}_{n^a})(u_\Pi). \qquad (3.107)$$

Note that $\hat{\gamma}_{i_1^a i_2^a} \circ \cdots \circ \hat{\gamma}_{i_{j-1}^a i_j^a}(V_{i_j^a}(x_{i_j^a}))$ is an arbitrary element of $\mathbb{V}_\Pi(x)$. Choose $\bar{\alpha}_\Pi$ as the minimum of the $\bar{\alpha}_{n^a}$'s corresponding to all the elements in $\mathbb{V}_\Pi(x)$. Then, $\bar{\alpha}_\Pi$ is a continuous and positive definite function and satisfies $(\mathrm{Id} - \bar{\alpha}_\Pi) \in \mathcal{K}_\infty$. Considering both Case 1 and Case 2, we have

$$V_\Pi(f(x, u)) \le (\mathrm{Id} - \bar{\alpha}_\Pi)(\max\{V_\Pi(x), u_\Pi\}), \qquad (3.108)$$

which is in the refined gain margin form defined by (3.87).

In the proof of Theorem 3.4, V_Π is shown to satisfy the gain margin formulation. An ISS-Lyapunov function in the dissipation form can be further constructed based on V_Π with a similar idea as in [131, Remark 3.3] and the proof of [23, Proposition 2.6]. Notice that α_Π in (3.108) should be of class $\mathcal{K}_\infty$ to apply these methods. This problem can be solved with [132, Lemma 2.8].

3.2.2   PROOF OF CYCLIC-SMALL-GAIN THEOREM FOR DISCRETE-TIME DYNAMIC NETWORKS IN GAIN MARGIN FORM

Properties (3.90) and (3.91) still hold for V_Π in this case.

As with the proof of Theorem 3.4, consider an *arbitrary* element in $\mathbb{V}_\Pi(x)$ that corresponds to a simple path $n^a = (V_{i_j^a}, \ldots, V_{i_2^a}, V_{i_1^a})$. With approaches similar to properties (3.95) and (3.96), we can ultimately obtain

$$\hat{\gamma}_{i_j^a l}(V_l(x_l)) \leq \hat{\gamma}_{i_{j-1}^a i_j^a}^{-1} \circ \cdots \circ \hat{\gamma}_{i_1^a i_2^a}^{-1}(V_\Pi(x)) \tag{3.109}$$

for all $l \in \{1, \ldots, N\} \setminus \{i_j^a\}$.

With (3.95), (3.96), and (3.109) satisfied, property (3.87) implies

$$V_{i_j^a}(f_{i_j^a}(x, u_{i_j^a})) \leq (\mathrm{Id} - \delta_{i_j^a}) \circ \hat{\gamma}_{i_{j-1}^a i_j^a}^{-1} \circ \cdots \circ \hat{\gamma}_{i_1^a i_2^a}^{-1} (\max\{V_\Pi(x), u_\Pi\}). \tag{3.110}$$

With Lemma C.4, by considering $\hat{\gamma}_{i_{j-1}^a i_j^a}^{-1} \circ \cdots \circ \hat{\gamma}_{i_1^a i_2^a}^{-1}$ as the χ and $\delta_{i_j^a}$ as the ε, there exists a continuous, positive definite $\bar{\delta}_{i_j^a}$ satisfying $(\mathrm{Id} - \bar{\delta}_{i_j^a}) \in \mathcal{K}_\infty$ such that

$$V_{i_j^a}(f_{i_j^a}(x, u_{i_j^a})) \leq \hat{\gamma}_{i_{j-1}^a i_j^a}^{-1} \circ \cdots \circ \hat{\gamma}_{i_1^a i_2^a}^{-1} \circ (\mathrm{Id} - \bar{\delta}_{i_j^a}) (\max\{V_\Pi(x), u_\Pi\}), \tag{3.111}$$

i.e.,

$$\hat{\gamma}_{i_1^a i_2^a} \cdots \circ \hat{\gamma}_{i_{j-1}^a i_j^a}(V_{i_j^a}(f_{i_j^a}(x, u_{i_j^a}))) \leq (\mathrm{Id} - \bar{\delta}_{i_j^a})(\max\{V_\Pi(x), u_\Pi\}). \tag{3.112}$$

Define $\bar{\delta}(s) = \min_{i \in \{1, \ldots, N\}}\{\bar{\delta}_i(s)\}$ for $s \geq 0$. It is clear that $\bar{\delta}$ is a continuous and positive definite function, and satisfies $(\mathrm{Id} - \bar{\delta}) \in \mathcal{K}_\infty$. Note that n^a corresponds to an arbitrary element in $\mathbb{V}_\Pi(x)$. It can be concluded that

$$V_\Pi(f(x, u)) \leq (\mathrm{Id} - \bar{\delta})(\max\{V_\Pi(x), u_\Pi\}). \tag{3.113}$$

Theorem 3.5 is proved.

We employ an example to show the construction of an ISS-Lyapunov function for a discrete-time dynamic network.

Example 3.5 *Consider a discrete-time dynamic network in the form of (3.76) with $N = 3$. The dynamics of the subsystems are defined as:*

$$f_1(x, u_1) = 0.6x_1 + \max\{0.36x_2^3, 3.2x_3^3, u_1\}, \tag{3.114}$$

$$f_2(x, u_2) = 0.4x_2 + \max\{0.6x_1^{1/3}, 1.2x_3, u_2\}, \tag{3.115}$$

$$f_3(x, u_3) = 0.2x_3 + \max\{0.36x_1^{1/3}, 0.36x_2, u_3\}. \tag{3.116}$$

Each subsystem is ISS with $V_i(x_i) = |x_i|$ as an ISS-Lyapunov function satisfying the dissipation formulation:

$$V_i(f_i(x, u_i)) - V_i(x_i) = -\alpha_i(V_i(x_i)) + \max_{j \neq i}\{\sigma_{ij}(V_j(x_j)), \sigma_{ui}(u_i)\}, \tag{3.117}$$

where

$$\begin{aligned}
\alpha_1(s) &= 0.4s, \quad \sigma_{12}(s) = 0.36s^3, \quad \sigma_{13}(s) = 3.2s^3, \quad \sigma_{u1}(s) = s, \\
\alpha_2(s) &= 0.6s, \quad \sigma_{21}(s) = 0.6s^{1/3}, \quad \sigma_{23}(s) = 1.2s, \quad \sigma_{u2}(s) = s, \quad (3.118) \\
\alpha_3(s) &= 0.8s, \quad \sigma_{31}(s) = 0.36s^{1/3}, \quad \sigma_{32}(s) = 0.36s, \quad \sigma_{u1}(s) = s,
\end{aligned}$$

for $s \in \mathbb{R}_+$.

Choose $\rho_1(s) = \rho_2(s) = \rho_3(s) = 0.02s$ *and define*

$$\begin{aligned}
\hat{\gamma}_{12}(s) &= 0.9184s^3, \quad \hat{\gamma}_{13}(s) = 8.1633s^3, \\
\hat{\gamma}_{21}(s) &= 1.0204s^{1/3}, \quad \hat{\gamma}_{23}(s) = 2.0408s, \quad (3.119) \\
\hat{\gamma}_{31}(s) &= 0.4592s^{1/3}, \quad \hat{\gamma}_{32}(s) = 0.4592s
\end{aligned}$$

for $s \in \mathbb{R}_+$. *The* $\hat{\gamma}_{(.)}$*'s are considered as the ISS gains of the subsystems. The gain digraph of the dynamic network is shown in Figure 3.8.*

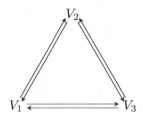

FIGURE 3.8   The gain digraph of the dynamic network in Example 3.5.

It can be directly checked that the dynamic network satisfies the cyclic-small-gain condition:

$$\begin{aligned}
\hat{\gamma}_{12} \circ \hat{\gamma}_{21} &< \text{Id}, \quad \hat{\gamma}_{23} \circ \hat{\gamma}_{32} < \text{Id}, \quad \hat{\gamma}_{31} \circ \hat{\gamma}_{13} < \text{Id}, \\
\hat{\gamma}_{12} \circ \hat{\gamma}_{23} \circ \hat{\gamma}_{31} &< \text{Id}, \quad \hat{\gamma}_{13} \circ \hat{\gamma}_{32} \circ \hat{\gamma}_{21} < \text{Id}, \quad (3.120)
\end{aligned}$$

and thus it is ISS.

In the gain digraph, $\mathcal{RS}(1) = \{1, 2, 3\}$. *By choosing* $\Pi = \{1\}$, *we construct the following ISS-Lyapunov function for the dynamic network:*

$$\begin{aligned}
V_\Pi(x) &= \max \left\{ \begin{array}{l} V_1(x_1), \hat{\gamma}_{12}(V_2(x_2)), \hat{\gamma}_{13} \circ \hat{\gamma}_{32}(V_2(x_2)), \\ \hat{\gamma}_{13}(V_3(x_3)), \hat{\gamma}_{12} \circ \hat{\gamma}_{23}(V_3(x_3)) \end{array} \right\} \\
&= \max\{V_1(x_1), 0.9184V_2^3(x_2), 8.1633V_3^3(x_3)\}. \quad (3.121)
\end{aligned}$$

Correspondingly,

$$\begin{aligned}
u_\Pi &= \max \left\{ \begin{array}{l} \sigma_{u1}(u_1), \hat{\gamma}_{12} \circ \sigma_{u2}(u_2), \hat{\gamma}_{13} \circ \hat{\gamma}_{32} \circ \sigma_{u2}(u_2), \\ \hat{\gamma}_{13} \circ \sigma_{u3}(u_3), \hat{\gamma}_{12} \circ \hat{\gamma}_{23} \circ \sigma_{u3}(u_3) \end{array} \right\} \\
&= \max\{2.5|u_1|, 4.2516|u_2|^3, 15.9439|u_3|^3\}. \quad (3.122)
\end{aligned}$$

Figure 3.9 shows the evolutions of V_Π *and* u_Π *with initial condition* $x(0) = [0.6, 0.87, 0.42]^T$ *and inputs* $u(T) = [0.1\sin(9T), 0.1\sin(11T), 0.1\sin(17T)]^T$. *From Figure 3.9,* V_Π *ultimately converges to the region determined by the magnitude of* $|u_\Pi|$. *This is in accordance with the theoretical result (3.108).*

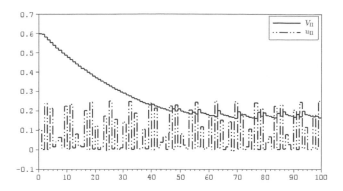

FIGURE 3.9 The evolutions of V_Π and u_Π of the dynamic network in Example 3.5.

3.3 HYBRID DYNAMIC NETWORKS

Based on the results for continuous-time dynamic networks and discrete-time dynamic networks, it is possible to develop a cyclic-small-gain result for hybrid dynamic networks, which involve both continuous-time and discrete-time dynamics. The hybrid dynamic network studied in this section is composed of N subsystems whose trajectories may be continuous, piecewise constant, or impulsive on the timeline. Define $\mathcal{N} = \{1, \dots, N\}$ as the set of indices of the subsystems. For $i \in \mathcal{N}$, each i-th subsystem of the dynamic network is modeled by

$$\dot{x}_i(t) = f_i(x(t), u_i(t)), \qquad t \in \mathbb{R}_+ \backslash \pi_i \tag{3.123}$$

$$x_i(t) = g_i(x(t^-), u_i(t^-)), \quad t \in \pi_i, \tag{3.124}$$

where $x_i \in \mathbb{R}^{n_i}$ is the state of the i-th subsystem, $x = [x_1^T, \dots, x_N^T]^T \in \mathbb{R}^n$ with $n := \sum_{i=1}^{N} n_i$ is the state of the dynamic network, $u_i : \mathbb{R}_+ \to \mathbb{R}^{m_i}$ is the input of the i-th subsystem, and $\pi_i \subset \mathbb{R}_+$ is the set of impulsive time instants of the i-th subsystem. For each $i \in \mathcal{N}$, it is assumed that $f_i : \mathbb{R}^{n+m_i} \to \mathbb{R}^{n_i}$ is locally Lipschitz and $f_i(0,0) = 0$; $g_i : \mathbb{R}^{n+m_i} \to \mathbb{R}^{n_i}$ is continuous and $g_i(0,0) = 0$. Denote $u = [u_1^T, \dots, u_N^T]^T$ as the input vector of the hybrid dynamic network. Assume that each u_i is piecewise continuous and bounded.

Note that differential equation (3.123) represents continuous-time dynamics, and difference equation (3.124) captures discrete-time dynamics. We consider three kinds of subsystems: the first kind is described by only continuous-time models (3.123) with $\pi_i = \emptyset$, the second kind is purely described by discrete-time models (3.124), and the third kind is described by a mix of (3.123) and (3.124). Assumption 3.1 is made on the ISS properties of the subsystems.

Assumption 3.1 *Each x_i-subsystem for $i \in \mathcal{N}$ has Lyapunov-based ISS properties. Specifically, for $i \in \mathcal{N}$, there exists a function $V_i : \mathbb{R}^{n_i} \to \mathbb{R}_+$ which is locally Lipschitz on $\mathbb{R}^{n_i} \backslash \{0\}$, positive definite and radially unbounded, and satisfies the following:*

1. *for each $i \in \mathcal{N}_C$, $\pi_i = \emptyset$, and there exist $\gamma_{ij}, \gamma_{u_i} \in \mathcal{K} \cup \{0\}$ with $j \neq i$ such that*

$$V_i(x_i) \geq \max_{j \neq i} \{\gamma_{ij}(V_j(x_j)), \gamma_{u_i}(|u_i|)\}$$

$$\Rightarrow \nabla V_i(x_i) f_i(x, u_i) \leq -\alpha_i(V_i(x_i)) \quad \text{a.e.}, \tag{3.125}$$

   where α_i is a continuous and positive definite function;

2. *for each $i \in \mathcal{N}_D$, $f_i \equiv 0$, and there exist $\gamma_{ij}, \gamma_{u_i} \in \mathcal{K} \cup \{0\}$ with $j \neq i$ such that*

$$V_i(g_i(x, u_i)) \leq (\text{Id} - \rho_i) \left(\max_{j \neq i} \{\gamma_{ij}(V_j(x_j)), V_i(x_i), \gamma_{u_i}(|u_i|)\} \right), \tag{3.126}$$

   where ρ_i is a continuous and positive definite function and satisfies $(\text{Id} - \rho_i) \in \mathcal{K}_\infty$;

3. *for each $i \in \mathcal{N}_H$, $\pi_i \neq \emptyset$, and there exist $\gamma_{ij}, \gamma_{u_i} \in \mathcal{K} \cup \{0\}$ with $j \neq i$ such that both properties (3.125) and (3.126) are satisfied.*

It should be noted that, if the π_i's are different, then the impulsive time instants of different subsystems are different. From this point of view, the hybrid dynamic network (3.123)–(3.124) is more general than the discrete-time dynamic network (3.76), even if $\mathcal{N}_C \cup \mathcal{N}_H = \emptyset$.

A mild assumption is made on the intervals between the impulsive time instants. For each $i \in \mathcal{N}_D$, π_i is of the form $\pi_i = \{t_{iw} > 0 : w \in \mathbb{Z}_+\}$, and there exist constants $\overline{\delta t}, \underline{\delta t} > 0$ such that for all $i \in \mathcal{N}_D$,

$$\underline{\delta t} \leq t_{i(w+1)} - t_{iw} \leq \overline{\delta t} \tag{3.127}$$

holds for all $w \in \mathbb{Z}_+$.

If for each i-th subsystem with $i \in \mathcal{N}_D$, there is an upper bound and a lower bound of the intervals between the impulsive time instants, then one can always find common bounds for all the discrete-time subsystems.

Under the conditions above, the existence and uniqueness of solutions of the dynamic network with subsystems in the form of (3.123)–(3.124) can be guaranteed in the sense of Carathéology [60]. We use $x(t, t_0, \xi, u) = [x_1^T(t, t_0, \xi, u), \ldots, x_N^T(t, t_0, \xi, u)]^T$ or simply $x(t)$ to denote the state trajectory of the dynamic network with initial condition $\xi \in \mathbb{R}^n$ at time t_0 and input u. For each $i \in \mathcal{N}_D \cap \mathcal{N}_H$, $x_i(t, t_0, \xi, u)$ is right-continuous on the time-line. It should be noted that the semi-group property is satisfied for the hybrid dynamic network because the impulsive time sets π_i's are fixed and do not depend on the initial condition [135]. With the state trajectories defined on continuous time, we still use Definition 1.8 for hybrid dynamic networks.

The main result of this section is that a hybrid dynamic network composed of subsystems (3.123)–(3.124) is ISS if it satisfies the cyclic-small-gain condition.

3.3.1 EQUIVALENCE BETWEEN CYCLIC-SMALL-GAIN AND GAINS LESS THAN THE IDENTITY

The proofs of the cyclic-small-gain theorems for continuous-time and discrete-time dynamic networks mainly deal with the simple cycles in the gain digraphs. For hybrid dynamic networks, the analysis of the cycles involving both continuous-time and discrete-time dynamics could be much more complicated. In this subsection, a result on the equivalence between cyclic-small-gain and gains less than Id is developed. Based on this observation, in the following subsection, the cyclic-small-gain theorem for hybrid dynamic networks is proved by showing that hybrid dynamic networks with interconnection gains less than Id are ISS.

The proof of the equivalence is based on the fact that, for a continuous-time or discrete-time system, if V is an ISS-Lyapunov function, then for any $\sigma \in \mathcal{K}_\infty$ being locally Lipschitz on $(0, \infty)$, $\sigma(V)$ is also an ISS-Lyapunov function. Consider a continuous-time system, for example. Assume that system $\dot{x} = f(x, u)$ with state $x \in \mathbb{R}^n$ and external input $u \in \mathbb{R}^m$ is ISS with $V : \mathbb{R}^n \to \mathbb{R}_+$ as an ISS-Lyapunov function satisfying

$$V(x) \geq \gamma_u(|u|) \Rightarrow \nabla V(x) f(x, u) \leq -\alpha(V(x)), \quad \text{a.e.,} \qquad (3.128)$$

where $\gamma \in \mathcal{K}$ and α is a continuous and positive definite function. Then, for any $\sigma \in \mathcal{K}_\infty$ being locally Lipschitz on $(0, \infty)$, $\bar{V} := \sigma(V)$ is also continuously differentiable almost everywhere, and there exists a continuous and positive definite function $\bar{\alpha}$ such that

$$\bar{V}(x) \geq \sigma \circ \gamma(|u|) \Rightarrow \nabla \bar{V}(x) f(x, u) \leq -\bar{\alpha}(\bar{V}(x)), \quad \text{a.e.} \qquad (3.129)$$

Such transformation is also valid for discrete-time systems.

Example 3.6 *Consider the continuous-time dynamic network (3.8). Define $\bar{V}_i = \sigma_i(V_i)$ for $i = 1, 2, 3$ with $\sigma_i \in \mathcal{K}_\infty$ being locally Lipschitz on $(0, \infty)$. Then, $\bar{V}_i$ is still an ISS-Lyapunov function of the x_i-subsystem and satisfies*

$$\bar{V}_i(x_i) \geq \max_{j \neq i}\{\bar{\gamma}_{ij}(\bar{V}_i(x_i))\} \Rightarrow \nabla V_i(x_i) f_i(x) \leq -\bar{\alpha}_i(V_i(x_i)), \quad \forall x, \qquad (3.130)$$

where $\bar{\alpha}_i$ is a continuous and positive definite function, and $\bar{\gamma}_{ij} = \sigma_i \circ \gamma_{ij} \circ \sigma_j^{-1}$.

Now we consider the special case where $\gamma_{12}, \gamma_{13}, \gamma_{21}, \gamma_{32}, \gamma_{31}$ are of class $\mathcal{K}_\infty$ and are continuously differentiable on $(0, \infty)$. Also assume that the cyclic-

small-gain condition is satisfied, i.e.,

$$\gamma_{13} \circ \gamma_{31} < \text{Id}, \tag{3.131}$$

$$\gamma_{13} \circ \gamma_{32} \circ \gamma_{21} < \text{Id}. \tag{3.132}$$

Then, by using Lemma C.1, there exists a continuous and positive definite δ such that

$$(\gamma_{13} + \delta) \circ \gamma_{31} < \text{Id}, \tag{3.133}$$

$$(\gamma_{13} + \delta) \circ \gamma_{32} \circ (\gamma_{21} + \delta) < \text{Id}. \tag{3.134}$$

By choosing $\sigma_1 = \gamma_{21} + \delta$, $\sigma_2 = \text{Id}$ and $\sigma_3 = (\gamma_{21} + \delta) \circ (\gamma_{13} + \delta)$, we can show that all the interconnection gains $\bar{\gamma}_{ij}$'s are less than Id.

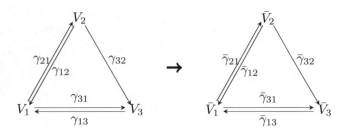

FIGURE 3.10 The equivalence between cyclic-small-gain and gains less than Id: $\bar{\gamma}_{ij} = \sigma_i \circ \gamma_{ij} \circ \sigma_j^{-1}$.

Under Assumption 3.1, the rest of this subsection shows that, if a hybrid dynamic network satisfies the cyclic-small-gain condition, then we can find a transformation $\sigma_i(V_i)$ for each V_i as done in Example 3.6, such that the interconnection gains in the dynamic network are less than Id.

The first step is to replace the ISS gains $\gamma_{(\cdot)}$'s with $\hat{\gamma}_{(\cdot)}$'s such that

1. $\hat{\gamma}_{(\cdot)} \in \mathcal{K}_\infty$ and $\hat{\gamma}_{(\cdot)} > \gamma_{(\cdot)}$ if $\gamma_{(\cdot)} \in \mathcal{K}$; $\hat{\gamma}_{(\cdot)} = 0$ if $\gamma_{(\cdot)} = 0$,
2. $\hat{\gamma}_{(\cdot)}$'s are locally Lipschitz on $(0, \infty)$, and
3. for each $r = 2, \ldots, N$,

$$\hat{\gamma}_{i_1 i_2} \circ \cdots \circ \hat{\gamma}_{i_r i_1} < \text{Id} \tag{3.135}$$

holds for all $1 \le i_j \le N$ and $i_j \ne i_{j'}$ if $j \ne j'$.

The existence of such $\hat{\gamma}_{(\cdot)}$'s can be guaranteed for hybrid dynamic networks, by reasoning similar to continuous-time dynamic networks; see Subsection 3.1.2. The only difference is that, here we consider a more general case in which $\hat{\gamma}_{(\cdot)}$'s are chosen to be locally Lipschitz on $(0, \infty)$.

Recall that $\mathcal{RS}(i)$ represents the reaching set of the i-th subsystem. For a set Ξ satisfying $\bigcup_{i\in\Xi}\mathcal{RS}(i)=\mathcal{N}$, define

$$\Gamma^{[q\to\Xi]}(s)=\bigcup_{p\in\Xi}\bigcup_{j\in\mathcal{N}}\Gamma_j^{[q\to p]}(s),\qquad(3.136)$$

where

$$\Gamma_j^{[q\to p]}(s)=\left\{\hat{\gamma}_{0i_1^{[q\to p]}}\circ\hat{\gamma}_{i_1^{[q\to p]}i_2^{[q\to p]}}\circ\cdots\circ\hat{\gamma}_{i_{j-1}^{[q\to p]}i_j^{[q\to p]}}(s)\right\}\qquad(3.137)$$

for $s\ge 0$, with $\hat{\gamma}_{0i_1^{[q\to p]}}=\mathrm{Id}$, $i_1^{[q\to p]}=q$, $i_k^{[q\to p]}\in\mathcal{N}$, $k\in\{1,\ldots,j\}$, $i_k^{[q\to p]}\ne i_{k'}^{[q\to p]}$ if $k\ne k'$, for $j\in\mathcal{N}$. Clearly, each $\Gamma_j^{[q\to p]}$ corresponds to a simple path from the x_q-subsystem to the x_p-subsystem, and $\Gamma^{q\to\Xi}$ corresponds to the simple paths from the x_q-subsystem to all the x_p-subsystems with $p\in\Xi$. A special case is $q=p$ and $j=1$.

Define

$$\bar{\mathbb{V}}_q^\Xi(x_q)=\Gamma^{[q\to\Xi]}(V_q(x_q))$$
$$\bar{V}_q^\Xi(x_q)=\hat{\gamma}^{[q\to\Xi]}(V_q(x_q))\qquad(3.138)$$

with $\hat{\gamma}^{[q\to\Xi]}(s)=\max\Gamma^{[q\to\Xi]}(s)$ for $s\ge 0$. Clearly, $\bar{V}_q^\Xi(x_q)=\max\bar{\mathbb{V}}_q^\Xi(x_q)$.

As there exists at least one subsystem in Ξ that is reachable from the x_q-subsystem and $\hat{\gamma}_{(\cdot)}\in\mathcal{K}_\infty\cup\{0\}$, it can be observed that for all $q\in\mathcal{N}$, $\hat{\gamma}^{[q\to\Xi]}\in\mathcal{K}_\infty$. Moreover, $\bar{V}_q^\Xi$ is positive definite and radially unbounded with respect to x_q, and is continuously differentiable almost everywhere. Correspondingly, to simplify the discussions, we define

$$\bar{u}_q^\Xi=\hat{\gamma}^{[q\to\Xi]}\circ\gamma_{u_q}(|u_q|)\qquad(3.139)$$

as the new input of the x_q-subsystem.

Proposition 3.1 presents the main result on the equivalence between cyclic-small-gain and interconnection gains less than Id.

Proposition 3.1 *A hybrid dynamic network composed of subsystems (3.123)–(3.124) which satisfy Assumption 3.1 and the cyclic-small-gain condition (3.22) can be reformulated as one with interconnection gains less than* Id *by considering $\bar{V}_q^\Xi$ defined in (3.138) as the new ISS-Lyapunov function and $\bar{u}_q^\Xi$ defined in (3.139) as the new input for each x_q-subsystem with $q\in\mathcal{N}$.*

Proposition 3.1 is proved below by studying both the continuous-time dynamics and the discrete-time dynamics.

With Lemma C.5, there exists a continuous and positive definite function δ satisfying $(\mathrm{Id}-\delta)\in\mathcal{K}_\infty$ such that $\hat{\gamma}_{ij}\circ(\mathrm{Id}-\delta)\ge\gamma_{ij}$ for all $\hat{\gamma}_{ij}\ne 0$. Thus, for all $i,j\in\mathcal{N}$, $i\ne j$, $\hat{\gamma}_{ij}\circ(\mathrm{Id}-\delta)\ge\gamma_{ij}$.

Furthermore, there exists a continuous and positive definite function δ' satisfying $(\mathrm{Id} - \delta') \in \mathcal{K}_\infty$ such that

$$(\mathrm{Id} - \delta') \circ \hat{\gamma}_{i_1^{[q \to p]} i_2^{[q \to p]}} \circ \cdots \circ \hat{\gamma}_{i_{j-1}^{[q \to p]} i_j^{[q \to p]}}$$

$$\geq \hat{\gamma}_{i_1^{[q \to p]} i_2^{[q \to p]}} \circ \cdots \circ \hat{\gamma}_{i_{j-1}^{[q \to p]} i_j^{[q \to p]}} \circ (\mathrm{Id} - \delta) \qquad (3.140)$$

for all $q \in \mathcal{N}$, all $p \in \Xi$, and all $j \in \mathcal{N}$, where $i_k^{[q \to p]} \in \mathcal{N}$ for $k \in \{1, \ldots, j\}$, and $i_k^{[q \to p]} \neq i_{k'}^{[q \to p]}$ if $k \neq k'$.

We now study the continuous-time dynamics and the discrete-time dynamics separately.

Continuous-Time Dynamics

For any $q^* \in \mathcal{N}_C \cup \mathcal{N}_H$, consider the case where

$$\bar{V}_{q^*}^\Xi(x_{q^*}) \geq \max_{j \neq q^*} \left\{ (\mathrm{Id} - \delta')(\bar{V}_j^\Xi(x_j)), \bar{u}_{q^*}^\Xi \right\}. \qquad (3.141)$$

Denote any one of the elements in $\bar{\mathbb{V}}_{q^*}^\Xi(x_{q^*})$ taking the value of $\bar{V}_{q^*}^\Xi(x_{q^*})$ as

$$\hat{\gamma}_{i_1^* i_2^*} \circ \cdots \circ \hat{\gamma}_{i_{j^*-1}^* i_{j^*}^*}(V_{i_{j^*}^*}(x_{i_{j^*}^*})), \qquad (3.142)$$

where $i_1^* = p^* \in \Xi$ and $i_{j^*}^* = q^*$.

For each $i_k^* \in \{i_1^*, \ldots, i_{j^*-1}^*\}$, $V_{i_k^*}$ satisfies

$$\hat{\gamma}_{i_1^* i_2^*} \circ \cdots \circ \hat{\gamma}_{i_{k-1}^* i_k^*} \circ (\mathrm{Id} - \delta)(V_{i_k^*}(x_{i_k^*}))$$

$$\leq (\mathrm{Id} - \delta') \circ \hat{\gamma}_{i_1^* i_2^*} \circ \cdots \circ \hat{\gamma}_{i_{k-1}^* i_k^*}(V_{i_k^*}(x_{i_k^*}))$$

$$\leq \max \bar{\mathbb{V}}_{i_k^*}^\Xi(x_{i_k^*})$$

$$\leq \bar{V}_{q^*}^\Xi(x_{q^*})$$

$$\leq \hat{\gamma}_{i_1^* i_2^*} \circ \cdots \circ \hat{\gamma}_{i_{j^*-1}^* i_{j^*}^*}(V_{i_{j^*}^*}(x_{i_{j^*}^*})) \qquad (3.143)$$

and thus,

$$(\mathrm{Id} - \delta)(V_{i_k^*}(x_{i_k^*})) \leq \hat{\gamma}_{i_k^* i_{k+1}^*} \circ \cdots \circ \hat{\gamma}_{i_{j^*-1}^* i_{j^*}^*}(V_{i_{j^*}^*}(x_{i_{j^*}^*})), \qquad (3.144)$$

which implies

$$\hat{\gamma}_{i_{j^*}^* i_k^*} \circ (\mathrm{Id} - \delta)(V_{i_k^*}(x_{i_k^*})) \leq \hat{\gamma}_{i_{j^*}^* i_k^*} \circ \hat{\gamma}_{i_k^* i_{k+1}^*} \circ \cdots \circ \hat{\gamma}_{i_{j^*-1}^* i_{j^*}^*}(V_{i_{j^*}^*}(x_{i_{j^*}^*})). \qquad (3.145)$$

With the cyclic-small-gain condition satisfied, one has

$$\hat{\gamma}_{i_{j^*}^* i_k^*} \circ \hat{\gamma}_{i_k^* i_{k+1}^*} \circ \cdots \circ \hat{\gamma}_{i_{j^*-1}^* i_{j^*}^*} < \mathrm{Id}. \qquad (3.146)$$

Then, (3.145) implies

$$V_{i^*_{j^*}}(x_{i^*_{j^*}}) \geq \hat{\gamma}_{i^*_{j^*} i^*_k} \circ (\mathrm{Id} - \delta)(V_{i^*_k}(x_{i^*_k})) \geq \gamma_{i^*_{j^*} i^*_k}(V_{i^*_k}(x_{i^*_k})) \qquad (3.147)$$

for all $i^*_k \in \{i^*_1, i^*_2, \ldots, i^*_{j^*-1}\}$.

For each $V_{i^*_k}$ with $i^*_k \in S \backslash \{i^*_1, i^*_2, \ldots, i^*_j\}$, it can be observed that

$$\hat{\gamma}_{i^*_1 i^*_2} \circ \cdots \circ \hat{\gamma}_{i^*_{j^*-1} i^*_{j^*}} \circ \hat{\gamma}_{i^*_{j^*} i^*_k} \circ (\mathrm{Id} - \delta)(V_{i^*_k}(x_{i^*_k}))$$

$$\leq (\mathrm{Id} - \delta') \circ \hat{\gamma}_{i^*_1 i^*_2} \circ \cdots \circ \hat{\gamma}_{i^*_{j^*-1} i^*_{j^*}} \circ \hat{\gamma}_{i^*_{j^*} i^*_k}(V_{i^*_k}(x_{i^*_k}))$$

$$\leq \max \bar{\mathbb{V}}^{\Xi}_{i^*_k}(x_{i^*_k})$$

$$\leq \bar{V}^{\Xi}_{q^*}(x_{q^*})$$

$$\leq \hat{\gamma}_{i^*_1 i^*_2} \circ \cdots \circ \hat{\gamma}_{i^*_{j^*-1} i^*_{j^*}}(V_{i^*_{j^*}}(x_{i^*_{j^*}})). \qquad (3.148)$$

Thus, it is achieved that

$$V_{i^*_{j^*}}(x_{i^*_{j^*}}) \geq \hat{\gamma}_{i^*_{j^*} i^*_k} \circ (\mathrm{Id} - \delta)(V_{i^*_k}(x_{i^*_k})) \geq \gamma_{i^*_{j^*} i^*_k}(V_{i^*_k}(x_{i^*_k})) \qquad (3.149)$$

for all $i^*_k \in S \backslash \{i^*_1, i^*_2, \ldots, i^*_j\}$.

Note that $i^*_{j^*} = q^*$. By combining (3.147) and (3.149) and considering $\bar{V}^{\Xi}_{q^*} \geq \bar{u}^{\Xi}_{q^*}$, one gets

$$V_{q^*}(x_{q^*}) \geq \max_{l \neq q^*} \left\{ \gamma_{q^* l}(V_l(x_l)), \gamma_{u_{q^*}}(|u_{q^*}|) \right\}, \qquad (3.150)$$

which implies that

$$\nabla V_{q^*}(x_{q^*}) f_{q^*}(x, u_{q^*}) \leq -\alpha_{q^*}(V_{q^*}(x_{q^*})) \qquad (3.151)$$

holds almost everywhere.

Recall that $\bar{V}^{\Xi}_{q^*}(x_{q^*}) = \hat{\gamma}^{[q^* \to \Xi]}(V_{q^*}(x_{q^*}))$. The continuous differentiability of the $\hat{\gamma}_{(\cdot)}$'s implies that the $\mathcal{K}_\infty$ function $\hat{\gamma}^{[q^* \to \Xi]}$ is continuously differentiable almost everywhere. Then, there exist continuous and positive definite functions $\check{\alpha}^{\Xi}_{q^*}$ and $\bar{\alpha}^{\Xi}_{q^*}$ such that

$$\nabla \bar{V}^{\Xi}_{q^*}(x_{q^*}) f_{q^*}(x, u_{q^*}) \leq -\check{\alpha}^{\Xi}_{q^*}(V_{q^*}(x_{q^*}))$$

$$= -\check{\alpha}^{\Xi}_{q^*} \circ (\hat{\gamma}^{[q^* \to \Xi]})^{-1}(\bar{V}^{\Xi}_{q^*}(x_{q^*}))$$

$$\leq -\bar{\alpha}^{\Xi}_{q^*}(\bar{V}^{\Xi}_{q^*}(x_{q^*})) \qquad (3.152)$$

holds almost everywhere.

Discrete-Time Dynamics

For any $q^* \in \mathcal{N}_D \cup \mathcal{N}_H$, using Lemma C.5, one can find a continuous and positive definite function ρ'_{q^*} satisfying $(\mathrm{Id} - \rho'_{q^*}) \in \mathcal{K}_\infty$ such that

$$
\bar{V}^\Xi_{q^*}(g_{q^*}(x, u_{q^*}))
$$

$$
\leq \hat{\gamma}^{[q^* \to \Xi]} \circ (\mathrm{Id} - \rho_{q^*})(\max_{j \neq q^*}\{\gamma_{q^* j}(V_j(x_j)), V_{q^*}(x_{q^*}), \gamma_{u_{q^*}}(|u_{q^*}|)\})
$$

$$
\leq (\mathrm{Id} - \rho'_{q^*})\Big(\max_{j \neq q^*}\{\hat{\gamma}^{[q^* \to \Xi]} \circ \gamma_{q^* j}(V_j(x_j)),
$$

$$
\hat{\gamma}^{[q^* \to \Xi]}(V_{q^*}(x_{q^*})), \hat{\gamma}^{[q^* \to \Xi]} \circ \gamma_{u_{q^*}}(|u_{q^*}|)\} \Big)
$$

$$
= (\mathrm{Id} - \rho'_{q^*})(\max_{j \neq q^*}\{\hat{\gamma}^{[q^* \to \Xi]} \circ \gamma_{q^* j}(V_j(x_j)), \bar{V}^\Xi_{q^*}(x_{q^*}), \bar{u}^\Xi_{q^*}\}). \tag{3.153}
$$

Denote any element in $\Gamma^{[q^* \to \Xi]}(s)$ as $\hat{\gamma}_{i_1^* i_2^*} \circ \cdots \circ \hat{\gamma}_{i_{j^*-1}^* i_{j^*}^*}(s)$ for $s \in \mathbb{R}_+$, with $i_1^* = p^* \in \Xi$ and $i_{j^*}^* = q^*$. For any $j \in \mathcal{N}\backslash\{q^*\}$, consider $\hat{\gamma}_{i_1^* i_2^*} \circ \cdots \circ \hat{\gamma}_{i_{j^*-1}^* i_{j^*}^*} \circ \gamma_{i_{j^*}^* j}(V_j(x_j))$.

If $j \in \{i_1^*, i_1^*, \ldots, i_{j^*-1}^*\}$, then denote $j = i_k^*$ with $i_k^* \in \{i_1^*, \ldots, i_{j^*-1}^*\}$. In this case, the following property holds

$$
\hat{\gamma}_{i_1^* i_2^*} \circ \cdots \circ \hat{\gamma}_{i_{j^*-1}^* i_{j^*}^*} \circ \gamma_{i_{j^*}^* j}(V_j(x_j))
$$

$$
= \hat{\gamma}_{i_1^* i_2^*} \circ \cdots \circ \hat{\gamma}_{i_k^* i_{k+1}^*} \circ \cdots \circ \hat{\gamma}_{i_{j^*-1}^* i_{j^*}^*} \circ \gamma_{i_{j^*}^* i_k^*}(V_{i_k^*}(x_{i_k^*}))
$$

$$
\leq \hat{\gamma}_{i_1^* i_2^*} \circ \cdots \circ \hat{\gamma}_{i_k^* i_{k+1}^*} \circ \cdots \circ \hat{\gamma}_{i_{j^*-1}^* i_{j^*}^*} \circ \hat{\gamma}_{i_{j^*}^* i_k^*} \circ (\mathrm{Id} - \delta)(V_{i_k^*}(x_{i_k^*}))
$$

$$
\leq \hat{\gamma}_{i_1^* i_2^*} \circ \cdots \hat{\gamma}_{i_{k-1}^* i_k^*} \circ (\mathrm{Id} - \delta)(V_{i_k^*}(x_{i_k^*}))
$$

$$
\leq (\mathrm{Id} - \delta') \circ \hat{\gamma}^{[j \to \Xi]}(V_j(x_j))
$$

$$
= (\mathrm{Id} - \delta')(\bar{V}^\Xi_j(x_j)), \tag{3.154}
$$

where the cyclic-small-gain condition $\hat{\gamma}_{i_k^* i_{k+1}^*} \circ \cdots \circ \hat{\gamma}_{i_{j^*-1}^* i_{j^*}^*} \circ \hat{\gamma}_{i_{j^*}^* i_k^*} < \mathrm{Id}$ is used for the second inequality.

If $j \in \mathcal{N}\backslash\{i_1^*, i_2^*, \ldots, i_{j^*}^*\}$, then it can be directly derived that

$$
\hat{\gamma}_{i_1^* i_2^*} \circ \cdots \circ \hat{\gamma}_{i_{j^*-1}^* i_{j^*}^*} \circ \gamma_{i_{j^*}^* j}(V_j(x_j))
$$

$$
\leq \hat{\gamma}_{i_1^* i_2^*} \circ \cdots \circ \hat{\gamma}_{i_{j^*-1}^* i_{j^*}^*} \circ \hat{\gamma}_{i_{j^*}^* j} \circ (\mathrm{Id} - \delta)(V_j(x_j))
$$

$$
\leq (\mathrm{Id} - \delta') \circ \hat{\gamma}^{[j \to \Xi]}(V_j(x_j))
$$

$$
= (\mathrm{Id} - \delta') \circ (\bar{V}^\Xi_j(x_j)). \tag{3.155}
$$

Combining (3.154) and (3.155), for any $j \in S\backslash\{q^*\} = \mathcal{N}\backslash\{i_{j^*}^*\}$ and for any $\hat{\gamma}_{i_1^* i_2^*} \circ \cdots \circ \hat{\gamma}_{i_{j^*-1}^* i_{j^*}^*}(s)$ in $\Gamma^{[q^* \to \Xi]}(s)$, one has

$$
\hat{\gamma}_{i_1^* i_2^*} \circ \cdots \circ \hat{\gamma}_{i_{j^*-1}^* i_{j^*}^*} \circ \gamma_{i_{j^*}^* j}(V_j(x_j)) \leq (\mathrm{Id} - \delta') \circ (\bar{V}^\Xi_j(x_j)). \tag{3.156}
$$

Recall that $\hat{\gamma}^{[q^* \to \Xi]}(s) = \max \Gamma^{[q^* \to \Xi]}(s)$ for $s \in \mathbb{R}_+$. Then, for any $j \in \mathcal{N} \backslash \{q^*\} = \mathcal{N} \backslash \{i_{j*}^*\}$, one has

$$\hat{\gamma}^{[q^* \to \Xi]} \circ \gamma_{i_{j*}^*, j}(V_j(x_j)) \leq (\mathrm{Id} - \delta') \circ (\bar{V}_j^\Xi(x_j)). \tag{3.157}$$

Then, from (3.153), it can be achieved that

$$\bar{V}_{q^*}^\Xi(g_{q^*}(x, u_{q^*})) \leq (\mathrm{Id} - \rho_{q^*}') \left(\max_{j \neq q^*} \left\{ (\mathrm{Id} - \delta')(\bar{V}_j^\Xi(x_j)), \bar{V}_{q^*}^\Xi(x_{q^*}), \bar{u}_i^\Xi \right\} \right)$$

$$\leq (\mathrm{Id} - \rho_{q^*}') \left(\max_{j \neq q^*} \left\{ \bar{V}_j^\Xi(x_j), \bar{V}_{q^*}^\Xi(x_{q^*}), \bar{u}_i^\Xi \right\} \right) \tag{3.158}$$

for any $q^* \in \mathcal{N}_D \cup \mathcal{N}_H$.

Properties (3.152) and (3.158) imply that a hybrid dynamic network of (3.123)–(3.124) satisfying the cyclic-small-gain condition (3.22) can be transformed into a network with interconnection gains less than Id by appropriately scaling the ISS-Lyapunov functions of the subsystems. Based on this observation, the cyclic-small-gain theorem for hybrid dynamic networks can be proved by checking the ISS of hybrid dynamic networks with interconnection gains less than Id.

3.3.2   CYCLIC-SMALL-GAIN THEOREM FOR HYBRID DYNAMIC NETWORKS

In this subsection, we first consider the dynamic networks composed of subsystems in the form of (3.123)–(3.124) with interconnection gains γ_{ij}'s ($i, j \in \mathcal{N}$, $i \neq j$) defined in (3.125) and (3.126) less than Id. Based on the equivalence result developed in Subsection 3.3.1, the Lyapunov function constructed for such systems is further used to validate the Lyapunov functions in the form of (3.31) for general dynamic networks.

For a continuous-time dynamic network composed of two subsystems, if the interconnection gains are less than Id, then one may choose $\sigma = \mathrm{Id}$ in (3.23), and construct an ISS-Lyapunov function as the maximum of the Lyapunov functions of the subsystems. Similarly, for the hybrid dynamic networks with interconnection gains less than the identity, we construct the following Lyapunov function candidate

$$V(x) = \max \mathbb{V}(x) \tag{3.159}$$

with

$$\mathbb{V}(x) = \{V_1(x_1), \ldots, V_N(x_N)\}. \tag{3.160}$$

Since for each $i \in \mathcal{N}$, $V_i(x_i)$ is positive definite and radially unbounded with respect to x_i, it can be verified that $V(x)$ is positive definite and radially unbounded with respect to x. Moreover, V is locally Lipschitz on $\mathbb{R}^n \backslash \{0\}$.

Correspondingly, we define

$$\bar{u} = \max \mathbb{U} \qquad (3.161)$$

with

$$\mathbb{U} = \{\gamma_{u_1}(|u_1|), \ldots, \gamma_{u_N}(|u_N|)\}. \qquad (3.162)$$

Denote $\pi = \bigcup_{i \in \mathcal{N}} \pi_i$ as the set of the impulsive time instants of the dynamic network. The following theorem shows that the hybrid dynamic network with gains less than Id is ISS with $V(x)$ defined in (3.159) as a weak Lyapunov function in the sense that $V(x(t))$ is, not necessarily strictly, decreasing along the solutions $x(t)$.

Theorem 3.6 *Consider the hybrid dynamic network composed of subsystems (3.123)–(3.124). Under Assumption 3.1, if all the interconnection gains γ_{ij}'s $(i, j \in \mathcal{N}, \ i \neq j)$ are less than Id, i.e., $\gamma_{ij} < \mathrm{Id}$, then $V(x)$ defined in (3.159) is a weak ISS-Lyapunov function and admits the following properties:*

1. for any ξ, u and $t_0 \geq 0$,

$$V(x(t, t_0, \xi, u)) \geq \bar{u}(t) \Rightarrow \dot{V}(x(t, t_0, \xi, u)) \leq 0 \qquad (3.163)$$

holds for almost all $t \in [t_0, \infty) \backslash \pi$;
2. for any ξ, u and $t_0 \geq 0$,

$$V(x(t, t_0, \xi, u)) \leq \max\{V(x(t^-, t_0, \xi, u)), \bar{u}(t^-)\} \qquad (3.164)$$

holds for all $t \in (t_0, \infty) \cap \pi$;
3. there exist a $\overline{\delta t}_D > 0$ and a positive definite function ρ^* satisfying $(\mathrm{Id} - \rho^*) \in \mathcal{K}_\infty$, such that for any ξ and any u,

$$V(x(t, t_0, \xi, u)) \leq \max\{(\mathrm{Id} - \rho^*)(V(\xi)), \|\bar{u}\|_{[t_0, t]}\} \qquad (3.165)$$

holds for any pair of nonnegative numbers (t, t_0) satisfying $t - t_0 \geq \overline{\delta t}_D$, and the hybrid dynamic network is ISS.

The proof of Theorem 3.6 is given in Appendix D.4.

Based on the observation that a hybrid dynamic network satisfying the cyclic-small-gain condition can be reformulated as one with interconnection gains less than the identity function, we develop a cyclic-small-gain theorem for hybrid dynamic networks.

If a hybrid dynamic network composed of subsystems (3.123)–(3.124) satisfies the cyclic-small-gain condition (3.22), based on Proposition 3.1 and Theorem 3.6, we can construct an ISS-Lyapunov function as

$$V^\Xi(x) = \max_{q \in \mathcal{N}}\{\bar{V}_q^\Xi(x_q)\}. \qquad (3.166)$$

Corresponding to V^Ξ, we define

$$\bar{u}^\Xi = \max_{q \in \mathcal{N}} \{\bar{u}_q^\Xi\} \tag{3.167}$$

as the new input of the hybrid dynamic network. Then, properties (3.163)–(3.165) hold for the hybrid dynamic network with V replaced by V^Ξ and $\bar{u}$ replaced by $\bar{u}^\Xi$. Our main theorem is as follows.

Theorem 3.7 *A hybrid dynamic network composed of subsystems (3.123)–(3.124) satisfying the cyclic-small-gain condition (3.22) is ISS with V^Ξ defined in (3.166) as a weak ISS-Lyapunov function.*

3.3.3 AN EXAMPLE

Consider a hybrid dynamic network in the form of (3.123)–(3.124) with $N = 3$ and each $x_i \in \mathbb{R}$.

The x_1-subsystem involves only continuous-time dynamics with

$$f_1(x, u_1) = -|x_1|x_1 + 0.3x_2 + 0.3x_3^2 + u_1^2. \tag{3.168}$$

The x_2-subsystem is defined on discrete time with

$$g_2(x, u_2) = 0.4x_2 + 0.25x_1^2 + 0.25x_3^2 + u_2 \tag{3.169}$$

and $\pi_2 = \mathbb{Z}_+ \backslash \{0\}$.

The x_3-subsystem is an impulsive system with

$$f_3(x, u_3) = -2|x_3|x_3 + 0.5x_1^2 + 0.5x_2 + 3u_3^2, \tag{3.170}$$

$$g_3(x, u_3) = 0.4x_3 + 0.4x_1 \tag{3.171}$$

and $\pi_3 = \{k + 0.2(k \bmod 2) : k \in \mathbb{Z}_+ \backslash \{0\}\}$.

Define $V_i(x_i) = |x_i|$ for $i = 1, 2, 3$. Then, each V_i is positive definite, radially unbounded, and continuously differentiable on $\mathbb{R} \backslash \{0\}$.

For the continuous-time dynamics, it can be verified that

$$V_1(x_1) \geq \max\{0.707V_2(x_2)^{1/2}, V_3(x_3), 5|u_1|\}$$
$$\Rightarrow \nabla V_1(x_1)f_1(x, u_1) \leq -0.06V_1(x_1)^2, \tag{3.172}$$
$$V_3(x_3) \geq \max\{0.9V_1(x_1), 0.707V_2(x_2)^{1/2}, 3|u_3|\}$$
$$\Rightarrow \nabla V_3(x_3)f_3(x, u_3) \leq -0.0494V_3(x_3)^2, \tag{3.173}$$

and for the discrete-time dynamics, it can be verified that

$$V_2(g_2(x, u_2)) \leq 0.8\max\{V_2(x_2), 1.875V_1(x_1)^2, 1.875V_3(x_3)^2, 7.5|u_2|\}, \tag{3.174}$$

$$V_3(g_3(x, u_3)) \leq 0.8889\max\{V_3(x_3), 0.9V_1(x_1)\}. \tag{3.175}$$

Define $\gamma_{12}(s) = 0.707s^{1/2}$, $\gamma_{13}(s) = s$, $\gamma_{21}(s) = 1.875s^2$, $\gamma_{23}(s) = 1.875s^2$, $\gamma_{31}(s) = 0.9s$, $\gamma_{32}(s) = 0.707s^{1/2}$, $\gamma_{u_1}(s) = 5s$, $\gamma_{u_2}(s) = 7.5s$, and $\gamma_{u_3}(s) = 3s$ for $s \in \mathbb{R}_+$. Then,

$$\gamma_{12} \circ \gamma_{21} < \mathrm{Id} \tag{3.176}$$

$$\gamma_{13} \circ \gamma_{31} < \mathrm{Id} \tag{3.177}$$

$$\gamma_{23} \circ \gamma_{32} < \mathrm{Id} \tag{3.178}$$

$$\gamma_{13} \circ \gamma_{32} \circ \gamma_{21} < \mathrm{Id} \tag{3.179}$$

$$\gamma_{12} \circ \gamma_{23} \circ \gamma_{31} < \mathrm{Id}. \tag{3.180}$$

Thus, the hybrid dynamic network composed of x_1-, x_2-, and x_3-subsystems satisfies the cyclic-small-gain condition, and from the cyclic-small-gain theorem, it is ISS with $[u_1, u_2, u_3]^T$ as input.

We employ the technique in Subsection 3.3.2 to construct an ISS-Lyapunov function for the hybrid dynamic network. Firstly, define $\hat{\gamma}_{12}(s) = 0.71s^{1/2}$, $\hat{\gamma}_{13}(s) = 1.01s$, $\hat{\gamma}_{21}(s) = 1.88s^2$, $\hat{\gamma}_{23}(s) = 1.88s^2$, $\hat{\gamma}_{31}(s) = 0.91s$, and $\hat{\gamma}_{32}(s) = 0.71s^{1/2}$ for $s \in \mathbb{R}_+$.

Define $\Xi = \{2\}$. Then, $\bigcup_{i \in \Xi} \mathcal{RS}(i) = \{1, 2, 3\}$. Define

$$\hat{\gamma}^{[1 \to \Xi]} = \max \Gamma^{[1 \to \Xi]}(s) = \max\{\hat{\gamma}_{21}(s), \hat{\gamma}_{23} \circ \hat{\gamma}_{31}(s)\} = 1.88s^2 \tag{3.181}$$

$$\hat{\gamma}^{[2 \to \Xi]} = \max \Gamma^{[2 \to \Xi]}(s) = s \tag{3.182}$$

$$\hat{\gamma}^{[3 \to \Xi]} = \max \Gamma^{[3 \to \Xi]}(s) = \max\{\hat{\gamma}_{23}(s), \hat{\gamma}_{21} \circ \hat{\gamma}_{13}(s)\} = 1.92s^2. \tag{3.183}$$

Define

$$V_1^{\Xi}(x_1) = \hat{\gamma}^{[1 \to \Xi]}(V_1(x_1)) = 1.88V_1(x_1)^2 \tag{3.184}$$

$$V_2^{\Xi}(x_2) = \hat{\gamma}^{[2 \to \Xi]}(V_2(x_2)) = V_2(x_2) \tag{3.185}$$

$$V_3^{\Xi}(x_3) = \hat{\gamma}^{[3 \to \Xi]}(V_3(x_3)) = 1.92V_3(x_3)^2 \tag{3.186}$$

$$\bar{u}_1^{\Xi} = \hat{\gamma}^{[1 \to \Xi]} \circ \gamma_{u_1}(|u_1|) = 47|u_1|^2 \tag{3.187}$$

$$\bar{u}_2^{\Xi} = \hat{\gamma}^{[2 \to \Xi]} \circ \gamma_{u_2}(|u_2|) = 7.5|u_2| \tag{3.188}$$

$$\bar{u}_3^{\Xi} = \hat{\gamma}^{[3 \to \Xi]} \circ \gamma_{u_3}(|u_3|) = 17.3|u_3|^2. \tag{3.189}$$

Then, for the continuous-time dynamics, it holds that

$$V_1^{\Xi}(x_1) \geq \max\{0.94V_2^{\Xi}(x_2), 0.98V_3^{\Xi}(x_3), \bar{u}_1^{\Xi}\}$$
$$\Rightarrow \nabla V_1^{\Xi}(x_1)f_1(x, u_1) \leq -0.089V_1^{\Xi}(x_1)^{3/2} \tag{3.190}$$

$$V_3^{\Xi}(x_3) \geq \max\{0.827V_1^{\Xi}(x_1), 0.96V_2^{\Xi}(x_2), \bar{u}_3^{\Xi}\}$$
$$\Rightarrow \nabla V_3^{\Xi}(x_3)f_3(x, u_3) \leq -0.0713V_3^{\Xi}(x_3)^{3/2}, \tag{3.191}$$

and for the discrete-time dynamics, it holds that

$$V_2^{\Xi}(g_2(x, u_2)) \leq 0.8 \max\{V_2^{\Xi}(x_2), 0.998V_1^{\Xi}(x_1), 0.98V_3^{\Xi}(x_3), \bar{u}_2^{\Xi}\} \tag{3.192}$$

$$V_3^{\Xi}(g_3(x, u_3)) \leq 0.79 \max\{V_3^{\Xi}(x_3), 0.83V_1^{\Xi}(x_1)\}. \tag{3.193}$$

In this way, the subsystems of the hybrid dynamic network are reformulated with new ISS-Lyapunov functions, with which the interconnection gains are less than the identity. Based on this achievement, we can construct the ISS-Lyapunov function of the hybrid dynamic network as:

$$V^{\Xi}(x) = \max\{V_1^{\Xi}(x_1), V_2^{\Xi}(x_2), V_3^{\Xi}(x_3)\}. \tag{3.194}$$

3.4 NOTES

Some recent extensions of the ISS small-gain theorem can be found in [221, 43, 231, 134, 121, 138, 137]. To the best of the authors' knowledge, Teel [259] stated an extension of the nonlinear small-gain theorem for the first time, for networks of discrete-time ISS systems. Shortly, the authors of [43, 44, 231] developed a matrix-small-gain criterion for networks with plus-type interconnections. In [134, 121], a more general cyclic-small-gain theorem for networks of IOS systems was developed. The corresponding Lyapunov formulations have been developed in [178, 179]. It should be noted that the matrix-small-gain condition is given by matrix inequalities of nonlinear functions, which is usually not easily checkable. As shown in this chapter, the cyclic-small-gain condition can be easily verified by directly testing specific compositions of ISS gains of the subsystems.

The small-gain methods have also been introduced in hybrid systems, which involve both continuous-time and discrete-time dynamics; see e.g., [167, 168, 135, 86, 211, 23, 42, 138]. In [86], the impulses are time-triggered and a (converse) dwell-time-based strategy is developed to evaluate the ISS property of impulsive systems. In [23], the discrete evolution is state triggered and both the continuous evolution and the discrete evolution are required to possess some stability property to guarantee the ISS of a hybrid system. ISS small-gain criteria for hybrid feedback systems and their corresponding Lyapunov formulations have also been developed by [167, 135, 211]. The interest in these results for quantized control, impulsive control, and networked control can be found in recent papers; see e.g., [86, 168]. Reference [84] considers nonlinear systems with discontinuous right-hand sides. References [138, 42] generalize the small-gain results to large-scale hybrid dynamic networks, based on vector Lyapunov functions and the matrix-small-gain theorem, respectively. One recent result on global stabilization of nonlinear systems based on vector-control Lyapunov functions can be found in [139]. It should be pointed out that, in [167, 211, 42], the impulses of the subsystems are supposed to be triggered at the same time. A cyclic-small-gain theorem for hybrid dynamic networks with the impulses of the subsystems triggered asynchronously is developed in [188]. A time-delay version of the cyclic-small-gain theorem can be found in [265].

This chapter has presented cyclic-small-gain results for continuous-time, discrete-time, and hybrid dynamic networks with the ISS property of the subsystems formulated by ISS-Lyapunov functions based on recent results in

[134, 121, 178, 179, 188].

The continuous-time dynamic networks considered in this chapter are modeled by differential equations. For discontinuous systems, i.e., systems with discontinuous dynamics, differential inclusions are used. In the discontinuous case, the cyclic-small-gain condition is still valid as long as the subsystems are ISS. See [84] for the extension of the original ISS small-gain theorem for discontinuous systems. A similar approach also applies to discrete-time dynamic networks.

The hybrid dynamic networks considered in this chapter are composed of ISS subsystems, whose motions may be continuous, piecewise constant, or impulsive on the timeline. In particular, the impulses of the subsystems are time-triggered and the impulsive time instants of different subsystems are allowed to be different. For hybrid dynamic networks, this chapter has only studied time-triggered impulsive events. Small-gain theorems for hybrid systems with state-triggered impulses were studied in [211, 42] for the $\mathcal{KLL}$ stability and input-to-state stability based on hybrid inclusions [71] and hybrid input-to-state stability in [23]. In these results, the impulses of different subsystems are triggered by the same state conditions. However, the impulses of different subsystems may be triggered under different state conditions in practical systems. Based on the achievements in this chapter, further effort may be devoted to a small-gain result for hybrid dynamic networks with the impulses of different subsystems triggered by different state conditions.

At a fundamental level, passivity and dissipativity concepts would appear to add extra flexibility to a reliance on just gain. However, there are other questions to explore. Firstly, throughout stability theory, the input–output and Lyapunov approaches have been linked by versions of the famous Kalman-Yakubovich (or Positive-Real) Lemma, from linear systems [5] to abstract systems [92]. Further, this lemma established the Lyapunov function in an additive form, which became the basic idea of dissipative systems, i.e., the Lyapunov functions are constructed by adding up the "storage functions" of all the subsystems. This was consistent with the study of large-scale system stability [205, 207] where Lyapunov functions took a weighted additive or vector form with discussions of their relative merits being a major point of interest. Some step towards exploring connections with ISS and the more general iISS has been taken recently in [41, 109]. Other features of the earlier stability theory were equivalences between passivity and small-gain theorems via transformations [4], the capability to study stability with different norms [48] (passivity goes naturally with Hilbert spaces where finite energy signals can be usefully studied) as noted for gain properties in [248, 249], the distinctions between the input–output and state stability concepts [91] (of interest in adaptive control where internal chaos was consistent with robust stabilization [198]), and instability theorem counterparts to the stability results [48, 93]. The last area includes ways to describe unstable systems within a gain/dissipativity framework. Some applications of passivity methods to nonlinear control can be found in [268].

This chapter has only considered the dynamic networks with stable subsystem dynamics. It is well known that appropriately switching between unstable dynamics may still lead to stable behaviors. This is often formulated by the "dwell-time" condition [86]. This issue was addressed in [169] for a Lyapunov-based small-gain theorem for hybrid systems. More effort is desired for the theoretical development of the cyclic-small-gain theorem in this research direction. Some extensions of passivity theory can be found in recent works [62, 63].

4 Control under Sensor Noise

The robust control problem for nonlinear uncertain systems with measurement feedback (i.e., in the presence of sensor noise) is challenging, yet important. The purpose of this chapter is to show that several nonlinear control problems can be studied in a unified framework of measurement feedback control. Figure 4.1 shows the block diagram of a measurement feedback control system.

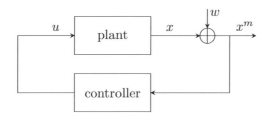

FIGURE 4.1 The block diagram of a measurement feedback control system: u is the control input, x is the state of the plant, and $x^m = x + w$ is the measurement of the state with w representing the additive sensor noise.

First, by means of an elementary example, we show that stabilization in the absence of sensor noise may not imply external stability or robustness in the presence of sensor noise.

Example 4.1 *Consider a first-order nonlinear system*

$$\dot{x} = x^2 + u, \tag{4.1}$$

where $x \in \mathbb{R}$ is the state and $u \in \mathbb{R}$ is the control input. If there is no sensor noise, we can design a feedback linearizing control law $u = -x^2 - 0.1x$ such that the closed-loop system is $\dot{x} = -0.1x$, which is asymptotically, and even exponentially in this case, stable at the origin. If the measurement of the state is subject to additive sensor noise, denoted by w, then the realizable feedback control law is $u = -(x+w)^2 - 0.1(x+w)$ and the resulting closed-loop system is

$$\dot{x} = -(0.1 + 2w)x - w^2 - 0.1w, \tag{4.2}$$

which clearly does not satisfy the Bounded-Input Bounded-State (BIBS) stability property when w is considered as the input.

This chapter contributes new cyclic-small-gain design methods to cope with the problems caused by sensor noise. In Section 4.1, we propose a new measurement feedback control design for nonlinear uncertain systems in the strict-feedback form. As an alternative to static state feedback, a dynamic state feedback control strategy is developed in Section 4.2 for measurement feedback control. A further extension of the design to decentralized control of nonlinear systems with output measurements is given in Section 4.3. The applications of the designs to event-triggered and self-triggered control, synchronization, and robust adaptive control are given in Sections 4.4, 4.5, and 4.6, respectively. With nontrivial modifications, the tools introduced in this chapter are also very useful in solving the problems in the following chapters.

4.1 STATIC STATE MEASUREMENT FEEDBACK CONTROL

In this section, we propose a small-gain design approach to robust control of nonlinear uncertain systems with disturbed measurement. As a design ingredient, a modified gain assignment lemma for measurement feedback control of first-order nonlinear systems is first proposed. Then, the measurement feedback control problem for nonlinear uncertain systems in the strict-feedback form is solved by recursively applying the modified gain assignment technique.

4.1.1 A MODIFIED GAIN ASSIGNMENT LEMMA

Consider the system

$$\dot{\eta} = \phi(\eta, w_1, \ldots, w_{n-2}) + \bar{\kappa} \tag{4.3}$$

$$\eta^m = \eta + w_{n-1} + \mathrm{sgn}(\eta)|w_n|, \tag{4.4}$$

where $\eta \in \mathbb{R}$ is the state, $\bar{\kappa} \in \mathbb{R}$ is the control input, $w_1, \ldots, w_n \in \mathbb{R}$ represent external disturbance inputs, and the function $\phi(\eta, w_1, \ldots, w_{n-2})$ is locally Lipschitz and satisfies

$$|\phi(\eta, w_1, \ldots, w_{n-2})| \leq \psi_\phi(|[\eta, w_1, \ldots, w_{n-2}]^T|), \quad \forall \eta, w_1, \ldots, w_{n-2} \tag{4.5}$$

with known $\psi_\phi \in \mathcal{K}_\infty$. The η^m defined by (4.4) is considered as the measurement of η. The case of $w_n = 0$ was considered in the past literature; see e.g., [130, 223, 123, 125]. The reason why we introduce this additional term $\mathrm{sgn}(\eta)|w_n|$ in (4.4) is that we need to develop a tool as stated in Lemma 4.1 for the development of robust small-based measurement feedback controllers for higher-dimensional nonlinear systems in Subsection 4.1.3. As will be clear later, the ISS-gain from w_n to η can always be made zero. Additionally, the gain from w_{n-1} to η has a specific form.

Lemma 4.1 shows that one can design a measurement feedback control law so that the closed-loop system is ISS with the disturbances as external inputs. Define $\alpha_V(s) = s^2/2$ for $s \in \mathbb{R}_+$.

Lemma 4.1 *Consider system* (4.3)–(4.4). *For any specified* $0 < c < 1$, $\epsilon > 0$, $\ell > 0$, *and* $\chi_\eta^{w_1}, \dots, \chi_\eta^{w_{n-2}} \in \mathcal{K}_\infty$, *one can find a continuously differentiable, odd, strictly decreasing, and radially unbounded function* $\kappa : \mathbb{R} \to \mathbb{R}$ *such that with control law*

$$\bar{\kappa} = \kappa(\eta^m), \tag{4.6}$$

$V_\eta(\eta) = \alpha_V(|\eta|)$ *is an ISS-Lyapunov function of the closed-loop system and satisfies*

$$V_\eta(\eta) \geq \max_{k=1,\dots,n-2} \left\{ \chi_\eta^{w_k}(|w_k|), \alpha_V\left(\frac{|w_{n-1}|}{c}\right), \epsilon \right\}$$
$$\Rightarrow \nabla V_\eta(\eta)(\phi(\eta, w_1, \dots, w_{n-2}) + \kappa(\eta^m)) \leq -\ell V_\eta(\eta), \quad \forall \eta, w_1, \dots, w_n. \tag{4.7}$$

If moreover, ψ_ϕ *is Lipschitz on compact sets and each* $\chi_\eta^{w_k}$ *for* $k = 1, \dots, n - 2$ *is chosen such that* $(\chi_\eta^{w_k})^{-1} \circ \alpha_V$ *is Lipschitz on compact sets, then an appropriate* κ *can be found such that* (4.7) *holds with* $\epsilon = 0$.

Proof. With (4.5) satisfied, one can find $\psi_\phi^\eta, \psi_\phi^{w_1}, \dots, \psi_\phi^{w_{n-2}} \in \mathcal{K}_\infty$ such that

$$|\phi(\eta, w_1, \dots, w_{n-2})| \leq \psi_\phi^\eta(|\eta|) + \sum_{k=1}^{n-2} \psi_\phi^{w_k}(|w_k|). \tag{4.8}$$

Since $\psi_\phi^\eta(s) + \sum_{k=1}^{n-2} \psi_\phi^{w_k} \circ (\chi_\eta^{w_k})^{-1} \circ \alpha_V(s) + \ell s/2$ is a class $\mathcal{K}_\infty$ function of s, from Lemma C.8 in the Appendix, for any constants $0 < c < 1$ and $\epsilon > 0$, one can find a $\nu : \mathbb{R}_+ \to \mathbb{R}_+$ which is positive, nondecreasing, and continuously differentiable on $(0, \infty)$ such that

$$(1 - c)\nu((1 - c)s)s \geq \psi_\phi^\eta(s) + \sum_{k=1}^{n-2} \psi_\phi^{w_k} \circ (\chi_\eta^{w_k})^{-1} \circ \alpha_V(s) + \frac{\ell}{2}s \tag{4.9}$$

for all $s \geq \sqrt{2\epsilon}$.

With the ν satisfying (4.9), define

$$\kappa(r) = -\nu(|r|)r \tag{4.10}$$

for $r \in \mathbb{R}$. Then, κ is continuously differentiable, odd, strictly decreasing, and radially unbounded.

With $V_\eta(\eta) = \alpha_V(|\eta|) = |\eta|^2/2$, we consider the case of

$$V_\eta(\eta) \geq \max_{k=1,\dots,n-2} \left\{ \chi_\eta^{w_k}(|w_k|), \alpha_V\left(\frac{|w_{n-1}|}{c}\right), \epsilon \right\}. \tag{4.11}$$

In this case, we have

$$|w_k| \leq (\chi_\eta^{w_k})^{-1} \circ \alpha_V(|\eta|), \quad k = 1, \dots, n - 2, \tag{4.12}$$

$$|w_{n-1}| \leq c\alpha_V^{-1}(V_\eta(\eta)) = c|\eta|, \tag{4.13}$$

$$|\eta| \geq \sqrt{2\epsilon}. \tag{4.14}$$

Recall the definition of η^m in (4.4). With $0 < c < 1$ and property (4.13), when $\eta \neq 0$, we have

$$\text{sgn}(\eta^m) = \text{sgn}(\eta), \tag{4.15}$$

$$|\eta^m| \geq (1-c)|\eta|. \tag{4.16}$$

In the case of (4.11), using (4.8)–(4.10) and (4.12)–(4.16), we have

$$
\begin{aligned}
&\nabla V_\eta(\eta)(\phi(\eta, w_1, \ldots, w_{n-2}) + \kappa(\eta^m)) \\
&= \eta(\phi(\eta, w_1, \ldots, w_{n-2}) - \nu(|\eta^m|)\eta^m) \\
&\leq |\eta||\phi(\eta, w_1, \ldots, w_{n-2})| - |\eta|\nu(|\eta^m|)|\eta^m| \\
&\leq |\eta|\left(\psi_\phi^\eta(|\eta|) + \sum_{k=1}^{n-2} \psi_\phi^{w_k}(|w_k|) - (1-c)\nu((1-c)|\eta|)|\eta|\right) \\
&\leq |\eta|\left(\psi_\phi^\eta(|\eta|) + \sum_{k=1}^{n-2} \psi_\phi^{w_k} \circ \left(\chi_\eta^{w_k}\right)^{-1} \circ \alpha_V(|\eta|) - (1-c)\nu((1-c)|\eta|)|\eta|\right) \\
&\leq -\frac{\ell}{2}|\eta|^2 = -\ell V_\eta(\eta). \tag{4.17}
\end{aligned}
$$

According to Lemma 4.1, if ψ_ϕ is Lipschitz on compact sets, then $\psi_\phi^\eta, \psi_\phi^{w_1}, \ldots, \psi_\phi^{w_{n-2}}$ can be chosen to be Lipschitz on compact sets. With $\left(\chi_\eta^{w_k}\right)^{-1} \circ \alpha_V$ being Lipschitz on compact sets, one can guarantee that the right-hand side of (4.9) is Lipschitz on compact sets. In this case, by using Lemma C.8 in the Appendix, one can find an appropriate ν, and thus κ, for $\epsilon = 0$. This ends the proof. $\diamond$

If w_{n-1} is bounded, then by using a set-valued map to cover the influence of w_{n-1}, the closed-loop system composed of (4.3), (4.4), and (4.6) can be represented with a differential inclusion:

$$
\begin{aligned}
\dot{\eta} &\in \{\phi(\eta, w_1, \ldots, w_{n-2}) + \kappa(\eta + a\bar{w}_{n-1} + \text{sgn}(\eta)|w_n|) : |a| \leq 1\} \\
&:= F(\eta, w_1, \ldots, w_{n-2}, \bar{w}_{n-1}, w_n), \tag{4.18}
\end{aligned}
$$

where $\bar{w}_{n-1}$ is an upper bound of $|w_{n-1}|$. Clearly, $0 \in F(0, \ldots, 0)$. With the differential inclusion formulation, property (4.7) can be equivalently represented by

$$
\begin{aligned}
V_\eta(\eta) &\geq \max_{k=1,\ldots,n-2} \left\{ \chi_\eta^{w_k}(|w_k|), \alpha_V\left(\frac{|\bar{w}_{n-1}|}{c}\right), \epsilon \right\} \\
&\Rightarrow \max_{f \in F(\eta, w_1, \ldots, w_{n-2}, \bar{w}_{n-1}, w_n)} \nabla V_\eta(\eta) f \leq -\ell V_\eta(\eta). \tag{4.19}
\end{aligned}
$$

Compared with the gain assignment result in Subsection 2.3.1, Lemma 4.1 takes into account sensor noise caused by w_{n-1} and w_n. Here, w_n is taken into account for the later recursive control design. The gain assignment techniques given in [223, 125] do not take into account the influence of $w_n \neq 0$.

It should be noted that with the proposed design, the ISS gain from w_n to η is zero. However, property (4.7) implies that the ISS gain from w_{n-1} to η is $1/c$, where the constant c should be chosen to satisfy $0 < c < 1$. Thus, the gain from w_{n-1} to η is larger than one. This means that the influence of the sensor noise cannot be attenuated to an arbitrarily small level. In fact, this is also the case if the system (4.3) is reduced to a linear system; see Example 4.2.

Example 4.2 *Consider a first-order linear time-invariant system in the form of (4.3)–(4.4) with $w_1, \ldots, w_{n-2}, w_n = 0$, and*

$$\phi(\eta, w_1, \ldots, w_{n-2}) = a\eta, \tag{4.20}$$

where a is an unknown constant satisfying $0 < a \le \bar{a}$ with known constant $\bar{a} > 0$. With a linear measurement feedback control law $u = -k\eta^m$ with constant $k > \bar{a}$, the closed-loop system is

$$\dot{\eta} = (a - k)\eta - kw_{n-1}, \tag{4.21}$$

which can be equivalently represented by transfer function

$$\frac{(\mathcal{L}\eta)(s)}{(\mathcal{L}w_{n-1})(s)} = \frac{-k}{s + (k - a)} := G(s), \tag{4.22}$$

where $\mathcal{L}$ represents the Laplace transform, and $s \in \mathbb{C}$. For the linear system, the gain from w_{n-1} to η can be calculated in the frequency domain as $\sup_{\omega^f \ge 0} |G(j\omega^f)| = k/(k-a)$, which is larger than one and can be designed to be arbitrarily close to one by choosing k large enough. This is in accordance with the modified gain assignment lemma.

With the gain assignment technique in Lemma 4.1, measurement feedback control can be solved for system (4.3) appended with ISS dynamic uncertainties [125]. Example 4.3 considers a simplified case to show the basic idea of this ISS small-gain design.

Example 4.3 *Consider the nonlinear system*

$$\dot{z} = q(z, x) \tag{4.23}$$
$$\dot{x} = f(x, z) + u \tag{4.24}$$
$$x^m = x + w, \tag{4.25}$$

where $[z, x] \in \mathbb{R}^2$ is the state, $u \in \mathbb{R}$ is the control input, x is considered as the output, x^m is the measurement of x with $w \in \mathbb{R}$ representing sensor noise, and $q, f : \mathbb{R}^2 \to \mathbb{R}$ are locally Lipschitz functions. Only the measurement x^m is available for feedback control design. For this system, the z-subsystem represents dynamic uncertainties.

Assume that the z-subsystem is ISS with x as the input and admits an ISS-Lyapunov function $V_z : \mathbb{R} \to \mathbb{R}_+$ such that

$$\underline{\alpha}_z(|z|) \leq V_z(z) \leq \overline{\alpha}_z(|z|), \quad \forall z, \tag{4.26}$$

$$V_z(z) \geq \chi_z^x(|x|) \Rightarrow \nabla V_z(z)q(z,x) \leq -\alpha_z(V_z(z)), \quad \forall z, x, \tag{4.27}$$

where $\underline{\alpha}_z, \overline{\alpha}_z \in \mathcal{K}_\infty$, $\chi_z^x \in \mathcal{K}$, and α_z is a continuous and positive definite function. Also assume that there exists a $\psi_f \in \mathcal{K}_\infty$ such that for all x, z,

$$|f(x,z)| \leq \psi_f(|[x,z]^T|). \tag{4.28}$$

Based on Lemma 4.1, we can design a control law $u = \bar{u}(x^m)$ such that the x-subsystem is ISS with $V_x(x) = \alpha_V(|x|) = x^2/2$ as an ISS-Lyapunov function. In particular, for any specific $\chi_x^z \in \mathcal{K}_\infty$, continuous and positive definite function α_x, and constant $\epsilon > 0$, the control law can be designed such that $V_x(x) = \alpha_V(|x|) = x^2/2$ satisfies

$$V_x(x) \geq \max\left\{\chi_x^z(|z|), \alpha_V\left(\frac{|w|}{c}\right), \epsilon\right\}$$
$$\Rightarrow \nabla V_x(x)(f(x,z) + \bar{u}(x^m)) \leq -\alpha_x(V_x(x)) \quad \forall x, z, w, \tag{4.29}$$

and thus

$$V_x(x) \geq \max\left\{\chi_x^z \circ \underline{\alpha}_z^{-1}(V_z(z)), \alpha_V\left(\frac{|w|}{c}\right), \epsilon\right\}$$
$$\Rightarrow \nabla V_x(x)(f(x,z) + \bar{u}(x^m)) \leq -\alpha_x(V_x(x)) \quad \forall x, z, w. \tag{4.30}$$

.

Also note that property (4.27) implies

$$V_z(z) \geq \chi_z^x \circ \alpha_V^{-1}(V_x(x)) \Rightarrow \nabla V_z(z)q(z,x) \leq -\alpha_z(V_z(z)). \tag{4.31}$$

With the design above, the closed-loop system is transformed into an interconnection of two ISS subsystems. The closed-loop system is ISS if χ_x^z is chosen to satisfy the small-gain condition:

$$\chi_x^z \circ \underline{\alpha}_z^{-1} \circ \chi_z^x \circ \alpha_V^{-1} < \mathrm{Id}. \tag{4.32}$$

By using the Lyapunov-based ISS small-gain theorem, we can also construct an ISS-Lyapunov function to analyze the influence of the sensor noise on the convergence of x.

If ψ_f is Lipschitz on compact sets, and there exists a $\chi_x^z \in \mathcal{K}_\infty$ such that $(\chi_x^z)^{-1} \circ \alpha_V$ is Lipschitz on compact sets and the small-gain condition (4.32) is satisfied, then the ϵ in (4.30) can be chosen to be zero.

4.1.2 PROBLEMS WITH HIGH-ORDER NONLINEAR SYSTEMS

By a repeated application of the gain assignment Lemma 4.1, this subsection develops a class of measurement feedback controllers for nonlinear uncertain systems in the strict-feedback form:

$$\dot{x}_i = x_{i+1} + \Delta_i(\bar{x}_i, d), \quad i = 1, \ldots, n-1 \tag{4.33}$$

$$\dot{x}_n = u + \Delta_n(\bar{x}_n, d) \tag{4.34}$$

$$x_i^m = x_i + w_i, \quad i = 1, \ldots, n, \tag{4.35}$$

where $[x_1, \ldots, x_n]^T := x \in \mathbb{R}^n$ is the state, $\bar{x}_i = [x_1, \ldots, x_i]^T$, $u \in \mathbb{R}$ is the control input, $d \in \mathbb{R}^{n_d}$ represents external disturbance inputs, x_i^m is the measurement of x_i with w_i being the corresponding sensor noise, and Δ_i's ($i = 1, \ldots, n$) are uncertain, locally Lipschitz functions.

The following assumptions are made on system (4.33)–(4.35).

Assumption 4.1 *For each $i = 1, \ldots, n$, there exists a known $\psi_{\Delta_i} \in \mathcal{K}_\infty$ such that for all $\bar{x}_i, d$,*

$$|\Delta_i(\bar{x}_i, d)| \le \psi_{\Delta_i}(|[\bar{x}_i^T, d^T]^T|). \tag{4.36}$$

Assumption 4.2 *There exists a constant $\bar{d} \ge 0$ such that*

$$|d(t)| \le \bar{d} \tag{4.37}$$

for $t \ge 0$.

Assumption 4.3 *For each $i = 1, \ldots, n$, there exists a constant $\bar{w}_i > 0$ such that*

$$|w_i(t)| \le \bar{w}_i \tag{4.38}$$

for $t \ge 0$.

A small-gain design has been developed for the stabilization of strict-feedback systems in the form of (4.33)–(4.34) and more general cascade systems with dynamic uncertainties [123]; see also the discussions for a simplified case in Subsection 2.3.2. If system (4.33)–(4.34) is free of sensor noise, i.e., $w_i = 0$, then it can be stabilized with a nonlinear controller in the form of

$$x_1^* = \check{\kappa}_1(x_1) \tag{4.39}$$

$$x_{i+1}^* = \check{\kappa}_i(x_i - x_i^*), \quad i = 2, \ldots, n-1 \tag{4.40}$$

$$u = \check{\kappa}_n(x_n - x_n^*), \tag{4.41}$$

where the $\breve{\kappa}_i$'s for $i = 1, \ldots, n$ are appropriately designed nonlinear functions, and u is the implementable control law. In this case, to analyze the stability property of the closed-loop system, we can define new state variables as

$$e_1 = x_1, \tag{4.42}$$

$$e_i = x_i - \breve{\kappa}_{i-1}(e_{i-1}), \quad i = 2, \ldots, n. \tag{4.43}$$

For continuous differentiability of the new state variables, the functions $\breve{\kappa}_i$ for $i = 1, \ldots, n-1$ are required to be continuously differentiable.

To take into account the influence of the sensor noise, we may consider replacing each x_i in (4.39)–(4.41) with x_i^m. We choose the new measurement feedback control law in the following form:

$$x_1^* = \kappa_1(x_1^m) \tag{4.44}$$

$$x_{i+1}^* = \kappa_i(x_i^m - x_i^*), \quad i = 2, \ldots, n-1 \tag{4.45}$$

$$u = \kappa_n(x_n^m - x_n^*), \tag{4.46}$$

where the κ_i's are not necessarily the same with the $\breve{\kappa}_i$'s in (4.39)–(4.41). For such a control law, the state transformation in the form of (4.42)–(4.43) may be modified as

$$e_1 = x_1^m, \tag{4.47}$$

$$e_i = x_i^m - \kappa_{i-1}(e_{i-1}), \quad i = 2, \ldots, n. \tag{4.48}$$

However, with such treatment, if the sensor noise is not differentiable, then the new e_i's for $i = 2, \ldots, n$ are not differentiable and one cannot use differential equations to represent the dynamics of the e_i-subsystems. Practically, it might be too restrictive to assume the continuous differentiability of sensor noise.

The main objective of this section is to develop a new small-gain design method which leads to nonlinear controllers that are:

1. robust to nondifferentiable and even discontinuous sensor noise, and moreover,
2. capable of attenuating the influence of the sensor noise on the control system to the largest extent possible.

4.1.3   RECURSIVE CONTROL DESIGN

This subsection employs set-valued maps to handle the problem caused by sensor noise. With the new design, the closed-loop system is transformed into an interconnection of ISS subsystems represented by differential inclusions. Specifically, the $[x_1, \ldots, x_n]^T$-system is transformed into a new $[e_1, \ldots, e_n]^T$-system through the following transformation:

$$e_1 = x_1 \tag{4.49}$$

$$e_i = \vec{d}(x_i, S_{i-1}(\bar{x}_{i-1})), \quad i = 2, \ldots, n, \tag{4.50}$$

where $S_i : \mathbb{R}^i \rightsquigarrow \mathbb{R}$ is an appropriately chosen set-valued map, and

$$\vec{d}(z, \Omega) := z - \underset{z' \in \Omega}{\arg\min}\{|z - z'|\} \qquad (4.51)$$

for any $z \in \mathbb{R}$ and any compact $\Omega \subset \mathbb{R}$. Basically, the set-valued maps are employed to cover the influence of the sensor noise and to represent the possible control laws in the control design procedure. A control law in the form of (4.44)–(4.46) is at last found as a selection of the set-valued map $S_n : \mathbb{R}^n \rightsquigarrow \mathbb{R}$.

For convenience of notation, denote $\bar{e}_i = [e_1, \dots, e_i]^T$ and $W_i = [\bar{w}_1, \dots, \bar{w}_i]^T$ for $i = 1, \dots, n$. Also denote $x_{n+1} = u$.

Initial Step: The e_1-subsystem

By taking the derivative of e_1, we have

$$\dot{e}_1 = x_2 + \Delta_1(\bar{x}_1, d)$$
$$= x_2 - e_2 + \Delta_1(\bar{x}_1, d) + e_2. \qquad (4.52)$$

Recall that $e_2 = \vec{d}(x_2, S_1(\bar{x}_1))$. Then,

$$x_2 - e_2 \in S_1(\bar{x}_1). \qquad (4.53)$$

Define the set-valued map S_1 as

$$S_1(\bar{x}_1) = \{\kappa_1(x_1 + a_1\bar{w}_1) : |a_1| \le 1\}, \qquad (4.54)$$

where $\kappa_1 : \mathbb{R} \to \mathbb{R}$ is a continuously differentiable, odd, strictly decreasing, and radially unbounded function.

Since κ_1 is strictly decreasing, $\max S_1(\bar{x}_1) = \kappa_1(x_1 - \bar{w}_1)$ and $\min S_1(\bar{x}_1) = \kappa_1(x_1 + \bar{w}_1)$. Set-valued map S_1 and the definition of e_2 are shown in Figure 4.2.

Intuitively, set-valued map S_1 is defined such that if x_2 is the control input of the e_1-subsystem, then the measurement feedback control law $x_2 = \kappa_1(x_1 + w_1) = \kappa_1(e_1 + w_1)$ with $|w_1| \le \bar{w}_1$ is a selection of $S_1(x_1)$. Moreover, by choosing a continuously differentiable κ_1, the boundaries of S_1 are continuously differentiable, and as shown below, the derivative of e_2 exists almost everywhere. As a result, the problem caused by the nondifferentiable sensor noise is solved.

Recursive Step: The e_i-subsystems for $i = 2, \dots, n$

For convenience, denote $S_0(\bar{x}_0) = \{0\}$. For each $k = 1, \dots, i - 1$, define set-valued map S_k as

$$S_k(\bar{x}_k) = \{\kappa_k(x_k - p_{k-1} + a_k\bar{w}_k) : p_{k-1} \in S_{k-1}(\bar{x}_{k-1}), |a_k| \le 1\}, \qquad (4.55)$$

where $\kappa_k : \mathbb{R} \to \mathbb{R}$ is a continuously differentiable, odd, strictly decreasing, and radially unbounded function.

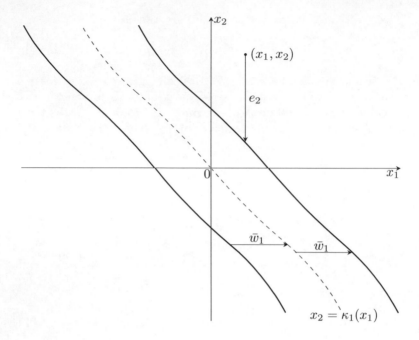

FIGURE 4.2 Boundaries of set-valued map S_1 and the definition of e_2.

Lemma 4.2 *Consider the $[x_1, \ldots, x_n]^T$-system defined by (4.33)–(4.35) with Assumptions 4.1–4.3 satisfied. If each S_k is defined in (4.55) for $k = 1, \ldots, i-1$, then when $e_i \neq 0$, the e_i-subsystem can be written in the form of*

$$\dot{e}_i = x_{i+1} + \phi_i^*(\bar{x}_i, d), \qquad (4.56)$$

where

$$|\phi_i^*(\bar{x}_i, d)| \leq \psi_{\phi_i^*}(|[\bar{e}_i^T, d^T, W_{i-1}^T]^T|) \qquad (4.57)$$

with known $\psi_{\phi_i^*} \in \mathcal{K}_\infty$. If, moreover, the ψ_{Δ_k}'s for $k = 1, \ldots, i$ are Lipschitz on compact sets, then one can find a $\psi_{\phi_i^*} \in \mathcal{K}_\infty$ which is Lipschitz on compact sets.

The proof of Lemma 4.2 is provided in Appendix E.1.

Define set-valued map S_i as

$$S_i(\bar{x}_i) = \left\{ \kappa_i(x_i - p_{i-1} + a_i \bar{w}_i) : p_{i-1} \in S_{i-1}(\bar{x}_{i-1}), |a_i| \leq 1 \right\}, \qquad (4.58)$$

where κ_i is a continuously differentiable, odd, strictly decreasing, and radially unbounded function.

Then, the e_i-subsystem can be rewritten as

$$\dot{e}_i = x_{i+1} - e_{i+1} + \phi_i^*(\bar{x}_i, d) + e_{i+1}, \qquad (4.59)$$

where

$$x_{i+1} - e_{i+1} \in S_i(\bar{x}_i) \tag{4.60}$$

according to the definition $e_{i+1} = \vec{d}(x_{i+1}, S_i(\bar{x}_i))$.

It can be observed that the $S_1(\bar{x}_1)$ defined in (4.54) is also in the form of (4.58) with $S_0(\bar{x}_0) = \{0\}$, and the e_1-subsystem defined in (4.52) is in the form of (4.59).

State Measurement Feedback Control Law and the Closed-Loop System

We design the measurement feedback control law as

$$p_1^* = \kappa_1(x_1^m), \tag{4.61}$$

$$p_i^* = \kappa_i(x_i^m - p_{i-1}^*), \quad i = 2, \dots, n-1 \tag{4.62}$$

$$u = \kappa_n(x_n^m - p_{n-1}^*). \tag{4.63}$$

Recall that $x_i^m = x_i + w_i$ for $i = 1, \dots, n$. It is directly checked that

$$p_1^* \in S_1(\bar{x}_1) \Rightarrow \dots \Rightarrow p_i^* \in S_i(\bar{x}_i) \Rightarrow \dots \Rightarrow u \in S_n(\bar{x}_n), \tag{4.64}$$

which means $e_{n+1} = 0$.

Considering $x_{i+1} - e_{i+1} \in S_i(\bar{x}_i)$ for $i = 1, \dots, n$ and $e_{n+1} = 0$, when $e_i \neq 0$, we can represent each e_i-subsystem for $i = 1, \dots, n$ with a differential inclusion:

$$\dot{e}_i \in \{p_i + \phi_i^*(\bar{x}_i, d) + e_{i+1} : p_i \in S_i(\bar{x}_i)\}$$
$$:= F_i(\bar{x}_i, e_{i+1}, d). \tag{4.65}$$

Thus, the closed-loop system with control law (4.61)–(4.63) is transformed into a network composed of the e_i-subsystems, each of which is represented by a differential inclusion.

Also, when $e_i \neq 0$, for each $p_{i-1} \in S_{i-1}(\bar{x}_{i-1})$, it holds that $|x_i - p_{i-1}| > |e_i|$ and $\text{sgn}(x_i - p_{i-1}) = \text{sgn}(e_i)$, which imply $\text{sgn}(x_i - e_i - p_{i-1}) = \text{sgn}(e_i)$. Thus, for $i = 1, \dots, n$, each $S_i(\bar{x}_i)$ in the form of (4.58) can be rewritten as

$$S_i(\bar{x}_i) = \{\kappa_i(e_i + \text{sgn}(e_i)|w_{i0}| + a_i\bar{w}_i) : |a_i| \leq 1\}, \tag{4.66}$$

where $w_{i0} = x_i - e_i - p_{i-1}$ with $p_{i-1} \in S_{i-1}(\bar{x}_{i-1})$.

By combining (4.65) and (4.66), we can recognize that each e_i-subsystem is in the form of (4.18). In the next subsection, we employ the gain assignment technique introduced in Subsection 4.1.1 to choose the κ_i's such that all the e_i-subsystems are ISS with the ISS gains satisfying the cyclic-small-gain condition.

Clearly, if there is no sensor noise, then the problem is reduced to the one considered in Section 2.3. In this case, the state transformation (4.49)–(4.50) is reduced to

$$e_1 = x_1 \tag{4.67}$$
$$e_i = x_i - \kappa_{i-1}(e_{i-1}), \quad i = 2, \dots, n, \tag{4.68}$$

and the control law (4.61)–(4.63) is reduced to

$$p_1^* = \kappa_1(x_1) \tag{4.69}$$
$$p_i^* = \kappa_i(x_i - p_{i-1}^*), \quad i = 2, \dots, n-1 \tag{4.70}$$
$$u = \kappa_n(x_n - p_{n-1}^*). \tag{4.71}$$

4.1.4   CYCLIC-SMALL-GAIN SYNTHESIS

Define

$$V_i(e_i) = \alpha_V(|e_i|) := \frac{1}{2}|e_i|^2 \tag{4.72}$$

as the ISS-Lyapunov function candidate for each e_i-subsystem. For convenience of discussions, denote $V_{n+1}(e_{n+1}) = \alpha_V(|e_{n+1}|)$.

Consider each e_i-subsystem (4.65) with S_i in the form of (4.66). By using Lemma 4.1, for any specified $\epsilon_i > 0$, $\ell_i > 0$, $0 < c_i < 1$, $\gamma_{e_i}^{e_k}, \gamma_{e_i}^{w_k} \in \mathcal{K}_\infty$ ($k = 1, \dots, i-1$), and $\gamma_{e_i}^{e_{i+1}}, \gamma_{e_i}^{d} \in \mathcal{K}_\infty$, we can find a continuously differentiable, odd, strictly decreasing, and radially unbounded function κ_i for S_i such that

$$V_i(e_i) \geq \max_{k=1,\dots,i-1} \left\{ \begin{array}{l} \gamma_{e_i}^{e_k} \circ \alpha_V(|e_k|), \gamma_{e_i}^{e_{i+1}} \circ \alpha_V(|e_{i+1}|), \\ \gamma_{e_i}^{w_k}(\bar{w}_k), \gamma_{e_i}^{w_i}(\bar{w}_i), \gamma_{e_i}^{d}(\bar{d}), \epsilon_i \end{array} \right\}$$
$$\Rightarrow \max_{f_i \in F_i(\bar{x}_i, e_{i+1}, d)} \nabla V_i(e_i) f_i \leq -\ell_i V_i(e_i) \tag{4.73}$$

and thus,

$$V_i(e_i) \geq \max_{k=1,\dots,i-1} \left\{ \begin{array}{l} \gamma_{e_i}^{e_k}(V_k(e_k)), \gamma_{e_i}^{e_{i+1}}(V_{i+1}(e_{i+1})), \\ \gamma_{e_i}^{w_k}(\bar{w}_k), \gamma_{e_i}^{w_i}(\bar{w}_i), \gamma_{e_i}^{d}(\bar{d}), \epsilon_i \end{array} \right\}$$
$$\Rightarrow \max_{f_i \in F_i(\bar{x}_i, e_{i+1}, d)} \nabla V_i(e_i) f_i \leq -\ell_i V_i(e_i), \tag{4.74}$$

where

$$\gamma_{e_i}^{w_i}(s) = \alpha_V\left(\frac{s}{c_i}\right) \tag{4.75}$$

for $s \in \mathbb{R}_+$.

Recall that $\bar{e}_i = [e_1, \dots, e_i]^T$. Denote $e = \bar{e}_n$. The gain digraph of the e-system is shown in Figure 4.3.

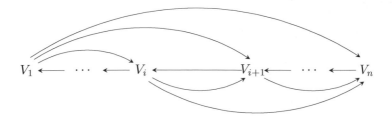

FIGURE 4.3 The gain digraph of the e-system.

According to the recursive design, given the $\bar{e}_{i-1}$-subsystem, by appropriately choosing set-valued map S_i for the e_i-subsystem, we can design the ISS gains $\gamma_{e_i}^{e_k}$ for $1 \le k \le i-1$ such that

$$
\begin{aligned}
\gamma_{e_1}^{e_2} \circ \gamma_{e_2}^{e_3} \circ \gamma_{e_3}^{e_4} \circ \cdots \circ \gamma_{e_{i-1}}^{e_i} \circ \gamma_{e_i}^{e_1} &< \mathrm{Id} \\
\gamma_{e_2}^{e_3} \circ \gamma_{e_3}^{e_4} \circ \cdots \circ \gamma_{e_{i-1}}^{e_i} \circ \gamma_{e_i}^{e_2} &< \mathrm{Id} \\
&\vdots \\
\gamma_{e_{i-1}}^{e_i} \circ \gamma_{e_i}^{e_{i-1}} &< \mathrm{Id}
\end{aligned}
\tag{4.76}
$$

By applying this reasoning repeatedly, we can guarantee (4.76) for all $2 \le i \le n$. In this way, the e-system satisfies the cyclic-small-gain condition.

With the Lyapunov-based ISS cyclic-small-gain theorem in Chapter 3, an ISS-Lyapunov function can be constructed for the e-system to evaluate the influence of the sensor noise:

$$
V(e) = \max_{i=1,\ldots,n} \{\sigma_i(V_i(e_i))\}
\tag{4.77}
$$

where $\sigma_1(s) = s$, $\sigma_i(s) = \gamma_{e_1}^{e_2} \circ \cdots \circ \hat{\gamma}_{e_{i-1}}^{e_i}(s)$ for $i = 2, \ldots, n$ for $s \in \mathbb{R}_+$. The $\hat{\gamma}_{(\cdot)}^{(\cdot)}$'s are class $\mathcal{K}_\infty$ functions that are continuously differentiable on $(0,\infty)$, slightly larger than the corresponding $\gamma_{(\cdot)}^{(\cdot)}$'s, and still satisfy the cyclic-small-gain condition. (Recall the construction in Subsection 3.1.2.)

The influence of w_i's, ϵ_i's, and d can be represented by

$$
\theta = \max_{i=1,\ldots,n} \left\{ \sigma_i \left(\max_{k=1,\ldots,i} \{\gamma_{e_i}^{w_k}(\bar{w}_k), \gamma_{e_i}^d(\bar{d}), \epsilon_i\} \right) \right\}.
\tag{4.78}
$$

Then, it holds that

$$
V(e) \ge \theta \Rightarrow \max_{f \in F(x,e,d)} \nabla V(e) f \le -\alpha(V(e)) \quad \text{a.e.,}
\tag{4.79}
$$

where $F(x,e,d) = [F_1(\bar{x}_1, e_2, d), \ldots, F_n(\bar{x}_n, e_{n+1}, d)]^T$ and α is a continuous and positive definite function.

By choosing $\gamma_{e_i}^{w_k}$ $(i = 2,\ldots,n,\ k = 1,\ldots,i-1)$, $\gamma_{e_i}^d$ $(i = 1,\ldots,n)$, ϵ_i $(i = 1,\ldots,n)$, and $\gamma_{e_i}^{e_{i+1}}$ $(i = 1,\ldots,n-1)$ small enough, we can make σ_i for $i = 2,\ldots,n$ small enough and get

$$\theta = \gamma_{e_1}^{w_1}(\bar{w}_1) = \alpha_V\left(\frac{\bar{w}_1}{c_1}\right). \tag{4.80}$$

Then, from (4.79), it is achieved that

$$V(e) \geq \alpha_V\left(\frac{\bar{w}_1}{c_1}\right) \Rightarrow \max_{f \in F(x,e,d)} \nabla V(e)f \leq -\alpha(V(e)) \tag{4.81}$$

holds wherever ∇V exists.

Property (4.81) implies that $V(e)$ ultimately converges to within the region $V(e) \leq \alpha_V(\bar{w}/c_1)$. Using the definitions of e_1, $V_1(e_1)$, $V(e)$ (see (4.72) and (4.77)), we have $|x_1| = |e_1| = \alpha_V^{-1}(V_1(e_1)) \leq \alpha_V^{-1}(V(e))$, which implies that x_1 ultimately converges to within the region $|x_1| \leq \bar{w}_1/c_1$. Note that constant c_1 can be arbitrarily chosen as long as $0 < c_1 < 1$. By choosing c_1 to be arbitrarily close to one, x_1 can be steered arbitrarily close to the region $|x_1| \leq \bar{w}_1$.

According to the design above, because of the nonzero ϵ_i terms, asymptotic stability of the closed-loop system cannot be guaranteed even if $w_i \equiv 0$ for $i = 1,\ldots,n$ and $d \equiv 0$. This problem can be solved if the ψ_{Δ_i}'s for $i = 1,\ldots,n$ in Assumption 4.1 are Lipschitz on compact sets. In this case, for each e_i-subsystem, we can find a $\psi_{\phi_i^*}$ that is Lipschitz on compact sets such that (4.57) holds. Then, according to Lemma 4.1, by choosing the $\gamma_{e_i}^{e_k}$ for $k = 1,\ldots,i-1$ and $\gamma_{e_i}^d$ such that $\left(\gamma_{e_i}^{e_k} \circ \alpha_V\right)^{-1} \circ \alpha_V$ and $\left(\gamma_{e_i}^d\right)^{-1} \circ \alpha_V$ are Lipschitz on compact sets, (4.73) and thus (4.74) can be realized with $\epsilon_i = 0$. Asymptotic stabilization is achieved if $w_i \equiv 0$ for $i = 1,\ldots,n$ and $d \equiv 0$.

The main result on state measurement feedback control of nonlinear uncertain systems is given in Theorem 4.1.

Theorem 4.1 *Consider system* (4.33)–(4.35). *Under Assumptions 4.1, 4.2 and 4.3, the closed-loop signals are bounded and, in particular, state x_1 can be steered arbitrarily close to the region $|x_1| \leq \bar{w}_1$ with the measurement feedback control law* (4.61)–(4.63). *If the system is disturbance-free, i.e., $w_i \equiv 0$ for $i = 1,\ldots,n$ and $d \equiv 0$, and the ψ_{Δ_i}'s for $i = 1,\ldots,n$ in Assumption 4.1 are Lipschitz on compact sets, then one can design the control law in the form of* (4.61)–(4.63) *such that x_1 asymptotically converges to the origin.*

It should be noted that the design proposed in this section can also be applied to strict-feedback systems with ISS dynamic uncertainties:

$$\dot{z} = g(z, x_1, d) \tag{4.82}$$

$$\dot{x}_i = x_{i+1} + \Delta_i(\bar{x}_i, z, d), \quad i = 1,\ldots,n-1 \tag{4.83}$$

$$\dot{x}_n = u + \Delta_n(\bar{x}_n, z, d) \tag{4.84}$$

$$x_i^m = x_i + w_i, \quad i = 1,\ldots,n, \tag{4.85}$$

where the z-subsystem with $z \in \mathbb{R}^m$ represents the dynamic uncertainties. If the z-subsystem is ISS with x_1, d as the inputs, then following the design procedure in this section, the closed-loop system can still be transformed into a network of ISS subsystems, and the control problem can then be solved by using the cyclic-small-gain theorem.

4.2 DYNAMIC STATE MEASUREMENT FEEDBACK CONTROL

In practical industrial applications, low-pass filters are often employed to attenuate high-frequency noise and to estimate the measured signals. Motivated by low-pass filters, in this section, we develop a dynamic state measurement feedback control structure for input-to-state stabilization of nonlinear systems under sensor noise.

We still consider nonlinear systems in the strict-feedback form:

$$\dot{x}_i = x_{i+1} + \Delta_i(\bar{x}_i), \quad i = 1, \ldots, n-1 \tag{4.86}$$

$$\dot{x}_n = u + \Delta_n(\bar{x}_n) \tag{4.87}$$

$$x_i^m = x_i + w_i, \quad i = 1, \ldots, n, \tag{4.88}$$

where $[x_1, \ldots, x_n]^T := x \in \mathbb{R}^n$ is the state, $u \in \mathbb{R}$ is the control input, $\bar{x}_i = [x_1, \ldots, x_i]^T$, $x_i^m \in \mathbb{R}$ is the disturbed measurement of x_i with sensor noise $w_i \in \mathbb{R}$, and Δ_i's for $i = 1, \ldots, n$ are unknown locally Lipschitz functions.

Assumption 4.4 *For each Δ_i with $i = 1, \ldots, n$ in (4.86)–(4.87), there exists a known $\psi_{\Delta_i} \in \mathcal{K}_\infty$ such that for all $\bar{x}_i$,*

$$|\Delta_i(\bar{x}_i)| \le \psi_{\Delta_i}(|\bar{x}_i|). \tag{4.89}$$

The objective is to design a dynamic state measurement feedback controller of the form

$$\dot{\zeta} = \varphi(\zeta, x^m) \tag{4.90}$$

$$u = \lambda(\zeta) \tag{4.91}$$

such that system (4.86)–(4.88) is made ISS with the w_i's as the inputs, and thus is IOS with the w_i's as the inputs and x_1 as the output. Moreover, it is desired that the IOS gain from w_1 to x_1 can be designed to be arbitrarily close to the identity function, and the IOS gains from $w_2, \ldots, w_n$ to x_1 can be designed to be arbitrarily small.

The gain assignment technique is still an ingredient for the design in this section. The basic idea is to transform the closed-loop system into an interconnection of first-order nonlinear systems in the following form:

$$\dot{\eta} = \phi(\eta, w_1, \ldots, w_m) + \bar{\kappa} \tag{4.92}$$

$$\eta^m = \eta + w_{m+1}, \tag{4.93}$$

where $\eta \in \mathbb{R}$ is the state, $\bar{\kappa} \in \mathbb{R}$ is the control input, $w_1, \ldots, w_{m+1} \in \mathbb{R}$ represent external inputs, $\eta^m \in \mathbb{R}$ is the measurement of η, the nonlinear function $\phi(\eta, w_1, \ldots, w_m)$ is locally Lipschitz and satisfies

$$|\phi(\eta, w_1, \ldots, w_m)| \leq \psi_\phi^\eta(|\eta|) + \sum_{k=1}^m \psi_\phi^{w_k}(|w_k|) \qquad (4.94)$$

with known $\psi_\phi^\eta, \psi_\phi^{w_1}, \ldots, \psi_\phi^{w_m} \in \mathcal{K}_\infty$.

Now, a new system with state $[\hat{e}_1, \tilde{e}_1, \ldots, \hat{e}_n, \tilde{e}_n]^T$ is constructed based on the x-system. Moreover, all the $\hat{e}_i$-subsystems and the $\tilde{e}_i$-subsystems are designed to be ISS, and the cyclic-small-gain theorem is used to check the stability of the closed-loop system.

4.2.1 DYNAMIC STATE MEASUREMENT FEEDBACK CONTROL DESIGN

In this subsection, we design the dynamic state measurement feedback controller through a recursive approach.

If there is no sensor noise, then we may design a controller to transform the closed-loop system into ISS e_i-subsystems with e_i defined by (4.67)–(4.68). In the presence of sensor noise, we use an estimate $\hat{e}_{i-1}$ to replace the e_{i-1} on the right-hand side of (4.68), and the new state transformation is

$$e_1 = x_1 \qquad (4.95)$$
$$e_i = x_i - \kappa_{(i-1)2}(\hat{e}_{i-1}), \quad i = 2, \ldots, n, \qquad (4.96)$$

where $\kappa_{(i-1)2} : \mathbb{R} \to \mathbb{R}$ is a continuously differentiable, odd, strictly decreasing and radially unbounded function.

Denote $x_{n+1} = x_{n+1}^m = u$. For $i = 1, \ldots, n$, each $\hat{e}_i$ is generated by the following estimator:

$$\dot{\hat{e}}_i = \kappa_{i1}(\hat{e}_i - e_i^m) + x_{i+1}^m, \qquad (4.97)$$

where $\kappa_{i1} : \mathbb{R} \to \mathbb{R}$ is an odd and strictly decreasing function, and

$$e_1^m = x_1^m \qquad (4.98)$$
$$e_i^m = x_i^m - \kappa_{(i-1)2}(\hat{e}_{i-1}), \quad i = 2, \ldots, n. \qquad (4.99)$$

The structure of the $\hat{e}_i$-subsystem is shown in Figure 4.4.

For $i = 1, \ldots, n$, define

$$\tilde{e}_i = \hat{e}_i - e_i \qquad (4.100)$$

as the estimation error for e_i. By taking the derivative of $\tilde{e}_i$ and using $x_i^m =$

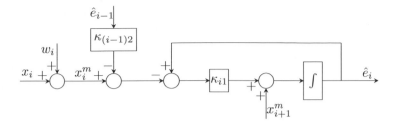

FIGURE 4.4 The estimator for e_i: the $\hat{e}_i$-subsystem.

$x_i + w_i$ and $x_{i+1}^m = x_{i+1} + w_{i+1}$, we have

$$
\begin{aligned}
\dot{\tilde{e}}_i &= \dot{\hat{e}}_i - \dot{e}_i \\
&= \dot{\hat{e}}_i - \dot{x}_i + \frac{\partial \kappa_{(i-1)2}(\hat{e}_{i-1})}{\partial \hat{e}_{i-1}} \dot{\hat{e}}_{i-1} \\
&= \kappa_{i1}(\tilde{e}_i - w_i) + w_{i+1} + x_{i+1} - \Delta_i(\bar{x}_i) - x_{i+1} \\
&\quad + \frac{\partial \kappa_{(i-1)2}(\hat{e}_{i-1})}{\partial \hat{e}_{i-1}} \dot{\hat{e}}_{i-1},
\end{aligned}
\tag{4.101}
$$

which can be represented in the form of

$$
\begin{aligned}
\dot{\tilde{e}}_i &= \Delta_{i1}^*(\hat{e}_1, \tilde{e}_1, \dots, \hat{e}_i, \tilde{e}_i, w_{i-1}, w_i, w_{i+1}) + \kappa_{i1}(\tilde{e}_i - w_i) \\
&:= f_{\tilde{e}_i}(\hat{e}_1, \tilde{e}_1, \dots, \hat{e}_i, \tilde{e}_i, w_{i-1}, w_i, w_{i+1})
\end{aligned}
\tag{4.102}
$$

with

$$
\begin{aligned}
&\Delta_{i1}^*(\hat{e}_1, \tilde{e}_1, \dots, \hat{e}_i, \tilde{e}_i, w_{i-1}, w_i, w_{i+1}) \\
&= w_{i+1} - \Delta_i(\bar{x}_i) + \frac{\partial \kappa_{(i-1)2}(\hat{e}_{i-1})}{\partial \hat{e}_{i-1}} \dot{\hat{e}}_{i-1}.
\end{aligned}
\tag{4.103}
$$

With Assumption 4.4 satisfied, we can find $\psi_{\Delta_{i1}^*}^{\hat{e}_k}, \psi_{\Delta_{i1}^*}^{\tilde{e}_k} \in \mathcal{K}_\infty$ for $k = 1, \dots, i$ and $\psi_{\Delta_{i1}^*}^{w_{i-1}}, \psi_{\Delta_{i1}^*}^{w_i}, \psi_{\Delta_{i1}^*}^{w_{i+1}} \in \mathcal{K}_\infty$ such that

$$
\begin{aligned}
&|\Delta_{i1}^*(\hat{e}_1, \tilde{e}_1, \dots, \hat{e}_i, \tilde{e}_i, w_{i-1}, w_i, w_{i+1})| \\
&\leq \sum_{k=1}^{i} \left(\psi_{\Delta_{i1}^*}^{\hat{e}_k}(|\hat{e}_k|) + \psi_{\Delta_{i1}^*}^{\tilde{e}_k}(|\tilde{e}_k|) \right) \\
&\quad + \psi_{\Delta_{i1}^*}^{w_{i-1}}(|w_{i-1}|) + \psi_{\Delta_{i1}^*}^{w_i}(|w_i|) + \psi_{\Delta_{i1}^*}^{w_{i+1}}(|w_{i+1}|).
\end{aligned}
\tag{4.104}
$$

By using $x_i^m = x_i + w_i$ and $x_{i+1}^m = x_{i+1} + w_{i+1}$, we can rewrite the $\hat{e}_i$-subsystem as

$$
\dot{\tilde{e}}_i = \kappa_{i1}(\tilde{e}_i - w_i) + w_{i+1} + x_{i+1}.
\tag{4.105}
$$

With $e_{i+1} = x_{i+1} - \kappa_{i2}(\hat{e}_i)$ according to (4.96), we have $x_{i+1} = e_{i+1} + \kappa_{i2}(\hat{e}_i)$, and thus, the $\hat{e}_i$-subsystem (4.105) can be rewritten as

$$
\begin{aligned}
\dot{\hat{e}}_i &= \Delta_{i2}^*(\tilde{e}_i, \tilde{e}_{i+1}, \hat{e}_{i+1}, w_i, w_{i+1}) + \kappa_{i2}(\hat{e}_i) \\
&:= f_{\hat{e}_i}(\tilde{e}_i, \tilde{e}_{i+1}, \hat{e}_i, \hat{e}_{i+1}, w_i, w_{i+1}),
\end{aligned}
\tag{4.106}
$$

where

$$
\Delta_{i2}^*(\tilde{e}_i, \tilde{e}_{i+1}, \hat{e}_{i+1}, w_i, w_{i+1}) = \kappa_{i1}(\tilde{e}_i - w_i) + w_{i+1} + \hat{e}_{i+1} - \tilde{e}_{i+1}.
\tag{4.107}
$$

Since κ_{i1} is odd and strictly decreasing, we can find $\psi_{\Delta_{i2}}^{\tilde{e}_i}, \psi_{\Delta_{i2}}^{\tilde{e}_{i+1}}, \psi_{\Delta_{i2}}^{\hat{e}_{i+1}}$, $\psi_{\Delta_{i2}}^{w_i}, \psi_{\Delta_{i2}}^{w_{i+1}} \in \mathcal{K}_\infty$ such that

$$
\begin{aligned}
&|\Delta_{i2}^*(\tilde{e}_i, \tilde{e}_{i+1}, \hat{e}_{i+1}, w_i, w_{i+1})| \\
&= \psi_{\Delta_{i2}}^{\tilde{e}_i}(|\tilde{e}_i|) + \psi_{\Delta_{i2}}^{\tilde{e}_{i+1}}(|\tilde{e}_{i+1}|) + \psi_{\Delta_{i2}}^{\hat{e}_{i+1}}(|\hat{e}_{i+1}|) + \psi_{\Delta_{i2}}^{w_i}(|w_i|) \\
&\quad + \psi_{\Delta_{i2}}^{w_{i+1}}(|w_{i+1}|).
\end{aligned}
\tag{4.108}
$$

Denote $\breve{e}_i = [\hat{e}_i, \tilde{e}_i]^T$ for $i = 1, \ldots, n$. The interconnection in the $\breve{e}_i$-subsystem is shown in Figure 4.5.

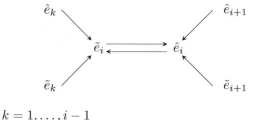

$$k = 1, \ldots, i - 1$$

FIGURE 4.5   The interconnection with each $\breve{e}_i$-system ($i = 1, \ldots, n$).

Define $\alpha_V(s) = s^2/2$ for $s \in \mathbb{R}_+$. For each $\tilde{e}_i$-subsystem and each $\hat{e}_i$-subsystem, we define the following ISS-Lyapunov function candidates, respectively:

$$
V_{\tilde{e}_i}(\tilde{e}_i) = \alpha_V(|\tilde{e}_i|), \quad i = 1, \ldots, n, \tag{4.109}
$$

$$
V_{\hat{e}_i}(\hat{e}_i) = \alpha_V(|\hat{e}_i|), \quad i = 1, \ldots, n. \tag{4.110}
$$

For convenience of notation, denote $\hat{e}_0 = e_0 = 0$ and $w_0 = 0$.

Denote $\breve{\bar{e}}_i = [\breve{e}_1^T, \ldots, \breve{e}_i^T]^T$ for $i = 1, \ldots, n$. Given the $\tilde{e}_i$-subsystem, we choose $\gamma_{\tilde{e}_i}^{\hat{e}_k}, \gamma_{\tilde{e}_i}^{\tilde{e}_k} \in \mathcal{K}_\infty$ with $k = 1, \ldots, i$ such that the compositions of the gain functions along all the simple loops through the $\tilde{e}_i$-subsystem in the $[\breve{\bar{e}}_{i-1}^T, \tilde{e}_i]^T$-subsystem are less than the identity function.

Consider the $\tilde{e}_i$-subsystem defined in (4.102). Using Lemma 4.1, for any specific constants $0 < c_{i1} < 1$, $\epsilon_{i1} > 0$ and $\ell_{i1} > 0$, any specific

$\gamma_{\tilde{e}_i}^{\hat{e}_i}, \gamma_{\tilde{e}_i}^{w_{i-1}}, \gamma_{\tilde{e}_i}^{w_i}, \gamma_{\tilde{e}_i}^{w_{i+1}} \in \mathcal{K}_\infty$, and the $\gamma_{\tilde{e}_i}^{\hat{e}_k}, \gamma_{\tilde{e}_i}^{\tilde{e}_k} \in \mathcal{K}_\infty$ for $k = 1, \ldots, i$ chosen above, we design κ_{i1} in the form of (4.6) such that $V_{\tilde{e}_i}$ satisfies

$$V_{\tilde{e}_i} \geq \max_{k=1,\ldots,i-1} \left\{ \begin{array}{l} \gamma_{\tilde{e}_i}^{\tilde{e}_k}(V_{\tilde{e}_k}), \gamma_{\tilde{e}_i}^{\hat{e}_k}(V_{\hat{e}_k}), \gamma_{\tilde{e}_i}^{\hat{e}_i}(V_{\hat{e}_i}), \\ \gamma_{\tilde{e}_i}^{w_{i-1}}(|w_{i-1}|), \gamma_{\tilde{e}_i}^{w_i}(|w_i|), \\ \gamma_{\tilde{e}_i}^{w_{i+1}}(|w_{i+1}|), \epsilon_{i1} \end{array} \right\}$$

$$\Rightarrow \nabla V_{\tilde{e}_i}(\tilde{e}_i) f_{\tilde{e}_i}(\hat{e}_1, \tilde{e}_1, \ldots, \hat{e}_i, \tilde{e}_i, w_{i-1}, w_i, w_{i+1}) \leq -\ell_{i1} V_{\tilde{e}_i}, \quad (4.111)$$

where

$$\gamma_{\tilde{e}_i}^{w_i} = \alpha_V \left(\frac{s}{c_{i1}} \right) \quad (4.112)$$

for $s \in \mathbb{R}_+$.

Given the $[\breve{e}_{i-1}, \tilde{e}_i]^T$-subsystem, we choose $\gamma_{\hat{e}_i}^{\tilde{e}_i} \in \mathcal{K}_\infty$ such that the compositions of the gain functions along all the simple loops through the $\hat{e}_i$-subsystem in the $\breve{e}_i$-subsystem are less than the identity function.

Consider the $\hat{e}_i$-subsystem defined in (4.106). Using Lemma 4.1, for any specified constants $\epsilon_{i2} > 0$ and $\ell_{i2} > 0$, any specified $\gamma_{\hat{e}_i}^{\tilde{e}_{i+1}}, \gamma_{\hat{e}_i}^{\hat{e}_{i+1}} \in \mathcal{K}_\infty$ and the $\gamma_{\hat{e}_i}^{\tilde{e}_i}$ chosen above, we design κ_{i2} in the form of (4.6) such that $V_{\hat{e}_i}$ satisfies

$$V_{\hat{e}_i}(\hat{e}_i) \geq \max \left\{ \begin{array}{l} \gamma_{\hat{e}_i}^{\tilde{e}_i}(V_{\tilde{e}_i}(\tilde{e}_i)), \gamma_{\hat{e}_i}^{\hat{e}_{i+1}}(V_{\hat{e}_{i+1}}(\hat{e}_{i+1})), \gamma_{\hat{e}_i}^{\tilde{e}_{i+1}}(V_{\tilde{e}_{i+1}}(\tilde{e}_{i+1})), \\ \gamma_{\hat{e}_i}^{w_i}(|w_i|), \gamma_{\hat{e}_i}^{w_{i+1}}(|w_{i+1}|), \epsilon_{i2} \end{array} \right\}$$

$$\Rightarrow \nabla V_{\hat{e}_i}(\hat{e}_i) f_{\hat{e}_i}(\tilde{e}_i, \tilde{e}_{i+1}, \hat{e}_i, \hat{e}_{i+1}, w_i, w_{i+1}) \leq -\ell_{i2} V_{\hat{e}_i}(\hat{e}_i). \quad (4.113)$$

In the case of $i = n$, $x_{i+1} = x_{i+1}^m = u$, $\hat{e}_{i+1} = e_{i+1} = 0$, and $w_{i+1} = 0$. The dynamic state measurement feedback control law is designed as

$$u = \kappa_{n2}(\hat{e}_n). \quad (4.114)$$

4.2.2 ISS OF THE CLOSED-LOOP SYSTEM

Define $e = [\breve{e}_1, \ldots, \breve{e}_n]^T$. The e-system can be represented by

$$\dot{e} = f_e(e, w_1, \ldots, w_n). \quad (4.115)$$

The gain digraph of the e-system is shown in Figure 4.6.

According to the recursive design, given the $\breve{e}_{i-1}$-system, by designing κ_{i1} for the $\tilde{e}_i$-subsystem, we can assign the ISS gains $\gamma_{\tilde{e}_i}^{\hat{e}_k}, \gamma_{\tilde{e}_i}^{\tilde{e}_k}$'s for $k = 1, \ldots, i-1$ such that all the simple loops in the $[\breve{e}_{i-1}^T, \tilde{e}_i]^T$-system through the $\tilde{e}_i$-subsystem satisfy the cyclic-small-gain condition. By designing κ_{i2} for the $\hat{e}_i$-subsystem, we can assign the ISS gain $\gamma_{\hat{e}_i}^{\tilde{e}_i}$ such that the simple loop in the $\breve{e}_i$-system through the $\tilde{e}_i$-subsystem satisfies the cyclic-small-gain condition. Through the recursive control design procedure, the e-system satisfies the cyclic-small-gain condition and is ISS.

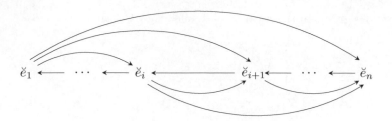

FIGURE 4.6 The gain digraph of the $[\breve{e}_1, \ldots, \breve{e}_n]^T$-system.

For the e-system, we construct the ISS-Lyapunov function candidate

$$V_e(e) = \max_{i=1,\ldots,n} \left\{ \sigma_{i1}(V_{\tilde{e}_i}(\tilde{e}_i)), \sigma_{i2}(V_{\hat{e}_i}(\hat{e}_i)) \right\}, \qquad (4.116)$$

where $\sigma_{12} = \mathrm{Id}$, and σ_{i1} with $i = 1, \ldots, n$ and σ_{i2} with $i = 2, \ldots, n$ are compositions of $\hat{\gamma}_{(\cdot)}^{(\cdot)}$'s which are continuously differentiable on $(0, \infty)$ and slightly larger than the corresponding $\gamma_{(\cdot)}^{(\cdot)}$'s, and still satisfy the cyclic-small-gain condition. Here, it is not necessary to give an explicit representation of the σ_{i1} and σ_{i2} to analyze the effect of the sensor noise.

Correspondingly, the influence from w_i and ϵ_i for $i = 1, \ldots, n$ can be represented as:

$$\theta = \max_{i=1,\ldots,n} \left\{ \begin{array}{c} \sigma_{i1} \circ \gamma_{\tilde{e}_i}^{w_{i-1}}(|w_{i-1}|), \sigma_{i1} \circ \gamma_{\tilde{e}_i}^{w_i}(|w_i|), \\ \sigma_{i1} \circ \gamma_{\tilde{e}_i}^{w_{i+1}}(|w_{i+1}|), \sigma_{i1}(\epsilon_{i1}), \\ \sigma_{i2} \circ \gamma_{\hat{e}_i}^{w_i}(|w_i|), \sigma_{i2} \circ \gamma_{\hat{e}_i}^{w_{i+1}}(|w_{i+1}|), \sigma_{i2}(\epsilon_{i2}) \end{array} \right\}. \qquad (4.117)$$

Then, it holds that

$$V_e(e) \geq \theta \Rightarrow \nabla V_e(e) f_e(e, w_1, \ldots, w_n) \leq -\alpha_e(V_e(e)) \qquad (4.118)$$

wherever ∇V_e exists, where α_e is a continuous and positive definite function.

By default, $\gamma_{\tilde{e}_{n+1}}^{w_n} := 0$, $\gamma_{\tilde{e}_0}^{w_1} := 0$, $\gamma_{\hat{e}_0}^{w_1} := 0$, $\sigma_{(n+1)1} := 0$, and $\sigma_{01} := 0$. Define

$$\gamma_e^{w_i}(s) = \max \left\{ \begin{array}{c} \sigma_{(i+1)1} \circ \gamma_{\tilde{e}_{i+1}}^{w_i}(s), \sigma_{i1} \circ \gamma_{\tilde{e}_i}^{w_i}(s), \\ \sigma_{(i-1)1} \circ \gamma_{\tilde{e}_{i-1}}^{w_i}(s), \sigma_{i2} \circ \gamma_{\hat{e}_i}^{w_i}(s), \\ \sigma_{(i-1)2} \circ \gamma_{\hat{e}_{i-1}}^{w_i}(s) \end{array} \right\} \qquad (4.119)$$

$$\epsilon = \max_{i=1,\ldots,n} \left\{ \sigma_{i1}(\epsilon_{i1}), \sigma_{i2}(\epsilon_{i2}) \right\}. \qquad (4.120)$$

Then, the θ defined in (4.117) can be equivalently represented by

$$\theta = \max_{i=1,\ldots,n} \left\{ \gamma_e^{w_i}(|w_i|), \epsilon \right\}. \qquad (4.121)$$

By choosing the $\gamma_{(\cdot)}^{(\cdot)}$'s small enough, we can make σ_{i1} for $i = 1, \ldots, n$ and σ_{i2} for $i = 2, \ldots, n$ small enough such that

$$\sigma_{i1} \circ \gamma_{\tilde{e}_i}^{w_i}(s) \geq \max \left\{ \begin{array}{l} \sigma_{(i+1)1} \circ \gamma_{\tilde{e}_{i+1}}^{w_i}(s), \sigma_{(i-1)1} \circ \gamma_{\tilde{e}_{i-1}}^{w_i}(s), \\ \sigma_{i2} \circ \gamma_{\tilde{e}_i}^{w_i}(s), \sigma_{(i-1)2} \circ \gamma_{\tilde{e}_{i-1}}^{w_i}(s) \end{array} \right\}. \tag{4.122}$$

Then, it is achieved that

$$\theta = \max_{i=1,\ldots,n} \left\{ \sigma_{i1} \circ \gamma_{\tilde{e}_i}^{w_i}(|w_i|), \epsilon \right\}. \tag{4.123}$$

Property (4.118) implies that there exists a $\beta_e \in \mathcal{KL}$ such that

$$V_e(e(t)) \leq \max \left\{ \beta_e(V_e(e(t_0)), t - t_0), \sup_{t_0 \leq \tau \leq t} (\theta(\tau)) \right\}, \tag{4.124}$$

where

$$\theta(\tau) = \max_{i=1,\ldots,n} \left\{ \sigma_{i1} \circ \gamma_{\tilde{e}_i}^{w_i}(|w_i(t)|), \epsilon \right\}. \tag{4.125}$$

From the definition of V_e in (4.118), using $\sigma_{12} = \mathrm{Id}$, we have

$$\begin{aligned} |x_1| = |e_1| = |\hat{e}_1 - \tilde{e}_1| &\leq |\hat{e}_1| + |\tilde{e}_1| \\ &= \alpha_V^{-1}(V_{\hat{e}_1}(\hat{e}_1)) + \alpha_V^{-1}(V_{\tilde{e}_1}(\tilde{e}_1)) \\ &\leq \alpha_V^{-1} \circ \sigma_{12}^{-1}(V_e(e)) + \alpha_V^{-1} \circ \sigma_{11}^{-1}(V_e(e)) \\ &= (\alpha_V^{-1} + \alpha_V^{-1} \circ \sigma_{11}^{-1})(V_e(e)). \end{aligned} \tag{4.126}$$

Define

$$\bar{\gamma}_{x_1}^{w_i} = (\alpha_V^{-1} + \alpha_V^{-1} \circ \sigma_{11}^{-1}) \circ \sigma_{i1} \circ \gamma_{\tilde{e}_i}^{w_i}, \quad i = 1, \ldots, n, \tag{4.127}$$

$$\bar{\beta}_{x_1} = (\alpha_V^{-1} + \alpha_V^{-1} \circ \sigma_{11}^{-1}) \circ \beta_e, \tag{4.128}$$

$$\bar{\epsilon}_{x_1} = (\alpha_V^{-1} + \alpha_V^{-1} \circ \sigma_{11}^{-1})(\epsilon). \tag{4.129}$$

Then, from (4.124) and (4.125), we obtain

$$\begin{aligned} |x_1(t)| &\leq (\alpha_V^{-1} + \alpha_V^{-1} \circ \sigma_{11}^{-1})(V_e(e(t))) \\ &\leq \max \left\{ \bar{\beta}_{x_1}(V_e(e(t_0)), t - t_0), \sup_{t_0 \leq \tau \leq t} \left(\max_{i=1,\ldots,n} \bar{\gamma}_{x_1}^{w_i}(|w_i(\tau)|) \right), \epsilon \right\}. \end{aligned} \tag{4.130}$$

Thus, the closed-loop system is IOS with x_1 as the output and the IOS gain from w_i to x_1 is $\bar{\gamma}_{x_1}^{w_i}$.

Recall that $\gamma_{\tilde{e}_1}^{w_1}(s) = \alpha_V(s/c_{11})$ for $s \in \mathbb{R}_+$. From the definition of $\bar{\gamma}_{x_1}^{w_1}$ in (4.127) with $i = 1$, we have

$$\begin{aligned} \bar{\gamma}_{x_1}^{w_1} &= (\alpha_V^{-1} + \alpha_V^{-1} \circ \sigma_{11}^{-1}) \circ \sigma_{11} \circ \gamma_{\tilde{e}_1}^{w_1} \\ &= (\mathrm{Id} + \alpha_V^{-1} \circ \sigma_{11} \circ \alpha_V) \left(\frac{s}{c_{11}} \right). \end{aligned} \tag{4.131}$$

Note that for $i = 1, \ldots, n$, each σ_{i1} is a composition of the $\hat{\gamma}^{(\cdot)}_{(\cdot)}$'s, which can be chosen to be arbitrarily small. Thus, the IOS gains $\bar{\gamma}^{w_i}_{x_1}$ for $i = 2, \ldots, n$ can be designed to be arbitrarily small. If we choose c_{11} to be arbitrarily close to one, and σ_{11} to be arbitrarily small, then $\bar{\gamma}^{w_1}_{x_1}$ is arbitrarily close to the identity function.

The main result of this section is summarized in the following theorem.

Theorem 4.2 *Under Assumption 4.4, system (4.86)–(4.88) can be input-to-state stabilized with the dynamic state measurement feedback control law defined in (4.96), (4.99), (4.97), and (4.114). Moreover, the closed-loop system is IOS with the sensor noise $w_1, \ldots, w_n$ as the inputs and x_1 as the output, the IOS gain from w_1 to x_1 can be designed to be arbitrarily close to the identity function, and the IOS gains from $w_2, \ldots, w_n$ to x_1 can be designed to be arbitrarily small.*

4.2.3 A DESIGN EXAMPLE

To verify the main result of this section, consider the following second-order nonlinear system:

$$\dot{x}_1 = x_2 \tag{4.132}$$

$$\dot{x}_2 = 0.2x_2^2 + u \tag{4.133}$$

$$x_1^m = x_1 + w_1 \tag{4.134}$$

$$x_2^m = x_2 + w_2. \tag{4.135}$$

For the sake of simplicity, we consider the case of $w_2 = 0$.

Define $e_1 = x_1$. Following the design procedure in Subsection 4.2.1, we have

$$\dot{\hat{e}}_1 = \kappa_{11}(\tilde{e}_1 - w_1) + \hat{e}_2 - \tilde{e}_2 + \kappa_{12}(\hat{e}_1) \tag{4.136}$$

$$\dot{\tilde{e}}_1 = \kappa_{11}(\tilde{e}_1 - w_1), \tag{4.137}$$

where $\hat{e}_1$ is the estimate of e_1, $\tilde{e}_1 = \hat{e}_1 - e_1$, and $\hat{e}_2$ and $\tilde{e}_2$ are defined later.

Consider the $\tilde{e}_1$-subsystem. Clearly, $\Delta^*_{11} = 0$. We choose $c_{11} = 0.8$ and $\ell_{11} = 0.02$. Then, the κ_{11} is designed in the form of $\kappa_{11}(r) = -\nu_{11}(|r|)r$ with ν_{11} satisfying

$$(1 - c_{11})\nu_{11}((1 - c_{11})s)s \geq 0.01s. \tag{4.138}$$

Then, we choose $\nu_{11}(s) = 0.05$ for $s \in \mathbb{R}_+$ and $\kappa_{11}(r) = -0.05r$ for $r \in \mathbb{R}$.

With κ_{11} designed, we have $\Delta^*_{12}(\tilde{e}_1, \tilde{e}_2, \hat{e}_2, w_1) = -0.05\tilde{e}_1 + 0.05w_1 + \hat{e}_2 - \tilde{e}_2$. Thus, $\psi^{\tilde{e}_1}_{\Delta^*_{12}}(s) = 0.05s$, $\psi^{w_1}_{\Delta^*_{12}}(s) = 0.05s$, $\psi^{\hat{e}_2}_{\Delta^*_{12}}(s) = s$, and $\psi^{\tilde{e}_2}_{\Delta^*_{12}}(s) = s$. Choose $\ell_{12} = 0.02$, $\gamma^{\tilde{e}_1}_{\hat{e}_1}(s) = s$, $\gamma^{\tilde{e}_2}_{\hat{e}_1}(s) = 0.99s$, $\gamma^{w_1}_{\hat{e}_1}(s) = 0.5s^2$, and $\gamma^{\hat{e}_2}_{\hat{e}_1}(s) = s$. Then, the κ_{12} is designed in the form of $\kappa_{12}(r) = -\nu_{12}(|r|)r$ with $\nu_{12}(s) = 2.11$ for $s \in \mathbb{R}_+$.

Define $e_2 = x_2 - \kappa_{12}(\hat{e}_1)$. The estimator for e_2 is designed in the following form:

$$\dot{\hat{e}}_2 = \kappa_{21}(\tilde{e}_2) + u, \tag{4.139}$$

where $\hat{e}_2$ is the estimate of e_2. Define $\tilde{e}_2 = \hat{e}_2 - e_2$. By directly taking the derivative of $\tilde{e}_2$, we have

$$\dot{\tilde{e}}_2 = \Delta_{21}^*(\tilde{e}_1, \hat{e}_1, \tilde{e}_2, \hat{e}_2, w_1) + \kappa_{21}(\tilde{e}_2), \tag{4.140}$$

where $|\Delta_{21}^*(\tilde{e}_1, \hat{e}_1, \tilde{e}_2, \hat{e}_2, w_1)|$ satisfies $|\Delta_{21}^*(\tilde{e}_1, \hat{e}_1, \tilde{e}_2, \hat{e}_2, w_1)| \leq 0.1055|\tilde{e}_1| + 1.7344|\hat{e}_1|^2 + 4.4521|\hat{e}_1| + 0.822|\tilde{e}_2|^2 + 2.11|\tilde{e}_2| + 0.822|\hat{e}_2|^2 + 2.11|\hat{e}_2| + 0.1055|w_1|$. Thus, we have $\psi_{\Delta_{21}^*}^{\hat{e}_2}(s) = 0.822s^2 + 2.11s$, $\psi_{\Delta_{21}^*}^{\tilde{e}_2}(s) = 0.822s^2 + 2.11s$, $\psi_{\Delta_{21}^*}^{\hat{e}_1}(s) = 1.7344s^2 + 4.4521s$, $\psi_{\Delta_{21}^*}^{\tilde{e}_1}(s) = 0.1055s$, and $\psi_{\Delta_{21}^*}^{w_1}(s) = 0.1055s$. We choose $\ell_{21} = 0.02$, $\gamma_{\tilde{e}_2}^{\hat{e}_1}(s) = 0.99s$, $\gamma_{\tilde{e}_2}^{\hat{e}_2}(s) = 0.99s$, $\gamma_{\tilde{e}_2}^{\tilde{e}_1}(s) = 0.99s$, and $\gamma_{\tilde{e}_2}^{w_1}(s) = 0.5s^2$. Then, the γ_{21} is designed in the form of $\kappa_{21}(r) = -\nu_{21}(|r|)r$ with $\nu_{21}(s) = 3.3784s + 8.7876$ for $s \in \mathbb{R}_+$.

With κ_{21} designed, we have $\Delta_{22}^*(\tilde{e}_2) = -3.3784|\tilde{e}_2|\tilde{e}_2 - 8.7876\tilde{e}_2$. Thus, $\psi_{\Delta_{22}^*}^{\tilde{e}_2}(s) = 3.3784s^2 + 8.7876s$. We choose $\ell_{22} = 0.02$ and $\gamma_{\hat{e}_2}^{\tilde{e}_2}(s) = 0.99s$. Then, the κ_{22} is designed in the form of $\kappa_{22}(r) = -\nu_{22}(|r|)r$ with $\nu_{22}(s) = 3.3784s + 8.7976$ for $s \in \mathbb{R}_+$.

In the construction of the ISS-Lyapunov function for the closed-loop system, we can choose all $\sigma_{(\cdot)} = \text{Id}$. With direct calculation following the procedure in Subsection 4.2.2, we have $\bar{\gamma}_{x_1}^{w_1}(s) = 2.5s$ for $s \in \mathbb{R}_+$.

Simulation results shown in Figures 4.7–4.8 are in accordance with the theoretical design.

4.3 DECENTRALIZED OUTPUT MEASUREMENT FEEDBACK CONTROL

Decentralized control problems arise from various engineering applications, such as power systems, transportation networks, water systems, chemical engineering, and telecommunication networks [239, 118]. Among the main characteristics of decentralized control are the dramatic reduction of computational complexity and the enhancement of robustness against uncertain interactions. This section studies decentralized output measurement feedback control of large-scale nonlinear systems with nonlinear dynamical interactions. It should be noted that, when reduced to single (centralized) systems, the observer-based design is still useful in handling the sensor noise. The discussions in this section neglect the possible influence of external disturbances, and focus on the impact of sensor noise to decentralized output-feedback control. The problems caused by external disturbances can be solved as in Section 4.1.

Consider the large-scale system with subsystems in the output-feedback

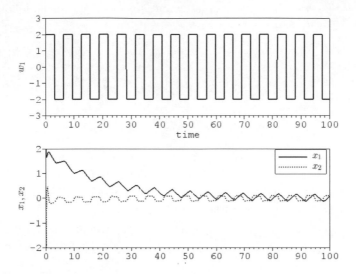

FIGURE 4.7 The sensor noise and the system states.

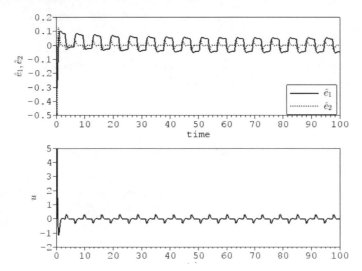

FIGURE 4.8 The estimator states and the control input.

form:

$$\dot{z}_i = \Delta_{i0}(z_i, y_i, w_i) \tag{4.141}$$

$$\dot{x}_{ij} = x_{i(j+1)} + \Delta_{ij}(y_i, z_i, w_i), \quad j = 1, \ldots, n_i - 1 \tag{4.142}$$

$$\dot{x}_{in_i} = u_i + \Delta_{in_i}(y_i, z_i, w_i) \tag{4.143}$$

$$w_i = [y_1, \ldots, y_{i-1}, y_{i+1}, y_N]^T \tag{4.144}$$

$$y_i = x_{i1} \tag{4.145}$$

$$y_i^m = y_i + d_i, \tag{4.146}$$

where, for each $i = 1, \ldots, N$, $[z_i^T, x_{i1}, \ldots, x_{in_i}]^T$ with $z_i \in \mathbb{R}^{n_{z_i}}$ and $x_{ij} \in \mathbb{R}$ $(j = 1, \ldots, n_i)$ is the state, $u_i \in \mathbb{R}$ is the control input, $y_i \in \mathbb{R}$ is the output, z_i and $[x_{i2}, \ldots, x_{in_i}]^T$ are the unmeasured portions of the state, $y_i^m \in \mathbb{R}$ is the measurement of the output with $d_i \in \mathbb{R}$ being sensor noise, and Δ_{ij}'s $(j = 1, \ldots, n_i)$ are unknown locally Lipschitz functions.

Figure 4.9 shows the block diagram including the i-th and the i'-th subsystems $(1 \le i, i' \le N,\ i \ne i')$ of the large-scale system.

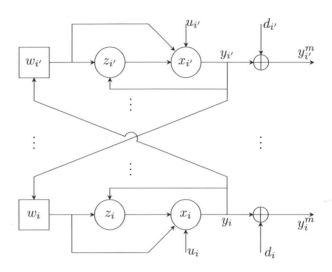

FIGURE 4.9   The block diagram of the large-scale system (4.141)–(4.146).

Assumptions 4.5, 4.6, and 4.7 are made on system (4.141)–(4.146).

Assumption 4.5 *For $i = 1, \ldots, N$, $j = 1, \ldots, n_i$, each Δ_{ij} satisfies*

$$|\Delta_{ij}(y_i, w_i)| \le \psi_{\Delta_{ij}}(|[y_i, w_i^T]^T|) \tag{4.147}$$

for all $[y_i, w_i^T]^T \in \mathbb{R}^N$, with a known $\psi_{\Delta_{ij}} \in \mathcal{K}_\infty$.

Assumption 4.6 *For each $i = 1, \ldots, N$, there exists a constant $\bar{d}_i \ge 0$, such that*

$$|d_i(t)| \le \bar{d}_i \tag{4.148}$$

for $t \ge 0$.

Assumption 4.7 *For $i = 1, \ldots, N$, each z_i-subsystem (4.141) with y_i and w_i as inputs admits a continuously differentiable ISS-Lyapunov function V_{z_i}, satisfying the following:*

1. *there exist* $\underline{\alpha}_{z_i}, \overline{\alpha}_{z_i} \in \mathcal{K}_\infty$ *such that for all* z_i,

$$\underline{\alpha}_{z_i}(|z_i|) \leq V_{z_i}(z_i) \leq \overline{\alpha}_{z_i}(|z_i|); \qquad (4.149)$$

2. *there exist* $\chi_{z_i}^{y_i}, \chi_{z_i}^{w_i} \in \mathcal{K}$ *and a continuous and positive definite function* α_{z_i} *such that for all* z_i, y_i, w_i,

$$V_{z_i}(z_i) \geq \max\{\chi_{z_i}^{y_i}(|y_i|), \chi_{z_i}^{w_i}(|w_i|)\}$$
$$\Rightarrow \nabla V_{z_i}(z_i)\Delta_{i0}(z_i, y_i, w_i) \leq -\alpha_{z_i}(V_{z_i}(z_i)). \qquad (4.150)$$

In the presence of the sensor noise, the objective of this section is to design a decentralized controller for the large-scale system composed of (4.141)–(4.146) by using the measurements y_i^m for $i = 1, \ldots, N$, such that the outputs y_i's $(i = 1, \ldots, N)$ are steered to within some small neighborhoods of the origin.

4.3.1 DECENTRALIZED REDUCED-ORDER OBSERVER

For each i-th subsystem, a decentralized reduced-order observer is designed to estimate the unmeasurable internal states by using the measurement y_i^m of the output:

$$\dot{\xi}_{ij} = \xi_{i(j+1)} + L_{i(j+1)}y_i^m - L_{ij}(\xi_{i2} + L_{i2}y_i^m), \quad j = 2, \ldots, n_i - 1 \qquad (4.151)$$

$$\dot{\xi}_{in_i} = u_i - L_{in_i}(\xi_{i2} + L_{i2}y_i^m), \qquad (4.152)$$

where ξ_{ij} is to be used as an estimate of $x_{ij} - L_{ij}y_i$ for each $j = 2, \ldots, n_i$. With $y_i^m = y_i + d_i$, observer (4.151)–(4.152) can be equivalently represented by

$$\dot{\xi}_{ij} = \xi_{i(j+1)} + L_{i(j+1)}y_i - L_{ij}(\xi_{i2} + L_{i2}y_i) + (L_{i(j+1)} - L_{ij}L_{i2})d_i,$$
$$j = 2, \ldots, n_i - 1 \qquad (4.153)$$

$$\dot{\xi}_{in_i} = u_i - L_{in_i}(\xi_{i2} + L_{i2}y_i) - L_{in_i}L_{i2}d_i. \qquad (4.154)$$

Define $\zeta_i = [x_{i2} - L_{i2}y_i - \xi_{i2}, \ldots, x_{in_i} - L_{in_i}y_i - \xi_{in_i}]^T$ as the observation error. Then, direct calculation yields:

$$\dot{\zeta}_i = A_i\zeta_i + \phi_{i0}(y_i, z_i, w_i, d_i)$$
$$:= f_{\zeta_i}(\zeta_i, y_i, z_i, w_i, d_i), \qquad (4.155)$$

where

$$A_i = \begin{bmatrix} -L_{i2} & & \\ \vdots & & I_{n_i-2} \\ -L_{i(n_i-1)} & & \\ -L_{in_i} & 0 & \cdots & 0 \end{bmatrix},$$

$$\phi_{i0}(y_i, z_i, w_i, d_i) = \begin{bmatrix} -L_{i2} & \\ \vdots & I_{n_i-1} \\ -L_{in_i} & \end{bmatrix} \begin{bmatrix} \Delta_{i1}(y_i, z_i, w_i) \\ \vdots \\ \Delta_{in_i}(y_i, z_i, w_i) \end{bmatrix}$$

$$+ \begin{bmatrix} L_{i2}^2 - L_{i3} \\ \vdots \\ L_{i(n_i-1)}L_{i2} - L_{in_i} \\ L_{in_i}L_{i2} \end{bmatrix} d_i.$$

Under Assumption 4.5, by using the definition of ϕ_{i0}, we can find $\psi_{\phi_{i0}}^{y_i}, \psi_{\phi_{i0}}^{z_i}, \psi_{\phi_{i0}}^{w_i}, \psi_{\phi_{i0}}^{d_i} \in \mathcal{K}_\infty$ such that $|\phi_{i0}(y_i, z_i, w_i, d_i)|^2 \leq \psi_{\phi_{i0}}^{y_i}(|y_i|) + \psi_{\phi_{i0}}^{z_i}(|z_i|) + \psi_{\phi_{i0}}^{w_i}(|w_i|) + \psi_{\phi_{i0}}^{d_i}(|d_i|)$ for all y_i, z_i, w_i, d_i.

The real constants L_{ij}'s $(j = 2, \ldots, n_i)$ are chosen so that A_i is a Hurwitz matrix, i.e., its eigenvalues have negative real parts. As a result, there exists a positive definite matrix $P_i = P_i^T \in \mathbb{R}^{(n_i-1)\times(n_i-1)}$ satisfying $P_i A_i + A_i^T P_i = -2I_{n_i-1}$. Define $V_{\zeta_i}(\zeta_i) = \zeta_i^T P_i \zeta_i$. Then, there exist $\underline{\alpha}_{\zeta_i}, \overline{\alpha}_{\zeta_i} \in \mathcal{K}_\infty$ such that $\underline{\alpha}_{\zeta_i}(|\zeta_i|) \leq V_{\zeta_i}(\zeta_i) \leq \overline{\alpha}_{\zeta_i}(|\zeta_i|)$ holds for all ζ_i. With direct calculation, we have

$$\nabla V_{\zeta_i}(\zeta_i) f_{\zeta_i}(\zeta_i, y_i, z_i, w_i, d_i)$$
$$= -2\zeta_i^T \zeta_i + 2\zeta_i^T P_i \phi_{i0}(y_i, z_i, w_i, d_i)$$
$$\leq -\zeta_i^T \zeta_i + |P_i|^2 |\phi_{i0}(y_i, z_i, w_i, d_i)|^2$$
$$\leq -\frac{1}{\lambda_{\max}(P_i)} V_{\zeta_i}(\zeta_i)$$
$$+ |P_i|^2 \left(\psi_{\phi_{i0}}^{y_i}(|y_i|) + \psi_{\phi_{i0}}^{z_i}(|z_i|) + \psi_{\phi_{i0}}^{w_i}(|w_i|) + \psi_{\phi_{i0}}^{d_i}(|d_i|) \right). \quad (4.156)$$

Define $\chi_{\zeta_i}^{y_i} = 4\lambda_{\max}(P_i)|P_i|^2 \psi_{\phi_{i0}}^{y_i}$, $\chi_{\zeta_i}^{z_i} = 4\lambda_{\max}(P_i)|P_i|^2 \psi_{\phi_{i0}}^{z_i}$, $\chi_{\zeta_i}^{w_i} = 4\lambda_{\max}(P_i)|P_i|^2 \psi_{\phi_{i0}}^{w_i}$, and $\chi_{\zeta_i}^{d_i} = 4\lambda_{\max}(P_i)|P_i|^2 \psi_{\phi_{i0}}^{d_i}$. Then, direct calculation yields:

$$V_{\zeta_i}(\zeta_i) \geq \max\{\chi_{\zeta_i}^{y_i}(|y_i|), \chi_{\zeta_i}^{z_i}(|z_i|), \chi_{\zeta_i}^{w_i}(|w_i|), \chi_{\zeta_i}^{d_i}(|d_i|)\}$$
$$\Rightarrow \nabla V_{\zeta_i}(\zeta_i) f_{\zeta_i}(\zeta_i, y_i, z_i, w_i, d_i) \leq -\alpha_{\zeta_i}(V_{\zeta_i}(\zeta_i)), \quad (4.157)$$

where $\alpha_{i0}(s) = s/4\lambda_{\max}(P_i)$ for $s \in \mathbb{R}_+$.

The reduced-order observer in this subsection is motivated by the design in [116]. Due to the unavailability of the accurate output y_i, the measurement y_i^m is used instead. The influence of the sensor noise d_i is represented by an ISS gain.

4.3.2 RECURSIVE CONTROL DESIGN

In this subsection, a new $[e_{i0}^T, e_{i1}, \ldots, e_{in_i}]^T$-system composed of ISS subsystems is recursively constructed based on the $[\zeta_i^T, z_i^T, y_i, \xi_{i2}, \ldots, \xi_{in_i}]^T$-system:

$$\dot{\zeta}_i = A_i \zeta_i + \phi_{i0}(y_i, z_i, w_i, d_i) \tag{4.158}$$

$$\dot{z}_i = \Delta_{i0}(z_i, y_i, w_i) \tag{4.159}$$

$$\dot{y}_i = \xi_{i2} + \phi_{i1}(\zeta_i, z_i, y_i, w_i) \tag{4.160}$$

$$\dot{\xi}_{ij} = \xi_{i(j+1)} + \phi_{ij}(y_i, \xi_{i2}, d_i), \qquad j = 2, \ldots, n_i - 1 \tag{4.161}$$

$$\dot{\xi}_{in_i} = u_i + \phi_{in_i}(y_i, \xi_{i2}, d_i), \tag{4.162}$$

where

$$\phi_{i1}(\zeta_i, z_i, y_i, w_i) = L_{i2} y_i + (x_{i2} - L_{i2} e_{i1} - \xi_{i2}) + \Delta_{i1}(y_i, z_i, w_i),$$

$$\phi_{ij}(y_i, \xi_{i2}, d_i) = L_{i(j+1)} y_i - L_{ij}(\xi_{i2} + L_{i2} y_i) + (L_{i(j+1)} - L_{ij} L_{i2}) d_i,$$

$$\phi_{in_i}(y_i, \xi_{i2}, d_i) = -L_{in_i}(\xi_{i2} + L_{i2} y_i) - L_{in_i} L_{i2} d_i.$$

The ϕ_{i1} is denoted as a function of ζ_i, z_i, y_i, w_i because $(x_{i2} - L_{i2} y_i - \xi_{i2})$ is the first element of vector ζ_i. Based on the observer design in Subsection 4.3.1, the large-scale system (4.141)–(4.146) is stabilized if (4.158)–(4.162) is stabilized. Note that the part composed of (4.160), (4.161) and (4.162) is in the strict-feedback form, and the part composed of (4.158) and (4.159) can be considered as dynamic uncertainty. The difference from the standard stabilization problem is that y_i is not available and y_i^m should be used instead.

Define $e_{i0} = [\zeta_i^T, z_i^T]^T$ and $e_{i1} = y_i$. The e_{ij} for $j = 2, \ldots, n_i$ are defined one by one in the following design procedure. The ISS-Lyapunov function candidates of the e_{ij}-subsystems $(j = 1, \ldots, n_i)$ are defined as

$$V_{ij}(e_{ij}) = \alpha_V(|e_{ij}|) = \frac{1}{2}|e_{ij}|^2. \tag{4.163}$$

In the following discussions in this subsection, we sometimes use V_{ij} instead of $V_{ij}(e_{ij})$ to simplify the notations. For convenience of notation, denote $\bar{e}_{ij} = [e_{i0}^T, e_{i1}, \ldots, e_{ij}]^T$ and $\bar{\xi}_{ij} = [\xi_{i2}, \ldots, \xi_{ij}]^T$.

The e_{i0}-subsystem

Recall that the ζ_i-subsystem and the z_i-subsystem are ISS. The e_{i0}-subsystem is a cascade connection of the ζ_i-subsystem and the z_i-subsystem, which satisfies the small-gain condition automatically. Define Lyapunov-based ISS gains $\gamma_{\zeta_i}^{e_{i1}} = \chi_{\zeta_i}^{y_i} \circ \alpha_V^{-1}$, $\gamma_{\zeta_i}^{z_i} = \chi_{\zeta_i}^{z_i} \circ \underline{\alpha}_{z_i}^{-1}$, and $\gamma_{z_i}^{e_{i1}} = \chi_{z_i}^{y_i} \circ \alpha_V^{-1}$. Then, from (4.157) and (4.150), we have

$$V_{\zeta_i} \geq \max\{\gamma_{\zeta_i}^{e_{i1}}(V_{i1}), \gamma_{\zeta_i}^{z_i}(V_{z_i}), \chi_{\zeta_i}^{w_i}(|w_i|), \chi_{\zeta_i}^{d_i}(|d_i|)\}$$

$$\Rightarrow \nabla V_{\zeta_i}(\zeta_i) f_{\zeta_i}(\zeta_i, y_i, z_i, w_i, d_i) \leq -\alpha_{\zeta_i}(V_{\zeta_i}), \tag{4.164}$$

$$V_{z_i} \geq \max\{\gamma_{z_i}^{e_{i1}}(V_{i1}), \chi_{z_i}^{w_i}(|w_i|)\}$$

$$\Rightarrow \nabla V_{z_i}(z_i) \Delta_{i0}(z_i, y_i, w_i) \leq -\alpha_{z_i}(V_{z_i}). \tag{4.165}$$

With the Lyapunov-based ISS cyclic-small-gain theorem, the ISS-Lyapunov function of the e_{i0}-subsystem is constructed as:

$$V_{i0}(e_{i0}) = \max\{V_{\zeta_i}(\zeta_i), \hat{\gamma}_{\zeta_i}^{z_i}(V_{z_i}(z_i))\},\qquad(4.166)$$

where $\hat{\gamma}_{\zeta_i}^{z_i} \in \mathcal{K}_\infty$ is slightly larger than $\gamma_{\zeta_i}^{z_i}$ and continuously differentiable on $(0, \infty)$. Then, there exist $\underline{\alpha}_{i0}, \overline{\alpha}_{i0} \in \mathcal{K}_\infty$ such that $\underline{\alpha}_{i0}(|e_{i0}|) \leq V_{i0}(e_{i0}) \leq \overline{\alpha}_{i0}(|e_{i0}|)$.

Define

$$\gamma_{e_{i0}}^{e_{i1}}(s) = \max\{\gamma_{\zeta_i}^{e_{i1}}(s), \hat{\gamma}_{\zeta_i}^{z_i} \circ \gamma_{z_i}^{e_{i1}}(s)\}\qquad(4.167)$$

$$\chi_{e_{i0}}^{w_i}(s) = \max\{\chi_{\zeta_i}^{w_i}(s), \hat{\gamma}_{\zeta_i}^{z_i} \circ \chi_{z_i}^{w_i}(s)\}\qquad(4.168)$$

$$\chi_{e_{i0}}^{d_i}(s) = \chi_{\zeta_i}^{d_i}(s)\qquad(4.169)$$

for $s \in \mathbb{R}_+$. Then, $\gamma_{e_{i0}}^{e_{i1}}$ is the ISS gain from V_{i1} to V_{i0}, $\chi_{e_{i0}}^{w_i}$ is the ISS gain from w_i to V_{i0}, and $\chi_{e_{i0}}^{d_i}$ is the ISS gain from d_i to V_{i0}.

With the cyclic-small-gain theorem, there exists a continuous and positive definite function α_{i0} such that

$$V_{i0} \geq \max\{\gamma_{e_{i0}}^{e_{i1}}(V_{i1}), \chi_{e_{i0}}^{w_i}(|w_i|), \chi_{e_{i0}}^{d_i}(|d_i|)\}$$
$$\Rightarrow \nabla V_{i0}(e_{i0})f_{e_{i0}}(e_{i0}, e_{i1}, w_i, d_i) \leq -\alpha_{i0}(V_{i0})\qquad(4.170)$$

wherever V_{i0} is differentiable. Here, $f_{e_{i0}}$ represents the dynamics of the e_{i0}-subsystem, i.e., $\dot{e}_{i0} = f_{e_{i0}}(e_{i0}, e_{i1}, w_i, d_i)$. It should be noted that V_{i0} is locally Lipschitz, and thus continuously differentiable almost everywhere.

The e_{i1}-subsystem

The design for the e_{i1}-subsystem (i.e., the y_i-subsystem) is quite similar to the e_1-subsystem in Subsection 4.1.3.

To deal with the sensor noise, we employ a set-valued map

$$S_{i1}(y_i) = \{\kappa_{i1}(y_i + \delta\bar{d}_i) : |\delta| \leq 1\}\qquad(4.171)$$

with κ_{i1} continuously differentiable, odd, strictly decreasing, and radially unbounded, to be determined later. Then, $\max S_{i1}(y_i) = \kappa_{i1}(y_i - \bar{d}_i)$ and $\min S_{i1}(y_i) = \kappa_{i1}(y_i + \bar{d}_i)$.

Recall the definition of $\vec{d}$ in (4.51). Define e_{i2} as

$$e_{i2} = \vec{d}(\xi_{i2}, S_{i1}(y_i)).\qquad(4.172)$$

Then, we have

$$\xi_{i2} - e_{i2} \in S_{i1}(y_i),\qquad(4.173)$$

and the e_{i1}-subsystem can be represented by

$$
\begin{aligned}
\dot{e}_{i1} &= \xi_{i2} - e_{i2} + e_{i2} + \phi_{i1}(\zeta_i, z_i, y_i, w_i) \\
&:= \xi_{i2} - e_{i2} + \phi_{i1}^*(\zeta_i, z_i, y_i, e_{i2}, w_i) \\
&\in \{\xi_{i2} - e_{i2} + \phi_{i1}^*(\zeta_i, z_i, y_i, e_{i2}, w_i) : \xi_{i2} - e_{i2} \in S_{i1}(y_i)\} \\
&:= F_{e_{i1}}(\zeta_i, z_i, y_i, e_{i2}, w_i).
\end{aligned}
\tag{4.174}
$$

From Assumption 4.5 and the definition of ϕ_{i1}, we can find a $\psi_{\phi_{i1}^*} \in \mathcal{K}_\infty$ such that $|\phi_{i1}^*(\zeta_i, z_i, y_i, e_{i2}, w_i)| \leq \psi_{\phi_{i1}^*}(\|[\bar{e}_{i2}^T, w_i^T]^T\|)$.

Note that $e_{i1} = y_i$. With Lemma 4.1, for any $0 < c_{i1} < 1$, $\epsilon_{i1} > 0$, $l_{i1} > 0$, $\gamma_{e_{i1}}^{e_{i0}}, \gamma_{e_{i1}}^{e_{i2}}, \chi_{e_{i1}}^{w_i} \in \mathcal{K}_\infty$, we can find a continuously differentiable, odd, strictly decreasing and radially unbounded κ_{i1} such that the e_{i1}-subsystem (4.174) with $\xi_{i2} - e_{i2}$ satisfying (4.173) is ISS with V_{i1} satisfying

$$
\begin{aligned}
&V_{i1}(e_{i1}) \geq \max\left\{\gamma_{e_{i1}}^{e_{i0}} \circ \underline{\alpha}_{i0}(|e_{i0}|), \gamma_{e_{i1}}^{e_{i2}} \circ \alpha_V(|e_{i2}|), \chi_{e_{i1}}^{w_i}(|w_i|), \chi_{e_{i1}}^{d_i}(\bar{d}_i), \epsilon_{i1}\right\} \\
&\Rightarrow \max_{f_{e_{i1}} \in F_{e_{i1}}(\zeta_i, z_i, y_i, e_{i2}, w_i)} \nabla V_{i1}(e_{i1}) f_{e_{i1}} \leq -l_{i1} V_{i1}(e_{i1})
\end{aligned}
\tag{4.175}
$$

where

$$
\chi_{e_{i1}}^{d_i}(s) = \alpha_V\left(\frac{1}{c_{i1}} s\right)
\tag{4.176}
$$

for $s \in \mathbb{R}_+$. Note that $V_{i0} \geq \underline{\alpha}_{i0}(|e_{i0}|)$ and $V_{i2} = \alpha_V(|e_{i2}|)$. Thus, with the appropriately designed κ_{i1}, we can achieve

$$
\begin{aligned}
&V_{i1}(e_{i1}) \geq \max\left\{\gamma_{e_{i1}}^{e_{i0}}(V_{i0}), \gamma_{e_{i1}}^{e_{i2}}(V_{i2}), \chi_{e_{i1}}^{w_i}(|w_i|), \chi_{e_{i1}}^{d_i}(\bar{d}_i), \epsilon_{i1}\right\} \\
&\Rightarrow \max_{f_{e_{i1}} \in F_{e_{i1}}(\zeta_i, z_i, y_i, e_{i2}, w_i)} \nabla V_{i1}(e_{i1}) f_{e_{i1}} \leq -l_{i1} V_{i1}(e_{i1}).
\end{aligned}
\tag{4.177}
$$

The e_{ij}-subsystem ($j = 2, \ldots, n_i$)

By convention, $S_{i1}(y_i, \bar{\xi}_{i1}) := S_{i1}(y_i)$. When $j = 3, \ldots, n_i$, for each $k = 2, \ldots, j-1$, a set-valued map S_{ik} is defined as

$$
S_{ik}(y_i, \bar{\xi}_{ik}) = \left\{\kappa_{ik}(\xi_{ik} - p_{ik}) : p_{ik} \in S_{i(k-1)}(y_i, \bar{\xi}_{i(k-1)})\right\},
\tag{4.178}
$$

where κ_{ik} is a continuously differentiable, odd, strictly decreasing and radially unbounded function, and $e_{i(k+1)}$ is defined as

$$
e_{i(k+1)} = \vec{d}(\xi_{i(k+1)}, S_{ik}(y_i, \bar{\xi}_{ik})).
\tag{4.179}
$$

Lemma 4.3 *Consider the $[\zeta_i^T, z_i^T, y_i, \xi_{i2}, \ldots, \xi_{in_i}]^T$-system defined by (4.158)–(4.162). If for $k = 1, \ldots, j-1$, S_{ik} and $e_{i(k+1)}$ are defined as (4.171), (4.172), (4.178) and (4.179), then when $e_{ij} \neq 0$, the e_{ij}-subsystem can be represented with*

$$
\dot{e}_{ij} = \xi_{i(j+1)} + \phi_{ij}'(z_i, y_i, \bar{\xi}_{ij}, w_i, d_i),
\tag{4.180}
$$

where

$$|\phi'_{ij}(y_i, \bar{\xi}_{ij}, w_i, d_i)| \leq \psi_{\phi'_{ij}}(|[\bar{e}_{ij}^T, w_i, d_i]^T|) \tag{4.181}$$

with $\psi_{\phi'_{ij}} \in \mathcal{K}_\infty$ known. Specifically, $\xi_{i(n_i+1)} = u_i$.

The proof of Lemma 4.3 is given in Appendix E.2.

Define a set-valued map S_{ij} as

$$S_{ij}(y_i, \bar{\xi}_{ij}) = \{\kappa_{ij}(\xi_{ij} - p_{ij}) : p_{ij} \in S_{i(j-1)}(y_i, \bar{\xi}_{i(j-1)})\} \tag{4.182}$$

with κ_{ij} continuously differentiable, odd, strictly decreasing, and radially unbounded, to be determined later. Define

$$e_{i(j+1)} = \vec{d}(\xi_{i(j+1)}, S_{ij}(y_i, \bar{\xi}_{ij})). \tag{4.183}$$

Then, $\xi_{i(j+1)} - e_{i(j+1)} \in S_{ij}(y_i, \bar{\xi}_{ij})$, and when $e_{ij} \neq 0$, the e_{ij}-subsystem can be rewritten as:

$$\begin{aligned}
\dot{e}_{ij} &= \xi_{i(j+1)} - e_{i(j+1)} + e_{i(j+1)} + \phi'_{ij}(z_i, y_i, \bar{\xi}_{ij}, w_i, d_i) \\
&:= \xi_{i(j+1)} - e_{i(j+1)} + \phi^*_{ij}(z_i, y_i, \bar{\xi}_{ij}, e_{i(j+1)}, w_i, d_i) \\
&\in \{\xi_{i(j+1)} - e_{i(j+1)} + \phi^*_{ij}(z_i, y_i, \bar{\xi}_{ij}, e_{i(j+1)}, w_i, d_i) : \\
&\qquad\qquad \xi_{i(j+1)} - e_{i(j+1)} \in S_{ij}(y_i, \bar{\xi}_{ij})\} \\
&:= F_{e_{ij}}(z_i, y_i, \bar{\xi}_{ij}, e_{i(j+1)}, w_i, d_i).
\end{aligned} \tag{4.184}$$

Clearly, under Assumption 4.5, there exists a $\psi_{\phi^*_{ij}} \in \mathcal{K}_\infty$ such that

$$\phi^*_{ij}(z_i, y_i, \bar{\xi}_{ij}, e_{i(j+1)}, w_i, d_i) \leq \psi_{\phi^*_{ij}}(|[\bar{e}_{i(j+1)}^T, w_i^T, d_i]^T|). \tag{4.185}$$

From the definition of e_{ij} in (4.179), in the case of $e_{ij} \neq 0$, for all $p_{ij} \in S_{i(j-1)}(y_i, \bar{\xi}_{i(j-1)})$, it holds that $|\xi_{ij} - p_{ij}| > |e_{ij}|$ and $\mathrm{sgn}(\xi_{ij} - p_{ij}) = \mathrm{sgn}(e_{ij})$, which implies $\mathrm{sgn}(\xi_{ij} - p_{ij} - e_{ij}) = \mathrm{sgn}(e_{ij})$, and thus $\xi_{ij} - p_{ij} = e_{ij} + (\xi_{ij} - p_{ij} - e_{ij}) = e_{ij} + \mathrm{sgn}(e_{ij})|\xi_{ij} - p_{ij} - e_{ij}|$. Then, we can rewrite the set-valued map S_{ij} as

$$\begin{aligned}
S_{ij}(y_i, \bar{\xi}_{ij}) = \{&\kappa_{ij}(e_{ij} + \mathrm{sgn}(e_{ij})|\xi_{ij} - p_{ij} - e_{ij}|) : \\
&p_{ij} \in S_{i(j-1)}(y_i, \bar{\xi}_{i(j-1)})\}.
\end{aligned} \tag{4.186}$$

With Lemma 4.1, for any $\epsilon_{ij} > 0$, $\ell_{ij} > 0$, $\gamma_{e_{ij}}^{e_{i0}}, \ldots, \gamma_{e_{ij}}^{e_{i(j-1)}}$, $\gamma_{e_{ij}}^{e_{i(j+1)}}, \chi_{e_{ij}}^{w_i}, \chi_{e_{ij}}^{d_i} \in \mathcal{K}_\infty$, we can find a continuously differentiable, odd, strictly decreasing, and radially unbounded κ_{ij} such that the e_{ij}-subsystem (4.184) with $\xi_{i(j+1)} - e_{i(j+1)} \in S_{ij}(y_i, \bar{\xi}_{ij})$ is ISS with V_{ij} satisfying

$$\begin{aligned}
V_{ij}(e_{ij}) &\geq \max_{k=1,\ldots,j-1,j+1} \left\{ \begin{array}{l} \gamma_{e_{ij}}^{e_{i0}} \circ \underline{\alpha}_{i0}(|e_{i0}|), \gamma_{e_{ij}}^{e_{ik}} \circ \alpha_V(|e_{ik}|), \\ \chi_{e_{ij}}^{w_i}(|w_i|), \chi_{e_{ij}}^{d_i}(\bar{d}_i), \epsilon_{ij} \end{array} \right\} \\
&\Rightarrow \max_{f_{e_{ij}} \in F_{e_{ij}}(z_i, y_i, \bar{\xi}_{ij}, e_{i(j+1)}, w_i, d_i)} \nabla V_{ij}(e_{ij}) f_{e_{ij}} \leq -\ell_{ij} V_{ij}(e_{ij}).
\end{aligned} \tag{4.187}$$

Notice that $V_{i0} \geq \underline{\alpha}_{i0}(|e_{i0}|)$ and $V_{ik} = \alpha_V(|e_{ik}|)$ for $k = 1, \ldots, j-1, j+1$. With the appropriately designed κ_{ij}, we can achieve

$$V_{ij}(e_{ij}) \geq \max_{k=0,\ldots,j-1,j+1} \left\{ \gamma_{e_{ij}}^{e_{ik}}(V_{ik}), \chi_{e_{ij}}^{w_i}(|w_i|), \chi_{e_{ij}}^{d_i}(\bar{d}_i), \epsilon_{ij} \right\}$$

$$\Rightarrow \max_{f_{e_{ij}} \in F_{e_{ij}}(z_i, y_i, \bar{\xi}_{ij}, e_{i(j+1)}, w_i, d_i)} \nabla V_{ij}(e_{e_{ij}}) f_{e_{ij}} \leq -\ell_{ij} V_{ij}(e_{ij}). \qquad (4.188)$$

Decentralized Control Law

In the case of $j = n_i$, the true control input u_i occurs and thus we can set $e_{i(n_i+1)} = 0$ and $V_{i(n_i+1)} = 0$. Indeed, our decentralized control law can be chosen as

$$p_{i2}^* = \kappa_{i1}(y_i + d_i) \qquad (4.189)$$

$$p_{ij}^* = \kappa_{i(j-1)}(\xi_{i(j-1)} - p_{i(j-1)}^*), \quad j = 3, \ldots, n_i \qquad (4.190)$$

$$u_i = \kappa_{in_i}(\xi_{in_i} - p_{in_i}^*). \qquad (4.191)$$

It is directly checked that

$$|d_i| \leq \bar{d}_i \Rightarrow p_{i2}^* \in S_{i1}(y_i) \Rightarrow \cdots \Rightarrow p_{in_i}^* \in S_{i(n_i-1)}(y_i, \bar{\xi}_{i(n_i-1)})$$
$$\Rightarrow u_i \in S_{in_i}(y_i, \bar{\xi}_{in_i}).$$

4.3.3   CYCLIC-SMALL-GAIN SYNTHESIS OF THE SUBSYSTEMS

Denote $\bar{e}_i = \bar{e}_{in_i}$. With the recursive control design, each $\bar{e}_i$-subsystem is an interconnection of ISS subsystems. With the cyclic-small-gain theorem, in this subsection, the decentralized controller designed above is fine-tuned, to yield the ISS property of each $\bar{e}_i$-subsystem.

According to the recursive design, given the $\bar{e}_{i(j-1)}$-subsystem, by appropriately choosing set-valued map S_{ij} for the e_{ij}-subsystem, we can design the ISS gains $\gamma_{e_{ij}}^{e_{ik}}$'s ($k = 0, \ldots, j-1$) such that

$$\gamma_{e_{ik}}^{e_{i(k+1)}} \circ \cdots \circ \gamma_{e_{i(j-1)}}^{e_{ij}} \circ \gamma_{e_{ij}}^{e_{ik}} < \mathrm{Id}, \quad k = 0, \ldots, j-1. \qquad (4.192)$$

By applying this reasoning repeatedly, we can guarantee (4.192) for all $j = 1, \ldots, n_i$. In this way, the $\bar{e}_i$-system satisfies the cyclic-small-gain condition in Chapter 3.

The gain interconnections between the subsystems can be represented by the gain digraph. In the gain digraph of the $\bar{e}_i$-subsystem, the e_{i1}-subsystem is reachable from the subsystems of $e_{i0}, e_{i2}, \ldots, e_{in_i}$. With the Lyapunov-based cyclic-small-gain theorem in Chapter 3, we construct an ISS-Lyapunov function for the $\bar{e}_i$-system as the "potential influence" from $V_{i0}(e_{i0}), \ldots, \ldots, V_{in_i}(e_{in_i})$ to $V_{i1}(e_{i1})$:

$$V_i(\bar{e}_i) = \max_{j=0,\ldots,n_i} \{\sigma_{ij}(V_{ij}(e_{ij}))\} \qquad (4.193)$$

with $\sigma_{i1}(s) = s$, $\sigma_{ij}(s) = \hat{\gamma}_{e_{i1}}^{e_{i2}} \circ \cdots \circ \hat{\gamma}_{e_{i(j-1)}}^{e_{ij}}(s)$ $(j = 2, \ldots, n_i)$, and $\sigma_{i0}(s) =$ $\max_{j=1,\ldots,n_i}\{\sigma_{ij} \circ \hat{\gamma}_{e_{ij}}^{e_{i0}}(s)\}$ for $s \in \mathbb{R}_+$, where the $\hat{\gamma}_{(\cdot)}^{(\cdot)}$'s are $\mathcal{K}_\infty$ functions continuously differentiable on $(0, \infty)$, slightly larger than the corresponding $\gamma_{(\cdot)}^{(\cdot)}$'s, and still satisfy (4.192) for all $j = 1, \ldots, n_i$.

For convenience, denote $\epsilon_{i0} = 0$. We represent "potential influence" from w_i, d_i, and ϵ_{ij}'s to V_{i1} as

$$\theta_i = \max_{j=0,\ldots,n_i} \left\{ \sigma_{ij} \circ \chi_{e_{ij}}^{w_i}(|w_i|), \sigma_{ij} \circ \chi_{e_{ij}}^{d_i}(\bar{d}_i), \sigma_{ij}(\epsilon_{ij}) \right\}. \tag{4.194}$$

Using the cyclic-small-gain theorem, we have that

$$V_i(\bar{e}_i) \geq \theta_i$$
$$\Rightarrow \max_{f_{\bar{e}_i} \in F_{\bar{e}_i}(z_i, y_i, \bar{\xi}_{in_i}, \bar{e}_i, w_i, d_i)} \nabla V_i(\bar{e}_i) f_{\bar{e}_i} \leq -\alpha_i(V_i(\bar{e}_i)) \tag{4.195}$$

holds wherever $\nabla V_i(\bar{e}_i)$ exists, where α_i is a continuous and positive definite function, and

$$F_{\bar{e}_i}(z_i, y_i, \bar{\xi}_{in_i}, \bar{e}_i, w_i, d_i) := \begin{bmatrix} \{f_{e_{i0}}(e_{i0}, e_{i1}, w_i, d_i)\} \\ F_{e_{i1}}(\zeta_i, z_i, y_i, e_{i2}, w_i) \\ \vdots \\ F_{e_{in_i}}(z_i, y_i, \bar{\xi}_{in_i}, e_{i(n_i+1)}, w_i, d_i) \end{bmatrix} \tag{4.196}$$

with $e_{n_i+1} = 0$.

From the definitions of w_i, e_{l1}'s, V_{l1}'s, and V_l's $(1 \leq l \leq N)$ and Assumption 4.6, we have

$$|w_i| \leq \sqrt{\sum_{1 \leq l \leq N, \, l \neq i} (\alpha_V^{-1}(V_l))^2 + \bar{d}_i^{e2}}$$
$$\leq \sqrt{\max_{1 \leq l \leq N, \, l \neq i} \{N \cdot (\alpha_V^{-1}(V_l))^2, N \cdot \bar{d}_i^{e2}\}}$$
$$= \max_{1 \leq l \leq N, \, l \neq i} \{\sqrt{N} \cdot (\alpha_V^{-1}(V_l)), \sqrt{N} \cdot \bar{d}_i^e\}. \tag{4.197}$$

Define $\gamma_{\bar{e}_i}^{\bar{e}_l}(s) = \max_{j=0,\ldots,n_i}\{\sigma_{ij} \circ \chi_{e_{ij}}^{w_i}(\sqrt{N} \cdot a_{il} \cdot \alpha_V^{-1}(s))\}$, $\chi_{\bar{e}_i}^{d_i^e}(s) =$ $\max_{j=0,\ldots,n_i}\{\sigma_{ij} \circ \chi_{e_{ij}}^{w_i}(\sqrt{N}s)\}$, $\chi_{\bar{e}_i}^{d_i}(s) = \max_{j=0,\ldots,n_i}\{\sigma_{ij} \circ \chi_{e_{ij}}^{d_i}(s)\}$ and $\epsilon_i =$ $\max_{j=0,\ldots,n_i}\{\sigma_{ij}(\epsilon_{ij})\}$ for $s \in \mathbb{R}_+$. Then, by substituting (4.197) into (4.194) and substituting (4.194) into (4.195), we get

$$V_i(\bar{e}_i) \geq \max_{1 \leq l \leq N, \, l \neq i} \left\{ \gamma_{\bar{e}_i}^{\bar{e}_l}(V_l(\bar{e}_l)), \chi_{\bar{e}_i}^{d_i^e}(\bar{d}_i^e), \chi_{\bar{e}_i}^{d_i}(\bar{d}_i), \epsilon_i \right\}$$
$$\Rightarrow \max_{f_{\bar{e}_i} \in F_{\bar{e}_i}(z_i, y_i, \bar{\xi}_{in_i}, \bar{e}_i, w_i, d_i)} \nabla V_i(\bar{e}_i) f_{\bar{e}_i} \leq -\alpha_i(V_i(\bar{e}_i)) \tag{4.198}$$

wherever $\nabla V_i(\bar{e}_i)$ exists, with α_i being a continuous and positive definite function.

According to the recursive design approach, we can design the $\gamma_{(\cdot)}^{(\cdot)}$'s (and thus the $\hat{\gamma}_{(\cdot)}^{(\cdot)}$'s) to be arbitrarily small to get arbitrarily small σ_{ij}'s ($j = 0, 2, \ldots, n_i$). We can also design the $\chi_{e_{ij}}^{w_i}$'s ($j = 1, \ldots, n_i$), the ϵ_{ij}'s ($j = 1, \ldots, n_i$), and the $\chi_{e_{ij}}^{d_i}$'s ($j = 2, \ldots, n_i$) to be arbitrarily small. Thus, from the definitions of $\gamma_{\bar{e}_i}^{\bar{e}_l}$ and $\chi_{\bar{e}_i}^{d_i^e}$, we can design the $\gamma_{\bar{e}_i}^{\bar{e}_l}$'s ($1 \leq l \leq N$, $l \neq i$), the $\chi_{\bar{e}_i}^{d_i^e}(\bar{d}_i^e)$, and the ϵ_i in (4.198) to be arbitrarily small. By designing the $\sigma_{ij} \circ \chi_{e_{ij}}^{d_i}$'s ($j = 0, 2, \ldots, n_i$) small enough, from the definitions of $\chi_{\bar{e}_i}^{d_i}$ and $\chi_{e_{i1}}^{d_i}$ (see (4.176)), it can be achieved that

$$\chi_{\bar{e}_i}^{d_i}(s) = \sigma_{i1} \circ \chi_{e_{i1}}^{d_i}(s) = \chi_{e_{i1}}^{d_i}(s) = \alpha_V\left(\frac{s}{c_{i1}}\right), \qquad (4.199)$$

where c_{i1} may be chosen to be any value satisfying $0 < c_{i1} < 1$.

4.3.4 ANALYSIS OF THE CLOSED-LOOP DECENTRALIZED SYSTEM

The closed-loop decentralized system is an interconnection of ISS $\bar{e}_i$-subsystems ($i = 1, \ldots, N$) with ISS-Lyapunov functions satisfying (4.198). As the discussion in Subsection 4.3.3 shows, we can design all the ISS gains $\gamma_{\bar{e}_i}^{\bar{e}_l}$'s ($1 \leq i, l \leq N$, $i \neq l$) to be arbitrarily small. Thus, the $\gamma_{\bar{e}_i}^{\bar{e}_l}$'s ($1 \leq i, l \leq N$, $i \neq l$) can be tuned such that all the simple loops in the closed-loop decentralized system satisfy the cyclic-small-gain condition. In this way, the ISS of the closed-loop decentralized system is achieved.

Consider the gain digraph of the $[\bar{e}_1^T, \ldots, \bar{e}_N^T]^T$-system. Recall the definition of reaching set, denoted by $\mathcal{RS}$, in Definition A.5 in Appendix A.

To analyze the effect of the disturbances on each $\bar{e}_i$-subsystem ($i = 1, \ldots, N$), we construct an ISS-Lyapunov function of the interconnected system composed of the $\bar{e}_r$-subsystems with $r \in \mathcal{RS}(i) \subseteq \{1, \ldots, N\}$ as:

$$V_{\bar{i}}(\bar{e}_{\bar{i}}) = \max_{r \in \mathcal{RS}(i)} \{\rho_r(V_r(\bar{e}_r))\}, \qquad (4.200)$$

where state $\bar{e}_{\bar{i}}$ consists of all the $\bar{e}_r$'s ($r \in \mathcal{RS}(i)$), $\rho_i = \mathrm{Id}$, and the ρ_r's ($r \in \mathcal{RS}(i) \backslash \{i\}$) are compositions of $\hat{\gamma}_{\bar{e}_r}^{\bar{e}_{r'}}$'s ($r, r' \in \mathcal{RS}(i)$, $r \neq r'$) which are continuously differentiable on $(0, \infty)$ and slightly larger than the corresponding $\gamma_{\bar{e}_r}^{\bar{e}_{r'}}$'s. Note that the dynamics of the $\bar{e}_{\bar{i}}$-system can be described by a differential inclusion $\dot{\bar{e}}_{\bar{i}} \in F_{\bar{i}}(\cdot)$.

Correspondingly, we can represent the influence of the disturbances to $V_{\bar{i}}(\bar{e}_{\bar{i}})$ as

$$\theta_{\bar{i}} = \max_{r \in \mathcal{RS}(i)} \{\rho_r \circ \chi_{\bar{e}_r}^{d_r^e}(\bar{d}_r^e), \rho_r \circ \chi_{\bar{e}_r}^{d_r^m}(\bar{d}_r), \rho_r(\epsilon_r)\}. \qquad (4.201)$$

Again, by using the ISS cyclic-small-gain theorem, we have

$$V_{\bar{i}}(\bar{e}_{\bar{i}}) \geq \theta_{\bar{i}} \Rightarrow \max_{f_{\bar{i}} \in F_{\bar{i}}(\cdot)} \nabla V_{\bar{i}}(\bar{e}_{\bar{i}}) f_{\bar{i}} \leq -\alpha_{\bar{i}}(V_{\bar{i}}(\bar{e}_{\bar{i}})) \qquad (4.202)$$

with $\alpha_{\bar{i}}$ positive definite, wherever $\nabla V_{\bar{i}}(\bar{e}_{\bar{i}})$ exists.

Note that, in (4.201), ρ_r ($r \in \mathcal{RS}(i)\backslash\{i\}$) can be designed to be arbitrarily small by designing $\gamma_{\bar{e}_r}^{\bar{e}_{r'}}$'s (and of course the $\hat{\gamma}_{\bar{e}_r}^{\bar{e}_{r'}}$'s) to be arbitrarily small, $\chi_{\bar{e}_r}^{d_r^e}$ and ϵ_r can be designed to be arbitrarily small, and $\chi_{\bar{e}_r}^{d_r^m}(s)$ can be designed to be $\alpha_V\left(\frac{s}{c_{r1}}\right)$. Thus, through an appropriate design, we can get

$$\theta_{\bar{i}} = \rho_i \circ \alpha_V\left(\frac{\bar{d}_i}{c_{i1}}\right) = \alpha_V\left(\frac{\bar{d}_i}{c_{i1}}\right). \qquad (4.203)$$

From (4.202) and (4.203), we can see that $V_{\bar{i}}(\bar{e}_{\bar{i}})$ ultimately converges to within the region $V_{\bar{i}}(\bar{e}_{\bar{i}}) \leq \alpha_V(\bar{d}_i/c_{i1})$. Using the definitions of e_{i1}, $V_{i1}(e_{i1})$, $V_i(\bar{e}_i)$, and $V_{\bar{i}}(\bar{e}_{\bar{i}})$ (see (4.163), (4.193), and (4.200)), we have $\alpha_V(|e_{i1}|) = V_{i1}(e_{i1}) \leq V_i(\bar{e}_i) \leq V_{\bar{i}}(\bar{e}_{\bar{i}})$, which implies that $y_i = e_{i1}$ ultimately converges to within the region $|y_i| \leq \bar{d}_i/c_{i1}$. By choosing c_{i1} to be arbitrarily close to one, the output y_i can be driven arbitrarily close to the region $|y_i| \leq \bar{d}_i$.

The main result of the section is summarized in Theorem 4.3.

Theorem 4.3 *Consider the large-scale system* (4.141)–(4.146). *Under Assumptions 4.5, 4.6, and 4.7, the closed-loop signals are bounded, and in particular, each output y_i ($i = 1, \ldots, N$) can be steered arbitrarily close to the region $|y_i| \leq \bar{d}_i$ with the decentralized controller composed of the decentralized reduced-order observer* (4.151)–(4.152) *and the decentralized control law* (4.189)–(4.191).

In the case of $N = 1$, system (4.141)–(4.146) is reduced to

$$\dot{z} = \Delta_0(z, y) \qquad (4.204)$$

$$\dot{x}_i = x_{i+1} + \Delta_i(y, z), \quad i = 1, \ldots, n-1 \qquad (4.205)$$

$$\dot{x}_n = u + \Delta_n(y, z) \qquad (4.206)$$

$$y = x_1 \qquad (4.207)$$

$$y^m = y + d, \qquad (4.208)$$

where $[z^T, x_1, \ldots, x_n]^T$ with $z \in \mathbb{R}^m$ and $x_i \in \mathbb{R}$ ($i = 1, \ldots, n$) is the state, $u \in \mathbb{R}$ is the control input, $y \in \mathbb{R}$ is the output, z and $[x_2, \ldots, x_n]^T$ are the unmeasured portions of the state, $y^m \in \mathbb{R}$ is the measurement of the output with $d \in \mathbb{R}$ being sensor noise, and Δ_i's ($i = 1, \ldots, n_i$) are unknown locally Lipschitz functions. In this case, the proposed decentralized controller is reduced to a centralized controller, and the design is still valid.

Example 4.4 *Consider the axial compressor model in Example 3.1. By defin-*
ing the transformations $z = R$, $x_1 = \phi$, $x_2 = -\psi$, $y = x_1$, and $u = (v-1)/\beta^2$
(which transform the control problem into a stabilization problem with respect
to the origin), one can rewrite system (3.1)–(3.3) in the output-feedback form
with y as the output and u as the control input:

$$\dot{z} = g(z, x_1) \tag{4.209}$$

$$\dot{x}_1 = x_2 + \Delta_1(x_1, z) \tag{4.210}$$

$$\dot{x}_2 = u + \Delta_2(x_1) \tag{4.211}$$

$$y = x_1, \tag{4.212}$$

where

$$\Delta_1(x_1, z) = -\frac{3}{2}x_1 + \frac{1}{2} - \frac{1}{2}(x_1 + 1)^3 - 3(x_1 + 1)z, \tag{4.213}$$

$$\Delta_2(x_1) = -\frac{1}{\beta^2}x_1. \tag{4.214}$$

The proposed measurement output-feedback design can be readily applied even
if Δ_1 and Δ_2 contain uncertainties.

Example 4.5 *We employ an interconnected system composed of two identical*
subsystems to demonstrate the control design procedure. Each i-th subsystem
($i = 1, 2$) is in the following form:

$$\begin{aligned}
\dot{z}_i &= -2z_i + w_i \\
\dot{x}_{i1} &= x_{i2} + \frac{1}{4}x_{i1}^2 \\
\dot{x}_{i2} &= u_i + \frac{\sqrt{2}}{8}x_{i1} + \frac{1}{4}x_{i1}^2 + \frac{\sqrt{2}}{8}x_{i1}z_i \\
w_i &= y_{(3-i)} \\
y_i &= x_{i1} \\
y_i^m &= y_i + d_i,
\end{aligned} \tag{4.215}$$

where $[z_i, x_{i1}, x_{i2}]^T \in \mathbb{R}^3$ is the state, $u_i \in \mathbb{R}$ is the control input, $y_i \in \mathbb{R}$
is the output, z_i and x_{i2} are unmeasured portions of the state, $y_i^m \in \mathbb{R}$ is
the measurement of output y_i, and d_i represents the sensor noise satisfying
$|d_i(t)| \le \bar{d}_i = \bar{d} = 0.1$ for $t \ge 0$. Define $\alpha_V(s) = 0.5s^2$ for $s \in \mathbb{R}_+$.
  Define $V_{z_i}(z_i) = 0.5z_i^2$. Then, we have

$$V_{z_i}(z_i) \ge \chi_{z_i}^{w_i}(|w_i|) \Rightarrow \nabla V_{z_i}(z_i)\dot{z}_i \le -V_{z_i}(z_i), \tag{4.216}$$

where $\chi_{z_i}^{w_i}(s) = 0.25s^2$ for $s \in \mathbb{R}_+$.
  For each i-th subsystem, the reduced-order observer is constructed as

$$\dot{\xi}_{i2} = u - \xi_{i2} - y_i^m. \tag{4.217}$$

Define $\zeta_i = \zeta_{i2} = x_{i2} - y_i - \xi_{i2}$. Then,

$$\dot{\zeta}_i = -\zeta_i + d_i + \frac{\sqrt{2}}{8}x_{i1} + \frac{\sqrt{2}}{8}z_i. \tag{4.218}$$

Define $V_{\zeta_i}(\zeta_i) = \zeta_i^2$. Then, we have

$$V_{\zeta_i}(\zeta_i) \geq \max\{\gamma_{\zeta_i}^{z_i}(V_{z_i}(z_i)), \gamma_{\zeta_i}^{e_{i1}}(V_{i1}(e_{i1}))\}$$
$$\Rightarrow \nabla V_{\zeta_i}(\zeta_i)\dot{\zeta}_i \leq -0.25 V_{\zeta_i}(\zeta_i), \qquad (4.219)$$

where $e_{i1} = y_i$, $V_{i1}(e_{i1}) = \alpha_V(|e_{i1}|)$, and $\gamma_{\zeta_i}^{z_i}(s) = s$ and $\gamma_{\zeta_i}^{e_{i1}}(s) = s$ for $s \in \mathbb{R}_+$.

Define $e_{i0} = [\zeta_i, z_i]^T$. Then, the ISS-Lyapunov function for the e_{i0}-subsystem can be constructed as

$$V_{i0}(e_{i0}) = \max\{V_{\zeta_i}(\zeta_i), \hat{\gamma}_{\zeta_i}^{z_i}(V_{z_i}(z_i))\}, \qquad (4.220)$$

where $\hat{\gamma}_{\zeta_i}^{z_i}(s) = 1.1s$ for $s \in \mathbb{R}_+$. Moreover, $V_{i0}(e_{i0})$ satisfies

$$V_{i0}(e_{i0}) \geq \max\{\gamma_{e_{i0}}^{e_{i1}}(V_{i1}(e_{i1})), \chi_{e_{i0}}^{w_i}(|w_i|), \chi_{e_{i0}}^{d_i}(\bar{d}_i)\}$$
$$\Rightarrow \nabla V_{i0}(e_{i0})\dot{e}_{i0} \leq -0.25 V_{i0}(e_{i0}), \qquad (4.221)$$

where $\gamma_{e_{i0}}^{e_{i1}}(s) = s$, $\chi_{e_{i0}}^{w_i}(s) = 0.275s^2$ and $\chi_{e_{i0}}^{d_i}(s) = 8s^2$.

The e_{i1}-subsystem can be rewritten as

$$\dot{e}_{i1} = \xi_{i2} - e_{i2} + (\zeta_i + e_{i1} + 0.25e_{i1}^2 + e_{i2}). \qquad (4.222)$$

Choose $c_{i1} = 0.5$ and $\gamma_{e_{i1}}^{e_{i2}}(s) = 0.9s$ for $s \in \mathbb{R}_+$. Select $\gamma_{e_{i1}}^{e_{i0}}(s) = 0.9s$ for $s \in \mathbb{R}_+$ such that $\gamma_{e_{i1}}^{e_{i0}} \circ \gamma_{e_{i0}}^{e_{i1}} < \mathrm{Id}$. With the gain assignment lemma, design $\kappa_{i1}(r) = -\nu_{i1}(|r|)r$ with $\nu_{i1}(s) = s + 6.5989$ for $s \in \mathbb{R}_+$. In this way, if $\kappa_{i1}(e_{i1} + \bar{d}_i) \leq \xi_{i2} - e_{i2} \leq \kappa_{i1}(e_{i1} - \bar{d}_i)$, then the implication

$$V_{i1}(e_{i1}) \geq \max\{\gamma_{e_{i1}}^{e_{i0}}(V_{i0}(e_{i0})), \gamma_{e_{i1}}^{e_{i2}}(V_{i2}(e_{i2})), \chi_{e_{i1}}^{d_i}(\bar{d}_i)\}$$
$$\Rightarrow \nabla V_{i1}(e_{i1})\dot{e}_{i1} \leq -V_{i1}(e_{i1}) \qquad (4.223)$$

holds, where $\chi_{e_{i1}}^{d_i}(s) = \alpha_V(s/c_{i1}) = 2s^2$.

The e_{i2}-subsystem can be rewritten as

$$\dot{e}_{i2} = u + \phi_{i2}^*(\zeta_i, e_{i1}, e_{i2}, \xi_{i2}, d_i), \qquad (4.224)$$

where $|\phi_{i2}^*(\zeta_i, e_{i1}, e_{i2}, \xi_{i2}, d_i)| \leq \psi_{\phi_{i2}^*}^{e_{i0}}(|e_{i0}|) + \psi_{\phi_{i2}^*}^{e_{i1}}(|e_{i1}|)\psi_{\phi_{i2}^*}^{e_{i2}}(|e_{i2}|) + \psi_{\phi_{i2}^*}^{d_i}(\bar{d}_i)$ with $\psi_{\phi_{i2}^*}^{e_{i0}}(s) = 6.5989s + 2s^2$, $\psi_{\phi_{i2}^*}^{e_{i1}}(s) = 44.5455s + 12.8489s^2 + 9.75s^3$, $\psi_{\phi_{i2}^*}^{e_{i2}}(s) = 7.5989s + 2s^2$, and $\psi_{\phi_{i2}^*}^{d_i}(s) = 36.9466s + 41.5934s^2 + 9.25s^3$ for $s \in \mathbb{R}_+$.

Choose $\chi_{e_{i2}}^{d_i}(s) = 2s^2$. Select $\gamma_{e_{i2}}^{e_{i0}}(s) = 0.9s$ and $\gamma_{e_{i2}}^{e_{i1}}(s) = s$ for $s \in \mathbb{R}_+$ such that $\gamma_{e_{i0}}^{e_{i1}} \circ \gamma_{e_{i1}}^{e_{i2}} \circ \gamma_{e_{i2}}^{e_{i0}} < \mathrm{Id}$ and $\gamma_{e_{i1}}^{e_{i2}} \circ \gamma_{e_{i2}}^{e_{i1}} < \mathrm{Id}$. For the e_{i2}-subsystem, design $\kappa_{i2}(r) = -\nu_{i2}(|r|)r$ with $\nu_{i2}(s) = 76.0.62 + 58.7526s + 10.9063s^2$ for $s \in \mathbb{R}_+$. Then, with $\min S_{i2}(y_i, \xi_{i2}) \leq u_i \leq \max S_{i2}(y_i, \xi_{i2})$, $V_{i2}(e_{i2})$ satisfies

$$V_{i2}(e_{i2}) \geq \max\{\gamma_{e_{i2}}^{e_{i0}}(V_{i0}(e_{i0})), \gamma_{e_{i2}}^{e_{i1}}(V_{i1}(e_{i1})), \chi_{e_{i2}}^{d_i}(\bar{d}_i)\}$$
$$\Rightarrow \nabla V_{i2}(e_{i2})\dot{e}_{i2} \leq -V_{i2}(e_{i2}). \qquad (4.225)$$

Denote $\bar{e}_i = [e_{i0}, e_{i1}, e_{i2}]^T$. Define $\hat{\gamma}_{e_{i1}}^{e_{i0}}(s) = 0.95s$, $\hat{\gamma}_{e_{i1}}^{e_{i2}}(s) = 0.95s$ and $\hat{\gamma}_{e_{i2}}^{e_{i0}}(s) = 0.95s$ for $s \in \mathbb{R}_+$. For each i-th subsystem, we can construct an ISS-Lyapunov function $V_i(\bar{e}_i) = \max\{\sigma_{i0}(V_{i0}(e_{i0})), \sigma_{i1}(V_{i1}(e_{i1})), \sigma_{i2}(V_{i2}(e_{i2}))\}$ with $\sigma_{i0}(s) = \max\{\hat{\gamma}_{e_{i1}}^{e_{i0}}(s), \hat{\gamma}_{e_{i1}}^{e_{i2}} \circ \hat{\gamma}_{e_{i2}}^{e_{i0}}(s)\} = 0.95s$, $\sigma_{i1}(s) = s$ and $\sigma_{i2}(s) = \hat{\gamma}_{e_{i1}}^{e_{i2}}(s) = 0.95s$ for $s \in \mathbb{R}_+$. We can also calculate $\chi_{\bar{e}_i}^{d_i}(s) = 2s^2$ for $s \in \mathbb{R}_+$.

Note that $|w_i| = |y_{(3-i)}| \le \alpha_V^{-1}(V_{(3-i)}(\bar{e}_{(3-i)}))$. Then, $\gamma_{\bar{e}_i}^{\bar{e}(3-i)}(s) = \sigma_{i0} \circ \chi_{e_{i0}}^{w_i} \circ \alpha_V^{-1}(s) = 0.5225$. Thus, $\gamma_{\bar{e}_i}^{\bar{e}(3-i)} \circ \gamma_{\bar{e}(3-i)}^{\bar{e}_i} < \mathrm{Id}$. Define $V_{\tilde{i}}(e_{\tilde{i}}) = \max\{V_i(\bar{e}_i), \hat{\gamma}_{\bar{e}_i}^{\bar{e}(3-i)}(V_{(3-i)}(\bar{e}_{(3-i)}))\}$ with $\hat{\gamma}_{\bar{e}_i}^{\bar{e}(3-i)}(s) = 0.55s$ for $s \in \mathbb{R}_+$. Then, we can calculate $\theta_{\tilde{i}} = \max\{\chi_{\bar{e}_i}^{d_i}(\bar{d}_i), \hat{\gamma}_{\bar{e}_i}^{\bar{e}(3-i)} \circ \chi_{\bar{e}(3-i)}^{d_{(3-i)}^m}(\bar{d}_{(3-i)})\} = \chi_{\bar{e}_i}^{d_i}(\bar{d}_i) = 2s^2$.

Based on the theory, the measurement feedback controller with κ_{i1} and κ_{i2} designed above could drive the output y_i to the region $|y_i| \le \bar{d}_i/c_{i1} = 2\bar{d}_i = 0.2$. Simulation results with disturbances $d_1^m(t) = 0.09\sin(5t) + 0.01\mathrm{sign}(\sin(30t))$ and $d_2^m(t) = 0.09\cos(5t) + 0.01\mathrm{sign}(\cos(30t))$, and initial conditions $z_1(0) = 0.5$, $x_{11}(0) = 0$, $x_{12}(0) = 0$, $\xi_{12}(0) = 0$, $z_2(0) = -0.5$, $x_{21}(0) = 0$, $x_{22}(0) = 0$, and $\xi_{22}(0) = 0$, shown in Figures 4.10 and 4.11, are in accordance with our theoretic results.

4.4 EVENT-TRIGGERED AND SELF-TRIGGERED CONTROL

An event-triggered control system is a sampled-data system in which the sampling time instants are determined by events generated by the real-time system state. By taking the advantage of the inter-sample behavior, event-triggered sampling may realize improved control performance over periodic sampling [160, 83].

An event-triggered state-feedback control system is generally in the following form:

$$\dot{x}(t) = f(x(t), u(t)) \tag{4.226}$$

$$u(t) = v(x(t_k)), \quad t \in [t_k, t_{k+1}), \ k \in \mathbb{S}, \tag{4.227}$$

where $x \in \mathbb{R}^n$ is the state, $u \in \mathbb{R}^m$ is the control input, $f : \mathbb{R}^n \times \mathbb{R}^m \to \mathbb{R}^n$ is a locally Lipschitz function representing system dynamics, $v : \mathbb{R}^n \to \mathbb{R}^m$ is a locally Lipschitz function representing the control law, t_k represents the sampling time instants, and $\mathbb{S} \subseteq \mathbb{Z}_+$ is the set of the indices of all the sampling time instants. It is assumed that $f(0, v(0)) = 0$. The sequence $\{t_k\}_{k \in \mathbb{S}}$ is determined online based on the measurement of the real-time system state. If there is an infinite number of sampling time instants, then $\mathbb{S} = \mathbb{Z}_+$; otherwise, $\mathbb{S}$ is in the form of $\{0, \ldots, k^*\}$ with $k^* \in \mathbb{Z}_+$ being the last sampling time instant. For convenience of notation, we denote $t_{k^*+1} = \infty$. Figure 4.12 shows the block diagram of an event-triggered control system.

Define

$$w(t) = x(t_k) - x(t), \quad t \in [t_k, t_{k+1}), \ k \in \mathbb{S} \tag{4.228}$$

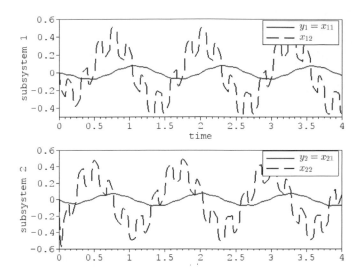

FIGURE 4.10   State trajectories of Example 4.5.

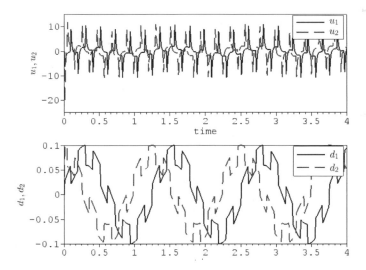

FIGURE 4.11   Control signals and disturbances of Example 4.5.

as the measurement error caused by data sampling, and rewrite

$$u(t) = v(x(t) + w(t)). \qquad (4.229)$$

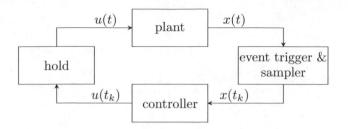

FIGURE 4.12 The block diagram of an event-triggered control system.

Then, by substituting (4.229) into (4.226), we have

$$\dot{x}(t) = f(x(t), v(x(t) + w(t)))$$
$$:= \bar{f}(x(t), x(t) + w(t)). \qquad (4.230)$$

Clearly, the event-triggered control problem is closely related to the measurement feedback control problem. However, through event-based triggering, the measurement error w caused by data sampling is adjustable, and the objective of event-triggered control is to adjust w online to asymptotically steer the system state $x(t)$ to the origin, if possible.

Based on the idea of robust control, a widely recognized approach to event-triggered control contains two steps:

1. Designing a continuous-time controller which guarantees the robustness of the closed-loop system with respect to the measurement error caused by data sampling;
2. Designing an appropriate event trigger to restrict the measurement error caused by data sampling to be within the margin of robust stability.

The block diagram of the system is shown in Figure 4.13.

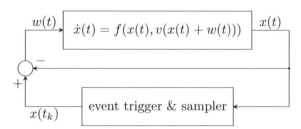

FIGURE 4.13 Event-triggered control problem as a robust control problem.

Due to its equivalence to robust stability, ISS has been used for event-triggered control of nonlinear systems. In [255, 160, 272], it is assumed that system (4.230) is ISS with w as the input and has an ISS-Lyapunov function

$V : \mathbb{R}^n \to \mathbb{R}_+$ satisfying

$$\nabla V(x)\bar{f}(x, x + w) \le -\alpha(V(x)) + \gamma(|w|) \tag{4.231}$$

for all x, w, where $\alpha \in \mathcal{K}_\infty$ and $\gamma \in \mathcal{K}$. Then, by designing the event trigger such that

$$|w(t)| \le \rho(V(x(t))) \tag{4.232}$$

always holds with $\rho \in \mathcal{K}$ satisfying

$$\alpha^{-1} \circ \gamma \circ \rho < \text{Id}, \tag{4.233}$$

asymptotic stability of the closed-loop system is achieved. In this case, V is a Lyapunov function of the closed-loop system.

Theoretically, a system is ISS if and only if it has an ISS-Lyapunov function. However, even if a nonlinear system has been designed to be ISS, the construction of an ISS-Lyapunov function may not be straightforward. Note that, given an ISS-Lyapunov function, one can easily determine the ISS characteristics of a system. By using the relationship between ISS and robust stability, the study in this section shows that a known ISS-Lyapunov function may not be necessary for the design of event-triggered control.

For physical realization of event-triggered sampling, a positive lower bound of the inter-sample periods should be guaranteed throughout the process of event-triggered control, i.e., $\inf_{k \in \mathbb{S}}(t_{k+1} - t_k) > 0$, to avoid infinitely fast sampling. A special case is the Zeno behavior, with which, $\lim_{k \to \infty} t_k < \infty$ [70].

In this section, we first present a trajectory-based ISS condition for asymptotic stabilization of general event-triggered control systems without infinitely fast sampling. Then, in Subsection 4.4.2, we discuss the event-triggered control problem in the presence of external disturbances. Subsection 4.4.3 gives a design for strict-feedback systems to fulfill the condition for event-triggered control.

4.4.1   AN ISS GAIN CONDITION FOR EVENT-TRIGGERED CONTROL

In this subsection, we assume that a measurement feedback controller exists for system (4.226) such that system (4.230) is ISS with the measurement error as the input.

Assumption 4.8 *System (4.230) is ISS with w as the input, that is, there exist $\beta \in \mathcal{KL}$ and $\gamma \in \mathcal{K}$ such that for any initial state $x(0)$ and any piecewise continuous, bounded w, it holds that*

$$|x(t)| \le \max\{\beta(|x(0)|, t), \gamma(\|w\|_\infty)\} \tag{4.234}$$

for all $t \ge 0$.

Under Assumption 4.8, with the robust stability property of ISS, if the event trigger is designed such that $|w(t)| \leq \rho(|x(t)|)$ for all $t \geq 0$ with $\rho \in \mathcal{K}$ satisfying

$$\rho \circ \gamma < \mathrm{Id}, \tag{4.235}$$

then $x(t)$ asymptotically converges to the origin. Based on this idea, the event trigger considered in this section is defined as: if $x(t_k) \neq 0$, then

$$t_{k+1} = \inf \left\{ t > t_k : H(x(t), x(t_k)) = 0 \right\}, \tag{4.236}$$

where $H : \mathbb{R}^n \times \mathbb{R}^n \to \mathbb{R}$ is defined by

$$H(x, x') = \rho(|x|) - |x - x'|. \tag{4.237}$$

If $x(t_k) = 0$ or $\{t > t_k : H(x(t), x(t_k)) = 0\} = \emptyset$, then the data sampling event is not triggered and in this case, $t_{k+1} = \infty$. Note that, under the assumption of $f(0, v(0)) = 0$, if $x(t_k) = 0$, then $u(t) = v(x(t_k)) = 0$ keeps the state at the origin for all $t \in [t_k, \infty)$.

With the event trigger proposed above, given t_k and $x(t_k) \neq 0$, t_{k+1} is the first time instant after t_k such that

$$\rho(|x(t_{k+1})|) - |x(t_{k+1}) - x(t_k)| = 0. \tag{4.238}$$

Since $\rho(|x(t_k)|) - |x(t_k) - x(t_k)| > 0$ for any $x(t_k) \neq 0$ and $x(t)$ is continuous on the timeline, the proposed event trigger guarantees that

$$\rho(|x(t)|) - |x(t) - x(t_k)| \geq 0 \tag{4.239}$$

for $t \in [t_k, t_{k+1})$. Recall the definition of $w(t)$ in (4.228). Property (4.239) implies that

$$|w(t)| \leq \rho(|x(t)|) \tag{4.240}$$

holds for $t \in [t_k, t_{k+1})$. At this stage, it cannot be readily guaranteed that (4.240) holds for all $t \geq 0$, as $\bigcup_{k \in \mathbb{S}} [t_k, t_{k+1})$ may not cover the whole timeline, i.e., $\mathbb{R}_+ \backslash \bigcup_{k \in \mathbb{S}} [t_k, t_{k+1}) \neq \emptyset$.

As mentioned above, for physical realization of (4.240) with event-triggered sampling, a positive lower bound of the inter-sample periods should be guaranteed throughout the event-triggered control procedure, i.e., $\inf_{k \in \mathbb{S}} \{t_{k+1} - t_k\} > 0$, to avoid infinitely fast sampling.

Theorem 4.4 presents a condition on the ISS gain γ to find a ρ such that $\inf_{k \in \mathbb{S}} \{t_{k+1} - t_k\} > 0$, and the closed-loop event-triggered control system is asymptotically stable at the origin.

Theorem 4.4 *Consider the event-triggered control system (4.230) with locally Lipschitz $\bar{f}$ satisfying $\bar{f}(0, 0) = 0$ and w defined in (4.228). If Assumption 4.8 is satisfied with a γ which is Lipschitz on compact sets, then one can find a $\rho \in \mathcal{K}_\infty$ such that*

- ρ satisfies (4.235), and
- ρ^{-1} is Lipschitz on compact sets.

Moreover, with the sampling time instants triggered by (4.236) with H defined in (4.237), for any specific initial state $x(0)$, system state $x(t)$ satisfies

$$|x(t)| \leq \breve{\beta}(|x(0)|, t) \tag{4.241}$$

for all $t \geq 0$, with $\breve{\beta} \in \mathcal{KL}$, and the inter-sample periods are lower bounded by a positive constant.

Proof. With a $\gamma \in \mathcal{K}$ being Lipschitz on compact sets, one can always find a $\bar{\gamma} \in \mathcal{K}_\infty$ being Lipschitz on compact sets such that $\bar{\gamma} > \gamma$. By choosing $\rho = \bar{\gamma}^{-1}$, we have $\rho \circ \gamma = \bar{\gamma} \circ \gamma < \bar{\gamma} \circ \bar{\gamma}^{-1} < \mathrm{Id}$, and $\rho^{-1} = \bar{\gamma}$ is Lipschitz on compact sets.

Along each trajectory of the closed-loop system, for each $k \in \mathbb{S}$ with state $x(t_k)$ at time instant t_k, define

$$\Theta_1(x(t_k)) = \left\{ x \in \mathbb{R}^n : |x - x(t_k)| \leq \rho \circ (\mathrm{Id} + \rho)^{-1}(|x(t_k)|) \right\}, \tag{4.242}$$

$$\Theta_2(x(t_k)) = \left\{ x \in \mathbb{R}^n : |x - x(t_k)| \leq \rho(|x|) \right\}. \tag{4.243}$$

By directly using Lemma C.6, it can be proved that $\Theta_1(x(t_k)) \subseteq \Theta_2(x(t_k))$. An illustration with $x = [x_1, x_2]^T \in \mathbb{R}^2$ is given in Figure 4.14.

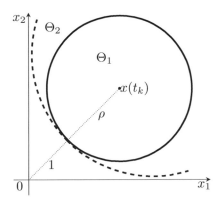

FIGURE 4.14   An illustration of $\Theta_1(x(t_k)) \subseteq \Theta_2(x(t_k))$.

Given a $\rho \in \mathcal{K}_\infty$ such that ρ^{-1} is Lipschitz on compact sets, it can be proved that $(\rho \circ (\mathrm{Id} + \rho)^{-1})^{-1} = (\mathrm{Id} + \rho) \circ \rho^{-1} = \rho^{-1} + \mathrm{Id}$ is Lipschitz on compact sets and there exists a continuous, positive function $\breve{\rho} : \mathbb{R}_+ \to \mathbb{R}_+$ such that $(\rho^{-1} + \mathrm{Id})(s) \leq \breve{\rho}(s)s := \hat{\rho}(s)$ for $s \in \mathbb{R}_+$. By using the definition of $\hat{\rho}$, one has $s = (\breve{\rho} \circ \hat{\rho}^{-1}(s)) \hat{\rho}^{-1}(s)$ and thus $\hat{\rho}^{-1}(s) = s / (\breve{\rho} \circ \hat{\rho}^{-1}(s)) := \bar{\rho}(s)s$. Here, it can be directly checked that $\bar{\rho} : \mathbb{R}_+ \to \mathbb{R}_+$ is continuous and positive. Thus,

$$\rho \circ (\mathrm{Id} + \rho)^{-1}(s) = (\rho^{-1} + \mathrm{Id})^{-1}(s) \geq \hat{\rho}^{-1}(s) = \bar{\rho}(s)s. \tag{4.244}$$

Property (4.244) implies that, if

$$|x - x(t_k)| \le \bar{\rho}(|x(t_k)|)|x(t_k)|, \tag{4.245}$$

then $x \in \Theta_1(x(t_k))$.

Also, for any $x \in \Theta_1(x(t_k))$, by using the locally Lipschitz property of $\bar{f}$, it holds that

$$
\begin{aligned}
|f(x, v(x(t_k)))| &= |\bar{f}(x, x(t_k))| \\
&= |\bar{f}(x - x(t_k) + x(t_k), x(t_k))| \\
&\le L_{\bar{f}} \left(|[x^T - x^T(t_k), x^T(t_k)]^T| \right) |[x^T - x^T(t_k), x^T(t_k)]^T| \\
&\le \bar{L}(|x(t_k)|)|x(t_k)|,
\end{aligned} \tag{4.246}
$$

where $L_{\bar{f}}, \bar{L}$ are continuous, positive functions defined on $\mathbb{R}_+$. Property (4.245) is used for the last inequality.

Then, the minimum time $T_k^{\min}$ needed for the state of the closed-loop system starting at $x(t_k)$ to go outside the region $\Theta_1(x(t_k))$ can be estimated by

$$T_k^{\min} \ge \frac{\bar{\rho}(|x(t_k)|)|x(t_k)|}{\bar{L}(|x(t_k)|)|x(t_k)|} = \frac{\bar{\rho}(|x(t_k)|)}{\bar{L}(|x(t_k)|)}, \tag{4.247}$$

which is well defined and strictly larger than zero for any $x(t_k)$. Since $\Theta_1(x(t_k)) \subseteq \Theta_2(x(t_k))$ and $x(t)$ is continuous on the timeline, the minimum interval needed for the state starting at $x(t_k)$ to go outside $\Theta_2(x(t_k))$ is not less than $T_k^{\min}$.

By directly using (4.247), one has

$$T_0^{\min} \ge \frac{\bar{\rho}(|x(0)|)}{\bar{L}(|x(0)|)}. \tag{4.248}$$

If $\mathbb{S} = \{0\}$, then $w(t)$ is continuous and (4.239) holds for $t \in [0, \infty)$. We now consider the case of $\mathbb{S} \ne \{0\}$. Suppose that for a specific $k \in \mathbb{Z}_+ \backslash \{0\}$, the event trigger (4.236) with H defined by (4.237) guarantees that for $t \in [0, t_k)$, $w(t)$ is piecewise continuous and (4.240) holds. Under condition (4.234), by using the robust stability property of ISS, one has

$$|x(t)| \le \check{\beta}(|x(0)|, t) \tag{4.249}$$

for all $t \in [0, t_k)$, with $\check{\beta} \in \mathcal{KL}$. Due to the continuity of $x(t)$ with respect to t, $x(t_k) = \lim_{t \to t_k^-} x(t)$. Thus, $|x(t_k)| \le \check{\beta}(|x(0)|, 0)$. This, together with property (4.247) implies that

$$T_k^{\min} \ge \min \left\{ \frac{\bar{\rho}(|x|)}{\bar{L}(|x|)} : |x| \le \check{\beta}(|x(0)|, 0) \right\}. \tag{4.250}$$

This means that for $t \in [0, t_{k+1})$, $w(t)$ is piecewise continuous and (4.240) holds. By induction, $w(t)$ is piecewise continuous and (4.239) holds for $t \in [0, t_{k+1})$ for any $k \in \mathbb{S}$. If $\mathbb{S}$ is an infinite set, then $\lim_{k \to \infty} t_{k+1} = \infty$ by using (4.248); if $\mathbb{S}$ is a finite set, say $\{0, \ldots, k^*\}$, then $t_{k^*+1} = \infty$. In both cases, $w(t)$ is piecewise continuous and (4.239) holds for $t \in [0, \infty)$.

With the robust stability property of ISS, property (4.249) holds for $t \in [0, \infty)$. This ends the proof. $\diamond$

The proof of Theorem 4.4 also naturally leads to a self-triggered sampling strategy, which computes t_{k+1} by using t_k and $x(t_k)$, and thus does not continuously monitor the trajectory of $x(t)$. Suppose that Assumption 4.8 is satisfied for the closed-loop system composed of (4.226) and (4.229) with locally Lipschitz $\bar{f}$. With property (4.247), given t_k and $x(t_k)$, t_{k+1} can be computed as

$$t_{k+1} = \frac{\bar{\rho}(|x(t_k)|)}{\bar{L}(|x(t_k)|)} + t_k \tag{4.251}$$

for $k \in \mathbb{Z}_+$. Based on the proof of Theorem 4.4, it can be directly verified that $\rho(|x(t)|) - |x(t) - x(t_k)| \geq 0$ holds for all $t \in [t_k, t_{k+1})$, $k \in \mathbb{Z}_+$, and all the inter-sample periods are lower bounded by a positive constant, given any specific initial state $x(0)$. With Assumption 4.8 satisfied, the state $x(t)$ ultimately converges to the origin.

Example 4.6 *The condition for event-triggered control without infinitely fast sampling can be readily fulfilled by the linear system*

$$\dot{x} = Ax + Bu \tag{4.252}$$

with $x \in \mathbb{R}^n$ as the state and $u \in \mathbb{R}^m$ as the control input, if the system is controllable. One can find a K such that $A - BK$ is Hurwitz and design $u = -K(x + w)$ with w being the measurement error caused by data sampling. Then,

$$\dot{x} = Ax - BK(x + w) = (A - BK)x - BKw. \tag{4.253}$$

With initial state $x(0)$ and input w, the solution of the closed-loop system is

$$x(t) = e^{(A-BK)t}x(0) - \int_0^t e^{(A-BK)(t-\tau)}BKw(\tau)d\tau. \tag{4.254}$$

for $t \geq 0$. It can be verified that $x(t)$ satisfies property (4.234) with $\beta(s,t) = (1 + 1/\delta)|e^{(A-BK)t}|s$ and $\gamma(s) = (1 + \delta)\left(\int_0^\infty |e^{(A-BK)\tau}|d\tau\right)s$, where δ can be selected as any positive constant. Clearly, γ is Lipschitz on compact sets.

4.4.2 EVENT-TRIGGERED CONTROL AND SELF-TRIGGERED CONTROL IN THE PRESENCE OF EXTERNAL DISTURBANCES

Theorem 4.4 does not take into account the influence of external disturbances. To study the influence of the disturbances, we consider the following system

$$\dot{x}(t) = f(x(t), u(t), d(t)), \qquad (4.255)$$

where $d(t) \in \mathbb{R}^{n_d}$ represents the external disturbances, and the other variables are defined as for (4.226). It is assumed that d is piecewise continuous and bounded. The control law and the event trigger are still in the form of (4.227) and (4.236), respectively. In this case, we still discuss the ISS condition for the realization of event-triggered control with guaranteed positive inter-sample periods.

With w defined in (4.228) as the measurement error caused by data sampling, the control law (4.227) can be rewritten as (4.229). By substituting (4.229) into (4.255), we have

$$\begin{aligned}
\dot{x}(t) &= f(x(t), v(x(t) + w(t)), d(t)) \\
&:= \bar{f}(x(t), x(t) + w(t), d(t)).
\end{aligned} \qquad (4.256)$$

Corresponding to Assumption 4.8 for the disturbance-free case, we make the following assumption on system (4.256).

Assumption 4.9 *System* (4.256) *is ISS with w and d as the inputs, that is, there exist $\beta \in \mathcal{KL}$ and $\gamma, \gamma^d \in \mathcal{K}$ such that for any initial state $x(0)$ and any piecewise continuous, bounded w and d, it holds that*

$$|x(t)| \le \max\left\{\beta(|x(0)|, t), \gamma(\|w\|_\infty), \gamma^d(\|d\|_\infty)\right\} \qquad (4.257)$$

for all $t \ge 0$.

Under Assumption 4.9, if the event trigger is still capable of guaranteeing (4.239) with $\rho \in \mathcal{K}$ such that $\rho \circ \gamma < \mathrm{Id}$, then by using the robust stability property of ISS, we can prove that

$$|x(t)| \le \max\left\{\breve{\beta}(|x(0)|, t), \breve{\gamma}^d(\|d\|_\infty)\right\} \qquad (4.258)$$

with $\breve{\beta} \in \mathcal{KL}$ and $\breve{\gamma}^d \in \mathcal{K}$. As x converges to the origin, the upper bound of $|w(t)| = |x(t_k) - x(t)|$ converges to zero according to (4.239). However, due to the presence of the external disturbance d, the system dynamics $f(x(t), v(x(t) + w(t)), d(t))$ may not converge to zero as x converges to the origin. This means that the inter-sample period $t_{k+1} - t_k$ could be arbitrarily small.

Event-Triggered Sampling with ϵ Modification

Inspired by the recent result [52], we modify the event trigger (4.236) by replacing H defined in (4.237) with

$$H(x, x') = \max\{\rho(|x|), \epsilon\} - |x - x'|, \qquad (4.259)$$

where ρ is a class $\mathcal{K}_\infty$ function satisfying $\rho \circ \gamma < \mathrm{Id}$ and constant $\epsilon > 0$. In this case, corresponding to (4.239), we have

$$|x(t) - x(t_k)| < \max\{\rho(|x(t)|), \epsilon\} \qquad (4.260)$$

for $t \in [t_k, t_{k+1})$, $k \in \mathbb{Z}_+$. With robust stability property of ISS, it holds that

$$|x(t)| \leq \max\left\{ \breve{\beta}(|x(0)|, t), \breve{\gamma}(\epsilon), \breve{\gamma}^d(\|d\|_\infty) \right\} \qquad (4.261)$$

with $\breve{\beta} \in \mathcal{KL}$ and $\breve{\gamma}, \breve{\gamma}^d \in \mathcal{K}$. It should be noted that, with $\epsilon > 0$, $t_{k+1} - t_k > 0$ is guaranteed for all $k \in \mathbb{S}$ and function ρ^{-1} is no longer required to be Lipschitz on compact sets. This result is summarized by Theorem 4.5 without proof.

Theorem 4.5 *Consider the event-triggered control system* (4.256) *with locally Lipschitz $\bar{f}$ and w defined in* (4.228). *If Assumption 4.9 is satisfied, with the sampling time instants triggered by* (4.236) *with H defined in* (4.259), *for any specific initial state $x(0)$, the system state $x(t)$ satisfies* (4.261) *for all $t \geq 0$, with $\breve{\beta} \in \mathcal{KL}$ and $\breve{\gamma}, \breve{\gamma}^d \in \mathcal{K}$, and the inter-sample periods are lower bounded by a positive constant.*

For such an event-triggered control system, even if $d \equiv 0$, only practical convergence can be guaranteed, that is, $x(t)$ can only be guaranteed to converge to within a neighborhood of the origin defined by $|x| \leq \breve{\gamma}(\epsilon)$. In the next subsection, we present a self-triggered sampling mechanism to overcome this obstacle, under the assumption of an *a priori* known upper bound of $\|d\|_\infty$.

Self-Triggered Sampling

We show that if an upper bound of $\|d\|_\infty$ is known *a priori*, then we can design a self-triggered sampling mechanism such that $x(t)$ is practically steered to within a neighborhood of the origin with size depending solely on $\|d\|_\infty$. Moreover, if $d(t)$ converges to zero, then $x(t)$ asymptotically converges to the origin.

Assumption 4.10 *There is a known constant $B^d \geq 0$ such that*

$$\|d\|_\infty \leq B^d. \qquad (4.262)$$

Lemma 4.4 on a property of locally Lipschitz functions is used in the following self-triggered control design.

Lemma 4.4 *For any locally Lipschitz function $h : \mathbb{R}^{n_1} \times \mathbb{R}^{n_2} \times \cdots \times \mathbb{R}^{n_m} \to \mathbb{R}^p$ satisfying $h(0, \ldots, 0) = 0$ and any $\varphi_1, \ldots, \varphi_m \in \mathcal{K}_\infty$ with $\varphi_1^{-1}, \ldots, \varphi_m^{-1}$ being Lipschitz on compact sets, there exists a continuous, positive, and nondecreasing function $L_h : \mathbb{R}_+ \to \mathbb{R}_+$ such that*

$$|h(z_1, \ldots, z_m)| \leq L_h \left(\max_{i=1,\ldots,m} \{|z_i|\} \right) \max_{i=1,\ldots,m} \{\varphi_i(|z_i|)\} \qquad (4.263)$$

for all z, where $z = [z_1^T, \ldots, z_m^T]^T$.

Proof. For a locally Lipschitz h satisfying $h(0, \ldots, 0) = 0$, one can always find a continuous, positive, and nondecreasing function $L_{h0} : \mathbb{R}_+ \to \mathbb{R}_+$ such that

$$|h(z_1, \ldots, z_m)| \leq L_{h0} \left(\max_{i=1,\ldots,m} \{|z_i|\} \right) \max_{i=1,\ldots,m} \{|z_i|\} \qquad (4.264)$$

for all z.

Define

$$\check{\varphi}(s) = \max_{i=1,\ldots,m} \{\varphi_i^{-1}(s)\} \qquad (4.265)$$

for $s \in \mathbb{R}_+$. Then, $\check{\varphi} \in \mathcal{K}_\infty$. Since $\varphi_1^{-1}, \ldots, \varphi_m^{-1}$ are Lipschitz on compact sets, $\check{\varphi}$ is Lipschitz on compact sets.

From the definition, one has

$$\check{\varphi} \left(\max_{i=1,\ldots,m} \{\varphi_i(|z_i|)\} \right) = \max_{i=1,\ldots,m} \{\check{\varphi} \circ \varphi_i(|z_i|)\}$$
$$\geq \max_{i=1,\ldots,m} \{\varphi_i^{-1} \circ \varphi_i(|z_i|)\}$$
$$= \max_{i=1,\ldots,m} \{|z_i|\}. \qquad (4.266)$$

With the $\check{\varphi}$ which is Lipschitz on compact sets, there exists a continuous, positive, and nondecreasing function $L_{\check{\varphi}} : \mathbb{R}_+ \to \mathbb{R}_+$ such that

$$\check{\varphi} \left(\max_{i=1,\ldots,m} \{\varphi_i(|z_i|)\} \right) \leq L_{\check{\varphi}} \left(\max_{i=1,\ldots,m} \{\varphi_i(|z_i|)\} \right) \max_{i=1,\ldots,m} \{\varphi_i(|z_i|)\}. \quad (4.267)$$

Lemma 4.4 is proved by substituting (4.266) and (4.267) into (4.264), and defining a continuous, positive, and nondecreasing L_h such that

$$L_h \left(\max_{i=1,\ldots,m} \{|z_i|\} \right) \geq L_{h0} \left(\max_{i=1,\ldots,m} \{|z_i|\} \right) L_{\check{\varphi}} \left(\max_{i=1,\ldots,m} \{\varphi_i(|z_i|)\} \right) \quad (4.268)$$

for all z. $\diamond$

Assume that $\bar{f}$ is locally Lipschitz and $\bar{f}(0, 0, 0) = 0$. Then, with Lemma 4.4, for any specific $\chi, \chi^d \in \mathcal{K}_\infty$ with $\chi^{-1}, (\chi^d)^{-1}$ being Lipschitz on compact sets, one can find a continuous, positive and nondecreasing $L_{\bar{f}}$ such that

$$|\bar{f}(x + w, x, d)| \leq L_{\bar{f}} \left(\max \{|x|, |w|, |d|\} \right) \max \{\chi(|x|), |w|, \chi^d(|d|)\} \quad (4.269)$$

for all x, w, d.

By choosing $\chi, \chi^d \in \mathcal{K}_\infty$ with $\chi^{-1}, (\chi^d)^{-1}$ being locally Lipchitz, the self-triggered sampling mechanism is designed as

$$t_{k+1} = t_k + \frac{1}{L_{\bar{f}} \left(\max \left\{ \bar{\chi}(|x(t_k)|), \bar{\chi}^d(B^d) \right\} \right)}, \tag{4.270}$$

where $\bar{\chi}(s) = \max\{\chi(s), s\}$ and $\bar{\chi}^d(s) = \max\{\chi^d(s), s\}$ for $s \in \mathbb{R}_+$.

Theorem 4.6 provides the main result of this subsection.

Theorem 4.6 *Consider the event-triggered control system* (4.256) *with locally Lipschitz $\bar{f}$ satisfying $\bar{f}(0,0,0) = 0$ and w defined in* (4.228). *If Assumption 4.9 holds with a γ being Lipschitz on compact sets, then one can find a $\rho \in \mathcal{K}_\infty$ such that*

* *ρ satisfies*

$$\rho \circ \gamma < \mathrm{Id}, \tag{4.271}$$

   and
* *ρ^{-1} is Lipschitz on compact sets.*

Moreover, under Assumption 4.10, by choosing $\chi = \rho \circ (\mathrm{Id}+\rho)^{-1}$ and $\chi^d \in \mathcal{K}_\infty$ with $(\chi^d)^{-1}$ being Lipschitz on compact sets for the self-triggered sampling mechanism (4.270), *for any specific initial state $x(0)$, the system state $x(t)$ satisfies*

$$|x(t)| \leq \max\{\breve{\beta}(|x(0)|, t), \breve{\gamma} \circ \chi^d(\|d\|_\infty), \breve{\gamma}^d(\|d\|_\infty)\} \tag{4.272}$$

for all $t \geq 0$, with $\breve{\beta} \in \mathcal{KL}$ and $\breve{\gamma}, \breve{\gamma}^d \in \mathcal{K}$, and the inter-sample periods are lower bounded by a positive constant.

Proof. Note that $\chi = \rho \circ (\mathrm{Id}+\rho)^{-1}$ implies $\chi^{-1} = \mathrm{Id}+\rho^{-1}$. If ρ^{-1} is Lipschitz on compact sets, then χ^{-1} is Lipschitz on compact sets. Also note that $(\chi^d)^{-1}$ is chosen to be Lipschitz on compact sets.

For the locally Lipschitz $\bar{f}$ satisfying $\bar{f}(0,0,0) = 0$, by using Lemma 4.4, one can find a continuous, positive, and nondecreasing $L_{\bar{f}}$ such that (4.269) holds.

We first prove that the self-triggered sampling mechanism achieves that

$$|x(t) - x(t_k)| \leq \max\{\chi(|x(t_k)|), \chi^d(\|d\|_\infty)\} \tag{4.273}$$

for $t \in [t_k, t_{k+1})$.

By taking the integration of both the sides of (4.256), one has

$$x(t) - x(t_k) = \int_{t_k}^{t} \bar{f}(x(t_k) + w(\tau), x(t_k), d(\tau)) d\tau, \tag{4.274}$$

and thus,

$$|x(t) - x(t_k)| \leq \int_{t_k}^{t} |\bar{f}(x(t_k) + w(\tau), x(t_k), d(\tau))| d\tau. \qquad (4.275)$$

Denote $\Omega(x(t_k), \|d\|_\infty)$ as the region of x such that $|x - x(t_k)| \leq \max\{\chi(|x(t_k)|), \chi^d(\|d\|_\infty)\}$. Then, the minimum time needed for $x(t)$ to go outside the region $\Omega(x(t_k), \|d\|_\infty)$ can be estimated by

$$\frac{\max\{\chi(|x(t_k)|), \chi^d(\|d\|_\infty)\}}{C(x(t_k), \|d\|_\infty)}$$

$$\geq \frac{\max\{\chi(|x(t_k)|), \chi^d(\|d\|_\infty)\}}{L_{\bar{f}}\left(\max\{\bar{\chi}(|x(t_k)|), \bar{\chi}^d(\|d\|_\infty)\}\right) \max\{\chi(|x(t_k)|), \chi^d(\|d\|_\infty)\}}$$

$$= \frac{1}{L_{\bar{f}}\left(\max\{\bar{\chi}(|x(t_k)|), \bar{\chi}^d(\|d\|_\infty)\}\right)}$$

$$\geq \frac{1}{L_{\bar{f}}\left(\max\{\bar{\chi}(|x(t_k)|), \bar{\chi}^d(B^d)\}\right)}, \qquad (4.276)$$

where $\bar{\chi}(s) = \max\{\chi(s), s\}$ and $\bar{\chi}^d(s) = \max\{\chi^d(s), s\}$ for $s \in \mathbb{R}_+$, and

$$C(x(t_k), \|d\|_\infty) = \max\{|\bar{f}(x(t_k) + w, x(t_k), d)| :$$
$$|w| \leq \max\{\chi(|x(t_k)|), \chi^d(\|d\|_\infty)\},$$
$$|d| \leq \|d\|_\infty\}. \qquad (4.277)$$

Thus, the proposed self-triggered sampling mechanism (4.270) guarantees (4.273).

With Lemma C.6, (4.273) implies

$$|w(t)| = |x(t) - x(t_k)| \leq \max\{\rho(|x(t)|), \chi^d(\|d\|_\infty)\} \qquad (4.278)$$

for $t \in [t_k, t_{k+1})$. Note that $\rho \circ \gamma < \mathrm{Id}$. Using the robust stability property of ISS and employing a similar induction procedure as for the proof of Theorem 4.4, one can prove that (4.272) holds for all $t \geq 0$. $\diamond$

With the asymptotic gain property of ISS, if $d(t)$ converges to zero, then $x(t)$ asymptotically converges to the origin.

4.4.3   EVENT-TRIGGERED CONTROL DESIGN FOR NONLINEAR UNCERTAIN SYSTEMS

The measurement feedback control design in Section 4.1 provides a solution to robust control of nonlinear systems in the presence of measurement errors. Since the key of event-triggered control is to deal with the measurement error caused by data sampling, this subsection presents a design to fulfill the requirements for event-triggered control for nonlinear uncertain systems by refining the design in Section 4.1. In this subsection, we consider the case where the systems are free of external disturbances. The proposed design can be directly extended to fulfill the requirement for self-triggered control of systems under external disturbances as discussed in Subsection 4.4.2.

Consider the following nonlinear system in the strict-feedback form:

$$\dot{x}_i(t) = x_{i+1}(t) + \Delta_i(\bar{x}_i(t)), \quad i = 1, \ldots, n-1 \tag{4.279}$$

$$\dot{x}_n(t) = u(t) + \Delta_n(\bar{x}_n(t)), \tag{4.280}$$

where $[x_1, \ldots, x_n]^T := x \in \mathbb{R}^n$ is the state, $u \in \mathbb{R}$ is the control input, and Δ_i's for $i = 1, \ldots, n$ with $\bar{x}_i = [x_1, \ldots, x_i]^T$ are unknown, locally Lipschitz functions.

Assumption 4.11 *For each $i = 1, \ldots, n$, there exists a known $\psi_{\Delta_i} \in \mathcal{K}_\infty$ which is Lipschitz on compact sets such that for all $\bar{x}_i$,*

$$|\Delta_i(\bar{x}_i)| \le \psi_{\Delta_i}(|\bar{x}_i|). \tag{4.281}$$

Define $w(t)$ as in (4.228) as the measurement error caused by data sampling. For convenience of notation, denote $w = [w_1, \ldots, w_n]^T$. In the design, we first assume the boundedness of w, i.e., the existence of $\|w\|_\infty$, denoted by w^∞. Equivalently, $\|w_i\|_\infty$, denoted by w_i^∞, exists for $i = 1, \ldots, n$. Denote $\bar{w}_i^\infty = [w_1^\infty, \ldots, w_i^\infty]^T$.

Following the approach in Section 4.1, one may design a control law such that the closed-loop system is robust to measurement error. To clarify the influence of data sampling on the closed-loop system, we slightly modify the design procedure in Section 4.1 as follows.

The basic idea of the control design in this subsection is still to transform the closed-loop system into an interconnection of ISS subsystems, and use the cyclic-small-gain theorem to guarantee the ISS of the closed-loop system. Specifically, the state variables of the ISS subsystems are defined as

$$e_1 = x_1 \tag{4.282}$$

$$e_i = \vec{d}(x_i, S_{i-1}(\bar{x}_{i-1}, \bar{w}_i^\infty)), \quad i = 2, \ldots, n, \tag{4.283}$$

and the control law is designed such that

$$u \in S_n(\bar{x}_n, \bar{w}_n^\infty), \tag{4.284}$$

where $\vec{d}(z, \Omega) := z - \arg\min_{z' \in \Omega}\{|z - z'|\}$ for any $z \in \mathbb{R}$ and any compact $\Omega \subset \mathbb{R}$, and for each $i = 1, \ldots, n$, $S_i : \mathbb{R}^i \times \mathbb{R}^i \rightsquigarrow \mathbb{R}$ is an appropriately designed set-valued map to cover the influence of the measurement errors.

Moreover, the set-valued maps are recursively defined as

$$S_1(\bar{x}_1, \bar{w}_1^\infty) = \{\kappa_1(x_1 + a_1 w_1^\infty) : |a_1| \le 1\} \tag{4.285}$$

$$S_i(\bar{x}_i, \bar{w}_i^\infty) = \{\kappa_i(x_i + a_i w_i^\infty - p_{i-1}) : |a_i| \le 1,$$
$$p_{i-1} \in S_{i-1}(\bar{x}_{i-1}, \bar{w}_{i-1}^\infty)\}$$
$$i = 2, \ldots, n, \tag{4.286}$$

where the κ_i's for $i = 1, \ldots, n$ are continuously differentiable, odd, and strictly decreasing functions.

It can be proved that for $i = 1, \ldots, n$, each e_i-subsystem can be represented by a differential inclusion as

$$\dot{e}_i \in S_i(\bar{x}_i, \bar{w}_i^\infty) + \Phi_i(\bar{x}_i, \bar{w}_i^\infty, e_{i+1}), \tag{4.287}$$

where Φ_i satisfies

$$|\Phi_i(\bar{x}_i, \bar{w}_i^\infty, e_{i+1})| \leq \psi_{\Phi_i}(|[\bar{x}_i^T, \bar{w}_i^{\infty T}, e_{i+1}]^T|) \tag{4.288}$$

for all $\bar{x}_i, \bar{w}_i^\infty, e_{i+1}$, with $\psi_{\Phi_i} \in \mathcal{K}_\infty$ being Lipschitz on compact sets.

As shown in Section 4.1, by designing the κ_i's, the e_i-subsystems can be designed to be ISS with ISS gains satisfying the cyclic-small-gain condition for the ISS of the closed-loop system with e as the state. Based on this design, we show that the closed-loop system with x as the state and w as the input is ISS with an ISS gain being Lipschitz on compact sets.

With Assumption 4.12 satisfied, as discussed in Section 4.1, by choosing the $\gamma_{e_i}^{e_k}$'s for $i = 1, \ldots, n$, $k = 1, \ldots, i-1, i+1$ to be Lipschitz on compact sets and choosing $\gamma_{e_i}^{w_k}$ for $k = 1, \ldots, i-1$ to be in the form of $\alpha_V \circ \breve{\gamma}_{e_i}^{w_k}$ with $\breve{\gamma}_{e_i}^{w_k}$ being Lipschitz on compact sets, (4.74) can be realized with $\epsilon_i = 0$ for each e_i-subsystem.

Then, one may construct an ISS-Lyapunov function in the form of (4.77) with σ_i's for $i = 1, \ldots, n$ being Lipschitz on compact sets, and represent the influence of the measurement error caused by data sampling with

$$\theta = \max_{i=1,\ldots,n} \left\{ \sigma_i \left(\max_{k=1,\ldots,i} \{ \gamma_{e_i}^{w_k}(\bar{w}_k) \} \right) \right\}. \tag{4.289}$$

With such treatment, property (4.79) still holds, which means

$$V(e(t)) \leq \max\{\beta(V(e(0)), t), \gamma(w^\infty)\}, \tag{4.290}$$

where β is a class $\mathcal{KL}$ function and

$$\gamma(s) := \max_{i=1,\ldots,n} \left\{ \sigma_i \left(\max_{k=1,\ldots,i} \{ \gamma_{e_i}^{w_k}(s) \} \right) \right\} \tag{4.291}$$

for $s \in \mathbb{R}_+$. Thus,

$$|e(t)| \leq \max\left\{ \alpha_V^{-1} \circ \beta\left(\alpha_V(|e(0)|), t\right), \alpha_V^{-1} \circ \gamma(w^\infty) \right\}. \tag{4.292}$$

According to the definitions of $e_1, \ldots, e_n$, it can be observed that the increase of w_i^∞'s for $i = 1, \ldots, n$ leads to the decrease of $|e_i|$'s for $i = 2, \ldots, n$. Notice that, in the case of $w_i^\infty = 0$ for $i = 1, \ldots, n$, $e_i = x_i - \kappa_{i-1}(e_{i-1})$ for $i = 2, \ldots, n$. Thus, if $w_i^\infty \geq 0$ for $i = 1, \ldots, n$, then

$$|e_i| \leq |x_i - \kappa_{i-1}(e_{i-1})| \leq |x_i| + |\kappa_{i-1}(e_{i-1})|. \tag{4.293}$$

Then, one can find an $\alpha_x \in \mathcal{K}_\infty$ such that

$$|e| \leq \alpha_x(|x|). \tag{4.294}$$

Also, from the definitions of $e_1, \ldots, e_n$, we have

$$|x_1| = |e_1|, \tag{4.295}$$

$$|x_i| \leq \max \left\{ | \max S_{i-1}(\bar{x}_{i-1}, \bar{w}_i^\infty) + e_i|, | \min S_{i-1}(\bar{x}_{i-1}, \bar{w}_i^\infty) - e_i| \right\},$$
$$i = 2, \ldots, n. \tag{4.296}$$

Due to the continuous differentiability of the κ_i's used for the definition of the set-valued maps S_i's, there exist functions $\alpha_e, \alpha_w \in \mathcal{K}_\infty$ which are Lipschitz on compact sets such that

$$|x| \leq \max\{\alpha_e(|e|), \alpha_w(|w^\infty|)\}. \tag{4.297}$$

By substituting (4.294) and (4.297) into (4.292), one achieves

$$|x(t)| \leq \max \left\{ \alpha_e \circ \alpha_V^{-1} \circ \beta \left(\alpha_V \circ \alpha_x(|x(0)|), t \right), \alpha_e \circ \alpha_V^{-1} \circ \gamma(w^\infty), \alpha_w(w^\infty) \right\}$$
$$:= \max\{\bar{\beta}(|x(0)|, t), \bar{\gamma}(w^\infty)\}. \tag{4.298}$$

It can be verified that $\bar{\beta} \in \mathcal{KL}$ and $\bar{\gamma} \in \mathcal{K}$.

Note that the design of the control law does not depend on $w_1^\infty, \ldots, w_n^\infty$. The control law guarantees (4.298) for all w^∞. This proves the ISS of the closed-loop system with x as the state and w as the input.

Since α_e and α_w are Lipschitz on compact sets, one can prove that $\bar{\gamma}$ is Lipschitz on compact sets by showing that $\alpha_v^{-1} \circ \gamma$ is Lipschitz on compact sets. This, according to the definition of γ, can be proved by proving that $\alpha_V^{-1} \circ \sigma_i \circ \gamma_{e_i}^{w_k}$ is Lipschitz on compact sets. Recall that each $\gamma_{e_i}^{w_k}$ for $k = 1, \ldots, i-1$ is chosen to be in the form of $\alpha_V \circ \breve{\gamma}_{e_i}^{w_k}$ with $\breve{\gamma}_{e_i}^{w_k}$ being Lipschitz on compact sets and each $\gamma_{e_i}^{w_i}$ is in the form of $\alpha_V(s/c_i)$; check (4.75). Then, $\alpha_V^{-1} \circ \sigma_i \circ \gamma_{e_i}^{w_k}$ is Lipschitz on compact sets as $\alpha_V^{-1} \circ \sigma_i \circ \alpha_V$ is Lipschitz on compact sets.

The design result in this subsection is summarized in Theorem 4.7.

Theorem 4.7 *Consider nonlinear uncertain system (4.279)–(4.280) with Assumption 4.12 satisfied. Then, one can design an event-triggered controller with the event trigger in the form of (4.236) with H defined by (4.237) and control law in the form of (4.61)–(4.63) such that infinitely fast sampling is avoided and the state x of the system is bounded and converges to the origin asymptotically.*

4.5 SYNCHRONIZATION UNDER SENSOR NOISE

The measurement feedback control design presented in Section 4.1 can also be extended for synchronization control of nonlinear uncertain systems in the presence of sensor noise.

We consider nonlinear systems in the strict-feedback form:

$$\dot{x}_{ij} = x_{i(j+1)} + f_j(X_{ij}), \quad j = 1, \ldots, n-1 \tag{4.299}$$

$$\dot{x}_{in} = u_i + f_n(X_{in}) \tag{4.300}$$

$$x_{ij}^m = x_{ij} + d_{ij}, \tag{4.301}$$

where, for $i = 1, 2$, $[x_{i1}, \ldots, x_{in}]^T := x_i \in \mathbb{R}^n$ and $u_i \in \mathbb{R}$ are the state and the control input, respectively, x_{ij}^m is the measurement of x_{ij}, d_{ij} is the measurement disturbance, $X_{ij} = [x_{i1}, \ldots, x_{ij}]^T$, f_j is an unknown locally Lipschitz function, and x_{i1} is referred to as the output of the x_i-system.

In absence of disturbances, accurate synchronization, i.e., $\lim_{t \to \infty}(x_1(t) - x_2(t)) = 0$ for any initial state $x_1(0), x_2(0)$, is usually expected. However, due to the sensor noise, perfect synchronization would not be realizable, and partial synchronization in the sense of ISS is practically meaningful.

Assumptions 4.12 and 4.13 are made on system (4.299)–(4.301).

Assumption 4.12 *For $j = 1, \ldots, n$, each f_j satisfies*

$$|f_j(X_{1j}) - f_j(X_{2j})| \le \varphi_j(|X_{1j}|, |X_{2j}|) \, \psi_j(|X_{1j} - X_{2j}|) \tag{4.302}$$

for $X_{1j}, X_{2j} \in \mathbb{R}^j$, where $\psi_j \in \mathcal{K}_\infty$ and $\varphi_j : \mathbb{R}_+ \times \mathbb{R}_+ \to \mathbb{R}_+$ is positive and nondecreasing with respect to $|X_{1j}|$ and $|X_{2j}|$.

Assumption 4.13 *For $i = 1, 2$, $j = 1, \ldots, n$, each d_{ij} satisfies*

$$|d_{ij}(t)| \le \bar{d}_{ij} \tag{4.303}$$

for $t \ge 0$, with $\bar{d}_{ij} \ge 0$ being a constant.

It should be noted that no global Lipschitz condition is assumed on the dynamics of the nonlinear systems.

For convenience of notation, denote $X_{ij}^m = [x_{i1}^m, \ldots, x_{ij}^m]^T$ for $i = 1, 2$, $j = 1, \ldots, n$.

4.5.1 RECURSIVE CONTROL DESIGN

For $j = 1, \ldots, n$, define $z_j = x_{1j} - x_{2j}$ as the synchronization error between the two systems, and define $\phi_j(X_j) = f_j(X_{1j}) - f_j(X_{2j})$ as the difference of the system dynamics. Then, each z_j-subsystem can be represented by

$$\dot{z}_j = z_{j+1} + \phi_j(X_{1j}, X_{2j}). \tag{4.304}$$

Specifically, $z_{n+1} = x_{1(n+1)} - x_{2(n+1)} = u_1 - u_2$. Define $\bar{d}_j = \bar{d}_{1j} + \bar{d}_{2j}$ for $j = 1, \ldots, n$. Then, Assumption 4.13 implies $|d_{1j} - d_{2j}| \le \bar{d}_j$. Due to the sensor noise, $z_j^m = x_{1j}^m - x_{2j}^m$ instead of z_j is available for feedback. Then, $|z_j^m - z_j| \le \bar{d}_j$.

Define $Z_j = [z_1, \ldots, z_j]^T$. Assumption 4.12 implies, for each $j = 1, \ldots, n$,

$$|\phi_j(X_{1j}, X_{2j})| \le \varphi_j(|X_{1j}|, |X_{2j}|) \psi_j(|Z_j|). \tag{4.305}$$

The z_j-subsystems ($j = 1, \ldots, n$) describe the dynamical behavior of the synchronization error system. By considering X_{1j} and X_{2j} as external inputs, the Z_n-system is in the lower-triangular form. Following a similar idea as in Section 4.1, we construct a new $[e_1, \ldots, e_n]^T$-system composed of ISS subsystems by recursively designing a nonlinear control law for the Z_n-system.

For convenience of notation, define $\bar{D}_{ij} = [\bar{d}_{i1}, \ldots, \bar{d}_{ij}]^T$ and $\bar{D}_j = [\bar{d}_1, \ldots, \bar{d}_j]^T$ for $i = 1, 2$, $j = 1, \ldots, n$, and denote $E_j = [e_1, \ldots, e_j]^T$ for $j = 1, \ldots, n$.

Initial Step: The e_1-subsystem

Define $e_1 = z_1$. Then, the e_1-subsystem can be written as

$$\dot{e}_1 = z_2 + \phi_1(X_{11}, X_{21}). \tag{4.306}$$

Define a set-valued map

$$S_1(X_{11}, X_{21}) = \Big\{ \kappa_1(X_{11} + \delta_{11}, X_{21} + \delta_{21}, z_1 + \delta_1) :$$
$$- \bar{D}_{11} \le \delta_{11} \le \bar{D}_{11}, -\bar{D}_{21} \le \delta_{21} \le \bar{D}_{21}, |\delta_1| \le \bar{d}_1 \Big\}, \tag{4.307}$$

where κ_1 is in the form of

$$\kappa_1(a_1, a_2, a_3) = \mu_1(|a_1|, |a_2|)\theta_1(a_3) \tag{4.308}$$

with $\mu_1 : \mathbb{R}_+ \times \mathbb{R}_+ \to \mathbb{R}_+$ being continuously differentiable on $(0, \infty) \times (0, \infty)$, positive and nondecreasing with respect to the two variables, and $\theta_1 : \mathbb{R} \to \mathbb{R}$ being continuously differentiable, odd, strictly decreasing, and radially unbounded. Both μ_1 and θ_1 are defined later.

Recall the definition of $\vec{d}$ in (4.51). Define

$$e_2 = \vec{d}(z_2, S_1(X_{11}, X_{21})), \tag{4.309}$$

and rewrite

$$\dot{e}_1 = z_2 - e_2 + \phi_1(X_{11}, X_{21}) + e_2, \tag{4.310}$$

where $z_2 - e_2 \in S_1(X_{11}, X_{21})$ based on (4.309).

Recursive Step: The e_j-subsystems

For convenience, $S_0(X_0) := \{0\}$. For each $k = 1, \ldots, j-1$, a set-valued map S_k is defined as

$$
\begin{aligned}
S_k(X_{1k}, X_{2k}) = \Big\{ &\kappa_k(X_{1k} + \delta_{1k}, X_{2k} + \delta_{2k}, z_k - p_{k-1} + \delta_k) : \\
&- \bar{D}_{1k} \leq \delta_{1k} \leq \bar{D}_{1k}, -\bar{D}_{2k} \leq \delta_{2k} \leq \bar{D}_{2k}, \\
&|\delta_k| \leq \bar{d}_k, p_{k-1} \in S_{k-1}(X_{1(k-1)}, X_{2(k-1)}) \Big\},
\end{aligned} \tag{4.311}
$$

where κ_k is in the form of

$$
\kappa_k(a_1, a_2, a_3) = \mu_k(|a_1|, |a_2|)\theta_k(a_3) \tag{4.312}
$$

with $\mu_k : \mathbb{R}_+ \times \mathbb{R}_+ \to \mathbb{R}_+$ being continuously differentiable on $(0, \infty) \times (0, \infty)$, positive and nondecreasing with respect to the two variables, and $\theta_k : \mathbb{R} \to \mathbb{R}$ being continuously differentiable, odd, strictly decreasing, and radially unbounded; and e_{k+1} is defined as

$$
e_{k+1} = \vec{d}(z_{k+1}, S_k(X_{1k}, X_{2k})). \tag{4.313}
$$

Lemma 4.5 *Consider the z_j-subsystems defined in (4.304). If S_k and e_{k+1} are defined as in (4.311) and (4.313) for each $k = 1, \ldots, j-1$, then, when $e_j \neq 0$, the e_j-subsystem can be represented by*

$$
\dot{e}_j \in \left\{ z_{j+1} + \phi_j^* : \phi_j^* \in \Phi_j^*(X_{1j}, X_{2j}) \right\}, \tag{4.314}
$$

where Φ_j^* is a convex, compact, and upper semi-continuous set-valued map, and for any $\phi_j^* \in \Phi_j^*(X_{1j}, X_{2j})$, it holds that

$$
|\phi_j^*| \leq \varphi_j^*(|X_{1j}|, |X_{2j}|)\psi_j^*(|[E_j^T, \bar{D}_{j-1}^T]^T|) \tag{4.315}
$$

with $\varphi_j^* : \mathbb{R}_+ \times \mathbb{R}_+ \to \mathbb{R}_+$ is positive and nondecreasing with respect to the two variables, and $\psi_j^* \in \mathcal{K}_\infty$.

The proof of Lemma 4.5 is in Appendix E.3.
Define a set-valued map

$$
\begin{aligned}
S_j(X_{1j}, X_{2j}) = \Big\{ &\kappa_j(X_{1j} + \delta_{1j}, X_{2j} + \delta_{2j}, z_j - p_{j-1} + \delta_j) : \\
&- \bar{D}_{1j} \leq \delta_{1j} \leq \bar{D}_{1j}, -\bar{D}_{2j} \leq \delta_{2j} \leq \bar{D}_{2j}, \\
&|\delta_j| \leq \bar{d}_j, p_{j-1} \in S_{j-1}(X_{1(j-1)}, X_{2(j-1)}) \Big\},
\end{aligned} \tag{4.316}
$$

where κ_k is in the form of

$$
\kappa_j(a_1, a_2, a_3) = \mu_j(|a_1|, |a_2|)\theta_j(a_3) \tag{4.317}
$$

with $\mu_j : \mathbb{R}_+ \times \mathbb{R}_+ \to \mathbb{R}_+$ being continuously differentiable on $(0, \infty) \times (0, \infty)$, positive and nondecreasing with respect to the two variables, and $\theta_j : \mathbb{R} \to \mathbb{R}$ being continuously differentiable, odd, strictly decreasing, and radially unbounded. Both μ_j and θ_j are defined later.

Define

$$e_{j+1} = \vec{d}(z_{j+1}, S_j(X_{1j}, X_{2j})). \tag{4.318}$$

Clearly, $z_{j+1} - e_{i+1} \in S_j(X_{1j}, X_{2j})$. Then, the e_j-subsystem can be represented by the differential inclusion:

$$\dot{e}_j \in \{z_{j+1} - e_{j+1} + \phi_j^* + e_{j+1} : \phi_j^* \in \Phi_j^*(X_{1j}, X_{2j})$$
$$z_{j+1} - e_{j+1} \in S_j(X_{1j}, X_{2j})\}$$
$$:= F_j(X_{1j}, X_{2j}, e_{j+1}). \tag{4.319}$$

It can be observed that $S_1(X_{11}, X_{21})$ defined in (4.307) is in the form of (4.316) with $S_0(X_{10}, X_{20}) = \{0\}$, and the e_1-subsystem defined in (4.310) is in the form of (4.319) with $\Phi_1^*(X_{11}, X_{21}) = \{\phi_1(X_{11}, X_{21})\}$.

With Lemma 4.5, the $[z_1, \ldots, z_n]^T$-system has been transformed into the $[e_1, \ldots, e_n]^T$-system with each e_j-subsystem $(j = 1, \ldots, n)$ in the form of (4.319).

4.5.2 ISS OF THE TRANSFORMED SUBSYSTEMS

Define $\alpha_V(s) = s^2/2$ for $s \in \mathbb{R}_+$. For $j = 1, \ldots, n$, each e_j-subsystem is designed to be ISS with ISS-Lyapunov function

$$V_j(e_j) = \alpha_V(|e_j|). \tag{4.320}$$

For convenience of notation, define $V_{n+1}(e_{n+1}) = \alpha_V(|e_{n+1}|)$. In the following discussions, we sometimes simply use V_j instead of $V_j(e_j)$.

Lemma 4.6 *Consider the e_j-subsystem defined in (4.319) and the set-valued map S_j defined in (4.316). For any specified $\epsilon_j > 0$, $\ell_j > 0$, $0 < c_j < 1$, $\gamma_{e_j}^{e_k}, \gamma_{e_j}^{d_k} \in \mathcal{K}_\infty$ $(k = 1, \ldots, j-1)$, and $\gamma_{e_j}^{e_{j+1}} \in \mathcal{K}_\infty$, one can find a $\mu_j : \mathbb{R}_+ \times \mathbb{R}_+ \to \mathbb{R}_+$ being continuously differentiable on $(0, \infty) \times (0, \infty)$, positive and nondecreasing, and a $\theta_j : \mathbb{R} \to \mathbb{R}$ being continuously differentiable, odd, strictly decreasing, and radially unbounded such that with $z_{j+1} - e_{j+1} \in S_j(X_{1j}, X_{2j})$, it holds that*

$$V_j \geq \max_{k=1,\ldots,j-1} \left\{ \gamma_{e_j}^{e_k}(V_k), \gamma_{e_j}^{e_{j+1}}(V_{j+1}), \gamma_{e_j}^{d_k}(\bar{d}_k), \gamma_{e_j}^{d_j}(\bar{d}_j), \epsilon_j \right\}$$
$$\Rightarrow \max_{f_j \in F_j(X_{1j}, X_{2j}, e_{j+1})} \nabla V_j(e_j) f_j \leq -\ell_j V_j(e_j), \tag{4.321}$$

where

$$\gamma_{e_j}^{d_j}(s) = \alpha_V\left(\frac{s}{c_j}\right) \tag{4.322}$$

for $s \in \mathbb{R}_+$.

The proof of Lemma 4.6 is in Appendix E.4.

4.5.3 REALIZABLE CONTROL LAWS

The true control inputs u_1 and u_2 occur in the form of $u_1 - u_2$ in the e_n-subsystem. We set $e_{n+1} = 0$ and $V_{n+1} = 0$, and choose the following control law such that $u_1 - u_2 \in S_n(X_{1n}, X_{2n})$:

$$p_1^* = \mu_1(|X_{11}^m|, |X_{21}^m|)\theta_1(z_1^m) \tag{4.323}$$

$$p_j^* = \mu_j(|X_{1j}^m|, |X_{2j}^m|)\theta_j(z_j^m - p_{j-1}^*) \tag{4.324}$$

$$u_1 - u_2 = \mu_n(|X_{1n}^m|, |X_{2n}^m|)\theta_n(z_n^m - p_{n-1}^*). \tag{4.325}$$

It is directly checked that

$$p_1^* \in S_1(X_{11}, X_{21}) \Rightarrow \cdots \Rightarrow p_j^* \in S_j(X_{1j}, X_{2j}) \Rightarrow \cdots$$
$$\Rightarrow u_1 - u_2 \in S_n(X_{1n}, X_{2n}). \tag{4.326}$$

4.5.4 CYCLIC-SMALL-GAIN SYNTHESIS

Define $e = [e_1, \ldots, e_n]^T$. With Lemma 4.6, the closed-loop system has been transformed into a network of ISS subsystems. Moreover, the gains can be designed to satisfy the cyclic-small-gain condition. Now consult the similar design in Section 4.1 for the cyclic-small-gain condition for the synchronization control problem. An ISS-Lyapunov function for the e-system can be constructed as

$$V(e) = \max_{j=1,\ldots,n} \{\sigma_j(V_j(e_j))\}, \tag{4.327}$$

where $\sigma_1(s) = s$, $\sigma_j(s) = \gamma_{e_1}^{e_2} \circ \cdots \circ \hat{\gamma}_{e_{j-1}}^{e_j}(s)$ $(j = 2, \ldots, n)$ for $s \in \mathbb{R}_+$, where the $\hat{\gamma}_{(\cdot)}^{(\cdot)}$'s are $\mathcal{K}_\infty$ functions being continuously differentiable on $(0, \infty)$, slightly larger than the corresponding $\gamma_{(\cdot)}^{(\cdot)}$'s, and still satisfying the cyclic-small-gain condition.

The influence of d_j's and ϵ_j's can be represented as

$$\theta = \max_{j=1,\ldots,n} \left\{ \sigma_j \left(\max_{k=1,\ldots,j} \left\{ \gamma_{e_j}^{d_k}(\bar{d}_k), \epsilon_j \right\} \right) \right\}. \tag{4.328}$$

Using the Lyapunov-based ISS cyclic-small-gain theorem, we have that

$$V(e) \geq \theta \Rightarrow \max_{f \in F(X_{1n}, X_{2n}, e)} \nabla V(e)f \leq -\alpha(V(e)) \tag{4.329}$$

holds wherever ∇V exists, where α is a continuous and positive definite function, and

$$F(X_{1n}, X_{2n}, e) := [F_1(X_{11}, X_{21}, e_2), \ldots, F_n(X_{1n}, X_{2n}, e_{n+1})]^T \tag{4.330}$$

with $e_{n+1} = 0$.

By choosing $\gamma_{e_j}^{d_k}$ $(j = 2, \ldots, n, \ k = 1, \ldots, j - 1)$, ϵ_j $(j = 1, \ldots, n)$ and $\gamma_{e_j}^{e_{j+1}}$ $(j = 1, \ldots, n - 1)$ (and thus σ_j $(j = 2, \ldots, n)$) small enough, we can achieve

$$\theta = \gamma_{e_1}^{d_1}(\bar{d}_1) = \alpha_V\left(\frac{\bar{d}_1}{c_1}\right). \tag{4.331}$$

Then, from (4.329), we have

$$V(e) \geq \alpha_V\left(\frac{\bar{d}_1}{c_1}\right) \Rightarrow \max_{f \in F(X_{1n}, X_{2n}, e)} \nabla V(e) f \leq -\alpha(V(e)) \tag{4.332}$$

holds wherever ∇V exists.

Property (4.332) implies that $V(e)$ ultimately converges to within the region $V(e) \leq \alpha_V(\bar{d}/c_1)$. Using the definitions of e_1, $V_1(e_1)$, $V(e)$ (see (4.320) and (4.327)), we have $|z_1| = |e_1| = \alpha_V^{-1}(V_1(e_1)) \leq \alpha_V^{-1}(V(e))$, which implies that z_1 ultimately converges to within the region $|z_1| \leq \bar{d}_1/c_1$. Notice that c_1 can be chosen arbitrarily from the interval $(0, 1)$ in the recursive design procedure. Recall that $z_1 = x_{11} - x_{21}$. By choosing c_1 to be arbitrarily close to one, $x_{11} - x_{21}$ can be steered arbitrarily close to the region $|x_{11} - x_{21}| \leq \bar{d}_1$.

The main result of this section is summarized in the following theorem.

Theorem 4.8 *Consider the two uncertain nonlinear systems defined in* (4.299)–(4.301). *Under Assumptions 4.12 and 4.13, one can design synchronization controllers in the form of* (4.323)–(4.325) *such that synchronization errors $z_j = x_{1j} - x_{2j}$ for $j = 1, \ldots, n$ are bounded and $z_1 = x_{11} - x_{21}$ can be steered arbitrarily close to the region $|x_{11} - x_{21}| \leq \bar{d}_1$.*

In this section, we only consider the synchronization problem of two nonlinear systems. It is certainly of interest and deserves more effort to generalize the design based on the cyclic-small-gain theorem to the synchronization problem of more than two systems. Notice that the agreement problem, which is closely related to the synchronization problem, is studied in Chapter 6.

4.6 APPLICATION: ROBUST ADAPTIVE CONTROL UNDER SENSOR NOISE

This chapter has mainly considered the measurement feedback control problem of two classes of nonlinear systems: the strict-feedback system and the output-feedback system. In these designs, the dynamics of the systems are assumed to be bounded by class $\mathcal{K}_\infty$ functions; see Assumptions 4.1, 4.4, 4.5, and 4.12. This means that the origin is an equilibrium of the considered systems if the inputs, including external disturbances and control inputs, are zero.

This section shows that if the origin is not an equilibrium of the input-free systems, the measurement feedback control problem can still be solved by a robust adaptive controller. A practical example is the fan speed control problem; see [66, 124].

Instead of Assumption 4.1, for strict-feedback system (4.33)–(4.35), we make the following assumption.

Assumption 4.14 *For each $i = 1, \ldots, n$, there exists a known $\psi_{\Delta_i} \in \mathcal{K}_\infty$ such that for all $\bar{x}_i, d$,*

$$|\Delta_i(\bar{x}_i, d) - \Delta_i(0, 0)| \leq \psi_{\Delta_i}(|[\bar{x}_i^T, d^T]^T|). \tag{4.333}$$

It is not restrictive to assume a known ψ_{Δ_i}. But the uncertainty of Δ_i means uncertainty of $\Delta_i(0, 0)$. By finding a known constant $\Delta_i^0 > 0$ such that $|\Delta_i(0, 0)| \leq \Delta_i^0$ and considering $\Delta_i(0, 0)$ as an external disturbance, the design proposed in Section 4.1 still works. However, since $\Delta_i(0, 0) \neq 0$, perfect convergence cannot be guaranteed even if the system is disturbance-free.

It is intuitive to introduce some adaptive mechanism to deal with the constant uncertainty $\Delta_i(0, 0)$. We propose an ISS-induced design for robust adaptive control of system (4.33)–(4.35) in the presence of sensor noise. For convenience of notation, denote $c_i = \Delta_i(0, 0)$.

First consider the x_1-subsystem. Introduce new variables x_2^* and v_1 such that $\dot{x}_2^* = v_1$. Define $e_1 = x_1$, $e_2 = x_2 - x_2^*$, and $\bar{x}_2^* = x_2^* + c_1$.

Then, we have

$$\dot{e}_1 = \bar{x}_2^* + \Delta_1(e_1, d) + e_2 \tag{4.334}$$

$$\dot{\bar{x}}_2^* = v_1. \tag{4.335}$$

It should be noted that $\bar{x}_2^*$ is unknown. By considering d and e_2 as external inputs, and e_1 as the output, the $(e_1, \bar{x}_2^*)$-system is in the output-feedback form. According to the method in Section 4.3, we employ a reduced-order observer

$$\dot{\xi}_1 = v_1 - L_1(\xi_1 + L_1(e_1 + w_1)) \tag{4.336}$$

to estimate $\bar{x}_2^* - L_1 e_1$. Here, the available $e_1 + w_1 = x_1^m$ is used. Define $\zeta_1 = \bar{x}_2^* - L_1 e_1 - \xi_1$ as the estimation error. Then, direct calculation yields:

$$\dot{\zeta}_1 = -L_1 \zeta_1 - L_1(e_2 + \Delta_1(e_1, d) - L_1 w_1), \tag{4.337}$$

which is ISS with e_1, e_2, d, w_1 as the inputs.

According to the definitions above, $\bar{x}_2^* = \zeta_1 + L_1 e_1 + \xi_1$. Consider the (e_1, ξ_1)-system

$$\dot{e}_1 = \xi_1 + \zeta_1 + L_1 e_1 + \Delta_1(e_1, d) + e_2 \tag{4.338}$$

$$\dot{\xi}_1 = v_1 - L_1(\xi_1 + L_1(e_1 + w_1)), \tag{4.339}$$

which is in the strict-feedback form. By using the method in Section 4.1, we design a control law $v_1 := v_1(e_1 + w_1, \xi_1)$ such that the (e_1, ξ_1)-system is ISS with ζ_1, d, e_2, w_1 as the inputs. Moreover, the ISS gains from ζ_1, d, e_2 can be designed to be arbitrarily small.

Define $e_2 = x_2 - x_2^*$. Then,

$$
\begin{aligned}
\dot{e}_2 &= \dot{x}_2 - \dot{x}_2^* \\
&= x_3 + \Delta_2(\bar{x}_2, d) + c_2 - v_1(x_1^m, \xi_1) \\
&:= x_3 + \Delta_2'(e_1, e_2, \zeta_1, d, w_1) + c_2',
\end{aligned}
\tag{4.340}
$$

which is in the form of the e_1-subsystem (4.334).

Recursive design can be performed until the true control input u occurs. Then, the closed-loop system is transformed into a network of ISS subsystems. The ISS gains can be appropriately designed to satisfy the cyclic-small-gain condition. A detailed proof is left to interested readers. It should be noted that this design is still valid for strict-feedback systems with ISS inverse dynamics (4.82)–(4.85).

Compared with strict-feedback systems, output-feedback systems in the form of (4.204)–(4.208) with unknown $\Delta_i(0,0,0)$ can be handled more easily.

Define

$$
\begin{aligned}
x_1' &= x_1 \tag{4.341} \\
x_{i+1}' &= x_{i+1} + \Delta_i(0,0,0), \quad \text{for } i = 1, \ldots, n-1 \tag{4.342} \\
x_{n+1}' &= u \tag{4.343}
\end{aligned}
$$

and introduce a dynamic compensator

$$
\dot{u} = v. \tag{4.344}
$$

Then, system (4.204)–(4.208) can be transformed into

$$
\begin{aligned}
\dot{z} &= \Delta_0(z, y, w) \tag{4.345} \\
\dot{x}_i' &= x_{i+1}' + \Delta_i(y, z, w) - \Delta_i(0,0,0), \quad i = 1, \ldots, n \tag{4.346} \\
\dot{x}_{n+1}' &= v \tag{4.347} \\
y &= x_1' \tag{4.348} \\
y^m &= y + d, \tag{4.349}
\end{aligned}
$$

the measurement feedback control problem of which can be readily solved with the design in Section 4.3.

4.7   NOTES

Despite its importance, the robust control of nonlinear systems in the presence of sensor noise has not received considerable attention in the present literature. Examples showing the difficulty of the problem can be found in [66, Chapter 6] and [64]. In [66], a controller is designed with set-valued maps and "flattened" Lyapunov functions following the backstepping methodology

such that the control system is ISS with respect to the measurement disturbances. However, in that result, the influence of the sensor noise grows with
the order of the system. Reference [125] studies nonlinear systems composed
of two subsystems; one is ISS and the other one is input-to-state stabilizable with respect to the measurement disturbance. In [125], the ISS of the
control system is guaranteed by the gain assignment technique introduced in
[130, 223, 123] and the nonlinear small-gain theorem proposed in [130, 126]. In
[159], it is shown that, for general nonlinear control systems under persistently
acting disturbances, the existence of smooth Lyapunov functions is equivalent
to the existence of (possibly discontinuous) feedback stabilizers which are robust with respect to small measurement errors and small additive external
disturbances. Discontinuous controllers are developed in [159] for nonlinear
systems such that the closed-loop system is insensitive to small measurement
errors. With a refined gain assignment technique, this chapter has studied the
measurement feedback control problem for nonlinear uncertain systems.

Gain assignment is a vital tool for small-gain-based nonlinear control designs [130, 223, 123]. This chapter has employed modified gain assignment
methods for robust control of nonlinear systems under sensor noise. Reference [125, Proposition 4.1] presents a gain assignment technique to guarantee
the ISS of the control system with respect to the measurement disturbance
and the gain from the measurement disturbance to the corresponding output
is assigned to be of class $\mathcal{K}_\infty$. Lemmas 4.1 and 4.6 in this chapter consider
more general cases in which the control laws are considered as selections of
appropriately designed set-valued maps.

Significant contributions have been made to the development of decentralized control theory (see e.g., [239, 129, 237, 274, 279, 147, 118] and references therein). For large-scale systems, with small-gain methods, the basic
idea is to design decentralized controllers to make the subsystems have desired gain properties. For example, in [94], the large-scale small-gain criteria
in the finite-gain setting [207] were used in stability analysis of decentralized
adaptive control. ISS small-gain methods have played an important role in
decentralized control of large-scale nonlinear systems [118]. In this chapter,
we have proposed an ISS cyclic-small-gain design for decentralized nonlinear control in the presence of sensor noise. The disturbed output-feedback
nonlinear system defined in (4.141)–(4.146) is a measurement-disturbed version of the system considered in [129, 148] and exists in mechanical systems,
e.g., the interconnected system of cart-inverted double pendulum [237]. Different from centralized systems, the input–output feedback linearization method
cannot be implemented in the decentralized system due to the dependence
of the Δ_{ij}'s on w_i. In particular, the z_i-subsystem is referred to as the nonlinear zero-dynamics systems, which forms nonlinear dynamical interactions
between the subsystems of the decentralized system. The nonlinearities in
(4.141)–(4.146) are simply assumed to be bounded by $\mathcal{K}_\infty$ functions, which
relaxes the polynomial-type growth conditions imposed in [237, 274, 279].

The decentralized reduced-order observer design in this chapter is motivated by the designs in [222, 129]. The only difference is that the measurements of the outputs are subject to sensor noise, the impact of which on the control performance has been well handled with the new cyclic-small-gain design in this chapter.

The impact of the observer-based design also extends to the estimator design for dynamic state-feedback control and the case study of robust adaptive control subject to sensor noise in this chapter. Notice that ISS has been used as a powerful tool [115] for robust adaptive control of nonlinear systems with no sensor noise. By employing neural network approximation, small-gain overcomes a circularity issue in adaptive control for non-affine pure-feedback systems in [270]. More discussions on robust control of nonlinear systems under sensor noise can be found in [187, 186, 177].

5 Quantized Nonlinear Control

In modern automatic control systems, signals are usually quantized before being transmitted via communication channels. Figure 5.1 shows the block diagram of a typical quantized control system.

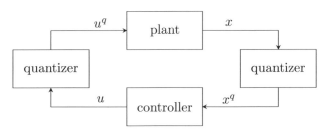

FIGURE 5.1 The block diagram of a quantized control system: u is the control input computed by the controller, x is the state of the plant, and u^q and x^q are the quantized signals of u and x, respectively.

A quantizer can be mathematically modeled as a discontinuous map from a continuous region to a discrete set of numbers. Two examples of commonly considered quantizers are shown in Figure 5.2.

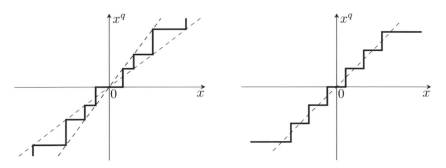

FIGURE 5.2 Two examples of quantizers.

The difference between the input and the output of a quantizer is called the quantization error. If the quantization errors are bounded, then by considering them as sensor noise, one may directly use the methods proposed in Chapter 4 to design quantized controllers. Usually, however, the quantization errors cannot be guaranteed to be bounded. For the quantizers shown in Figure

5.2, the quantization errors go to infinity as the inputs of the quantizers go to infinity. With nontrivial modifications of the methods in Chapter 4, this chapter resolves the quantized control problems for nonlinear systems.

Based on ISS cyclic-small-gain methods, in Section 5.1, a sector bound approach is developed for quantized control of nonlinear systems with static quantization. Due to the finite word-length of digital devices, practical quantizers have finite quantization levels. Section 5.2 introduces a dynamic quantization strategy such that the quantization levels can be dynamically adjusted during the control procedure for semiglobal quantized stabilization. In Section 5.3, the quantized output-feedback control problem is studied for a class of nonlinear systems taking the generalized output-feedback form. Section 5.4 gives some notes and references on quantized nonlinear control.

5.1 STATIC QUANTIZATION: A SECTOR BOUND APPROACH

This section considers the strict-feedback system as the plant:

$$\dot{x}_i = x_{i+1} + \Delta_i(\bar{x}_i), \quad 1 \le i \le n-1 \tag{5.1}$$

$$\dot{x}_n = u + \Delta_n(\bar{x}_n) \tag{5.2}$$

$$x_i^q = q_i(x_i), \quad 1 \le i \le n, \tag{5.3}$$

where $x = [x_1, \ldots, x_n]^T \in \mathbb{R}^n$ is the state, $u \in \mathbb{R}$ is the control input, $\bar{x}_i = [x_1, \ldots, x_i]^T$, x_i^q is the quantization of x_i, q_i's are state quantizers, each of which is a map from $\mathbb{R}$ to some discrete set Ω_{q_i}, and Δ_i's are unknown, locally Lipschitz functions.

Assumption 5.1 *For $1 \le i \le n$, the map $q_i : \mathbb{R} \to \Omega_i$ of the i-th quantizer is piecewise constant, and there exist known constants $0 \le b_i < 1$ and $a_i \ge 0$ such that for all $x_i \in \mathbb{R}$,*

$$|q_i(x_i) - x_i| \le b_i|x_i| + (1 - b_i)a_i. \tag{5.4}$$

One example of a quantizer which satisfies condition (5.4) is the truncated logarithmic quantizer defined as

$$q_i(x_i) = \begin{cases} \frac{(1+b_i)^{k+1}a_i}{(1-b_i)^k}, & \text{if } \frac{(1+b_i)^k a_i}{(1-b_i)^k} < x_i \le \frac{(1+b_i)^{k+1}a_i}{(1-b_i)^{k+1}}, \quad k \in \mathbb{Z}_+; \\ 0, & \text{if } 0 \le x_i \le a_i; \\ -q_i(-x_i), & \text{if } x_i < 0. \end{cases} \tag{5.5}$$

See also Figure 5.3. It should be noted that condition (5.4) can be satisfied by more general quantizers as long as their maps are bounded by sectors with an offset.

We make the following assumption on the system dynamics.

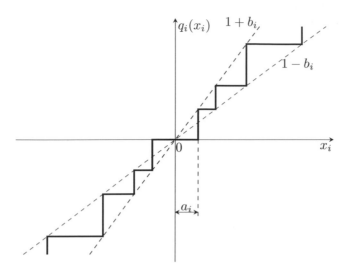

FIGURE 5.3 A truncated logarithmic quantizer: $|q_i(x_i) - x_i| \leq b_i|x_i| + (1 - b_i)a_i$ for all $x_i \in \mathbb{R}$ with $a_i \geq 0$ and $0 \leq b_i < 1$.

Assumption 5.2 *For each Δ_i $(1 \leq i \leq n)$, there exists a known $\psi_{\Delta_i} \in \mathcal{K}_\infty$ such that for all $\bar{x}_i \in \mathbb{R}^i$,*

$$|\Delta_i(\bar{x}_i)| \leq \psi_{\Delta_i}(|\bar{x}_i|). \tag{5.6}$$

The objective of this section is to find, if possible, a static quantized state-feedback controller in the form of

$$u = u(x^q) \tag{5.7}$$

with $x^q := [x_1^q, \ldots, x_n^q]^T$, for system (5.1)–(5.3) such that the closed-loop signals are bounded and the state $x_1(t)$ converges to some neighborhood of the origin whose size depends on the quantization error near the origin.

To convey the basic approach to quantized nonlinear control, this section does not consider the influence of the external disturbances. However, thanks to the natural robustness of the cyclic-small-gain design, quantized control can still be realized for the systems subject to dynamic uncertainty and external disturbance:

$$\dot{z} = g(z, x_1) \tag{5.8}$$

$$\dot{x}_i = x_{i+1} + \Delta_i(\bar{x}_i, d), \quad 1 \leq i \leq n - 1 \tag{5.9}$$

$$\dot{x}_n = u + \Delta_n(\bar{x}_n, d) \tag{5.10}$$

$$x_i^q = q_i(x_i), \quad 1 \leq i \leq n, \tag{5.11}$$

where $d \in \mathbb{R}^{n_d}$ represents external disturbance inputs, $z \in \mathbb{R}^{n_z}$ represents the state of the inverse dynamics representing dynamic uncertainties and is not measurable, and the other variables are defined the same as for system (5.1)–(5.3).

This section only considers the case where the state measurement is quantized. Actuator quantization (i.e., the control signal u is quantized before it is applied to the plant) is surely of interest. Based on the proposed design, interested readers may study quantized nonlinear control with both state quantization and actuator quantization. It should be noted that the actuator quantization error depends on the magnitude of the control signal, and it cannot be trivially treated as an external disturbance.

By considering the influence of the quantization error as an uncertain term, the closed-loop quantized system can be represented with the block diagram shown in Figure 5.4, where the operator Λ satisfies condition (5.4). For linear systems, it is standard to employ robust control designs to solve the problem. If the system is linear, or more generally, globally Lipschitz, then one may conjecture that there exists some appropriate b_i such that the closed-loop quantized system can be made robust to the quantization error. In the following discussions, global Lipschitz continuity is not assumed on the dynamics of system (5.1)–(5.3).

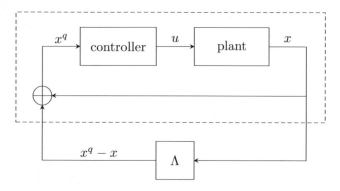

FIGURE 5.4 Quantized control as a robust control problem.

Example 5.1 shows that the quantized control problem can be solved for first-order systems based on a modification of the gain assignment technique.

Example 5.1 *Consider the system*

$$\dot{\eta} = \kappa + \phi(\eta) \tag{5.12}$$

$$\eta^q = q(\eta), \tag{5.13}$$

where $\eta \in \mathbb{R}$ is the state, $\kappa \in \mathbb{R}$ is the control input, $q : \mathbb{R} \to \mathbb{R}$ is the state quantizer, and $\phi : \mathbb{R} \to \mathbb{R}$ is an uncertain, locally Lipschitz function. Assume that there exists a known locally Lipschitz $\psi_\phi \in \mathcal{K}_\infty$ such that $|\phi(x)| \le \psi_\phi(|x|)$ for all $x \in \mathbb{R}$. Also assume the sector bound property on the quantizer, i.e., $|q(\eta) - \eta| \le b|\eta| + (1 - b)a$ with $a \ge 0$ and $0 \le b < 1$ being constants.

With Lemma C.8, for the locally Lipschitz $\psi_\phi \in \mathcal{K}_\infty$ and any specified $b \geq 0$ and $0 < c < 1$, one can always find a continuous, positive, and nondecreasing $\nu : \mathbb{R}_+ \to \mathbb{R}_+$ such that

$$(1 - b)(1 - c)\nu((1 - b)(1 - c)s)s \geq \frac{1}{2}s + \psi_\phi(s) \qquad (5.14)$$

for all $s \in \mathbb{R}_+$. The quantized control law is designed as

$$\kappa = \bar{\kappa}(\eta^q) := -\nu(|\eta^q|)\eta^q. \qquad (5.15)$$

For the closed-loop quantized system, we define a Lyapunov function candidate as $V(\eta) = \eta^2/2$. Consider the case of $V(\eta) \geq a^2/2c^2$. In this case, $|\eta| \geq a/c$. By also using $|q(\eta) - \eta| \leq b|\eta| + (1 - b)a$, it can be verified that

$$\text{sgn}(\eta^q) = \text{sgn}(\eta), \qquad (5.16)$$
$$|\eta^q| \geq (1 - b)(1 - c)|\eta|. \qquad (5.17)$$

Then,

$$\begin{aligned}
\nabla V(\eta)(\bar{\kappa}(\eta^q) + \phi(\eta)) &= \eta(-\nu(|\eta^q|)\eta^q + \phi(\eta)) \\
&\leq -\nu(|\eta^q|)|\eta^q||\eta| + |\eta||\phi(\eta)| \\
&= -|\eta|(-\nu(|\eta^q|)|\eta^q| + |\phi(\eta)|) \\
&\leq -|\eta|\Big(-\nu((1 - b)(1 - c)|\eta|)(1 - b)(1 - c)|\eta| \\
&\qquad\qquad + \psi_\phi(|\eta|)\Big) \\
&\leq -\frac{1}{2}|\eta|^2. \qquad (5.18)
\end{aligned}$$

That is,

$$V(\eta) \geq \frac{a^2}{2c^2} \Rightarrow \nabla V(\eta)(\kappa + \phi(\eta)) \leq -\frac{1}{2}|\eta|^2, \qquad (5.19)$$

which means that η can be steered to within the region $|\eta| \leq a/c$, where constant a represents the quantization error around the origin and parameter c can be chosen to be arbitrarily close to one. Theoretically, if a $= 0$, then asymptotic convergence can be achieved. Under mild conditions, such a design can also guarantee robustness to external disturbances.

This example shows that the influence of the quantization error can be attenuated through a modified gain assignment design.

This section extends the design in Example 5.1 in a recursive manner to solve the quantized control problem for system (5.1)–(5.3).

The main result of this section is summarized in Theorem 5.1.

Theorem 5.1 *Consider system* (5.1)–(5.3) *under Assumptions 5.1 and 5.2. A quantized control law in the form of* (5.7) *can be designed such that the closed-loop signals are bounded. Moreover, if $a_1 \neq 0$, then given any $0 < c_1 < 1$, the state x_1 can be steered to within the region $|x_1| \leq a_1/c_1$; if $a_1 = 0$, then given an arbitrarily small $\delta > 0$, the state x_1 can be steered to within the region $|x_1| \leq \delta$.*

The basic idea of designing the quantized control law is to transform the $[x_1, \dots, x_n]^T$-system into a new $[e_1, \dots, e_n]^T$-system through a recursive control design procedure. We employ set-valued maps to cover the discontinuity caused by the quantizers. By appropriately choosing the set-valued maps, each e_i-subsystem is designed to be ISS with the other states e_j's ($j \neq i$) as the inputs. Then, the cyclic-small-gain theorem is used to guarantee the ISS of the $[e_1, \dots, e_n]^T$-system. For convenience of discussions, denote $\bar{e}_i = [e_1, \dots, e_i]^T$ and $\bar{a}_i = [a_1, \dots, a_i]^T$.

5.1.1 RECURSIVE CONTROL DESIGN

Initial Step: The e_1-subsystem

Let $e_1 = x_1$. Rewrite the e_1-subsystem as

$$\dot{e}_1 = x_2 - e_2 + \phi_1^*(x_1, e_2), \tag{5.20}$$

where e_2 is a new state variable to be defined later and $\phi_1^*(x_1, e_2) := \Delta_1(e_1) + e_2$. Under condition (5.6), one can find a $\psi_{\phi_1^*} \in \mathcal{K}_\infty$ such that

$$|\phi_1^*(x_1, e_2)| \leq \psi_{\phi_1^*}(|\bar{e}_2|). \tag{5.21}$$

Inspired by the set-valued map design for measurement feedback control, we use set-valued maps to cover the discontinuity of quantization. With (5.4) satisfied, define set-valued maps $\check{S}_1$ and S_1 as

$$\check{S}_1(x_1) = \left\{ \kappa_1(x_1 + d_{11}) : |d_{11}| \leq b_1|x_1| + (1 - b_1)a_1 \right\} \tag{5.22}$$

$$S_1(x_1) = \left\{ d_{12}p_1 : p_1 \in \check{S}_1(x_1), \frac{1}{1 + b_2} \leq d_{12} \leq \frac{1}{1 - b_2} \right\}, \tag{5.23}$$

where κ_1 is a continuously differentiable, odd, strictly decreasing, and radially unbounded function, to be determined later.

Recall that $\vec{d}(z, \Omega) := z - \arg\min_{z' \in \Omega}\{|z - z'|\}$ for any $z \in \mathbb{R}$ and any compact $\Omega \subset \mathbb{R}$. Define

$$e_2 = \vec{d}(x_2, S_1(x_1)). \tag{5.24}$$

Then, we have $x_2 - e_2 \in S_1(x_1)$.

In the definition of $\check{S}_1$, the term d_{11} represents the impact of the quantization error $q_1(x_1) - x_1$, which satisfies (5.4). If x_2 is accurately available, following the idea of measurement feedback control in Chapter 4, one may design $\breve{S}_1$ such that if $x_2 \in \breve{S}_1(x_1)$, then the e_1-subsystem is ISS. However, $q_2(x_2)$ instead of x_2 is available. In this case, the new set-valued map $S_1(x_1)$ is employed to deal with the quantization error $q_2(x_2) - x_2$.

With a continuously differentiable κ_1, the boundaries of $S_1(x_1)$, i.e., $\max S_1(x_1)$ and $\min S_1(x_1)$, are continuously differentiable almost everywhere and the derivative of e_2 exists almost everywhere. Then, one may use a differential inclusion to represent the dynamics of the e_2-subsystem. An example of $S_1(x_1)$ is shown in Figure 5.5. Here, κ_1 is chosen to be continuously differentiable for simplicity of discussions. The design procedure is still valid if κ_1 is only locally Lipschitz, which means, κ_1 is continuously differentiable almost everywhere.

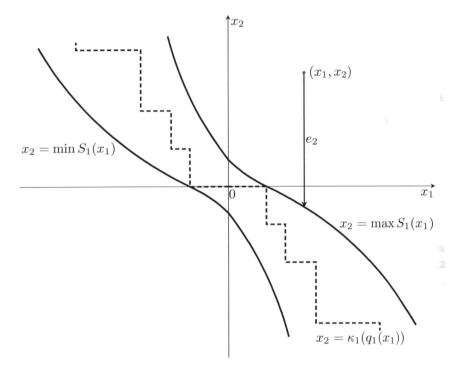

FIGURE 5.5 Set-valued map S_1 and the definition of e_2.

If $a_1 = b_1 = b_2 = 0$, then $q_1(x_1) = x_1$, $\breve{S}_1(x_1) = S_1(x_1) = \{\kappa_1(x_1)\}$ and $e_2 = x_2 - \kappa_1(x_1)$, and the set-valued map design is reduced to the function-based design in Section 2.3.

In the following procedure, the new e_i-subsystems $(2 \le i \le n)$ are derived in a recursive manner and are represented by differential inclusions.

Recursive Step: The e_i-subsystem ($2 \leq i \leq n$)

By default, $\check{S}_0(\bar{x}_0) := \{0\}$. With condition (5.4) satisfied, for each $k = 1, \ldots, i-1$, define set-valued maps $\check{S}_k$ and S_k as

$$\check{S}_k(\bar{x}_k) = \Big\{ \kappa_k(x_k - p_{k-1} + d_{k1}) : p_{k-1} \in \check{S}_{k-1}(\bar{x}_{k-1}),$$

$$|d_{k1}| \leq b_k |x_k| + (1 - b_k)a_k \Big\} \quad (5.25)$$

$$S_k(\bar{x}_k) = \Big\{ d_{k2} p_k : p_k \in \check{S}_k(\bar{x}_k), \frac{1}{1 + b_{k+1}} \leq d_{k2} \leq \frac{1}{1 - b_{k+1}} \Big\}, \quad (5.26)$$

where κ_k is continuously differentiable, odd, strictly decreasing, and radially unbounded.

For each $k = 1, \ldots, i-1$, define e_{k+1} as

$$e_{k+1} = \vec{d}(x_{k+1}, S_k(\bar{x}_k)). \quad (5.27)$$

It can be observed that e_k is defined as the directed distance from x_k to $S_{k-1}(\bar{x}_{k-1})$. However, e_k cannot be directly used for feedback. Alternatively, $x_{k+1} - p_k$ with $p_k \in \check{S}_k(\bar{x}_k)$ is employed to define set-valued map $\check{S}_k$ and thus S_k. With this treatment, if $p_k \in S_k(\bar{x}_k)$, then x_{k+1} can be steered to within $S_k(\bar{x}_k)$ by using the control design below.

Lemma 5.1 shows that with the set-valued maps defined above, each e_i-subsystem can be transformed into an appropriate form for gain assignment design.

Lemma 5.1 *Consider system* (5.1)–(5.3) *satisfying Assumptions 5.1 and 5.2. With the definitions in* (5.25)–(5.27) *for* $1 \leq k \leq i-1$, *for any variable* e_{i+1}, *when* $e_i \neq 0$, *the dynamics of each* e_i-*subsystem can be represented by the differential inclusion*

$$\dot{e}_i \in \{x_{i+1} - e_{i+1} + \phi_i^* : \phi_i^* \in \Phi_i^*(\bar{x}_i, e_{i+1})\}, \quad (5.28)$$

where $\Phi_i^*(\bar{x}_i, e_{i+1})$ *is a convex, compact, and upper semi-continuous set-valued map, and there exists a* $\psi_{\Phi_i} \in \mathcal{K}_\infty$ *such that for any* $\bar{x}_i, e_{i+1}$ *and any* $\phi_i^* \in \Phi_i^*(\bar{x}_i, e_{i+1})$,

$$|\phi_i^*| \leq \psi_{\Phi_i^*}(|[\bar{e}_{i+1}^T, \bar{a}_{i-1}^T]^T|). \quad (5.29)$$

The proof of Lemma 5.1 is in Appendix F.1. In the case without quantization, i.e., $a_i = 0$ and $b_i = 0$ for $1 \leq i \leq n$, we have $\max S_{i-1} = \min S_{i-1} = \kappa_{i-1}(e_{i-1})$. In this case, the differential inclusion (5.28) can be equivalently represented by a differential equation.

Define set-valued maps $\check{S}_i$ and S_i as in (5.25) and (5.26) with $k = i$, respectively. Specifically, in the case of $i = n$, $b_{i+1} = b_{n+1} = 0$ and

$S_i(\bar{x}_i) = S_n(\bar{x}_n) = \check{S}_n(\bar{x}_n)$. Define e_{i+1} as in (5.27) with $k = i$. Then, it always holds that

$$x_{i+1} - e_{i+1} \in S_i(\bar{x}_i). \tag{5.30}$$

By default, denote $x_{n+1} = u$.

Then, when $e_i \neq 0$, we can further represent the e_i-subsystem as

$$\dot{e}_i \in S_i(\bar{x}_i) + \Phi_i^*(\bar{x}_i, e_{i+1}). \tag{5.31}$$

Define $\Phi_1^*(x_1, e_2) = \{\phi_1^*(x_1, e_2)\}$ and $\psi_{\Phi^*} = \psi_{\phi^*}$. Then, the e_1-subsystem is also in the form of (5.31) with $\bar{a}_0 = 0$ and Φ_1^* satisfying (5.29). With Lemma 5.1, through the recursive design, the $[x_1, \ldots, x_n]^T$-system has been transformed into the $[e_1, \ldots, e_n]^T$-system with each e_i-subsystem ($1 \leq i \leq n$) in the form of (5.31).

The extended Filippov solution of each e_i-subsystem can be defined with differential inclusion (5.31) with the convex, compact, and upper semi-continuous set-valued map $S_i(\bar{x}_i) + \Phi_i^*(\bar{x}_i, e_{i+1})$.

5.1.2 QUANTIZED CONTROLLER

At Step $i = n$, the true control input u occurs, and thus we can set $e_{n+1} = 0$. Indeed, the desired quantized controller u can be chosen as follows:

$$p_1^* = \kappa_1(q_1(x_1)) \tag{5.32}$$
$$p_i^* = \kappa_i(q_i(x_i) - p_{i-1}^*), \quad 2 \leq i \leq n-1 \tag{5.33}$$
$$u = \kappa_n(q_n(x_n) - p_{n-1}^*). \tag{5.34}$$

It is directly checked that

$$p_1^* \in \check{S}_1(x_1) \Rightarrow \cdots \Rightarrow p_i^* \in \check{S}_i(\bar{x}_i) \Rightarrow \cdots \Rightarrow u \in S_n(\bar{x}_n), \tag{5.35}$$

which implies $e_{n+1} = 0$.

5.1.3 ISS OF THE TRANSFORMED SUBSYSTEMS AND CYCLIC-SMALL-GAIN THEOREM-BASED SYNTHESIS

Define $\alpha_V(s) = s^2/2$ for $s \in \mathbb{R}_+$. In this subsection, we refine the gain assignment technique and show that each e_i-subsystem ($1 \leq i \leq n$) can be rendered ISS with the ISS-Lyapunov function

$$V_i(e_i) = \alpha_V(|e_i|) \tag{5.36}$$

by appropriately choosing the κ_i's for the $\check{S}_i$'s and the S_i's.

Lemma 5.2 *Consider the e_i-subsystem ($1 \leq i \leq n$) in (5.31) with Φ_i^* satisfying (5.29). Under condition (5.4), for any specified $\epsilon_i > 0$, $0 < c_i < 1$, $\ell_i > 0$, $\gamma_{e_i}^{e_1}, \ldots, \gamma_{e_i}^{e_{i-1}}, \gamma_{e_i}^{e_{i+1}}, \chi_{e_i}^{a_1}, \ldots, \chi_{e_i}^{a_{i-1}} \in \mathcal{K}_\infty$, one can find a continuously differentiable, odd, strictly decreasing, and radially unbounded function κ_i for $S_i(\bar{x}_i)$ such that $V_i(e_i)$ satisfies*

$$V_i(e_i) \geq \max_{k=1,\ldots,i-1} \left\{ \gamma_{e_i}^{e_k}(V_k(e_k)), \gamma_{e_i}^{e_{i+1}}(V_{i+1}(e_{i+1})), \chi_{e_i}^{a_k}(a_k), \alpha_V\left(\frac{a_i}{c_i}\right), \epsilon_i \right\}$$

$$\Rightarrow \max_{f_i \in (S_i(\bar{x}_i) + \Phi_i^*(\bar{x}_i, e_{i+1}))} \nabla V_i(e_i) f_i \leq -\ell_i V_i(e_i), \tag{5.37}$$

where $V_{n+1}(e_{n+1}) = 0$.

Proof. With (5.29) satisfied, there exist $\psi_{\Phi_i^*}^{e_1}, \ldots, \psi_{\Phi_i^*}^{e_{i+1}}, \psi_{\Phi_i^*}^{a_1}, \ldots, \psi_{\Phi_i^*}^{a_{i-1}} \in \mathcal{K}_\infty$ such that for any $\phi_i^* \in \Phi_i^*(\bar{x}_i, e_{i+1})$,

$$|\phi_i^*| \leq \sum_{k=1}^{i+1} \psi_{\Phi_i^*}^{e_k}(|e_k|) + \sum_{k=1}^{i-1} \psi_{\Phi_i^*}^{a_k}(a_k). \tag{5.38}$$

Under Assumption 5.1, $0 \leq b_i, b_{i+1} < 1$. From Lemma C.8, for any $\epsilon_i > 0$ and any $0 < c_i < 1$, one can find a $\nu_i : \mathbb{R}_+ \to \mathbb{R}_+$ which is positive, nondecreasing and continuously differentiable on $(0, \infty)$, and satisfies

$$\frac{(1 - b_i)(1 - c_i)}{1 + b_{i+1}} \nu_i((1 - b_i)(1 - c_i)s)s$$

$$\geq \frac{\ell_i}{2} s + \psi_{\Phi_i^*}^{e_i}(s) + \sum_{k=1,\ldots,i-1,i+1} \psi_{\Phi_i^*}^{e_k} \circ \alpha_V^{-1} \circ (\gamma_{e_i}^{e_k})^{-1} \circ \alpha_V(s)$$

$$+ \sum_{k=1,\ldots,i-1} \psi_{\Phi_i^*}^{a_k} \circ (\chi_{e_i}^{a_k})^{-1} \circ \alpha_V(s) \tag{5.39}$$

for $s \geq \sqrt{2\epsilon_i}$.

With this kind of ν_i, define $\kappa_i(r) = -\nu_i(|r|)r$ for $r \in \mathbb{R}$. Then, κ_i is odd, strictly decreasing, radially unbounded, and continuously differentiable.

Recall that $V_i(e_i) = \alpha_V(|e_i|) = \frac{1}{2}|e_i|^2$. Consider the case of

$$V_i(e_i) \geq \max_{k=1,\ldots,i-1,i+1} \left\{ \gamma_{e_i}^{e_k}(V_k(e_k)), \chi_{e_i}^{a_k}(a_k), \alpha_V\left(\frac{a_i}{c_i}\right), \epsilon_i \right\}. \tag{5.40}$$

In this case, we have

$$|e_k| \leq \alpha_V^{-1} \circ (\gamma_{e_i}^{e_k})^{-1} \circ \alpha_V(|e_i|), \quad k = 1, \ldots, i-1, i+1 \tag{5.41}$$

$$|e_i| \geq \sqrt{2\epsilon_i} \tag{5.42}$$

$$e_i \neq 0 \tag{5.43}$$

$$a_i \leq c_i|e_i| \tag{5.44}$$

$$a_k \leq (\chi_{e_i}^{a_k})^{-1} \circ \alpha_V(|e_i|). \tag{5.45}$$

We simply use $\check{S}_k$ and S_k to denote $\check{S}_k(\bar{x}_k)$ and $S_k(\bar{x}_k)$ for $1 \leq k \leq n$. From the definition of S_{i-1}, we have

$$\max S_{i-1} \geq \max \left\{ \max \left(\frac{1}{1+b_i} \check{S}_{i-1} \right), \max \left(\frac{1}{1-b_i} \check{S}_{i-1} \right) \right\}$$

$$\min S_{i-1} \leq \min \left\{ \min \left(\frac{1}{1+b_i} \check{S}_{i-1} \right), \min \left(\frac{1}{1-b_i} \check{S}_{i-1} \right) \right\}.$$

Consider the following cases for $x_i - p_{i-1} + d_{i1}$ with $p_{i-1} \in \check{S}_{i-1}$ and $d_{i1} \leq b_i |x_i| + (1-b_i)a_i$:

a) $x_i > \max S_{i-1}$ (i.e., $e_i > 0$) and $x_i \geq 0$:

$$x_i - p_{i-1} + d_{i1} \geq (1-b_i)x_i - \max \check{S}_{i-1} - (1-b_i)a_i$$

$$= (1-b_i) \left(x_i - \max \left(\frac{1}{1-b_i} \check{S}_{i-1} \right) - a_i \right)$$

$$\geq (1-b_i)(x_i - \max S_{i-1} - a_i) = (1-b_i)(e_i - a_i).$$

b) $x_i < \min S_{i-1}$ (i.e., $e_i < 0$) and $x_i \geq 0$:

$$x_i - p_{i-1} + d_{i1} \leq (1+b_i)x_i - \min \check{S}_{i-1} + (1-b_i)a_i$$

$$= (1+b_i) \left(x_i - \min \left(\frac{1}{1+b_i} \check{S}_{i-1} \right) \right) + (1-b_i)a_i$$

$$\leq (1-b_i)(x_i - \min S_{i-1} + a_i) = (1-b_i)(e_i + a_i).$$

c) $x_i < \min S_{i-1}$ (i.e., $e_i < 0$) and $x_i \leq 0$:

$$x_i - p_{i-1} + d_{i1} \leq (1-b_i)x_i - \min \check{S}_{i-1} + (1-b_i)a_i$$

$$= (1-b_i) \left(x_i - \min \left(\frac{1}{1-b_i} \check{S}_{i-1} \right) + a_i \right)$$

$$\leq (1-b_i)(x_i - \min S_{i-1} + a_i) = (1-b_i)(e_i + a_i).$$

d) $x_i > \max S_{i-1}$ (i.e., $e_i > 0$) and $x_i \leq 0$:

$$x_i - p_{i-1} + d_{i1} \geq (1+b_i)x_i - \max \check{S}_{i-1} - (1-b_i)a_i$$

$$= (1+b_i) \left(x_i - \max \left(\frac{1}{1+b_i} \check{S}_{i-1} \right) \right) - (1-b_i)a_i$$

$$\geq (1-b_i)(x_i - \max S_{i-1} - a_i) = (1-b_i)(e_i - a_i).$$

Then, using (5.44), we have

$$|x_i - p_{i-1} + d_{i1}| \geq (1-b_i)(1-c_i)|e_i|, \qquad (5.46)$$

$$\text{sign}(x_i - p_{i-1} + d_{i1}) = \text{sign}(e_i). \qquad (5.47)$$

In the case of (5.40), for any $|d_{i1}| \leq b_i|x_i| + (1 - b_i)a_i$, $p_{i-1} \in \check{S}_{i-1}$, $1/(1 + b_{i+1}) \leq d_{i2} \leq 1/(1 - b_{i+1})$, and $\phi_i^* \in \Phi_i^*(\bar{x}_i, e_{i+1})$, using (5.39)–(5.47), we successfully achieve

$$\nabla V_i(e_i)(d_{i2}\kappa_i(x_i - p_{i-1} + d_{i1}) + \phi_i^*)$$
$$= e_i\big(-d_{i2}\nu_i(|x_i - p_{i-1} + d_{i1}|)(x_i - p_{i-1} + d_{i1}) + \phi_i^*\big)$$
$$\leq -d_{i2}\nu_i(|x_i - p_{i-1} + d_{i1}|)|x_i - p_{i-1} + d_{i1}||e_i| + |e_i||\phi_i^*|$$
$$\leq |e_i|\Big(-\frac{1}{1 + b_{i+1}}\nu_i((1 - b_i)(1 - c_i)|e_i|)(1 - b_i)(1 - c_i)|e_i|$$
$$+ \sum_{k=1}^{i+1} \psi_{\Phi_i^*}^{e_k}(|e_k|) + \sum_{k=1}^{i-1} \psi_{\Phi_i^*}^{a_k}(a_k)\Big)$$
$$\leq |e_i|\Big(-\frac{(1 - b_i)(1 - c_i)}{1 + b_{i+1}}\nu_i((1 - b_i)(1 - c_i)|e_i|)|e_i|$$
$$+ \sum_{k=1,\dots,i-1,i+1} \psi_{\Phi_i^*}^{e_k} \circ \alpha_V^{-1} \circ (\gamma_{e_i}^{e_k})^{-1} \circ \alpha_V(|e_i|)$$
$$+ \psi_{\Phi_i^*}^{e_i}(|e_i|) + \sum_{k=1}^{i-1} \psi_{\Phi_i^*}^{a_k} \circ (\chi_{e_i}^{a_k})^{-1} \circ \alpha_V(|e_i|)\Big)$$
$$\leq -\frac{\ell_i}{2}|e_i|^2 = -\ell_i V_i(e_i). \tag{5.48}$$

As a result, we obtain that

$$\max \nabla V_i(e_i)(S_i(\bar{x}_i) + \Phi_i^*(\bar{x}_i, e_{i+1})) \leq -\ell_i V_i(e_i) \tag{5.49}$$

holds almost everywhere.   ◇

Thus, the e_i-subsystem is ISS with $e_1, \dots, e_{i-1}, e_{i+1}$, a_k $(1 \leq k \leq i)$, and ϵ_i as the inputs. With Lemma 5.2, the e_i-subsystems can be rendered to be ISS one-by-one in the recursive design procedure. Furthermore, the ISS gains $\gamma_{(\cdot)}^{(\cdot)}$'s and $\chi_{(\cdot)}^{(\cdot)}$'s can be designed to be arbitrarily small or small enough to satisfy the cyclic-small-gain condition.

With the help of the cyclic-small-gain theorem, the quantized controller designed above can be fine-tuned to yield the ISS property of the closed-loop quantized system.

Denote $e = [e_1, \dots, e_n]^T$. For convenience, denote $\dot{e} = F(e, x)$. With the recursive control design procedure, the e-system is an interconnection of ISS subsystems with $a_1, \dots, a_n, \epsilon_1, \dots, \epsilon_n$ as inputs. The gain interconnection graph of the closed-loop quantized system is shown in 5.6.

In the recursive design procedure, given the $\bar{e}_{i-1}$-subsystem, by designing the set-valued maps $\check{S}_i$ and S_i for the e_i-subsystem, we can assign the ISS gains $\gamma_{e_i}^{e_k}$ for $1 \leq k \leq i-1$ to satisfy the cyclic-small-gain condition (4.76). By applying this reasoning repeatedly, we can guarantee (4.76) for all $2 \leq i \leq n$. In this way, the e-system satisfies the cyclic-small-gain condition.

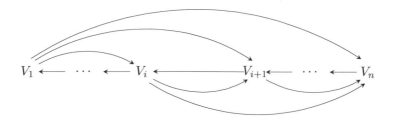

FIGURE 5.6 The gain interconnection graph of the closed-loop quantized system.

With the Lyapunov-based ISS cyclic-small-gain result in Chapter 3, an ISS-Lyapunov function can be constructed for the e-system as:

$$V(e) = \max_{1 \leq i \leq n} \{\sigma_i(V_i(e_i))\} \tag{5.50}$$

with $\sigma_1(s) = s$ and $\sigma_i(s) = \hat{\gamma}_{e_1}^{e_2} \circ \cdots \circ \hat{\gamma}_{e_{i-1}}^{e_i}(s)$ $(2 \leq i \leq n)$ for $s \in \mathbb{R}_+$, where the $\hat{\gamma}_{(\cdot)}^{(\cdot)}$'s are $\mathcal{K}_\infty$ functions being continuously differentiable on $(0, \infty)$, slightly larger than the corresponding $\gamma_{(\cdot)}^{(\cdot)}$'s and still satisfying (4.76) for all $2 \leq i \leq n$. Clearly, V is differentiable almost everywhere.

The influence of $a_1, \ldots, a_n, \epsilon_1, \ldots, \epsilon_n$ can be represented by

$$\vartheta = \max_{1 \leq i \leq n} \left\{ \sigma_i \circ \alpha_V \left(\frac{a_i}{c_i} \right), \max_{k=1,\ldots,i-1} \{\sigma_i \circ \chi_{e_i}^{a_k}(a_k)\}, \sigma_i(\epsilon_i) \right\}$$

$$:= \max \left\{ \vartheta_0, \sigma_1 \circ \alpha_V \left(\frac{a_1}{c_1} \right) \right\} = \max \left\{ \vartheta_0, \alpha_V \left(\frac{a_1}{c_1} \right) \right\}. \tag{5.51}$$

By using the Lyapunov-based cyclic-small-gain theorem (Theorem 3.1), it holds that

$$V(e) \geq \vartheta \Rightarrow \max \nabla V(e) F(e, x) \leq -\alpha(V(e)) \tag{5.52}$$

wherever $\nabla V(e)$ exists, where α is a continuous and positive definite function.

Property (5.52) implies that $V(e)$ ultimately converges to within the region $V(e) \leq \vartheta$. Recall the definitions of V_1 and V (see (5.36) and (5.50)). One can see $V(e) \geq \sigma_1(V_1(e_1)) = V_1(e_1) = \alpha_V(|e_1|)$. Thus, $x_1 = e_1$ ultimately converges to within the region $|x_1| \leq \alpha_V^{-1}(\vartheta)$.

In the recursive design procedure, we can design the $\gamma_{(\cdot)}^{(\cdot)}$'s (and thus the $\hat{\gamma}_{(\cdot)}^{(\cdot)}$'s) to be arbitrarily small to get arbitrarily small σ_i's $(2 \leq i \leq n)$. We can also design the $\chi_{e_i}^{a_k}$'s $(1 \leq k \leq i-1, 1 \leq i \leq n)$ and the ϵ_i's $(1 \leq i \leq n)$ to be arbitrarily small. In this way, if $a_1 \neq 0$, then we can design ϑ_0 to be arbitrarily small such that $\vartheta = \alpha_V(a_1/c_1)$. By choosing c_1 arbitrarily close to one, x_1 can be driven arbitrarily close to the region $|x_1| \leq a_1$. If $a_1 = 0$, then x_1 ultimately converges to within the region $|x_1| \leq \alpha_V^{-1}(\vartheta_0)$, where ϑ_0 can be designed to be arbitrarily small.

5.1.4   A NUMERICAL EXAMPLE

We employ a simple example to demonstrate the control design approach. Consider the system

$$\dot{x}_1 = x_2 + 0.5x_1 \tag{5.53}$$

$$\dot{x}_2 = u + 0.5x_2^2 \tag{5.54}$$

$$x_1^q = q_1(x_1) \tag{5.55}$$

$$x_2^q = q_2(x_2), \tag{5.56}$$

where $[x_1, x_2]^T \in \mathbb{R}^2$ is the state, $u \in \mathbb{R}$ is the control input, q_1 and q_2 are state quantizers satisfying (5.4) with $b_1 = b_2 = 0.1$ and $a_1 = a_2 = 0.2$.

Define $e_1 = x_1$. Then, the e_1-subsystem can be written as

$$\dot{e}_1 = x_2 - e_2 + (0.5e_1 + e_2), \tag{5.57}$$

where e_2 is defined as (5.24) with S_1 defined as (5.23). Define $\Phi_1^*(e_1, e_2) = \{0.5e_1 + e_2\}$. Then, $\psi_{\Phi_1^*}^{e_1}(s) = 0.5s$ and $\psi_{\Phi_1^*}^{e_2}(s) = s$. Choose $\gamma_{e_1}^{e_2}(s) = 0.95s$, $c_1 = 0.2$, and $\ell_1 = 0.89$. Choose $\nu_1(s) = 3.06$ according to (5.39). Then, $\kappa_1(r) = -\nu_1(|r|)r = -3.06r$.

Following the design procedure provided in the proof of Lemma 5.1 in Appendix F.1, the e_2-subsystem is in the following form:

$$\dot{e}_2 \in \{u + \phi_2^* : \phi_2^* \in \Phi_2^*(x_1, x_2)\}. \tag{5.58}$$

Here, we only present the calculation of $\Phi_2^*(x_1, x_2)$ in the case of $x_2 > \max S_1(x_1)$. The calculation in the case of $x_2 < \max S_1(x_1)$ is quite similar. Firstly, $\max \breve{S}_1(x_1)$ can be calculated as:

$$\max \breve{S}_1(x_1) = -3.06(x_1 - 0.1|x_1| - 0.18)$$

$$= \begin{cases} -3.06(0.9x_1 - 0.18), & \text{if } x_1 \geq 0; \\ -3.06(1.1x_1 - 0.18), & \text{if } x_1 < 0. \end{cases} \tag{5.59}$$

Then, it can be observed that $\max \breve{S}_1(x_1) > 0$ if $x_1 < 0.2$ and $\max \breve{S}_1(x_1) \leq 0$ if $x_1 \geq 0.2$. Then, we have

$$\max S_1(x_1) = \begin{cases} \frac{1}{0.9} \max \breve{S}_1(x_1), & \text{if } x_1 < 0.2; \\ \frac{1}{1.1} \max \breve{S}_1(x_1), & \text{if } x_1 \geq 0.2. \end{cases} \tag{5.60}$$

Combining (5.59) and (5.60), direct calculation yields:

$$\max S_1(x_1) = \begin{cases} -2.5036x_1 + 0.5007, & \text{if } x_1 \geq 0.2; \\ -3.06x_1 + 0.612, & \text{if } 0 \leq x_1 < 0.2; \\ -3.74x_1 + 0.612, & \text{if } x_1 < 0. \end{cases} \tag{5.61}$$

Thus, we get

$$\partial \max S_1(x_1) = \begin{cases} \{2.5036\}, & \text{if } x_1 > 0.2; \\ [2.5036, 3.06], & \text{if } x_1 = 0.2; \\ \{3.06\}, & \text{if } 0 < x_1 < 0.2; \\ [3.06, 3.74], & \text{if } x_1 = 0; \\ \{3.74\}, & \text{if } x_1 < 0. \end{cases} \quad (5.62)$$

Define $\phi_{21}(x_1, x_2) = -1.2518x_1 - 2.5036x_2 + 0.5x_2^2$, $\phi_{22}(x_1, x_2) = -1.53x_1 - 3.06x_2 + 0.5x_2^2$ and $\phi_{23}(x_1, x_2) = -1.87x_1 - 3.74x_2 + 0.5x_2^2$. Then, in the case of $x_2 > \max S_1(x_1)$, we can calculate

$$\Phi_2^*(x_1, x_2) = \begin{cases} \{\phi_{21}(x_1, x_2)\}, & \text{if } x_1 > 0.2; \\ \overline{\text{co}}\{\phi_{21}(x_1, x_2), \phi_{22}(x_1, x_2)\}, & \text{if } x_1 = 0.2; \\ \{\phi_{22}(x_1, x_2)\}, & \text{if } 0 < x_1 < 0.2; \\ \overline{\text{co}}\{\phi_{22}(x_1, x_2), \phi_{23}(x_1, x_2)\}, & \text{if } x_1 = 0; \\ \{\phi_{23}(x_1, x_2)\}, & \text{if } x_1 < 0. \end{cases} \quad (5.63)$$

Then, it can be verified that for any $\phi_2^* \in \Phi_2^*(x_1, x_2)$, $|\phi_2^*| \leq \psi_{\Phi_2^*}^{e_1}(|e_1|) + \psi_{\Phi_2^*}^{e_2}(|e_2|) + \psi_{\Phi_2^*}^{a_1}(a_1)$ with $\psi_{\Phi_2^*}^{e_1}(s) = 15.8576s$, $\psi_{\Phi_2^*}^{e_2}(s) = 3.74s + s^2$ and $\psi_{\Phi_2^*}^{a_1}(s) = 11.44s$ for $s \in \mathbb{R}_+$.

To satisfy the cyclic-small-gain condition, choose $\gamma_{e_2}^{e_1}(s) = s$. Choose $\chi_{e_2}^{a_1}(s) = \alpha_V(s/0.3)$ and $c_2 = 0.3$. Choose $\nu_2(s) = 23.5 + s$ according to (5.39). Then, $\kappa_2(r) = -\nu_2(|r|)r$.

The quantized controller is designed as

$$p_1^* = -3.06q_1(x_1) \quad (5.64)$$
$$u = -(q_2(x_2) - p_1^*)(23.5 + |q_2(x_2) - p_1^*|). \quad (5.65)$$

With the design method in Subsection 5.1.3, x_1 would ultimately converge to within the region $|x_1| \leq a_1/c_1 = 1$. Simulation results with initial conditions $x_1(0) = 1$ and $x_2(0) = -3$, shown in Figures 5.7 and 5.8, are in accordance with the theoretical results.

5.2 DYNAMIC QUANTIZATION

Practically, due to the finite word-length of digital devices, quantizers have only finite numbers of quantization levels. An example is the finite-level uniform quantizer, as shown in Figure 5.9. If the input of the quantizer is within the quantization range $M\mu$, then the quantization error is less than μ; otherwise, the output of the quantizer is saturated. Clearly, the sector bound condition assumed in Section 5.1 cannot be satisfied by the finite-level quantizers.

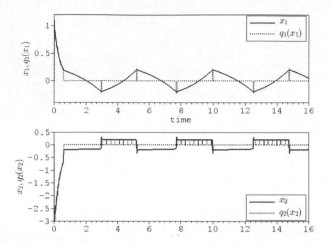

FIGURE 5.7   State trajectories of the example in Subsection 5.1.4.

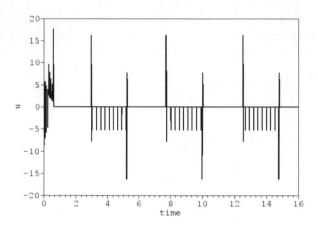

FIGURE 5.8   Control input of the example in Section 5.1.4.

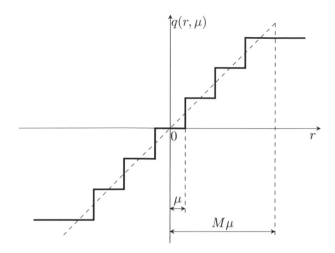

FIGURE 5.9 A uniform quantizer q with a finite number of levels: μ represents the quantization error within the quantization range $M\mu$, i.e., $|r| \le M\mu \Rightarrow |q(r,\mu) - r| \le \mu$, with M being a positive integer.

By considering μ as a variable of the quantizer, the basic idea of dynamic quantization is to dynamically update μ during the quantized control procedure for improved quantized control performance, e.g., semiglobal asymptotic stabilization. Example 5.2 shows the basic idea.

Example 5.2 *Consider a closed-loop quantized system*

$$\dot{x} = f(x, \kappa(q(x, \mu))) \qquad (5.66)$$

where $x \in \mathbb{R}$ is the state, $f : \mathbb{R}^2 \to \mathbb{R}$ is a locally Lipschitz function, $\kappa : \mathbb{R} \to \mathbb{R}$ is the control law, $q : \mathbb{R} \times \mathbb{R}_+ \to \mathbb{R}$ is the quantizer as shown in Figure 5.9 with parameter $M > 0$, and $\mu \in \mathbb{R}_+$ is the variable of the quantizer.

By defining quantization error $d(x, \mu) = q(x, \mu) - x$, the closed-loop quantized system can be rewritten as

$$\dot{x} = f(x, \kappa(x + d(x, \mu))). \qquad (5.67)$$

Assume that system (5.67) is ISS with d as the input and admits an ISS-Lyapunov function $V : \mathbb{R} \to \mathbb{R}_+$ satisfying

$$\underline{\alpha}(|x|) \le V(x) \le \overline{\alpha}(|x|) \qquad (5.68)$$

$$V(x) \ge \gamma(|d|) \Rightarrow \nabla V(x) f(x, \kappa(x + d)) \le -\alpha(V(x)), \qquad (5.69)$$

where $\underline{\alpha}, \overline{\alpha} \in \mathcal{K}_\infty$, $\gamma \in \mathcal{K}$, and α is a continuous and positive definite function. Also assume that $\underline{\alpha}$, γ, and M satisfy

$$\underline{\alpha}^{-1} \circ \gamma(\mu) \le M\mu \qquad (5.70)$$

for all $\mu \in \mathbb{R}_+$.

Consider the case of $\underline{\alpha}(M\mu) \geq V(x) \geq \gamma(\mu)$. Direct calculation yields:

$$V(x) \leq \underline{\alpha}(M\mu) \Rightarrow |x| \leq M\mu \Rightarrow |d| \leq \mu \atop V(x) \geq \gamma(\mu) \bigg\} \Rightarrow V(x) \geq \gamma(|d|). \qquad (5.71)$$

Then, by using (5.69), we have

$$\underline{\alpha}(M\mu) \geq V(x) \geq \gamma(\mu) \Rightarrow \nabla V(x)f(x,\kappa(x+d)) \leq -\alpha(V(x)). \qquad (5.72)$$

Suppose that an upper bound of $V(x(0))$ is known a priori. By choosing $\mu(0)$ such that $\underline{\alpha}(M\mu(0)) \geq V(x(0))$ and reducing μ on the timeline slowly, as shown in Figure 5.10, asymptotic stabilization can be achieved. In fact, property (5.72) defines nested invariant sets of the quantized control system, which play a central role in dynamically quantized control of nonlinear systems. The process of reducing μ is usually known as the "zooming-in" stage of dynamic quantization.

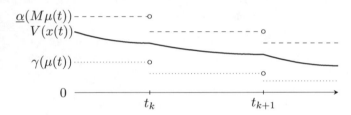

FIGURE 5.10 Basic idea of dynamic quantization.

In the case where the upper bound of $V(x(0))$ is unknown, semiglobal stabilization can be achieved for forward complete systems by employing a "zooming-out" stage, i.e., increasing μ and thus $\underline{\alpha}(M\mu)$ fast enough that $\underline{\alpha}(M\mu(t^*)) \geq V(x(t^*))$ at some finite time t^*. Very detailed discussions of this idea can be found in [164]. The dynamically quantized control design for high-order nonlinear systems in the following sections are based on these "zooming" ideas.

5.2.1 PROBLEM FORMULATION

The objective of this section is to design a new class of quantized controllers for stabilization of high-order nonlinear uncertain systems with dynamic quantization. The influence of dynamic uncertainty is also taken into account. Specifically, we consider the strict-feedback system with dynamic uncertainties:

$$\dot{z} = g(z, x_1) \qquad (5.73)$$
$$\dot{x}_i = x_{i+1} + \Delta_i(\bar{x}_i, z), \quad i = 1, \ldots, n-1 \qquad (5.74)$$
$$\dot{x}_n = u + \Delta_n(\bar{x}_n, z), \qquad (5.75)$$

where $[x_1, \ldots, x_n]^T := x \in \mathbb{R}^n$ is the measurable state, $z \in \mathbb{R}^{n_z}$ represents the state of the inverse dynamics representing dynamic uncertainties and is not measurable, $u \in \mathbb{R}$ is the control input, $\bar{x}_i = [x_1, \ldots, x_i]^T$, and Δ_i's $(i = 1, \ldots, n)$ are unknown locally Lipschitz continuous functions. We consider the general case in which both the measurement x and the control input u are quantized.

To realize quantized control with partial-state feedback, we assume that system (5.73)–(5.75) is unboundedness observable (UO) and small-time final-state norm-observable with x as the output. The definition of UO is given in Definition 1.12. The notion of small-time final-state norm-observability is recalled from [85] and is closely linked to the notion of UO [128]. It is of interest to note that a dynamic system which is small-time final-state norm-observable is automatically UO, but the converse statement is not true. See also [247] for detailed discussions of observability notions for nonlinear systems in the framework of ISS.

Definition 5.1 *Consider a dynamic system $\dot{x} = f(x)$, $y = h(x)$ where $x \in \mathbb{R}^n$ is the state, $y \in \mathbb{R}^m$ is the output, $f : \mathbb{R}^n \to \mathbb{R}^n$ is a locally Lipschitz function with $f(0) = 0$, and $h : \mathbb{R}^n \to \mathbb{R}^m$ is a continuous function with $h(0) = 0$. The system is said to be small-time final-state norm-observable if for any $\tau > 0$, there exists $\gamma \in \mathcal{K}_\infty$ such that*

$$|x(\tau)| \leq \gamma(\|y\|_{[0,\tau]}), \quad \forall x(0) \in \mathbb{R}^n. \tag{5.76}$$

Throughout the section, the following assumptions are made on system (5.73)–(5.75).

Assumption 5.3 *System (5.73)–(5.75) with $u = 0$ is forward complete and small-time final-state norm-observable with x as the output, i.e., for $u = 0$,*

$$\forall t_d > 0 \ \exists \varphi \in \mathcal{K}_\infty \text{ such that}$$
$$|X(t_d)| \leq \varphi(\|x\|_{[0,t_d]}), \quad \forall X(0) \in \mathbb{R}^{n+n_z}, \tag{5.77}$$

where $X := [z^T, x^T]^T$.

Assumption 5.3 is needed for semiglobal quantized stabilization. However, it is important to mention that Assumption 5.3 is not needed if the bounds of the initial state of system (5.73)–(5.75) are known *a priori*. See the discussions on dynamic quantization in Subsection 5.2.8.

Assumptions 5.4 and 5.5 are made on the system dynamics.

Assumption 5.4 *For each Δ_i with $i = 1, \ldots, n$, there exists a known $\lambda_{\Delta_i} \in \mathcal{K}_\infty$ such that for all $\bar{x}_i$, z,*

$$|\Delta_i(\bar{x}_i, z)| \leq \lambda_{\Delta_i}(|(\bar{x}_i, z)|). \tag{5.78}$$

Assumption 5.5 *The z-subsystem* (5.73) *with x_1 as the input admits an ISS-Lyapunov function $V_0 : \mathbb{R}^{n_z} \to \mathbb{R}_+$ which is locally Lipschitz on $\mathbb{R}^{n_z} \setminus \{0\}$ and satisfies the following:*

1. *there exist $\underline{\alpha}_0, \overline{\alpha}_0 \in \mathcal{K}_\infty$ such that*

$$\underline{\alpha}_0(|z|) \leq V_0(z) \leq \overline{\alpha}_0(|z|), \quad \forall z \in \mathbb{R}^{n_z}; \tag{5.79}$$

2. *there exist a $\chi_z^{x_1} \in \mathcal{K}$ and a continuous and positive definite α_0 such that*

$$V_0(z) \geq \chi_z^{x_1}(|x_1|) \Rightarrow \nabla V_0(z) g(z, x_1) \leq -\alpha_0(V_0(z)) \tag{5.80}$$

wherever ∇V_0 exists.

Under Assumption 5.5, system (5.73)–(5.75) represents an important class of minimum-phase nonlinear systems, which have been studied extensively by many researchers in the context of (non-quantized) robust and adaptive nonlinear control [153].

5.2.2   QUANTIZATION

This subsection provides a more detailed description of the quantizer shown in Figure 5.9. A quantizer $q(r, \mu)$ is defined as $q(r, \mu) = \mu q^o(r/\mu)$, where $r \in \mathbb{R}$ is the input of the quantizer, $\mu > 0$ is a variable to be explained later, and $q^o : \mathbb{R} \to \mathbb{R}$ is a piecewise constant function. Specifically, there exists a constant $M > 0$ such that

$$|q^o(a) - M| \leq 1, \quad \text{if } a > M; \tag{5.81}$$
$$|q^o(a) - a| \leq 1, \quad \text{if } |a| \leq M; \tag{5.82}$$
$$|q^o(a) + M| \leq 1, \quad \text{if } a < -M; \tag{5.83}$$
$$q^o(0) = 0. \tag{5.84}$$

Then, quantizer $q(r, \mu)$ satisfies:

$$|q(r, \mu) - M\mu| \leq \mu, \quad \text{if } r > M\mu; \tag{5.85}$$
$$|q(r, \mu) - r| \leq \mu, \quad \text{if } |r| \leq M\mu; \tag{5.86}$$
$$|q(r, \mu) + M\mu| \leq \mu, \quad \text{if } r < -M\mu; \tag{5.87}$$
$$q(0, \mu) = 0. \tag{5.88}$$

$M\mu$ is the quantization range of quantizer $q(r, \mu)$, and μ represents the largest quantization error when $|r| \leq M\mu$. Clearly, the quantizer shown in Figure 5.9 satisfies properties (5.85)–(5.88).

In several existing quantized control results (see e.g., [164]), two positive parameters, say M', δ', are used to formulate a quantizer q' as:

$$|q'(r, \mu') - r| \leq \delta'\mu', \quad \text{if } |r| \leq M'\mu'; \tag{5.89}$$
$$|q'(r, \mu')| > (M' - \delta')\mu', \quad \text{if } |r| > M'\mu'. \tag{5.90}$$

Actually, a quantizer satisfying (5.85)–(5.87) has such properties if the variables are appropriately defined. Indeed, by defining $M = M'/\delta'$, $\mu = \delta'\mu'$ and a new quantizer $q(r, \mu) = q'(r, \mu/\delta')$, properties (5.85)–(5.87) hold for the new quantizer q. Moreover, properties (5.85) and (5.87) explicitly represent the saturation property of the quantizer, and are quite useful in realizing the recursive design in this section.

As shown in Example 5.2, given fixed M, the basic idea of dynamic quantization is to dynamically update μ (and thus $M\mu$) to improve the control performance. Increasing μ, referred to as zooming-out, enlarges μ and thus the quantization range $M\mu$. Decreasing μ, referred to as zooming-in, reduces μ and $M\mu$. According to the literature, μ is called the zooming variable. With the design in this section, the zooming variables are updated in discrete time.

5.2.3 QUANTIZED CONTROLLER STRUCTURE AND CONTROL OBJECTIVE

We introduce a new quantized control structure, which is a natural extension of the ISS small-gain design without quantization. With the ISS small-gain design method in Chapter 2, we can recursively design a non-quantized controller for system (5.73)–(5.75) as

$$v_i = \breve{\kappa}_i(x_i - v_{i-1}), \quad i = 1, \ldots, n - 1 \tag{5.91}$$

$$u = \breve{\kappa}_n(x_n - v_{n-1}), \tag{5.92}$$

where $v_0 = 0$ and $\breve{\kappa}_i$'s for $i = 1, \ldots, n$ are appropriately chosen continuous functions. The maps defined in (5.91) are usually called virtual control laws, and (5.92) defines the true control law.

Our solution to the quantized control problem for system (5.73)–(5.75) is to add quantizers before and after each (virtual) control law defined in (5.91)–(5.92). Based on this idea, the quantized controller is in the form of

$$v_i = q_{i2}(\kappa_i(q_{i1}(x_i - v_{i-1}, \mu_{i1})), \mu_{i2}), \quad i = 1, \ldots, n - 1 \tag{5.93}$$

$$u = q_{n2}(\kappa_n(q_{n1}(x_n - v_{n-1}, \mu_{n1})), \mu_{n2}), \tag{5.94}$$

where $v_0 = 0$, the q_{ij}'s are quantizers with zooming variables μ_{ij}'s for $i = 1, \ldots, n$, $j = 1, 2$, and the κ_i's for $i = 1, \ldots, n$ are nonlinear functions. Each κ_i in (5.93)–(5.94) is not necessarily the same as the corresponding $\breve{\kappa}_i$ in (5.91)–(5.92) due to the implementation of the quantizers. The block diagram of the proposed quantized control system is shown in Figure 5.11.

In Assumption 5.6, each quantizer q_{ij} is assumed to have properties in the form of (5.85)–(5.88).

Assumption 5.6 *For $i = 1, \ldots, n$, $j = 1, 2$, each quantizer q_{ij} with zooming*

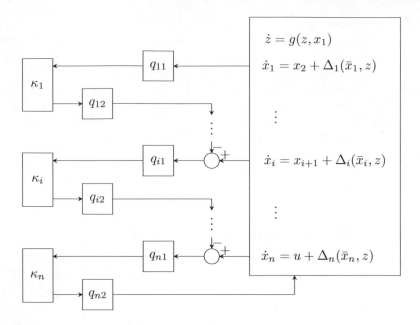

FIGURE 5.11 The quantized control structure for high-order nonlinear systems.

variable μ_{ij} satisfies

$$|q_{ij}(r, \mu_{ij}) - M_{ij}\mu_{ij}| \le \mu_{ij}, \quad \text{if } r > M_{ij}\mu_{ij}; \tag{5.95}$$

$$|q_{ij}(r, \mu_{ij}) - r| \le \mu_{ij}, \quad \text{if } |r| \le M_{ij}\mu_{ij}; \tag{5.96}$$

$$|q_{ij}(r, \mu_{ij}) + M_{ij}\mu_{ij}| \le \mu_{ij}, \quad \text{if } r < -M_{ij}\mu_{ij}; \tag{5.97}$$

$$q_{ij}(0, \mu_{ij}) = 0, \tag{5.98}$$

where $M_{i1} > 2$ and $M_{i2} > 1$.

The assumption on the parameters M_{i1} and M_{i2} is not restrictive. From Figure 5.12, it can be observed that the simplest three-level quantizer satisfies (5.85)–(5.88) with $M = 3$.

In the dynamic quantization design in this section, the zooming variables μ_{ij} for $i = 1, \dots, n$, $j = 1, 2$ are piecewise constant and updated in discrete-time. Without loss of generality, they are assumed to be right-continuous on the timeline. The dynamic quantization is composed of two stages: zooming-out and zooming-in. To simplify the discussions, the time sequences for the updates of all the zooming variables are designed to be the same and denoted by $\{t_k\}_{k \in \mathbb{Z}_+}$, in which $t_{k+1} - t_k = t_d$ with constant $t_d > 0$.

The update law of each μ_{ij} is expected to be in the following form:

$$\mu_{ij}(t_{k+1}) = Q_{ij}(\mu_{ij}(t_k)), \quad k \in \mathbb{Z}_+. \tag{5.99}$$

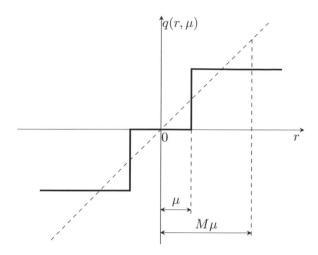

FIGURE 5.12   Three-level uniform quantizer with $M = 3$.

In the zooming-out stage, $Q_{ij} = Q_{ij}^{\text{out}}$; in the zooming-in stage, $Q_{ij} = Q_{ij}^{\text{in}}$.

The goal is to design a quantized controller in the form of (5.93)–(5.94) with dynamic quantization (5.99) to semiglobally stabilize system (5.73)–(5.75) such that all the signals including the state x in the closed-loop quantized system are bounded, and moreover, to steer x_1 to an arbitrarily small neighborhood of the origin.

5.2.4   RECURSIVE CONTROL DESIGN WITH SET-VALUED MAPS FOR STATIC QUANTIZATION

A fundamental technical obstacle for quantized feedback control design is that the quantized control system in question must be made robust with respect to the quantization errors. The nonlinearity and dimensionality of system (5.73)–(5.75) and the saturation and discontinuity of quantization together cause the major difficulties.

This subsection develops a recursive design procedure for κ_i in (5.93)–(5.94) by taking into account the effects of *static* quantization, such that the closed-loop quantized system admits nested invariant sets for further dynamic quantization designs. Set-valued maps are still used to handle the discontinuity of quantization. With appropriately designed set-valued maps, the closed-loop quantized system is transformed into a network of e_i-subsystems represented by differential inclusions. Moreover, each e_i-subsystem is designed to be ISS with gains satisfying the cyclic-small-gain condition. More importantly, it is shown in Subsection 5.2.6 that the design guarantees the existence of nested invariant sets for the closed-loop quantized system to realize dynamic quantization.

This subsection focuses on the influence of quantization error, and assumes that the zooming variables are constant.

Assumption 5.7 *For $i = 1, \ldots, n$, $j = 1, 2$, each zooming variable μ_{ij} is constant on the timeline.*

Note that Assumption 5.7 is removed for dynamic quantization design in Subsection 5.2.8.

Initial Step: The e_1-subsystem

Let $e_1 = x_1$. The e_1-subsystem is in the following form:

$$\dot{e}_1 = x_2 + \Delta_1(\bar{x}_1, z). \tag{5.100}$$

Define a set-valued map S_1 as

$$S_1(\bar{x}_1, \mu_{11}, \mu_{12}) = \left\{ \kappa_1(x_1 + b_{11}) + b_{12} : |b_{11}| \leq \max\{c_{11}|e_1|, \mu_{11}\}, |b_{12}| \leq \mu_{12} \right\}, \tag{5.101}$$

where κ_1 is a continuously differentiable, odd, strictly decreasing, and radially unbounded function and $0 < c_{11} < 1$ is a constant, both of which are determined later. It should be noted that b_{11}, b_{12} defined in (5.101) are used as auxiliary variables to define set-valued map S_1.

Recall that $\vec{d}(\xi, \Omega) := \xi - \arg\min_{\xi' \in \Omega}\{|\xi - \xi'|\}$ for any $\xi \in \mathbb{R}$ and any $\Omega \subset \mathbb{R}$.

Define

$$e_2 = \vec{d}(x_2, S_1(\bar{x}_1, \mu_{11}, \mu_{12})). \tag{5.102}$$

Rewrite the e_1-subsystem (5.100) as

$$\dot{e}_1 = x_2 - e_2 + \Delta_1(\bar{x}_1, z) + e_2, \tag{5.103}$$

where $x_2 - e_2 \in S_1(\bar{x}_1, \mu_{11}, \mu_{12})$ due to (5.102).

It is necessary to give a detailed description of the set-valued map S_1. Consider the first-order nonlinear system $\dot{e}_1 = x_2 + \Delta_1(\bar{x}_1, z)$. Recall the definition $e_1 = x_1$. With the gain assignment technique in Subsection 2.3, one can design a control law $x_2 = \kappa_1(x_1)$ to stabilize the e_1-system. In the existence of quantization errors, control law $x_2 = \kappa_1(x_1)$ should be modified as $x_2 = q_{12}(\kappa_1(q_{11}(x_1, \mu_{11})), \mu_{12})$. The set-valued map S_1 takes into account the quantization errors of both the quantizers q_{11} and q_{12}. As shown below, the motion of the new variable e_2 defined based on S_1 can be represented by a differential inclusion, and the problem caused by the discontinuity of quantization is solved. In the control design procedure below, we still use set-valued maps to deal with the discontinuity of quantization.

Recursive Step: The e_i-subsystems

Denote $\bar{\mu}_{i1} = [\mu_{11}, \ldots, \mu_{i1}]^T$ and $\bar{\mu}_{i2} = [\mu_{12}, \ldots, \mu_{i2}]^T$ for $i = 1, \ldots, n$. For each $i = 2, \ldots, n$, define a set-valued map S_i as

$$S_i(\bar{x}_i, \bar{\mu}_{i1}, \bar{\mu}_{i2}) = \Big\{ \kappa_i(x_i - \varsigma_{i-1} + b_{i1}) + b_{i2} : \varsigma_{i-1} \in S_{i-1}(\bar{x}_{i-1}, \bar{\mu}_{(i-1)1}, \bar{\mu}_{(i-1)2}),$$

$$|b_{i1}| \leq \max\{c_{i1}|e_i|, \mu_{i1}\}, |b_{i2}| \leq \mu_{i2} \Big\}, \tag{5.104}$$

where κ_i is a continuously differentiable, odd, strictly decreasing, and radially unbounded function and $0 < c_{i1} < 1$ is a constant. Both κ_i and c_{i1} are determined later. The definition of S_i guarantees its convexity, compactness, and upper semi-continuity of the set-valued map S_i. Here, b_{i1}, b_{i2} are auxiliary variables used to define the set-valued map S_i.

It can be observed that $S_1(\bar{x}_1, \bar{\mu}_{11}, \bar{\mu}_{12})$, defined in (5.101), is also in the form of (5.104) with $S_0(\bar{x}_0, \bar{\mu}_{01}, \bar{\mu}_{02}) := \{0\}$.

For each $i = 2, \ldots, n$, define e_{i+1} as

$$e_{i+1} = \vec{d}(x_{i+1}, S_i(\bar{x}_i, \bar{\mu}_{i1}, \bar{\mu}_{i2})). \tag{5.105}$$

Lemma 5.3 shows that with the recursive definitions of set-valued maps in (5.104) and new state variables in (5.105), the e_i-subsystems with $i = 1, \ldots, n$ can be represented by differential inclusions with specific properties.

Lemma 5.3 *Consider the $(x_1, \ldots, x_n)$-system in (5.74)–(5.75). Under Assumptions 5.4 and 5.7, with the definitions in (5.101), (5.102), (5.104), and (5.105), each e_i-subsystem for $1 \leq i \leq n$ can be represented with the differential inclusion:*

$$\dot{e}_i \in S_i(\bar{x}_i, \bar{\mu}_{i1}, \bar{\mu}_{i2}) + \Phi_i^*(e_{i+1}, \bar{x}_i, \bar{\mu}_{(i-1)1}, \bar{\mu}_{(i-1)2}, z), \tag{5.106}$$

where Φ_i^* is a convex, compact and upper semi-continuous set-valued map, and there exists a $\lambda_{\Phi_i^*} \in \mathcal{K}_\infty$ such that for all $(e_{i+1}, \bar{x}_i, \bar{\mu}_{(i-1)1}, \bar{\mu}_{(i-1)2}, z)$, any $\phi_i^* \in \Phi_i^*(e_{i+1}, \bar{x}_i, \bar{\mu}_{(i-1)1}, \bar{\mu}_{(i-1)2}, z)$ satisfies

$$|\phi_i^*| \leq \lambda_{\Phi_i^*}(|(\bar{e}_{i+1}, \bar{\mu}_{(i-1)1}, \bar{\mu}_{(i-1)2}, z)|), \tag{5.107}$$

where $\bar{e}_i := [e_1, \ldots, e_i]^T$.

The proof of Lemma 5.3 is in Appendix F.

With Lemma 5.3, through the recursive design approach, the $(x_1, \ldots, x_n)$-system has been transformed into the new $(e_1, \ldots, e_n)$-system with each e_i-subsystem $(i = 1, \ldots, n)$ in the form of (5.106). The extended Filippov solution of each e_i-subsystem can be defined with differential inclusion (5.106), because set-valued maps S_i and Φ_i^* are convex, compact, and upper semi-continuous.

ISS of the Subsystems

Denote $e_0 = z$. Then,

$$\dot{e}_0 = g(e_0, e_1). \tag{5.108}$$

Define $\gamma_{e_0}^{e_1}(s) = \chi_z^{x_1} \circ \alpha_V^{-1}(s)$ for $s \in \mathbb{R}_+$. Under Assumption 5.5, it holds that

$$V_0(e_0) \geq \gamma_{e_0}^{e_1}(V_1(e_1)) \Rightarrow \nabla V_0(e_0) g(e_0, e_1) \leq -\alpha_0(V_0(e_0)) \tag{5.109}$$

wherever ∇V_0 exists.

For each e_i-subsystem with $i = 1, \ldots, n$, define the following ISS-Lyapunov function candidate

$$V_i(e_i) = \alpha_V(|e_i|), \tag{5.110}$$

where $\alpha_V(s) = s^2/2$ for $s \in \mathbb{R}_+$. For convenience of notation, define $V_{n+1}(e_{n+1}) = \alpha_V(|e_{n+1}|)$.

Lemma 5.4 states that, for $i = 1, \ldots, n$, by appropriately choosing κ_i, each e_i-subsystem can be rendered to be ISS with V_i defined in (5.110) as an ISS-Lyapunov function. Furthermore, the ISS gains from μ_{i1} and μ_{i2} to e_i satisfy specific conditions to guarantee the existence of nested invariant sets.

Lemma 5.4 *Consider the e_i-subsystem $(i = 1, \ldots, n)$ in the form of (5.106) with S_i defined in (5.101) and (5.104). Under Assumptions 5.4 and 5.7, for any specified constants $\epsilon_i > 0$, $\iota_i > 0$, $0 < c_{i1}, c_{i2} < 1$, $\gamma_{e_i}^{e_k} \in \mathcal{K}_\infty$ for $k = 0, \ldots, i-1, i+1$, and $\gamma_{e_i}^{\mu_{k1}}, \gamma_{e_i}^{\mu_{k2}} \in \mathcal{K}_\infty$ for $k = 1, \ldots, i-1$, one can find a continuously differentiable, odd, strictly decreasing, and radially unbounded κ_i for the set-valued map S_i such that the e_i-subsystem is ISS with $V_i(e_i) = \alpha_V(|e_i|)$ as an ISS-Lyapunov function satisfying*

$$V_i(e_i) \geq \max_{k=1,\ldots,i-1} \left\{ \begin{array}{l} \gamma_{e_i}^{e_0}(V_0(e_0)), \gamma_{e_i}^{e_k}(V_k(e_k)), \gamma_{e_i}^{e_{i+1}}(V_{i+1}(e_{i+1})), \\ \gamma_{e_i}^{\mu_{k1}}(\mu_{k1}), \gamma_{e_i}^{\mu_{k2}}(\mu_{k2}), \gamma_{e_i}^{\mu_{i1}}(\mu_{i1}), \gamma_{e_i}^{\mu_{i2}}(\mu_{i2}), \epsilon_i \end{array} \right\}$$
$$\Rightarrow \max_{\psi_i \in \Psi_i(e_{i+1}, \bar{x}_i, \bar{\mu}_{i1}, \bar{\mu}_{i2}, z)} \nabla V_i(e_i) \psi_i \leq -\iota_i V_i(e_i), \tag{5.111}$$

where

$$\gamma_{e_i}^{\mu_{i1}}(s) = \alpha_V\left(\frac{1}{c_{i1}} s\right) \tag{5.112}$$

$$\gamma_{e_i}^{\mu_{i2}}(s) = \alpha_V\left(\frac{1}{1-c_{i1}} \bar{\kappa}_i^{-1}\left(\frac{1}{c_{i2}} s\right)\right) \tag{5.113}$$

$$\Psi_i(e_{i+1}, \bar{x}_i, \bar{\mu}_{i1}, \bar{\mu}_{i2}, z) := S_i(\bar{x}_i, \bar{\mu}_{i1}, \bar{\mu}_{i2}) + \Phi_i^*(e_{i+1}, \bar{x}_i, \bar{\mu}_{(i-1)1}, \bar{\mu}_{(i-1)2}, z) \tag{5.114}$$

with $\bar{\kappa}_i(s) = |\kappa_i(s)|$ for $s \in \mathbb{R}_+$.

The proof of Lemma 5.4 is in Appendix F.

Recall the definition $V_i(e_i) = \alpha_V(|e_i|)$. Consider the $\gamma_{e_i}^{\mu_{i1}}$ and $\gamma_{e_i}^{\mu_{i2}}$ defined in (5.112) and (5.113), respectively. It can be observed that the ISS gain from the quantization error μ_{i1} through the quantized control system to the signal e_i is s/c_{i1}. Direct calculation yields that the ISS gain from the quantization error μ_{i2} through the quantized control system to the signal $\kappa_i(e_i)$ is $\bar{\kappa}_i\left(\bar{\kappa}_i^{-1}(s/c_{i2})/(1-c_{i1})\right)$, which may not be linear, but is closely related to the linear function s/c_{i2}. As clarified later, properties (5.112) and (5.113) play a crucial role for the implementation of dynamic quantization in the quantized control system.

5.2.5 QUANTIZED CONTROLLER

In Subsection 5.2.4, set-valued maps are used to transform the closed-loop quantized system into a network of ISS subsystems described by differential inclusions. In this subsection, it is shown that the quantized control law u in the form of (5.93)–(5.94) with the κ_i's defined above belongs to the set-valued map S_n under realizable conditions. In this way, the closed-loop quantized system with the proposed quantized control law u can be represented as a dynamic network composed of ISS subsystems.

Recall that $\bar{\kappa}_i(s) = |\kappa_i(s)|$ for $s \in \mathbb{R}_+$. Lemma 5.5 provides conditions under which the quantized control law u in the form of (5.93)–(5.94) belongs to the set-valued map S_n.

Lemma 5.5 *Under Assumption 5.6, if*

$$\frac{1}{M_{i1}} < c_{i1} \leq 0.5, \quad \frac{1}{M_{i2}} < c_{i2} < 1 \tag{5.115}$$

for all $i = 1, \ldots, n$ and if

$$|e_i| \leq M_{i1}\mu_{i1}, \tag{5.116}$$

$$\bar{\kappa}_i((1-c_{i1})|e_i|) \leq M_{i2}\mu_{i2} \tag{5.117}$$

for all $i = 1, \ldots, n$, then v_i for $i = 1, \ldots, n-1$ and u defined in (5.93)–(5.94) satisfy

$$v_i \in S_i(\bar{x}_i, \bar{\mu}_{i1}, \bar{\mu}_{i2}), \quad i = 1, \ldots, n-1, \tag{5.118}$$

$$u \in S_n(\bar{x}_n, \bar{\mu}_{n1}, \bar{\mu}_{n2}). \tag{5.119}$$

The proof of Lemma 5.5 is given in Appendix F by fully using the properties of the quantizers and the set-valued maps.

Conditions (5.116) and (5.117) imply that the signals $|e_i|$ and $|\bar{\kappa}_i((1-c_i)|e_i|)|$ should be covered by the quantization ranges $M_{i1}\mu_{i1}$ and $M_{i2}\mu_{i2}$, respectively, such that the quantized control law u belongs to the set-valued map S_n. This is because of the saturation property (see (5.95) and (5.97)) of the quantizers.

5.2.6 SMALL-GAIN-BASED SYNTHESIS AND NESTED INVARIANT SETS OF THE CLOSED-LOOP QUANTIZED SYSTEM

Recall that $e = [e_0^T, e_1, \ldots, e_n]^T$. The gain digraph of the e-system is shown in Figure 5.13. The purpose of this subsection is to design the ISS gains to yield the ISS property of the closed-loop quantized system with e as the state by using the cyclic-small-gain theorem.

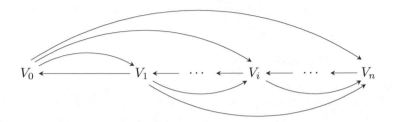

FIGURE 5.13   The gain digraph of the e-system.

Recall that $\bar{e}_i = [e_1, \ldots, e_i]^T$. For each $(e_0, \bar{e}_i)$-subsystem $(i = 1, \ldots, n)$, given the $(e_0, \bar{e}_{i-1})$-subsystem, by designing the set-valued map S_i for the e_i-subsystem, the ISS gains from states $e_0, \ldots, e_{i-1}$ to state e_i can be assigned. With the recursive design, the ISS gains $\gamma_{e_i}^{e_k}$ for $k = 0, \ldots, i-1$ can be designed such that

$$
\begin{aligned}
\gamma_{e_0}^{e_1} \circ \gamma_{e_1}^{e_2} \circ \gamma_{e_2}^{e_3} \circ \cdots \circ \gamma_{e_{i-1}}^{e_i} \circ \gamma_{e_i}^{e_0} &< \mathrm{Id} \\
\gamma_{e_1}^{e_2} \circ \gamma_{e_2}^{e_3} \circ \cdots \circ \gamma_{e_{i-1}}^{e_i} \circ \gamma_{e_i}^{e_1} &< \mathrm{Id} \\
&\vdots \\
\gamma_{e_{i-1}}^{e_i} \circ \gamma_{e_i}^{e_{i-1}} &< \mathrm{Id}
\end{aligned}
\tag{5.120}
$$

By applying this reasoning repeatedly, (5.120) can be guaranteed for all $i = 1, \ldots, n$. In this way, the e-system satisfies the cyclic-small-gain condition.

In the gain digraph of the e-system shown in Figure 5.13, the e_1-subsystem is reachable from the subsystems of $e_0, e_2, \ldots, e_n$, i.e., there are sequences of directed arcs from the subsystems of $e_0, e_2, \ldots, e_n$ to the e_1-subsystem.

By using the Lyapunov-based cyclic-small-gain theorem in Chapter 3, an ISS-Lyapunov function can be constructed for the e-system as

$$
V(e) = \max_{i=0,\ldots,n} \{\sigma_i(V_i(e_i))\}
\tag{5.121}
$$

with $\sigma_1(s) = s$, $\sigma_i(s) = \hat{\gamma}_{e_1}^{e_2} \circ \cdots \circ \hat{\gamma}_{e_{i-1}}^{e_i}(s)$ $(i = 2, \ldots, n)$, and $\sigma_0(s) = \max_{i=1,\ldots,n}\{\sigma_i \circ \hat{\gamma}_{e_i}^{e_0}(s)\}$ for $s \in \mathbb{R}_+$, where the $\hat{\gamma}_{(\cdot)}^{(\cdot)}$'s are $\mathcal{K}_\infty$ functions continuously differentiable on $(0, \infty)$ and slightly larger than the corresponding $\gamma_{(\cdot)}^{(\cdot)}$'s, and still satisfy the cyclic-small-gain condition.

The following lemma states that by appropriately choosing the κ_i's for the set-valued maps S_i's, the cyclic-small-gain condition (5.120) can be satisfied and the closed-loop quantized system with state e admits specific ISS properties.

Lemma 5.6 *Consider the e-system composed of the e_i-subsystems in the form of (5.108) and (5.106) satisfying (5.109) and (5.111), respectively. If the ISS gains defined in (5.109) and (5.111) satisfy (5.120) for all $i = 1,\ldots,n$ and if $u \in S_n(\bar{x}_n, \bar{\mu}_{n1}, \bar{\mu}_{n2})$, then the ISS-Lyapunov function candidate V defined in (5.121) for the e-system satisfies*

$$V(e) \geq \theta(\bar{\mu}_{n1}, \bar{\mu}_{n2}, \bar{\epsilon}_n) \Rightarrow \max_{\psi \in \Psi(e, x, \bar{\mu}_{n1}, \bar{\mu}_{n2})} \nabla V(e)\psi \leq -\alpha(V(e)) \quad (5.122)$$

wherever ∇V exists, where α is a continuous and positive definite function, and

$$\theta(\bar{\mu}_{n1}, \bar{\mu}_{n2}, \bar{\epsilon}_n) := \max_{i=1,\ldots,n} \left\{ \sigma_i \left(\max_{k=1,\ldots,i} \{\gamma_{e_i}^{\mu_{k1}}(\mu_{k1}), \gamma_{e_i}^{\mu_{k2}}(\mu_{k2}), \epsilon_i\} \right) \right\} \quad (5.123)$$

$$\Psi(e, x, \bar{\mu}_{n1}, \bar{\mu}_{n2}) := [\{g^T(e_0, e_1)\}, \Psi_1(e_2, \bar{x}_1, \bar{\mu}_{11}, \bar{\mu}_{12}), \ldots, \Psi_n(0, \bar{x}_n, \bar{\mu}_{n1}, \bar{\mu}_{n2})]^T \quad (5.124)$$

with $\bar{\epsilon}_n := [\epsilon_1, \ldots, \epsilon_n]^T$.

Proof. In the case of $u \in S_n(\bar{x}_n, \bar{\mu}_{n1}, \bar{\mu}_{n2})$, it holds that $e_{n+1} = 0$ and thus $V_{n+1}(e_{n+1}) = 0$. With the cyclic-small-gain condition (5.120) satisfied for all $i = 1, \ldots, n$, (5.122) can be proved. $\diamond$

For specified σ_i for $i = 1, \ldots, n$, by designing the $\gamma_{e_i}^{\mu_{k1}}$'s $(k = 1, \ldots, i-1)$ and the $\gamma_{e_i}^{\mu_{k2}}$'s $(k = 1, \ldots, i-1)$ small enough, we can achieve

$$\theta(\bar{\mu}_{n1}, \bar{\mu}_{n2}, \bar{\epsilon}_n) = \max_{i=1,\ldots,n} \left\{ \sigma_i \circ \gamma_{e_i}^{\mu_{i1}}(\mu_{i1}), \sigma_i \circ \gamma_{e_i}^{\mu_{i2}}(\mu_{i2}), \sigma_i(\epsilon_i) \right\} \quad (5.125)$$

for all $\mu_{i1}, \mu_{i2}, \epsilon_i > 0$ for $i = 1, \ldots, n$.

Motivated by [164], the dynamic quantization design in this section is based on nested invariant sets with sizes depending on zooming variables $\bar{\mu}_{n1}$ and $\bar{\mu}_{n2}$.

Define

$$B_1(\bar{\mu}_{n1}, \bar{\mu}_{n2}) = \max_{i=1,\ldots,n} \left\{ \begin{array}{l} \sigma_i \circ \alpha_V(M_{i1}\mu_{i1}), \\ \sigma_i \circ \alpha_V\left(\frac{1}{1-c_{i1}}\bar{\kappa}_i^{-1}(M_{i2}\mu_{i2})\right) \end{array} \right\}, \quad (5.126)$$

$$B_2(\bar{\mu}_{n1}, \bar{\mu}_{n2}) = \max_{i=1,\ldots,n} \left\{ \begin{array}{l} \sigma_i \circ \alpha_V\left(\frac{1}{c_{i1}}\mu_{i1}\right), \\ \sigma_i \circ \alpha_V\left(\frac{1}{1-c_{i1}}\bar{\kappa}_i^{-1}\left(\frac{1}{c_{i2}}\mu_{i2}\right)\right) \end{array} \right\}. \quad (5.127)$$

Lemma 5.7 summarizes this section by showing the existence of the nested invariant sets, defined by B_1 and B_2, for the closed-loop quantized system designed based on Lemmas 5.3–5.6.

Lemma 5.7 *Consider the quantized control system consisting of the plant* (5.73)–(5.75) *and the quantized control law* (5.93)–(5.94). *Under Assumptions 5.4, 5.5, 5.6 and 5.7, the closed-loop quantized system can be transformed into a large-scale system composed of e_i-subsystems in the form of* (5.108) *and* (5.106), *and for specified constants c_{i1}, c_{i2} satisfying* (5.115) *for $i = 1, \ldots, n$, specified ISS gains $\gamma_{e_k}^{e_{k'}}$ ($k \neq k'$) satisfying the cyclic-small-gain condition* (5.120) *for all $i = 1, \ldots, n$, specified ISS gains $\gamma_{e_i}^{\mu_{k1}}, \gamma_{e_i}^{\mu_{k2}}$ for $i = 1, \ldots, n$, $k = 1, \ldots, i - 1$ satisfying* (5.125), *and specified arbitrarily small constants ϵ_i for $i = 1, \ldots, n$, we can find continuously differentiable, odd, strictly decreasing, and radially unbounded functions κ_i for $i = 1, \ldots, n$ such that* (5.111) *holds for $i = 1, \ldots, n$. Moreover, if*

$$\sigma_i \circ \alpha_V (M_{i1} \mu_{i1}) = \sigma_i \circ \alpha_V \left(\frac{1}{1 - c_{i1}} \bar{\kappa}_i^{-1} (M_{i2} \mu_{i2}) \right)$$

$$= \sigma_j \circ \alpha_V (M_{j1} \mu_{j1}) = \sigma_j \circ \alpha_V \left(\frac{1}{1 - c_{j1}} \bar{\kappa}_j^{-1} (M_{j2} \mu_{j2}) \right) \qquad (5.128)$$

for all $i, j = 1, \ldots, n$ and if

$$B_1(\bar{\mu}_{n1}, \bar{\mu}_{n2}) \geq \theta_0 \qquad (5.129)$$

with $\theta_0 = \max_{i=1,\ldots,n} \{\sigma_i(\epsilon_i)\}$, then the ISS-Lyapunov function candidate V defined in (5.121) *satisfies*

$$B_1(\bar{\mu}_{n1}, \bar{\mu}_{n2}) \geq V(e) \geq \max\{B_2(\bar{\mu}_{n1}, \bar{\mu}_{n2}), \theta_0\} \qquad (5.130)$$

$$\Rightarrow \max_{\psi \in \Psi(e, x, \bar{\mu}_{n1}, \bar{\mu}_{n2})} \nabla V(e)\psi \leq -\alpha(V(e)), \qquad (5.131)$$

where Ψ is defined in (5.124).

Proof. Under Assumptions 5.4 and 5.7, with Lemma 5.3, we can transform the closed-loop quantized system into a large-scale system with state e composed of e_i-subsystems in the form of (5.108) and (5.106).

Under Assumptions 5.4, 5.5, 5.6, and 5.7, by directly using Lemma 5.4, for any specified constants c_{i1}, c_{i2} satisfying (5.115) for $i = 1, \ldots, n$, any ISS gains $\gamma_{e_k}^{e_{k'}}$ ($k \neq k'$) satisfying the cyclic-small-gain condition (5.120) for all $i = 1, \ldots, n$, any specified ISS gains $\gamma_{e_i}^{\mu_{k1}}, \gamma_{e_i}^{\mu_{k2}}$ for $i = 1, \ldots, n$, $k = 1, \ldots, i - 1$ satisfying (5.125) and specified arbitrarily small constants ϵ_i for $i = 1, \ldots, n$, we can find continuously differentiable, odd, strictly decreasing and radially unbounded functions κ_i for $i = 1, \ldots, n$ such that (5.111) holds for $i = 1, \ldots, n$.

The satisfaction of (5.115) by appropriately choosing the κ_i's for $i = 1, \ldots, n$ guarantees that $B_1(\bar{\mu}_{n1}, \bar{\mu}_{n2}) > B_2(\bar{\mu}_{n1}, \bar{\mu}_{n2})$ for all positive zooming variables μ_{i1}, μ_{i2}. By using (5.129), we have $B_1(\bar{\mu}_{n1}, \bar{\mu}_{n2}) \geq \max\{B_2(\bar{\mu}_{n1}, \bar{\mu}_{n2}), \theta_0\}$. Recall the definitions of $V_i(e_i)$ in (5.110) and $V(e)$

in (5.121). The equalities in (5.128) and the left inequality in (5.130) guarantee (5.116)–(5.117). Under Assumption 5.6, with (5.115) satisfied, by using Lemma 5.5, we have $u \in S_n(\bar{x}_n, \bar{\mu}_{n1}, \bar{\mu}_{n2})$.

Note that (5.125) is satisfied by appropriately choosing κ_i for $i = 1, \ldots, n$. By virtue of (5.112), (5.113), (5.125) and (5.127), $\theta(\bar{\mu}_{n1}, \bar{\mu}_{n2}, \bar{\epsilon}_n) = \max\{B_2(\bar{\mu}_{n1}, \bar{\mu}_{n2}), \theta_0\}$. With the cyclic-small-gain condition (5.120) satisfied by appropriately choosing κ_i for $i = 1, \ldots, n$ and $u \in S_n(\bar{x}_n, \bar{\mu}_{n1}, \bar{\mu}_{n2})$, Lemma 5.6 guarantees the implication in (5.130) and (5.131). $\diamond$

In the following subsection, based on Lemma 5.7, the invariant sets are used to design dynamic quantization.

5.2.7 A GUIDELINE FOR QUANTIZED CONTROL LAW DESIGN

To clarify the design procedure, we provide a guideline to choosing the functions κ_i for the quantized control law (5.93)–(5.94) such that the closed-loop quantized system satisfies property (5.130)–(5.131). The guideline includes two major steps:

1. *Choose the ISS parameters of the e_i-subsystems.*
   a. Choose constants c_{i1}, c_{i2} to satisfy (5.115) for $i = 1, \ldots, n$.
   b. Choose ISS gains $\gamma_{e_i}^{e_j} \in \mathcal{K}_\infty$ ($j \neq i$) and the corresponding functions $\hat{\gamma}_{e_i}^{e_j} > \gamma_{e_i}^{e_j}$ to satisfy the cyclic-small-gain condition (5.120) for all $i = 1, \ldots, n$, and calculate σ_i for $i = 1, \ldots, n$ in (5.121).
   c. Choose ISS gains $\gamma_{e_i}^{\mu_{k1}}, \gamma_{e_i}^{\mu_{k2}}$ for $i = 1, \ldots, n$, $k = 1, \ldots, i - 1$ such that (5.125) holds for all $\mu_{i1}, \mu_{i2}, \epsilon_i > 0$ for $i = 1, \ldots, n$.
   d. Choose specified $\epsilon_i, \iota_i > 0$ for $i = 1, \ldots, n$.
2. *Choose κ_i's based on Lemma 5.4 with the ISS parameters chosen in Step 1.*

In Step 1, it is only required that Step (c) is after Step (b), because condition (5.125) in Step (c) depends on the σ_i calculated in Step (b). Under Assumptions 5.4–5.7, if the ISS parameters and the κ_i are chosen according to the guideline and if conditions (5.128) and (5.129) are satisfied, then from Lemma 5.7, the nested invariant sets exist.

5.2.8 DYNAMIC QUANTIZATION

Because of the saturation property of the quantizers, the quantized control law designed in Subsection 5.2.4 can only guarantee local stabilization; see (5.130)–(5.131). In this subsection, based on the nested invariant sets given in Lemma 5.7, we design a dynamic quantization logic in the form of (5.99), composed of a zooming-in stage and a zooming-out stage, to dynamically adjust the zooming variables μ_{ij} ($i = 1, \ldots, n$, $j = 1, 2$) such that the closed-loop quantized system is semiglobally stabilized. In this design, the zooming variables $\mu_{ij}(t)$ are piecewise constant signals, and are adjusted on a discrete time sequence $\{t_k\}_{k \in \mathbb{Z}_+}$, where $t_{k+1} - t_k = t_d$ with constant $t_d > 0$.

To satisfy condition (5.128) in Lemma 5.7, we design dynamic quantization such that for all $t \in \mathbb{R}_+$,

$$\sigma_i \circ \alpha_V(M_{i1}\mu_{i1}(t)) = \sigma_i \circ \alpha_V\left(\frac{1}{1-c_{i1}}\bar{\kappa}_i^{-1}(M_{i2}\mu_{i2}(t))\right)$$

$$:= \Theta(t) \tag{5.132}$$

for $i = 1, \ldots, n$. Equivalently, it is required that

$$\mu_{i1}(t) = \frac{1}{M_{i1}}\alpha_V^{-1} \circ \sigma_i^{-1}(\Theta(t)) := \Upsilon_{i1}(\Theta(t)), \tag{5.133}$$

$$\mu_{i2}(t) = \frac{1}{M_{i2}}\bar{\kappa}_i\left((1-c_{i1})\alpha_V^{-1} \circ \sigma_i^{-1}(\Theta(t))\right) := \Upsilon_{i2}(\Theta(t)) \tag{5.134}$$

for $i = 1, \ldots, n$. According to the definitions, Υ_{i1} and Υ_{i2} are invertible for $i = 1, \ldots, n$. Thus, the dynamic quantization logic (5.99) can be designed by choosing an appropriate update law for Θ, which may reduce the design complexity for all the zooming variables μ_{ij} ($i = 1, \ldots, n$, $j = 1, 2$). The update law for Θ is expected to be in the following form:

$$\Theta(t_{k+1}) = Q(\Theta(t_k)), \quad k \in \mathbb{Z}_+. \tag{5.135}$$

In the zooming-out stage, $Q = Q^{\text{out}}$; in the zooming-in stage, $Q = Q^{\text{in}}$. With Q^{out} and Q^{in} designed, we can design the dynamic quantization logic (5.99) for μ_{ij} by choosing

$$Q_{ij}^{\text{out}} = \Upsilon_{ij} \circ Q^{\text{out}} \circ \Upsilon_{ij}^{-1}, \tag{5.136}$$

$$Q_{ij}^{\text{in}} = \Upsilon_{ij} \circ Q^{\text{in}} \circ \Upsilon_{ij}^{-1}. \tag{5.137}$$

Using the definition of B_1 in (5.126), we also have

$$\Theta(t) = B_1(\bar{\mu}_{n1}(t), \bar{\mu}_{n2}(t)). \tag{5.138}$$

Before designing dynamic quantization, the relation between zooming variables $\bar{\mu}_{n1}, \bar{\mu}_{n2}$ and control error e should be clarified. For $i = 1, \ldots, n$, using the definitions of S_i in (5.104), the strictly decreasing property of κ_i implies

$$\max S_i(\bar{x}_i, \bar{\mu}_{i1}, \bar{\mu}_{i2})$$
$$= \kappa_i(x_i - \max S_{i-1}(\bar{x}_{i-1}, \bar{\mu}_{(i-1)1}, \bar{\mu}_{(i-1)2}) - \max\{c_{i1}|e_i|, \mu_{i1}\}) + \mu_{i2}, \tag{5.139}$$

and

$$\min S_i(\bar{x}_i, \bar{\mu}_{i1}, \bar{\mu}_{i2})$$
$$= \kappa_i(x_i - \min S_{i-1}(\bar{x}_{i-1}, \bar{\mu}_{(i-1)1}, \bar{\mu}_{(i-1)2}) + \max\{c_{i1}|e_i|, \mu_{i1}\}) - \mu_{i2}. \tag{5.140}$$

Recall that $e = [e_0^T, e_1, \ldots, e_n]^T$. Given the definitions of e_i for $i = 2, \ldots, n$ in (5.105), denote

$$e = e(X, \bar{\mu}_{n1}, \bar{\mu}_{n2}) \tag{5.141}$$

with $X = [z^T, x^T]^T \in \mathbb{R}^{n+n_z}$. It can be observed that e is a continuous function of $X, \bar{\mu}_{n1}, \bar{\mu}_{n2}$. Clearly, the piecewise updates of $\bar{\mu}_{n1}, \bar{\mu}_{n2}$ cause jumps of e on the timeline. This should be carefully handled in the design.

Zooming-Out Stage

The purpose of the zooming-out stage in this subsection is to increase the zooming variables μ_{ij} such that at some finite time t_{k^*}, the state of the closed-loop quantized system is restricted to be in the larger invariant set corresponding to B_1 in (5.130). In this stage, the components κ_i's for $i = 1, \ldots, n$ of the controller are set to be zero, i.e., $u = 0$.

The small-time norm-observability assumed in Assumption 5.3 guarantees that for $t_d > 0$, there exists a $\varphi \in \mathcal{K}_\infty$ such that

$$|X(t_k + t_d)| \le \varphi(\|x\|_{[t_k, t_k + t_d]}) \tag{5.142}$$

for any $k \in \mathbb{Z}_+$. Considering the definitions of V and e in (5.121) and (5.141), for $t_d > 0$, property (5.142) can be represented with the Lyapunov function V as

$$|V(e(X(t_k + t_d), 0, 0))| \le \bar{\varphi}(\|x\|_{[t_k, t_k + t_d]}) \tag{5.143}$$

for any $k \in \mathbb{Z}_+$, where $\bar{\varphi} \in \mathcal{K}_\infty$.

With the forward completeness property assumed in Assumption 5.3, we design a zooming-out logic $Q^{\text{out}} : \mathbb{R}_+ \to \mathbb{R}_+$ to increase Θ fast enough to dominate the growth rate of $\bar{\varphi}(|x|)$ such that at some finite time $t_{k^*} > 0$ with $k^* \in \mathbb{Z}_+$, it holds that

$$M_{i1}\mu_{i1}(t_{k^*}) \ge |x_i(t_{k^*})|, \quad i = 1, \ldots, n, \tag{5.144}$$

$$\Theta(t_{k^*}) \ge \bar{\varphi}(\|x\|_{[t_{k^*} - t_d, t_{k^*}]}). \tag{5.145}$$

Due to the saturation of the quantizer, if the input signal of a quantizer is outside the range of the quantizer, then one cannot estimate the bound of the signal without using additional information. In the zooming-out stage, the κ_i's are set to be zero, and the input of the quantizer q_{i1} is x_i; see control law (5.93)–(5.94). Inequality (5.144) means that at some finite time t_{k^*}, x_i is in the quantization range of q_{i1}. Then, we can estimate the bound of $|x_i(t_{k^*})|$.

Using (5.143) and (5.145), we have

$$\Theta(t_{k^*}) \ge \max\{V(e(X(t_{k^*}), 0, 0)), \theta_0\}. \tag{5.146}$$

From the definitions of $\max S_i(\bar{x}_i, \bar{\mu}_{i1}, \bar{\mu}_{i2})$, $\min S_i(\bar{x}_i, \bar{\mu}_{i1}, \bar{\mu}_{i2})$, and e_{i+1}, one observes that increase of $\bar{\mu}_{n1}, \bar{\mu}_{n2}$ leads to increase of $\max S_i(\bar{x}_i, \bar{\mu}_{i1}, \bar{\mu}_{i2})$, decrease of $\min S_i(\bar{x}_i, \bar{\mu}_{i1}, \bar{\mu}_{i2})$ and thus decrease or hold of $|e_{i+1}|$ for $i = 1, \ldots, n-1$. Thus, with the zooming-out logic Q^{out}, we achieve that, at time $t_{k^*} > 0$ with $k^* \in \mathbb{Z}_+$, it holds that

$$\Theta(t_{k^*}) \geq \max\{V(e(X(t_{k^*}), \bar{\mu}_{n1}(t_{k^*}), \bar{\mu}_{n2}(t_{k^*}))), \theta_0\}. \tag{5.147}$$

With Q^{out} designed, we can design the zooming-out logic Q^{out}_{ij} for $i = 1, \ldots, n$, $j = 1, 2$ according to (5.136)–(5.137).

It should be noted that, if a bound of the initial state $X(0)$ is known *a priori*, then we can directly set $\Theta(t_{k^*})$ to satisfy (5.146) with $t_{k^*} = 0$. In this case, the zooming-out stage is not necessary and Assumption 5.3 is not required.

Zooming-In Stage

The zooming-out stage achieves (5.147) at time t_{k^*} with $k^* \in \mathbb{Z}_+$. Suppose that at some $t_k > 0$ with $k \geq k^*$, it is achieved that

$$\Theta(t_k) \geq \max\{V(e(X(t_k), \bar{\mu}_{n1}(t_k), \bar{\mu}_{n2}(t_k))), \theta_0\}. \tag{5.148}$$

We first design a $Q^{\text{in}} : \mathbb{R}_+ \to \mathbb{R}_+$ for the zooming-in stage such that

$$\Theta(t_{k+1}) = Q^{\text{in}}(\Theta(t_k))$$
$$\geq \max\{V(e(X(t_{k+1}), \bar{\mu}_{n1}(t_{k+1}), \bar{\mu}_{n2}(t_{k+1}))), \theta_0\}. \tag{5.149}$$

This objective is achievable by using Lemmas 5.8 and 5.9. Then, we show the convergence property of the update law (5.135) for Θ in the zooming-in stage by Lemma 5.10.

Recall that if (5.149) is achieved based on (5.148), then one can recursively guarantee that the state e of the closed-loop quantized system is always in the larger invariant set represented by B_1 in spite of the discontinuous update of Θ; see (5.130) and (5.138).

Lemma 5.8 describes the decreasing property of V during the time interval $[t_k, t_{k+1})$, based on which we design the zooming-in update law Q^{in} for Θ.

Lemma 5.8 *Consider the closed-loop quantized system with V satisfying property* (5.130)–(5.131). *If* (5.148) *holds at time t_k with $k \in \mathbb{Z}_+$, then there exists a continuous and positive definite function $\bar{\rho}$ such that*

$$(\text{Id} - \bar{\rho}) \in \mathcal{K}_\infty, \tag{5.150}$$
$$V(e(X(t_{k+1}), \bar{\mu}_{n1}(t_k), \bar{\mu}_{n2}(t_k))) \leq \max\{(\text{Id} - \bar{\rho})(\Theta(t_k)), \theta_0\}. \tag{5.151}$$

The proof of Lemma 5.8 is in Appendix F.

From the definition in (5.141), the piecewise constant update of the zooming variables $\bar{\mu}_{n1}, \bar{\mu}_{n2}$ causes jumps of e and thus jumps of V. Based on (5.151), we design the zooming-in logic Q^{in} to achieve (5.149) by taking into account the jumps.

For convenience of notation, define

$$W(\xi, s) = V(e(\xi, \bar{\Upsilon}_{n1}(s), \bar{\Upsilon}_{n2}(s))) \tag{5.152}$$

for $\xi \in \mathbb{R}^{n+n_z}$ and $s \in \mathbb{R}_+$, where

$$\bar{\Upsilon}_{n1}(s) = [\Upsilon_{11}(s), \dots, \Upsilon_{n1}(s)]^T, \tag{5.153}$$

$$\bar{\Upsilon}_{n1}(s) = [\Upsilon_{12}(s), \dots, \Upsilon_{n2}(s)]^T. \tag{5.154}$$

Then, $W(\xi, s)$ is a continuous function of (ξ, s).

Consider (ξ, s) satisfying

$$0 \leq s \leq \Theta(t_{k^*}) \tag{5.155}$$

$$W(\xi, s) \leq \Theta(t_{k^*}). \tag{5.156}$$

From the definitions of V and W in (5.121) and (5.152), we can find a compact set $\Omega^o \subset \mathbb{R}^{n+n_z} \times \mathbb{R}_+$ such that all the (ξ, s) satisfying (5.155)–(5.156) belong to Ω^o. By using the property of continuous functions, we can find a continuous and positive definite function $\rho^o < \mathrm{Id}$ such that for all $(\xi, s) \in \Omega^o$ and all $h \geq 0$, it holds that

$$|W(\xi, s - \rho^o(h)) - W(\xi, s)| \leq h. \tag{5.157}$$

We propose the following update law for Θ in the zooming-in stage:

$$Q^{\mathrm{in}}(\Theta) = \Theta - \rho^o \left(\frac{\Theta - \max\{\Xi(\Theta), \theta_0\}}{2} \right), \tag{5.158}$$

where $\Xi = (\mathrm{Id} - \bar{\rho})$. In the following procedure, we use Lemma 5.9 to guarantee the achievement of objective (5.149) and employ Lemma 5.10 to show the convergence of Θ with the update law defined in (5.158).

Lemma 5.9 shows that property (5.149) can be achieved with the zooming-in update law (5.158) for Θ, given that (5.147) and (5.148) are satisfied.

Lemma 5.9 *Consider the closed-loop quantized system with V satisfying property (5.130)–(5.131). Suppose that condition (5.147) holds at some finite time t_{k^*} and condition (5.148) holds at some time t_k with $k \geq k^*$. Then, property (5.149) is satisfied at time t_{k+1} with the update law $\Theta(t_{k+1}) = Q^{\mathrm{in}}(\Theta(t_k))$ with Q^{in} defined in (5.158).*

Proof. With $0 < \rho^o < \mathrm{Id}$, it can be guaranteed that

$$\Theta(t_{k+1}) \leq \Theta(t_k) \tag{5.159}$$

and

$$\Theta(t_{k+1}) = \Theta(t_k) - \rho^o \left(\frac{\Theta(t_k) - \max\{\Xi(\Theta(t_k)), \theta_0\}}{2} \right)$$

$$\geq \frac{\Theta(t_k) + \max\{\Xi(\Theta(t_k)), \theta_0\}}{2} \tag{5.160}$$

for $k \geq k^*$. Thus, $0 < \Theta(t_k) \leq \Theta(t_{k^*})$ for $k \geq k^*$.

From Lemma 5.8, (5.151) holds. Using (5.147), (5.151), and (5.152), we have

$$W(X(t_{k+1}), \Theta(t_k)) \leq \max\{\Xi(\Theta(t_k)), \theta_0\} \leq \Theta(t_{k^*}) \qquad (5.161)$$

for $k \geq k^*$. Hence, $(X(t_{k+1}), \Theta(t_k)) \in \Omega^o$ for $k \geq k^*$. Given $(X(t_{k+1}), \Theta(t_k)) \in \Omega^o$, from (5.158) and (5.161), we obtain

$$
\begin{aligned}
&W(X(t_{k+1}), \Theta(t_{k+1})) \\
&\leq W(X(t_{k+1}), \Theta(t_k)) + |W(X(t_{k+1}), \Theta(t_{k+1})) - W(X(t_{k+1}), \Theta(t_k))| \\
&\leq W(X(t_{k+1}), \Theta(t_k)) + |W(X(t_{k+1}), Q^{\mathrm{in}}(\Theta(t_k))) - W(X(t_{k+1}), \Theta(t_k))| \\
&\leq \max\{\Xi(\Theta(t_k)), \theta_0\} + \frac{\Theta(t_k) - \max\{\Xi(\Theta(t_k)), \theta_0\}}{2} \\
&= \frac{\Theta(t_k) + \max\{\Xi(\Theta(t_k)), \theta_0\}}{2}.
\end{aligned}
\qquad (5.162)
$$

From (5.148), we have $\theta_0 \leq \Theta(t_k)$, which implies

$$\theta_0 \leq \frac{\Theta(t_k) + \max\{\Xi(\Theta(t_k)), \theta_0\}}{2}. \qquad (5.163)$$

Properties (5.160), (5.162), and (5.163) together with the definition of W in (5.152) guarantee (5.149).   ◊

Lemma 5.10 shows the convergence property of the update law (5.135) for Θ with $Q = Q^{\mathrm{in}}$ defined in (5.158).

Lemma 5.10 *Suppose that at some $t_{k^*} > 0$ with $k^* \in \mathbb{Z}_+$, $\Theta(t_{k^*}) \geq \theta_0$. Then with Q^{in} defined in (5.158), update law $\Theta(t_{k+1}) = Q^{\mathrm{in}}(\Theta(t_k))$ achieves*

$$\lim_{k \to \infty} \Theta(t_k) = \theta_0. \qquad (5.164)$$

Proof. Consider the following two cases.

• $\Xi(\Theta(t_k)) \geq \theta_0$. From the definition of Ξ, one can find a continuous and positive definite function ρ_1^* such that $\Xi \leq \mathrm{Id} - \rho_1^*$. Then, one can find a continuous and positive definite function ρ_2^* such that $\rho^o\left(\frac{s - \Xi(s)}{2}\right) \geq \rho_2^*(s)$ for $s \in \mathbb{R}_+$. In the case of $\Xi(\Theta(t_k)) \geq \theta_0$, we have

$$
\begin{aligned}
\Theta(t_{k+1}) &= \Theta(t_k) - \rho^o\left(\frac{\Theta(t_k) - \Xi(\Theta(t_k))}{2}\right) \\
&\leq \Theta(t_k) - \rho_2^*(\Theta(t_k)),
\end{aligned}
\qquad (5.165)
$$

which guarantees that there exists a $t_{k^o} > t_k$ with $k^o \in \mathbb{Z}_+$ such that $\Xi(\Theta(t_{k^o})) < \theta_0$.

- $0 \leq \Xi(\Theta(t_k)) < \theta_0$. Define $\Theta'(t_k) = \Theta(t_k) - \theta_0$ for $k \in \mathbb{Z}_+$. Then, we obtain

$$\Theta'(t_{k+1}) = \Theta'(t_k) - \rho^o \left(\frac{\Theta'(t_k)}{2} \right), \tag{5.166}$$

which is an asymptotically stable first-order discrete-time system [131].

Recall the definition of Ξ. We can see $\Xi^{-1} > \mathrm{Id}$, $\Xi^{-1}(\theta_0)$ is larger than θ_0 and $\Theta < \Xi^{-1}(\theta_0)$ is an invariant set of system (5.166). Thus, $\lim_{k\to\infty} \Theta'(t_k) = 0$ and equivalently $\lim_{k\to\infty} \Theta(t_k) = \theta_0$. $\diamondsuit$

The motions of $\Theta(t)$ and $W(X(t), \Theta(t))$ are illustrated in Figure 5.14.

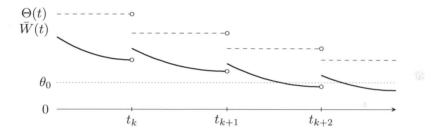

FIGURE 5.14   Motions of $\Theta(t)$ and $\bar{W}(t) = W(X(t), \Theta(t))$ in the zooming-in stage.

The zooming-in update law for Θ can be designed by finding the $\bar{\rho}$ with Lemma 5.8 and ρ^o by using the continuity of W. Lemmas 5.9 and 5.10 are used to prove the effectiveness of the zooming-in update law Q^{in} defined in (5.158). With Q^{in} designed, we can design the zooming-in logic Q_{ij}^{in} for $i = 1, \ldots, n$, $j = 1, 2$ according to (5.136)–(5.137).

Main Result

Based on the design above, the main result of quantized control is summarized in Theorem 5.2.

Theorem 5.2 *For system (5.73)–(5.75), under Assumptions 5.3–5.6, by choosing constants c_{i1}, c_{i2} satisfying (5.115) for $i = 1, \ldots, n$, ISS gains $\gamma_{e_k}^{e_{k'}}$ ($k \neq k'$) satisfying the cyclic-small-gain condition (5.120) for all $i = 1, \ldots, n$, ISS gains $\gamma_{e_i}^{\mu_{k1}}, \gamma_{e_i}^{\mu_{k2}}$ for $i = 1, \ldots, n$, $k = 1, \ldots, i-1$ satisfying (5.125) and constants $\epsilon_i > 0$ for $i = 1, \ldots, n$, one can design the functions κ_i for $i = 1, \ldots, n$ in (5.93)–(5.94) and the dynamic quantization logic Q_{ij} for $i = 1, \ldots, n$, $j = 1, 2$ in (5.99) such that the closed-loop solutions z and x are bounded. Moreover, by choosing the constants $\epsilon_i > 0$ for $i = 1, \ldots, n$ to be arbitrarily small, the output $x_1(t)$ can be steered to within an arbitrarily small neighborhood of origin.*

Proof. With Assumption 5.3 satisfied, at some time $t_{k^*} > 0$ with $k^* \in \mathbb{Z}_+$, (5.147) can be achieved by the zooming-out logic Q_{ij}^{out}.

With Assumptions 5.4–5.6 satisfied, using Lemma 5.7, by appropriately designing the functions κ_i for $i = 1, \ldots, n$ such that the ISS parameters satisfy the conditions (5.115), (5.120), and (5.125), the closed-loop quantized system has the nested invariant sets defined in (5.131).

Using Lemmas 5.8 and 5.9, (5.149) can be guaranteed with the designed zooming-in logic, and it holds that $V(e(X(t_k), \bar{\mu}_{n1}(t_k), \bar{\mu}_{n2}(t_k))) \leq \Theta(t_k)$ for $k \geq k^*$. Moreover, $\overline{\lim}_{k \to \infty} V(e(X(t_k), \bar{\mu}_{n1}(t_k), \bar{\mu}_{n2}(t_k))) \leq \theta_0$ according to Lemma 5.10. Recall the definition of V in (5.121). The closed-loop signal x_1 is driven to within the region $|x_1| \leq \alpha_V^{-1}(\theta_0)$. Recall the definition of θ_0 in Lemma 5.7. By designing ϵ_i ($i = 1, \ldots, n$) to be arbitrarily small, the state x_1 can be steered to within an arbitrarily small neighborhood of the origin. $\diamond$

If there are no inverse dynamics (i.e., the z-subsystem does not exist) in system (5.73)–(5.75), the assumption on small-time norm-observability in Assumption 5.3 is not needed.

5.3 QUANTIZED OUTPUT-FEEDBACK CONTROL

By designing a quantized observer, this section studies quantized output-feedback control of nonlinear systems. The main result shows that the output of the quantized control system can be steered to within an arbitrarily small neighborhood of the origin even with a three-level uniform quantizer.

Consider the disturbed output-feedback nonlinear system with quantized output:

$$\dot{x}_i = x_{i+1} + f_i(y, d), \quad i = 1, \ldots, n-1 \tag{5.167}$$

$$\dot{x}_n = u + f_n(y, d) \tag{5.168}$$

$$y = x_1 \tag{5.169}$$

$$y^q = q(y, \mu), \tag{5.170}$$

where $[x_1, \ldots, x_n]^T \in \mathbb{R}^n$ is the state, $u \in \mathbb{R}$ is the control input, $d \in \mathbb{R}^{n_d}$ represents external disturbance inputs, $y \in \mathbb{R}$ is the output, $q(y, \mu)$ is the output quantizer with variable $\mu > 0$, $y^q \in \mathbb{R}$ is the quantized output, $[x_2, \ldots, x_n]^T$ is the unmeasured portion of the state, and f_i's ($i = 1, \ldots, n$) are uncertain locally Lipschitz continuous functions.

The output quantizer is assumed to satisfy

$$|y| \leq M\mu \Rightarrow |q(y, \mu) - y| \leq \mu, \tag{5.171}$$

where $M > 0$, $M\mu$ is the quantization range and $\mu > 0$ is the maximum quantization error when $|y| \leq M\mu$.

In this section, the basic idea of dynamic quantization is still to appropriately update the zooming variable μ during the quantized control procedure for improved control performance. It is assumed that μ is right-continuous with respect to time and is updated in discrete-time as:

$$\mu(t_{k+1}) = Q(\mu(t_k)), \quad k \in \mathbb{Z}_+, \tag{5.172}$$

where $Q : \mathbb{R}_+ \to \mathbb{R}_+$ represents the dynamic quantization logic, and $t_k \geq 0$ with $k \in \mathbb{Z}_+$ are updating time instants satisfying $t_{k+1} - t_k = d_t$ with $d_t > 0$.

Assumptions 5.8–5.11 are made throughout this section.

Assumption 5.8 *System (5.167)–(5.169) with $u = 0$ is forward complete and small-time norm-observable with y as the output.*

Assumption 5.9 *For each $f_i(y, d)$ $(i = 1, \ldots, n)$ in (5.167)–(5.168), there exists a known $\psi_{f_i} \in \mathcal{K}_\infty$ such that for all y, d,*

$$|f_i(y, d)| \leq \psi_{f_i}(|[y, d^T]^T|). \tag{5.173}$$

Assumption 5.10 *There exists a $\bar{d} \geq 0$ such that*

$$|d(t)| \leq \bar{d} \tag{5.174}$$

for $t \geq 0$.

Assumption 5.11 *Quantizer $q(y, \mu)$ satisfies (5.171) with $M > 1$.*

5.3.1   REDUCED-ORDER OBSERVER DESIGN

For convenience of notation, denote $w = y^q - y$.

Inspired by the reduced-order observer used for measurement feedback control in Section 4.3, we design the following reduced-order observer which uses the quantized output y^q:

$$\dot{\xi}_i = \xi_{i+1} + L_{i+1}y^q - L_i(\xi_2 + L_2y^q), \ i = 2, \ldots, n-1 \tag{5.175}$$

$$\dot{\xi}_n = u - L_n(\xi_2 + L_2y^q), \tag{5.176}$$

where ξ_i is an estimate for the unmeasured state $x_i - L_iy$ for each $i = 2, \ldots, n$. Define $e_0 = [x_2 - L_2y - \xi_2, \ldots, x_n - L_ny - \xi_n]^T$ as the observation error. Then, from (5.167)–(5.170) and (5.175)–(5.176), the observation error system is

$$\dot{e}_0 = Ae_0 + \phi_0(y, d, w)$$
$$:= f_{e_0}(e_0, y, d, w), \tag{5.177}$$

where

$$A = \begin{bmatrix} -L_2 & & \\ \vdots & & I_{n-2} \\ -L_{n-1} & & \\ -L_n & 0 & \cdots & 0 \end{bmatrix}, \qquad (5.178)$$

$$\phi_0(y,d,w) = \begin{bmatrix} -L_2 & \\ \vdots & I_{n-1} \\ -L_n & \end{bmatrix} \begin{bmatrix} f_1(y,d) \\ \vdots \\ f_n(y,d) \end{bmatrix} + \begin{bmatrix} L_2^2 - L_3 \\ \vdots \\ L_{n-1}L_2 - L_n \\ L_n L_2 \end{bmatrix} w.$$

$$(5.179)$$

The real constants L_i's in (5.178) are chosen so that A is Hurwitz, and thus there exists a matrix $P = P^T > 0$ satisfying $PA + A^T P = -2I_{n-1}$. For ϕ_0 defined in (5.179), using Assumption 5.9, we can find $\psi_{\phi_0}^y, \psi_{\phi_0}^d, \psi_{\phi_0}^w \in \mathcal{K}_\infty$ such that $|\phi_0(y,d,w)|^2 \leq \psi_{\phi_0}^y(|y|) + \psi_{\phi_0}^d(|d|) + \psi_{\phi_0}^w(|w|)$ holds for all y, d, w.

Define $V_0(e_0) = e_0^T P e_0$. Define $\underline{\alpha}_0(s) = \lambda_{\min}(P)s^2$ and $\overline{\alpha}_0(s) = \lambda_{\max}(P)s^2$ for $s \in \mathbb{R}_+$. Then, $\underline{\alpha}_0(|e_0|) \leq V_0(e_0) \leq \overline{\alpha}_0(|e_0|)$ holds for all e_0. Direct computation yields:

$$\begin{aligned} \nabla V_0(e_0) f_{e_0}(e_0, y, d, w) &= -2e_0^T e_0 + 2e_0^T P \phi_0(y, d, w) \\ &\leq -e_0^T e_0 + |P|^2 |\phi_0(y, d, w)|^2 \\ &\leq -\frac{1}{\lambda_{\max}(P)} V_0(e_0) + |P|^2 \Big(\psi_{\phi_0}^y(|y|) \\ &\quad + \psi_{\phi_0}^d(|d|) + \psi_{\phi_0}^w(|w|) \Big). \end{aligned} \qquad (5.180)$$

Define $\chi_0^y = 4\lambda_{\max}(P)|P|^2 \psi_{\phi_0}^y$, $\chi_0^d = 4\lambda_{\max}(P)|P|^2 \psi_{\phi_0}^d$, and $\chi_0^\mu = 4\lambda_{\max}(P)|P|^2 \psi_{\phi_0}^w$. Then, we have

$$\begin{aligned} V_0(e_0) &\geq \max\{\chi_0^y(|y|), \chi_0^d(|d|), \chi_0^\mu(|w|)\} \\ &\Rightarrow \nabla V_0(e_0) f_{e_0}(e_0, y, d, w) \leq -\alpha_0(V_0(e_0)), \end{aligned} \qquad (5.181)$$

where $\alpha_0(s) = s/4\lambda_{\max}(P)$ for $s \in \mathbb{R}_+$.

5.3.2 QUANTIZED CONTROL DESIGN

The gain assignment technique still plays a central role in quantized output-feedback control design. Consider the following first-order system:

$$\dot{\eta} = \phi(\eta, \omega_1, \ldots, \omega_{n-2}) + \bar{\kappa}, \qquad (5.182)$$

where $\eta \in \mathbb{R}$ is the state, $\bar{\kappa} \in \mathbb{R}$ is the control input, $\omega_1, \ldots, \omega_{n-2} \in \mathbb{R}$ represent external disturbance inputs, and the nonlinear function $\phi(\eta, \omega_1, \ldots, \omega_{n-2})$ is locally Lipschitz and satisfies

$$|\phi(\eta, \omega_1, \ldots, \omega_{n-2})| \leq \psi_\phi(|[\eta, \omega_1, \ldots, \omega_{n-2}]^T|), \qquad (5.183)$$

with $\psi_\phi \in \mathcal{K}_\infty$ known. Define $\alpha_V(s) = \frac{1}{2}s^2$ for $s \in \mathbb{R}_+$. Notice that sgn denotes the standard sign function.

Lemma 5.11 *Consider system (5.182). For any specified $0 < c < 1$, $\epsilon > 0$, $\ell > 0$, and $\chi_\eta^{\omega_1}, \ldots, \chi_\eta^{\omega_{n-2}} \in \mathcal{K}_\infty$, one can find a continuously differentiable, odd, strictly decreasing, and radially unbounded κ such that if $\bar{\kappa}$ in (5.182) satisfies*

$$\bar{\kappa} \in \{\kappa(\eta + \text{sgn}(\eta)|\omega_n|) + \delta|\omega_{n-1}| : |\delta| \leq 1\}, \qquad (5.184)$$

where $\omega_{n-1}, \omega_n \in \mathbb{R}$ represent measurement disturbances, then it holds that

$$V_\eta(\eta) \geq \max_{k=1,\ldots,n-2} \left\{ \chi_\eta^{\omega_k}(|\omega_k|), \alpha_V\left(\frac{|\omega_{n-1}|}{c}\right), \epsilon \right\}$$
$$\Rightarrow \nabla V_\eta(\eta)(\phi(\eta, \omega_1, \ldots, \omega_{n-2}) + \bar{\kappa}) \leq -\ell V_\eta(\eta). \qquad (5.185)$$

Lemma 5.11 is a set-valued map version of Lemma 4.1 and can be proved in the same way.

Define $e_1 = y$. Consider the $[e_0^T, e_1, \xi_2, \ldots, \xi_n]^T$-system:

$$\dot{e}_0 = Ae_0 + \phi_0(e_1, d, w) \qquad (5.186)$$
$$\dot{e}_1 = \xi_2 + \phi_1(e_0, e_1, d) \qquad (5.187)$$
$$\dot{\xi}_i = \xi_{i+1} + \phi_i(e_1, \xi_2, w), \quad i = 2, \ldots, n-1 \qquad (5.188)$$
$$\dot{\xi}_n = u + \phi_n(e_1, \xi_2, w), \qquad (5.189)$$

where

$$\phi_1(e_0, e_1, d) = L_2 y + (x_2 - L_2 y - \xi_2) + f_1(y, d)$$
$$\phi_i(e_1, \xi_2, w) = L_{i+1} y^q - L_i(\xi_2 + L_2 y^q), \quad i = 2, \ldots, n-1$$
$$\phi_n(e_1, \xi_2, w) = -L_n(\xi_2 + L_2 y^q).$$

We get (5.187) from the x_1-subsystem (5.167) using the fact that $(x_2 - L_2 e_1 - \xi_2)$ is the first element of vector e_0. We get (5.188) and (5.189) from (5.175) and (5.176) by using $y^q = y + w = e_1 + w$.

We construct a new $[e_0^T, e_1, \ldots, e_n]^T$-system consisting of ISS subsystems obtained through a recursive design of the $[e_0^T, e_1, \xi_2, \ldots, \xi_n]^T$-system. The ISS-Lyapunov function V_0 for the e_0-subsystem is defined in Subsection 5.3.1. For $i = 1, \ldots, n$, each e_i-subsystem is designed with an ISS-Lyapunov function candidate

$$V_i(e_i) = \alpha_V(|e_i|), \qquad (5.190)$$

where $\alpha_V(s) = s^2/2$ for $s \in \mathbb{R}_+$. Denote $\bar{e}_i = [e_0^T, e_1, \ldots, e_i]^T$ and $\bar{\xi}_i = [\xi_2, \ldots, \xi_i]^T$.

In this subsection, we suppose that μ is constant and consider only the case of $|e_1| = |y| \leq M\mu$. From (5.171), this means $|w| = |y^q - y| \leq \mu$.

The e_0-subsystem

Define $\gamma_0^1 = \chi_0^y \circ \alpha_V^{-1}$ and $\chi_0^\mu = \chi_0^w$. Then, from (5.181), we have

$$V_0(e_0) \geq \max\{\gamma_0^1(V_1(e_1)), \chi_0^d(|d|), \chi_0^\mu(\mu)\}$$
$$\Rightarrow \nabla V_0(e_0) f_{e_0}(e_0, y, d, w) \leq -\alpha_0(V_0(e_0)). \tag{5.191}$$

The e_1-subsystem

The e_1-subsystem can be rewritten as

$$\dot{e}_1 = \xi_2 - e_2 + (\phi_1(e_0, e_1, d) + e_2)$$
$$:= \xi_2 - e_2 + \phi_1^*(\bar{e}_2, d)$$
$$:= f_{e_1}(\bar{e}_2, \xi_2, d) \tag{5.192}$$

with the new state variable e_2 to be defined below. From Assumption 5.9 and the definition of ϕ_1, we can find a $\psi_{\phi_1^*} \in \mathcal{K}_\infty$ such that $|\phi_1^*(\bar{e}_2, d)| \leq \psi_{\phi_1^*}(|[\bar{e}_2^T, d^T]^T|)$.

Define a set-valued map S_1 as

$$S_1(e_1, \mu) = \{\kappa_1(e_1 + a\mu) : |a| \leq 1\} \tag{5.193}$$

with κ_1 continuously differentiable, odd, strictly decreasing, and radially unbounded, to be determined later. State variable e_2 is defined as

$$e_2 = \vec{d}(\xi_2, S_1(e_1, \mu)). \tag{5.194}$$

Then, we have $\xi_2 - e_2 \in S_1(e_1, \mu)$.

For any $\gamma_1^0, \gamma_1^2 \in \mathcal{K}_\infty$, choose $\chi_1^0 = \gamma_1^0 \circ \alpha_0$ and $\chi_1^2 = \gamma_1^2 \circ \alpha_V$. Then, $\gamma_1^0(V_0) = \chi_1^0 \circ \alpha_0^{-1}(V_0) \geq \chi_1^0(|e_0|)$ and $\gamma_1^2(V_2) = \chi_1^2 \circ \alpha_V^{-1}(V_2) = \chi_1^2(|e_2|)$. With Lemma 5.11, for any specified $0 < c_1 < 1$, $\epsilon_1 > 0$, $\ell_1 > 0$, $\gamma_1^0, \gamma_1^2, \chi_1^d \in \mathcal{K}_\infty$, we can find a continuously differentiable, odd, strictly decreasing, and radially unbounded κ_1 such that the e_1-subsystem with $\xi_2 - e_2 \in S_1(e_1, \mu)$ is ISS with V_1 satisfying

$$V_1 \geq \max\{\gamma_1^0(V_0), \gamma_1^2(V_2), \chi_1^d(|d|), \chi_1^\mu(\mu), \epsilon_1\}$$
$$\Rightarrow \nabla V_1(e_1) f_{e_1}(\bar{e}_2, \xi_2, d) \leq -\ell_1 V_1, \tag{5.195}$$

where $\chi_1^\mu(s) = \alpha_V(s/c_1)$ for $s \in \mathbb{R}_+$.

The definition of set-valued map S_1 is quite similar with the definition of the S_1 in Section 4.1 and can be represented by Figure 4.2 with the $\bar{w}_1$ replaced by μ.

The e_i-subsystem ($i = 2, \ldots, n$)

When $i = 3, \ldots, n$, for each $k = 2, \ldots, i-1$, a set-valued map S_k is defined as

$$S_k(e_1, \bar{\xi}_k, \mu) = \{\kappa_k(\xi_k - p_k) : p_k \in S_{k-1}(e_1, \bar{\xi}_{k-1}, \mu)\}, \tag{5.196}$$

where κ_k is continuously differentiable, odd, strictly decreasing, and radially unbounded; and the new state variable e_{k+1} is defined as

$$e_{k+1} = \vec{d}(\xi_{k+1}, S_k(e_1, \bar{\xi}_k, \mu)). \tag{5.197}$$

It is worth noting that, since κ_k is strictly decreasing, it holds that

$$\max S_k(e_1, \bar{\xi}_k, \mu) = \kappa_k(\xi_k - \max S_{k-1}(e_1, \bar{\xi}_{k-1}, \mu)), \tag{5.198}$$

$$\min S_k(e_1, \bar{\xi}_k, \mu) = \kappa_k(\xi_k - \min S_{k-1}(e_1, \bar{\xi}_{k-1}, \mu)). \tag{5.199}$$

Lemma 5.12 *Consider the $[e_0^T, e_1, \xi_2, \ldots, \xi_n]^T$-system in (5.186)–(5.189) with $|e_1| \leq M\mu$. With $S_k(e_k, \mu)$ and e_{k+1} defined in (5.193), (5.194), (5.196), and (5.197) for $k = 1, \ldots, i-1$, for any variable e_{i+1}, when $e_i \neq 0$, the e_i-subsystem can be represented as*

$$\dot{e}_i = \xi_{i+1} - e_{i+1} + \phi_i^*(\bar{e}_{i+1}, d, \mu, w, \bar{\xi}_i), \tag{5.200}$$

where

$$|\phi_i^*(\bar{e}_{i+1}, d, \mu, w, \bar{\xi}_i)| \leq \psi_{\phi_i^*}(\|[\bar{e}_{i+1}^T, d^T, \mu]^T\|) \tag{5.201}$$

with $\psi_{\phi_i^*} \in \mathcal{K}_\infty$. Specifically, $\xi_{n+1} = u$.

With the quantized observer designed above, the system in the output-feedback form has been transformed into the strict-feedback form and Lemma 5.12 can be proved in the same way as for the strict-feedback system in Section 5.2.

Define a set-valued map S_i as

$$S_i(e_1, \bar{\xi}_i, \mu) = \{\kappa_i(\xi_i - p_i) : p_i \in S_{i-1}(e_1, \bar{\xi}_{i-1}, \mu)\} \tag{5.202}$$

with κ_i continuously differentiable, odd, strictly decreasing, and radially unbounded, to be defined later. Define e_{i+1} as

$$e_{i+1} = \vec{d}(\xi_{i+1}, S_i(e_1, \bar{\xi}_i, \mu)). \tag{5.203}$$

Then, we have $\xi_{i+1} - e_{i+1} \in S_i(e_1, \bar{\xi}_i, \mu)$.

From the definition of e_i (i.e., e_{k+1} with $k = i-1$) in (5.197), in the case of $e_i \neq 0$, for all $p_i \in S_{i-1}(e_1, \bar{\xi}_{i-1}, \mu)$, it holds that $|\xi_i - p_i| \geq |e_i|$ and $\text{sgn}(\xi_i - p_i) = \text{sgn}(e_i)$, which means $\text{sgn}(\xi_i - p_i - e_i) = \text{sgn}(e_i)$, and thus $\xi_i - p_i = e_i + (\xi_i - p_i - e_i) = e_i + \text{sgn}(e_i)|\xi_i - p_i - e_i|$. Note that

$\xi_{i+1} - e_{i+1} \in S_i(e_1, \bar{\xi}_i, \mu)$. There always exists a $p_i \in S_{i-1}(e_1, \bar{\xi}_{i-1}, \mu)$ such that $\xi_{i+1} - e_{i+1} = \kappa_i(\xi_i - p_i) = \kappa_i(e_i + \text{sgn}(e_i)|\xi_i - p_i - e_i|)$.

With Lemma 5.11, for any $\epsilon_i > 0$, $\ell_i > 0$, $\gamma_i^0, \ldots, \gamma_i^{i-1}, \gamma_i^{i+1}, \chi_i^d, \chi_i^\mu \in \mathcal{K}_\infty$, we can find a continuously differentiable, odd, strictly decreasing, and radially unbounded κ_i such that the e_i-subsystem with $\xi_{i+1} - e_{i+1} \in S_i(e_1, \bar{\xi}_i, \mu)$ is ISS with V_i satisfying

$$V_i \geq \max_{k=0,\ldots,i-1,i+1} \left\{ \gamma_i^k(V_k), \chi_i^d(|d|), \chi_i^\mu(\mu), \epsilon_i \right\}$$
$$\Rightarrow \nabla V_i(e_i) f_{e_i}(\bar{e}_{i+1}, \bar{\xi}_{i+1}, d, \mu, w) \leq -\ell_i V_i. \qquad (5.204)$$

Here, f_{e_i} represents the dynamics of the e_i-subsystem, i.e., $\dot{e}_i = f_{e_i}(\bar{e}_{i+1}, \bar{\xi}_{i+1}, d, \mu, w)$. By default, $V_{n+1} := \alpha_V(|e_{n+1}|)$. The true control input $u = \xi_{n+1}$ occurs with the e_n-subsystem, and we set $e_{n+1} = 0$.

Realizable Quantized Controller

From (5.204) with $i = n$, our desired quantized controller u can be chosen in the following form:

$$p_2^* = \kappa_1(y^q) \qquad (5.205)$$
$$p_i^* = \kappa_{i-1}(\xi_{i-1} - p_{i-1}^*), \quad i = 3, \ldots, n \qquad (5.206)$$
$$u = \kappa_n(\xi_n - p_n^*). \qquad (5.207)$$

In the case of $|y| \leq M\mu$, we have $|w| = |y^q - y| \leq \mu$ and thus $\kappa_1(y^q) = \kappa_1(y + w) = \kappa_1(e_1 + w) \in S_1(e_1, \mu)$. It is then directly checked that

$$p_2^* \in S_1(e_1, \mu) \Rightarrow \cdots \Rightarrow p_i^* \in S_{i-1}(e_1, \bar{\xi}_{i-1}, \mu)$$
$$\Rightarrow \cdots \Rightarrow u = \xi_{n+1} - e_{n+1} \in S_n(e_1, \bar{\xi}_n, \mu),$$

where $e_{n+1} = 0$. Thus, if $|y| \leq M\mu$, then the quantized control law (5.205)–(5.207) guarantees (5.195) and (5.204).

5.3.3   CYCLIC-SMALL-GAIN SYNTHESIS

Denote $e = \bar{e}_n$ and $\xi = \bar{\xi}_n$. For $i = 0, \ldots, n$, each e_i-subsystem has been made ISS (or more precisely, practically ISS). In this subsection, we choose the ISS-gains such that the e-system satisfies the cyclic-small-gain condition. The gain digraph of the e-system is still in the form shown in Figure 5.13.

According to the recursive design, given the $\bar{e}_{i-1}$-subsystem, by designing the set-valued map S_i for the e_i-subsystem, we assign the ISS gains γ_i^k ($k = 1, \ldots, i - 1$) such that

$$\gamma_k^{k+1} \circ \gamma_{k+1}^{k+2} \circ \cdots \circ \gamma_{i-2}^{i-1} \circ \gamma_{i-1}^i \circ \gamma_i^k < \text{Id}. \qquad (5.208)$$

Applying this reasoning repeatedly, the e-system satisfies the cyclic-small-gain condition.

An ISS-Lyapunov function is constructed as:

$$V(e) = \max_{i=0,\ldots,n} \{\sigma_i(V_i(e_i))\} \tag{5.209}$$

with $\sigma_1(s) = s$, $\sigma_i(s) = \hat{\gamma}_1^2 \circ \cdots \circ \hat{\gamma}_{i-1}^i(s)$ $(i = 2,\ldots,n)$, and $\sigma_0(s) = \max_{i=1,\ldots,n}\{\sigma_i \circ \hat{\gamma}_i^0(s)\}$ for $s \in \mathbb{R}_+$, where the $\hat{\gamma}_{(\cdot)}^{(\cdot)}$'s are $\mathcal{K}_\infty$ functions continuously differentiable on $(0,\infty)$ and slightly larger than the corresponding $\gamma_{(\cdot)}^{(\cdot)}$'s and still satisfy the cyclic-small-gain condition.

Recall that $|d| \le \bar{d}$. Denote $\epsilon_0 = 0$. We represent the maximal influence of d, μ, and ϵ_i $(i = 1,\ldots,n)$ as

$$\theta = \max_{i=0,\ldots,n} \left\{\sigma_i \circ \chi_i^d(\bar{d}), \sigma_i \circ \chi_i^\mu(\mu), \sigma_i(\epsilon_i)\right\}. \tag{5.210}$$

Using the Lyapunov-based cyclic-small-gain theorem, we achieve that if $|y| \le M\mu$, then the e-system with quantized control law (5.205)–(5.207) satisfies

$$V(e) \ge \theta \Rightarrow \nabla V(e) f_e(e, \xi, d, \mu, w) \le -\alpha(V(e)) \tag{5.211}$$

holds wherever $\nabla V(e)$ exists, with α positive definite. Note that $\nabla V(e)$ exists almost everywhere. Here, f_e represents the dynamics of the e-system, i.e., $\dot{e} = f_e(e, \xi, d, \mu, w)$.

In the recursive design approach, we can make the $\gamma_{(\cdot)}^{(\cdot)}$'s (and thus the $\hat{\gamma}_{(\cdot)}^{(\cdot)}$'s) arbitrarily small to get arbitrarily small σ_i's $(i = 0, 2, \ldots, n)$. We can also select the χ_i^d's $(i = 0, \ldots, n)$, the ϵ_i's $(i = 1, \ldots, n)$, and the χ_i^μ's $(i = 0, 2, \ldots, n)$ to be arbitrarily small. In this way, for arbitrarily small $\theta_0 > 0$, we can design the gains such that $\max_{i=1,\ldots,n} \left\{\sigma_i \circ \chi_i^d(\bar{d}), \sigma_i(\epsilon_i)\right\} \le \theta_0$ and $\max_{i=0,2,\ldots,n} \left\{\sigma_i \circ \chi_i^\mu(\mu)\right\} \le \theta_0$.

Recall that $\chi_1^\mu(s) = \alpha_V(s/c_1)$ for $s \in \mathbb{R}_+$ defined in (5.195). If $|y| \le M\mu$, then quantized control law (5.205)–(5.207) guarantees

$$V(e) \ge \max\{\alpha_V(\mu/c_1), \theta_0\} \Rightarrow \nabla V(e) f_e(e, \xi, d, \mu, w) \le -\alpha(V(e)) \tag{5.212}$$

wherever $\nabla V(e)$ exists.

5.3.4   DYNAMIC QUANTIZATION AND MAIN RESULT

Define $\Theta = \alpha_V(M\mu)$. Then, $\mu = \alpha_V^{-1}(\Theta)/M$. The update law for μ can be determined by designing an update law for Θ. Denote $x = [x_1, \ldots, x_n]^T$ and $\xi = [\xi_2, \ldots, \xi_n]^T$. Recall that $e = [e_0, \ldots, e_n]^T$ and the definition of e_i for $i = 0, \ldots, n$. The transformed state variable e can be considered as a continuous function of x, ξ, μ. By using $\mu = \alpha_V^{-1}(\Theta)/M$, we can denote $e = e(x, \xi, \Theta)$. In dynamic quantization, Θ is piecewise updated on the timeline and denoted as $\Theta(t)$. Clearly, the piecewise update of Θ leads to jumps of e.

Some of the results in this section can be considered as special cases of Subsection 5.2.8 and are presented without detailed proofs.

Zooming-Out Stage

In this stage, the control input u and the state ξ of the observer are set to be zero. The small-time norm-observability assumed in Assumption 5.8 guarantees that for $d_t > 0$, there exists a $\varphi \in \mathcal{K}_\infty$ such that

$$|x(t_k + d_t)| \leq \varphi(\|y\|_{[t_k, t_k+d_t]}) \tag{5.213}$$

for all $k \in \mathbb{Z}_+$. Considering the definitions of V and e, for $d_t > 0$, there exists a $\bar\varphi \in \mathcal{K}_\infty$ such that

$$|V(e(x(t_k + d_t), 0, 0))| \leq \bar\varphi(\|y\|_{[t_k, t_k+d_t]}) \tag{5.214}$$

for all $k \in \mathbb{Z}_+$.

The forward completeness assumed in Assumption 5.8 guarantees that we can increase Θ fast enough to dominate the growth rate of $\bar\varphi(|y|)$. Thus, we can design the zooming-out logic to increase Θ (and thus μ) fast enough such that at some time $t_{k^*} > 0$ with $k^* \in \mathbb{Z}_+$, it holds that

$$\Theta(t_{k^*}) \geq \bar\varphi(\|y\|_{[t_{k^*}-d_t, t_{k^*}]}) \geq \max\{V(e(x(t_{k^*}), 0, 0)), \theta_0\}. \tag{5.215}$$

From the definition of S_i in (5.193) and (5.202), it can be observed that an increase of μ (and thus Θ) leads to an increase of $\max S_i$ and a decrease of $\min S_i$. Using the definition of e_{i+1}, an increase of Θ leads to a decrease or hold of $|e_{i+1}|$ (and thus a decrease or hold of $V(e)$). Note that $\xi(t_{k^*}) = 0$. From (5.215), we achieve

$$\Theta(t_{k^*}) \geq \max\{V(e(x(t_{k^*}), \xi(t_{k^*}), \Theta(t_{k^*}))), \theta_0\}. \tag{5.216}$$

Zooming-In Stage

With the help of Assumption 5.11, in the recursive control design procedure, we can choose c_1 satisfying $1/M < c_1 < 1$. Then, one can find a positive definite ρ_1^z such that

$$\alpha_V(\mu/c_1) \leq (\mathrm{Id} - \rho_1^z)(\Theta). \tag{5.217}$$

Suppose that at some time $t_k > 0$ with $k \in \mathbb{Z}_+$, it holds that

$$\Theta(t_k) \geq \max\{V(e(x(t_k), \xi(t_k), \Theta(t_k))), \theta_0\}. \tag{5.218}$$

We want to find a $Q_{\mathrm{in}}^\Theta : \mathbb{R}_+ \to \mathbb{R}_+$ such that $\Theta(t_{k+1}) = Q_{\mathrm{in}}^\Theta(\Theta(t_k))$ satisfies

$$\Theta(t_{k+1}) \geq \max\{V(e(x(t_{k+1}), \xi(t_{k+1}), \Theta(t_{k+1}))), \theta_0\}, \tag{5.219}$$

where $t_{k+1} - t_k = d_t$.

One can find a positive definite ρ_2^z such that $(\mathrm{Id} - \rho_2^z) \in \mathcal{K}_\infty$ and $(\mathrm{Id} - \rho_2^z)(s) \geq \max\{(\mathrm{Id} - \rho_1^z)(s), s - d_t \cdot \min_{(\mathrm{Id}-\rho_1^z)(s) \leq v \leq s} \alpha(V)\}$ for $s \in \mathbb{R}_+$. Define

$$\Xi = \mathrm{Id} - \rho_2^z. \tag{5.220}$$

Condition (5.218) implies that $V(e(x(t_k), \xi(t_k), \Theta(t_k))) \leq \alpha_V(M\mu(t_k))$. From (5.212) and (5.217), if (5.218) holds, then

$$V(e(x(t_{k+1}), \xi(t_{k+1}), \Theta(t_k))) \leq \max\{\Xi(\Theta(t_k)), \theta_0\}. \qquad (5.221)$$

Using the property of continuous functions, we can find a positive definite $\rho_3^z < \mathrm{Id}$ such that for all $x \in \mathbb{R}^n$, $\xi \in \mathbb{R}^{n-1}$, $\Theta > 0$, and $h \geq 0$, it holds that

$$|V(e(x, \xi, \Theta - \rho_3^z(h))) - V(e(x, \xi, \Theta))| \leq h. \qquad (5.222)$$

Define

$$\Theta_{\mathrm{in}}^{\Theta}(\Theta) = \Theta - \rho_3^z\left(\frac{\Theta - \max\{\Xi(\Theta), \theta_0\}}{2}\right). \qquad (5.223)$$

Then, (5.221), (5.222), and (5.223) imply

$$V(e(x(t_{k+1}), \xi(t_{k+1}), \Theta(t_{k+1}))) \leq \frac{\Theta(t_k) + \max\{\Xi(\Theta(t_k)), \theta_0\}}{2}, \qquad (5.224)$$

and (5.218) and (5.223) imply

$$\Theta(t_{k+1}) \geq \frac{\Theta(t_k) + \max\{\Xi(\Theta(t_k)), \theta_0\}}{2} \geq \theta_0. \qquad (5.225)$$

Properties (5.224) and (5.225) together guarantee (5.219).

Lemma 5.13 *Suppose that $\Theta(t_{k^*}) \geq \theta_0$ with $k^* \in \mathbb{Z}_+$. Then, with zooming-in logic $\Theta(t_{k+1}) = Q_{\mathrm{in}}^{\Theta}(\Theta(t_k))$ for $k \in \mathbb{Z}_+$ with Q_{in}^{Θ} defined in (5.223), it holds that*

$$\lim_{k \to \infty} \Theta(t_k) = \theta_0. \qquad (5.226)$$

Lemma 5.13 can be proved in the same way as Lemma 5.10.

With the appropriately designed zooming-in logic in (5.223), it always holds that $V(e(x(t), \xi(t), \Theta(t))) \leq \Theta(t)$. Thus, the closed-loop signals are bounded. By using (5.226), we have

$$\varlimsup_{k \to \infty} V(e(x(t), \xi(t), \Theta(t))) = \theta_0. \qquad (5.227)$$

Recall the definition of V in (5.209). It can be observed that $y = x_1 = e_1$ ultimately converges to within the region $|y| \leq \alpha_V^{-1}(\theta_0)$. By choosing θ_0 to be arbitrarily small, output y can be steered to within an arbitrarily small neighborhood of the origin.

Recall that $\Theta = \alpha_V(M\mu)$. With Q_{in}^{Θ} defined in (5.223), the zooming-in logic for μ is designed as

$$Q(\mu) = Q_{\mathrm{in}}(\mu) = \frac{1}{M}\alpha_V^{-1} \circ Q_{\mathrm{in}}^{\Theta} \circ \alpha_V(M\mu). \qquad (5.228)$$

The main result of this section is summarized in Theorem 5.3.

Theorem 5.3 *Consider system* (5.167)–(5.170) *with output quantization satisfying* (5.171). *Under Assumptions 5.8–5.11, the closed-loop signals are bounded, and in particular, the output y can be steered to within an arbitrarily small neighborhood of the origin with the quantized output-feedback controller composed of reduced-order observer* (5.175)–(5.176), *control law* (5.205)–(5.207), *and dynamic quantization in the form of* (5.172) *with zooming-in dynamics* $Q = Q_{in}$ *defined in* (5.228).

The block diagram of the quantized output-feedback control system designed in this section is shown in Figure 5.15. Interested readers may try designing quantized controllers with actuator quantization by combining the designs in this section and Section 5.2. Notice that the recent paper [191] presents a result on output-feedback control of nonlinear systems with actuator quantization.

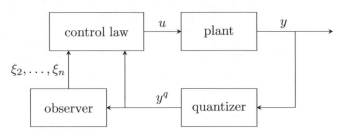

FIGURE 5.15   Quantized output-feedback control.

5.4   NOTES

Recent years have seen considerable efforts devoted to quantized control of linear and nonlinear systems. Reference [55] studied quantized stabilization of single-input single-output (SISO) linear systems with the coarsest quantizers, and showed that the coarsest quantizer should follow a logarithmic law for quadratic stabilization. By characterizing the coarsest quantizer as a sector bounded uncertainty, the authors of [67] considered the quantized control of multi-input multi-output (MIMO) linear systems and analyzed the robustness of the quantized control systems. An early result on quantized control of nonlinear systems with logarithmic quantizers appeared in [175], in which the general idea of using (robust) control Lyapunov functions to design (robust) quantized controllers is employed. Reference [26] studied the conditions under which a logarithmic quantizer does not cancel the stabilizing effect of a continuous feedback control law, for quantized control of dissipative nonlinear systems. In [26], set-valued maps are employed to overcome the problems caused by the discontinuity of the quantizers.

Based on the idea of scaling quantization levels, the authors of [19, 163, 168, 164] studied quantized control of linear and nonlinear systems with dynamic quantization. If the quantizer admits a finite number of levels, then the quantization error is large if the original signal is outside the range of the quantizer. To deal with this problem, a growth condition should be satisfied by the quantization error and the quantized signal [209, 163]. Reference [46] presents a semiglobal stabilization result for nonlinear systems in the feedforward form. Reference [163] clarifies the relation between ISS and quantized control. ISS with respect to quantization error appears to be fundamental in several results on quantized control; see e.g., Liberzon and Nešić's results [163, 165, 166, 168]. Reference [209] established a unified framework for control design of nonlinear systems with quantization and time scheduling via an emulation-like approach, with the ISS small-gain theorem [130] (see also Chapter 2) as a tool. For the systems with partial state information, a quantized output-feedback control strategy was developed [166].

Because of the discontinuity of the quantizers, a closed-loop quantized system is basically a discontinuous system and can be modeled by differential inclusions. Then, the extended Filippov solution introduced in [84] can be used to represent the motion of such systems. Based on the extended Filippov solution, the cyclic-small-gain theorem proposed in Chapter 3 can be directly generalized to dynamic networks described by differential inclusions. Detailed discussions on discontinuous systems and the Filippov solution can be found in [60, 35]. Appendix B gives a brief introduction to the related notions and the basic results.

The sector bound approach presented in this chapter for static quantization is motivated by [67, 26], which directly assume the sector bound property of the quantizers instead of discussing the nonlinearity and the discontinuity of the quantizers in details. In this chapter, set-valued maps have been employed to cover the sector bound of the quantizers such that the closed-loop quantized system is transformed into a network of ISS subsystems. Then, the quantized control design is finalized with the cyclic-small-gain theorem, and the influence of the quantization errors are explicitly represented by ISS gains.

System (5.73)–(5.75) represents an important class of minimum-phase nonlinear systems, which have been studied extensively by many authors in the context of (non-quantized) robust and adaptive nonlinear control. The reader may consult [153, 222, 262] and references therein for the details. For semiglobal quantized stabilization of system (5.73)–(5.75), it is assumed that the uncontrolled system is final-state norm-observable such that the quantized controller can estimate an upper bound of the internal state in some finite time at the zooming-out stage. Reference [85] discussed the equivalent characterizations of initial-state norm-observability, final-state norm-observability, and $\mathcal{KL}$ norm-observability for forward complete (or unboundedness observable) nonlinear systems with external inputs. By means of dynamic quantization, an n-dimensional strict-feedback nonlinear system with measurement and actuator quantization can be semiglobally stabilized (with global convergence of the closed-loop state signals) by a quantized controller with $2n$ three-level dynamic quantizers. When the strict-feedback system is reduced to the output-feedback form, quantized output-feedback control is solved by introducing a quantized observer. This result is introduced in Section 5.3.

In spite of the obtained results, several related problems should be addressed in the future research:

- Quantized control is closely related to other network control problems such as sampled-data control and control with time-delays. How to deal with more complicated network behaviors in a systematic way, in particular those hybrid/switching systems satisfying only a weak semigroup property (see [137]), should be studied in greater detail.
- The cyclic-small-gain theorem was originally developed for large-scale systems. It is thus very natural to ask whether decentralized quantized controllers can be developed for a class of large-scale nonlinear systems.
- Controllers are expected to possess adaptive capabilities to cope with "large" system uncertainties. A further extension of the presented methodology to quantized adaptive control is of practical interest for engineering applications.

More discussions on quantized control of nonlinear uncertain systems can be found in [122, 191, 189, 190] as well as Chapter 7.

6 Distributed Nonlinear Control

The spatially distributed structure of complex systems motivates the idea of distributed control. In a distributed control system, the subsystems are controlled by local controllers through information exchange with neighboring agents for coordination purposes. Formation control of mobile robots is an example. The major difficulties of distributed control are due to complex characteristics such as nonlinearity, dimensionality, uncertainty, and information constraints. This chapter develops small-gain methods for distributed control of nonlinear systems.

The discussion in this chapter starts with an example of a multi-vehicle formation control system in which each vehicle is modeled by an integrator. In the case of leader-following with fixed topology, it is shown that the problem can be transformed into the stability problem of a specific dynamic network composed of ISS subsystems. This motivates a cyclic-small-gain result in digraphs, which is given in Section 6.1. It is shown that the new result is extremely useful for distributed control of nonlinear systems. Specifically, Section 6.2 presents a cyclic-small-gain design for distributed output-feedback control of nonlinear systems. In Section 6.3, we study the distributed formation control problem of nonholonomic mobile robots with a fixed information exchange topology. An extension to the case of flexible topology is developed in Section 6.4.

Example 6.1 *Consider a group of $N + 1$ vehicles (multi-vehicle system) as shown in Figure 6.1, with each vehicle modeled by an integrator:*

$$\dot{x}_i = v_i, \quad i = 0, \ldots, N, \tag{6.1}$$

where $x_i \in \mathbb{R}$ is the position and $v_i \in \mathbb{R}$ is the velocity of the i-th vehicle. The vehicle with index 0 is the leader while the other vehicles are the followers. The objective is to control the follower vehicles to specific positions relative to the leader by adjusting the velocities v_i for $i = 1, \ldots, N$. More specifically, it is required that

$$\lim_{t \to \infty} (x_i(t) - x_j(t)) = d_{ij}, \quad i, j = 0, \ldots, N, \tag{6.2}$$

where constants d_{ij} represent the desired relative positions. Clearly, to define the problem well, $d_{ij} = d_{ik} + d_{kj}$ for any $i, j, k = 0, \ldots, N$ and $d_{ij} = -d_{ji}$ for any $i, j = 0, \ldots, N$. Also, by default, $d_{ii} = 0$ for any $i = 0, \ldots, N$. In the literature of distributed control, the vehicles are usually considered as agents and the multi-vehicle system is studied as a multi-agent system.

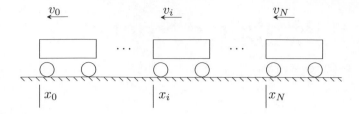

FIGURE 6.1 A multi-vehicle system.

Compared with global positions, relative positions between the vehicles are often easily measurable in practice, and are used for feedback in this example. Considering the position information exchange, agent j is called a neighbor of agent i if $(x_i - x_j)$ is available to agent i, and $\mathcal{N}_i \subseteq \{0, \dots, N\}$ is used to denote the set of agent i's neighbors. We consider the case where each vehicle only uses the position differences with the vehicles right before and after it, i.e., $\mathcal{N}_i = \{i-1, i+1\}$ for $i = 1, \dots, N-1$ and $\mathcal{N}_N = \{N-1\}$.

Define $\tilde{x}_i = x_i - x_0 - d_{i0}$ and $\tilde{v}_i = v_i - v_0$. By taking the derivative of $\tilde{x}_i$, we have

$$\dot{\tilde{x}}_i = \tilde{v}_i, \quad i = 1, \dots, N. \tag{6.3}$$

According to the definition of $\tilde{x}_i$, $\tilde{x}_i - \tilde{x}_j = x_i - x_j - d_{ij}$. Thus, the control objective is achieved if $\lim_{t \to \infty}(\tilde{x}_i - \tilde{x}_0) = 0$. Also, $(\tilde{x}_i - \tilde{x}_j)$ is available to the control of the $\tilde{x}_i$-subsystem if $(x_i - x_j)$ is available to agent i. This problem is normally known as the consensus problem. If the position information exchange topology has a spanning tree with agent 0 as the root, then the following distributed control law is effective:

$$\tilde{v}_i = k_i \sum_{j \in \mathcal{N}_i} (\tilde{x}_j - \tilde{x}_i), \tag{6.4}$$

where k_i is a positive constant. Moreover, if the velocities v_i are required to be bounded, one may modify (6.4) as

$$\tilde{v}_i = \varphi_i \left(\sum_{j \in \mathcal{N}_i} (\tilde{x}_j - \tilde{x}_i) \right), \tag{6.5}$$

where $\varphi_i : \mathbb{R} \to [\underline{v}_i, \overline{v}_i]$ with constants $\underline{v}_i < 0 < \overline{v}_i$ is a continuous, strictly increasing function satisfying $\varphi_i(0) = 0$. With control law (6.5), $v_i \in [v_0 + \underline{v}_i, v_0 + \overline{v}_i]$. The validity of the control laws defined by (6.4) and (6.5) can be directly verified by using the state agreement result in [172].

With control law (6.5), each $\tilde{x}_i$-subsystem can be rewritten as

$$\dot{\tilde{x}}_i = \varphi_i \left(\sum_{j \in \mathcal{N}_i} \tilde{x}_j - N_i \tilde{x}_i \right) := f_i(\tilde{x}), \tag{6.6}$$

where N_i is the size of $\mathcal{N}_i$ and $\tilde{x} = [\tilde{x}_0, \ldots, \tilde{x}_N]^T$. Define $V_i(\tilde{x}_i) = |\tilde{x}_i|$ as an ISS-Lyapunov function candidate for the $\tilde{x}_i$-subsystem for $i = 1, \ldots, N$. It can be verified that for any $\delta > 0$, there exists a continuous, positive definite α such that

$$V_i(\tilde{x}_i) \geq \frac{1}{(1-\delta_i)N_i} \sum_{j \in \mathcal{N}_i} V_j(\tilde{x}_j) \Rightarrow \nabla V_i(\tilde{x}_i)f_i(\tilde{x}) \leq -\alpha_i(V_i(\tilde{x}_i)) \quad a.e., \quad (6.7)$$

where, for convenience of notation, $V_0(\tilde{x}_0) = 0$. This shows the ISS of each $\tilde{x}_i$-subsystem with $i = 1, \ldots, N$. If the network of ISS subsystems is asymptotically stable, then the control objective is achieved.

We employ a digraph $\mathcal{G}_f$ to represent the underlying interconnection structure of the dynamic network. The vertices of the digraph correspond to agents $1, \ldots, N$, and for $i, j = 1, \ldots, N$, directed edge (j, i) exists in the graph if and only if $\tilde{x}_j$ is an input of the x_i-subsystem. We use $\overline{\mathcal{N}}_i$ to represent the set of neighbors of agent i in $\mathcal{G}_f$. Then, it is directly verified that $\overline{\mathcal{N}}_i = \mathcal{N}_i \backslash \{0\}$. Recall that $V_0(\tilde{x}_0) = 0$. Then, the $\mathcal{N}_i$ in (6.7) can be directly replaced by $\overline{\mathcal{N}}_i$. Figure 6.2 shows the digraph $\mathcal{G}_f$ for the case in which each follower vehicle uses the position differences with the vehicles right before and after it.

$$1 \rightleftarrows 2 \rightleftarrows 3 \rightleftarrows \cdots \rightleftarrows N$$

FIGURE 6.2   An example of information exchange digraph $\mathcal{G}_f$, for which each vehicle uses the position differences with the vehicles right before and after it. In this figure, $\overline{\mathcal{N}}_i = \{i-1, i+1\}$ for $i = 2, \ldots, N-1$, $\overline{\mathcal{N}}_1 = \{2\}$ and $\overline{\mathcal{N}}_N = \{N-1\}$.

Notice that for any positive constants $a_1, \ldots, a_n$ satisfying $\sum_{i=1}^n 1/a_i \leq n$, it holds that $\sum_{i=1}^n d_i = \sum_{i=1}^n (1/a_i)a_i d_i \leq n \max_{i=1,\ldots,n}\{a_i d_i\}$ for all $d_1, \ldots, d_n \geq 0$. Then, property (6.7) implies

$$V_i(\tilde{x}_i) \geq \frac{\overline{N}_i}{(1-\delta_i)N_i} \max_{j \in \overline{\mathcal{N}}_i}\{a_{ij}V_j(\tilde{x}_j)\} \Rightarrow \nabla V_i(\tilde{x}_i)f_i(\tilde{x}) \leq -\alpha_i(V_i(\tilde{x}_i)), \quad (6.8)$$

where $\overline{N}_i$ is the size of $\overline{\mathcal{N}}_i$ and a_{ij} are positive constants satisfying $\sum_{j \in \overline{\mathcal{N}}_i} 1/a_{ij} \leq \overline{N}_i$. It can be observed that $N_i = \overline{N}_i + 1$ if $0 \in \mathcal{N}_i$ and $N_i = \overline{N}_i$ if $0 \notin \mathcal{N}_i$.

Given specific $a_{ij} > 0$, one can test the stability property of the closed-loop system by directly checking whether the cyclic-small-gain condition is satisfied. But, for a specific $\mathcal{G}_f$, can we find appropriate coefficients a_{ij} to satisfy the cyclic-small-gain condition, and how?

It should be noted that the effectiveness of control law (6.5) can be proved by using the result in [172]. Here, our objective is to transform the problem into a stability problem of dynamic networks, and develop a result which is hopefully useful for more general distributed control problems.

To answer the question in Example 6.1, a cyclic-small-gain result in digraphs is developed in Section 6.1.

6.1   A CYCLIC-SMALL-GAIN RESULT IN DIGRAPHS

Consider a digraph $\mathcal{G}_f$ which has N vertices. For $i = 1, \ldots, N$, define $\overline{\mathcal{N}}_i$ such that if there is a directed edge (j, i) from the j-th vertex to the i-th vertex, then $j \in \overline{\mathcal{N}}_i$. Each edge (j, i) is assigned a positive variable a_{ij}. For a simple cycle $\mathcal{O}$ of $\mathcal{G}_f$, denote $A_\mathcal{O}$ as the product of the positive values assigned to the edges of the cycle. For $i = 1, \ldots, N$, denote $\mathcal{C}(i)$ as the set of simple cycles of $\mathcal{G}_f$ through the i-th vertex.

Lemma 6.1 *If the digraph $\mathcal{G}_f$ has a spanning tree $\mathcal{T}_f$ with vertices $i_1^*, \ldots, i_q^*$ as the roots, then for any $\epsilon > 0$, there exist $a_{ij} > 0$ for $i = 1, \ldots, N$, $j \in \overline{\mathcal{N}}_i$, such that*

$$\sum_{j \in \overline{\mathcal{N}}_i} \frac{1}{a_{ij}} \leq \overline{N}_i, \quad i = 1, \ldots, N \tag{6.9}$$

$$A_\mathcal{O} < 1 + \epsilon, \quad \mathcal{O} \in \mathcal{C}(i_1^*) \cup \cdots \cup \mathcal{C}(i_q^*) \tag{6.10}$$

$$A_\mathcal{O} < 1, \quad \mathcal{O} \in \left(\bigcup_{i=1,\ldots,N} \mathcal{C}(i) \right) \setminus \left(\mathcal{C}(i_1^*) \cup \cdots \cup \mathcal{C}(i_q^*) \right), \tag{6.11}$$

where $\overline{N}_i$ is the size of $\overline{\mathcal{N}}_i$.

Proof. We only consider the case of $q = 1$. The case of $q \geq 2$ can be proved similarly. Denote i^* as the root of the tree.

Define $a_{ij}^0 = 1$ for $1 \leq i \leq N$, $j \in \overline{\mathcal{N}}_i$. If $a_{ij} = a_{ij}^0$ for $1 \leq i \leq N$, $j \in \overline{\mathcal{N}}_i$, then

$$\sum_{j \in \overline{\mathcal{N}}_i} \frac{1}{a_{ij}^0} \leq \overline{N}_i, \quad i = 1, \ldots, N \tag{6.12}$$

$$A_\mathcal{O} = 1, \quad \mathcal{O} \in \bigcup_{i=1,\ldots,N} \mathcal{C}(i). \tag{6.13}$$

Consider one of the paths leading from root i^* in the spanning tree $\mathcal{T}_f$. Denote the path as $(p_1, \ldots, p_m)$ with $p_1 = i^*$.

One can find $a_{p_2 p_1}^1 = a_{p_2 p_1}^0 + \epsilon_{p_2 p_1}^0 > 0$ with $\epsilon_{p_2 p_1}^0 > 0$ and $a_{p_2 j}^1 = a_{p_2 j}^0 - \epsilon_{p_2 j}^0 > 0$ with $\epsilon_{p_2 j}^0 > 0$ for $j \in \overline{\mathcal{N}}_{p_2} \setminus \{p_1\}$ such that if $a_{ij} = a_{ij}^1$ for $i = p_2$ and $a_{ij} = a_{ij}^0$ for $i \neq p_2$, then (6.12) is satisfied, and also

$$A_\mathcal{O} < 1 + \epsilon' \text{ for } \mathcal{O} \in \mathcal{C}(p_1), \tag{6.14}$$

$$A_\mathcal{O} < 1 \text{ for } \mathcal{O} \in \mathcal{C}(p_2) \setminus \mathcal{C}(p_1) \tag{6.15}$$

with $0 < \epsilon' < \epsilon$.

Then, one can find $a_{p_3 p_2}^1 = a_{p_3 p_2}^0 + \epsilon_{p_3 p_2}^0 > 0$ with $\epsilon_{p_3 p_2}^0 > 0$ and $a_{p_3 j}^1 = a_{p_3 j}^0 - \epsilon_{p_3 j}^0 > 0$ with $\epsilon_{p_3 j}^0 > 0$ for $j \in \overline{\mathcal{N}}_{p_3} \setminus \{p_2\}$ such that if $a_{ij} = a_{ij}^1$ for

$i \in \{p_2, p_3\}$, and $a_{ij} = a_{ij}^0$ for $i \notin \{p_2, p_3\}$, then (6.12) is satisfied, and also

$$A_{\mathcal{O}} < 1 + \epsilon'' \text{ for } \mathcal{O} \in \mathcal{C}(p_1), \tag{6.16}$$

$$A_{\mathcal{O}} < 1 \text{ for } \mathcal{O} \in (\mathcal{C}(p_2) \cup \mathcal{C}(p_3)) \backslash \mathcal{C}(p_1) \tag{6.17}$$

with $0 < \epsilon' \le \epsilon'' < \epsilon$.

By doing this for $i = p_2, \ldots, p_m$, we can find $a_{ij}^1 > 0$ for $i \in \{p_2, \ldots, p_m\}$, $j \in \overline{\mathcal{N}}_i$, such that

$$A_{\mathcal{O}} < 1 + \epsilon_1 \text{ for } \mathcal{O} \in \mathcal{C}(p_1), \tag{6.18}$$

$$A_{\mathcal{O}} < 1 \text{ for } \mathcal{O} \in (\mathcal{C}(p_2) \cup \cdots \cup \mathcal{C}(p_m)) \backslash \mathcal{C}(p_1) \tag{6.19}$$

with $0 < \epsilon_0 < \epsilon$.

By considering each path leading from the root i^* in the spanning tree one-by-one, we can find $a_{ij}^1 > 0$ for $i \in \{1, \ldots, N\}$, $j \in \overline{\mathcal{N}}_i$, such that if $a_{ij} = a_{ij}^1$ for $i \in \{1, \ldots, N\}$, $j \in \overline{\mathcal{N}}_i$, then (6.12) and (6.11) are satisfied and

$$A_{\mathcal{O}} < 1 + \epsilon^1 \text{ for } \mathcal{O} \in \mathcal{C}(i_1^*) \cup \cdots \cup \mathcal{C}(i_q^*), \tag{6.20}$$

where $0 < \epsilon^1 < \epsilon$.

Note that the left-hand sides of inequalities (6.9), (6.10), and (6.11) continuously depend on a_{ij} for $i \in \{1, \ldots, N\}$, $j \in \overline{\mathcal{N}}_i$. One can find $a_{ij}^2 > 0$ for $i \in \{1, \ldots, N\}$, $j \in \overline{\mathcal{N}}_i$, such that if $a_{ij} = a_{ij}^2$ for $i \in \{1, \ldots, N\}$, $j \in \overline{\mathcal{N}}_i$, then conditions (6.9), (6.10), and (6.11) are satisfied. $\Diamond$

Example 6.2 *Continue Example 6.1. Define* $\mathcal{L} = \{i \in \{1, \ldots, N\} : 0 \in \mathcal{N}_i\}$. *Considering the relation between* N_i *and* $\overline{N}_i$, *and* $\overline{N}_i \le N$, *the cyclic-small-gain condition can be satisfied by the network of ISS subsystems with property* (6.8) *if*

$$A_{\mathcal{O}} < \frac{(1 - \bar{\delta})^N (N + 1)}{N}, \quad \mathcal{O} \in \bigcup_{i \in \mathcal{L}} \mathcal{C}(i), \tag{6.21}$$

$$A_{\mathcal{O}} < (1 - \bar{\delta})^N, \quad \mathcal{O} \in \left(\bigcup_{i \in \{1, \ldots, N\}} \mathcal{C}(i) \right) \backslash \left(\bigcup_{i \in \mathcal{L}} \mathcal{C}(i) \right), \tag{6.22}$$

where $\bar{\delta} = \max_{i=1, \ldots, N} \{\delta_i\}$.

By using Lemma 6.1, if graph $\mathcal{G}_f$ *has a spanning tree with the agents belonging to* $\mathcal{L}$ *as the roots, one can find a constant* $\bar{\delta} > 0$ *and constants* $a_{ij} > 0$ *satisfying* $\sum_{j \in \overline{\mathcal{N}}_i} 1/a_{ij} \le \overline{N}_i$ *such that conditions (6.21) and (6.22) are satisfied. The graph shown in Figure 6.2 satisfies this condition.*

Lemma 6.1 proves very useful in constructing distributed controllers for nonlinear agents to achieve convergence of their outputs to an agreement value. It provides for a form of gain assignment in the network coupling.

6.2 DISTRIBUTED OUTPUT-FEEDBACK CONTROL

In this section, the basic idea of cyclic-small-gain design for distributed control is generalized to high-order nonlinear systems. Consider a group of N nonlinear agents, of which each agent i $(1 \leq i \leq N)$ is in the output-feedback form:

$$\dot{x}_{ij} = x_{i(j+1)} + \Delta_{ij}(y_i, w_i), \quad 1 \leq j \leq n_i - 1 \tag{6.23}$$

$$\dot{x}_{in_i} = u_i + \Delta_{in_i}(y_i, w_i) \tag{6.24}$$

$$y_i = x_{i1}, \tag{6.25}$$

where $[x_{i1}, \ldots, x_{in_i}]^T := x_i \in \mathbb{R}^{n_i}$ with $x_{ij} \in \mathbb{R}$ $(1 \leq j \leq n_i)$ is the state, $u_i \in \mathbb{R}$ is the control input, $y_i \in \mathbb{R}$ is the output, $[x_{i2}, \ldots, x_{in_i}]^T$ is the unmeasured portion of the state, $w_i \in \mathbb{R}^{n_{w_i}}$ represents external disturbances, and Δ_{ij}'s $(1 \leq j \leq n_i)$ are unknown locally Lipschitz functions.

The objective of this section is to develop a new class of distributed controllers for the multi-agent system based on available information such that the outputs y_i for $1 \leq i \leq N$ converge to the same desired agreement value y_0. This problem is called the output agreement problem in this book.

In Section 4.3, decentralized control was developed such that a group of nonlinear systems can be stabilized despite the nonlinear interconnections between them. Different from decentralized control, the major objective of distributed control is to control the agents in a coordinated way for some desired group behavior. For the output agreement problem, the objective is to control the agents so that the outputs converge to a desired common value. Information exchange between the agents is required for coordination purposes. In practice, the information exchange is subject to constraints. As considered in Example 6.1, the position x_0 of the leader vehicle is only available to some of the follower vehicles, and the formation control objective is achieved through information exchange between the neighboring vehicles.

For distributed control of the multi-agent nonlinear system (6.23)–(6.25), we employ a digraph $\mathcal{G}^c$ to represent the information exchange topology between the agents. Digraph $\mathcal{G}^c$ contains N vertices corresponding to the N agents and M directed edges corresponding to the information exchange links. Specifically, if $y_i - y_k$ is available to the local controller design of agent i, then there is a directed link from agent k to agent i and agent k is called a neighbor of agent i; otherwise, there is no link from agent k to agent i. Set $\mathcal{N}_i \subseteq \{1, \ldots, N\}$ is used to represent agent i's neighbors. In this section, an agent is not considered as a neighbor of itself and thus $i \notin \mathcal{N}_i$ for $1 \leq i \leq N$. Agent i is called an informed agent if it has access to the knowledge of the agreement value y_0 for its local controller design. Let $\mathcal{L} \subseteq \{1, \ldots, N\}$ represent the set of all the informed agents.

The following assumption is made on the agreement value and system (6.23)–(6.25).

Assumption 6.1 *There exists a nonempty set* $\Omega \subseteq \mathbb{R}$ *such that*

1. $y_0 \in \Omega$;
2. for each $1 \le i \le N$, $1 \le j \le n_i$,

$$|\Delta_{ij}(y_i, w_i) - \Delta_{ij}(z_i, 0)| \le \psi_{\Delta_{ij}}(|[y_i - z_i, w_i^T]^T|) \qquad (6.26)$$

for all $[y_i, w_i^T]^T \in \mathbb{R}^{1+n_{w_i}}$ and all $z_i \in \Omega$, where $\psi_{\Delta_{ij}} \in \mathcal{K}_\infty$ is Lipschitz on compact sets and known.

It should be noted that *a priori* information on the bounds of y_0 (and thus Ω) is usually known in practice. In this case, condition 2 in Assumption 6.1 can be guaranteed if for each z_i, there exists a $\psi_{\Delta_{ij}}^{z_i} \in \mathcal{K}_\infty$ that is Lipschitz on compact sets such that

$$
\begin{aligned}
|\Delta_{ij}(y_i, w_i) - \Delta_{ij}(z_i, 0)| &= |\Delta_{ij}((y_i - z_i) + z_i, w_i) - \Delta_{ij}(z_i, 0)| \\
&\le \psi_{\Delta_{ij}}^{z_i}(|[y_i - z_i, w_i^T]^T|). \qquad (6.27)
\end{aligned}
$$

Then, $\psi_{\Delta_{ij}}$ can be defined as $\psi_{\Delta_{ij}}(s) = \sup_{z_i \in \Omega} \psi_{\Delta_{ij}}^{z_i}(s)$ for $s \in \mathbb{R}_+$. In fact, there always exists a $\psi_{\Delta_{ij}}^{z_i} \in \mathcal{K}_\infty$ that is Lipschitz on compact sets to fulfill condition (6.27) if Δ_{ij} is locally Lipschitz.

It is also assumed that the external disturbances are bounded.

Assumption 6.2 For each $i = 1, \ldots, N$, there exists a $\bar{w}_i \ge 0$ such that

$$|w_i(t)| \le \bar{w}_i \qquad (6.28)$$

for all $t \ge 0$.

The basic idea is to design observer-based local controllers for the agents such that each controlled agent i is IOS, and moreover, has the UO property. Then, the cyclic-small-gain theorem in digraphs can be used to guarantee the IOS of the closed-loop multi-agent system and then the achievement of output agreement.

By introducing a dynamic compensator

$$\dot{u}_i = v_i \qquad (6.29)$$

and defining $x'_{i1} = y_i - y_0$ and $x'_{i(j+1)} = x_{i(j+1)} + \Delta_{ij}(y_0, 0)$ for $1 \le j \le n_i$, we can transform each agent i defined by (6.23)–(6.25) into the form of

$$
\begin{aligned}
\dot{x}'_{ij} &= x'_{i(j+1)} + \Delta_{ij}(y_i, w_i) - \Delta_{ij}(y_0, 0), \quad 1 \le j \le n_i + 1 \qquad &(6.30) \\
\dot{x}'_{in_i} &= v_i + \Delta_{in_i}(y_i, w_i) - \Delta_{in_i}(y_0, 0) \qquad &(6.31) \\
y'_i &= x'_{i1} \qquad &(6.32)
\end{aligned}
$$

with the output tracking error $y'_i = y_i - y_0$ as the new output and v_i as the new control input.

Moreover, the dynamic compensator (6.29) guarantees that the origin is an equilibrium of the transformed agent system (6.30)–(6.32) if it is disturbance-free, and the distributed control objective can be achieved if the equilibrium at the origin of each transformed agent system is stabilized.

The local controller for each agent i is designed by directly using the available y_i^m, defined as follows:

$$y_i^m = \frac{1}{N_i + 1} \left(\sum_{k \in \mathcal{N}_i} (y_i - y_k) + (y_i - y_0) \right), \quad i \in \mathcal{L} \tag{6.33}$$

$$y_i^m = \frac{1}{N_i} \sum_{k \in \mathcal{N}_i} (y_i - y_k), \quad i \in \{1, \ldots, N\} \backslash \mathcal{L}, \tag{6.34}$$

where N_i is the size of $\mathcal{N}_i$. For convenience of discussions, we represent y_i^m with the new outputs as

$$y_i^m = y_i' - \mu_i \tag{6.35}$$

with

$$\mu_i = \frac{1}{N_i + 1} \sum_{k \in \mathcal{N}_i} y_k', \quad i \in \mathcal{L} \tag{6.36}$$

$$\mu_i = \frac{1}{N_i} \sum_{k \in \mathcal{N}_i} y_k', \quad i \in \{1, \ldots, N\} \backslash \mathcal{L}. \tag{6.37}$$

6.2.1 DISTRIBUTED OUTPUT-FEEDBACK CONTROLLER

Owing to the output-feedback structure, we design a local observer for each transformed agent system (6.30)–(6.32):

$$\dot{\xi}_{i1} = \xi_{i2} + L_{i2}\xi_{i1} + \rho_{i1}(\xi_{i1} - y_i^m) \tag{6.38}$$

$$\dot{\xi}_{ij} = \xi_{i(j+1)} + L_{i(j+1)}\xi_{i1} - L_{ij}(\xi_{i2} + L_{i2}\xi_{i1}), \quad 2 \le j \le n_i \tag{6.39}$$

$$\dot{\xi}_{i(n_i+1)} = v_i - L_{i(n_i+1)}(\xi_{i2} + L_{i2}\xi_{i1}), \tag{6.40}$$

where $\rho_{i1} : \mathbb{R} \to \mathbb{R}$ is an odd and strictly decreasing function, and $L_{i2}, \ldots, L_{in_i}$ are positive constants. In the observer, ξ_{i1} is an estimate of y_i', and ξ_{ij} is an estimate of $x_{ij}' - L_{ij}y_i'$ for $2 \le j \le n_i + 1$.

Here, equation (6.38) is constructed to estimate y_i' by using y_i^m which is influenced by the outputs y_k' ($k \in \mathcal{N}_i$) of the neighbor agents (see (6.35)). The nonlinear function ρ_{i1} in (6.38) is used to assign an appropriate *nonlinear* gain to the observation error system. As shown later, it is the key to making each controlled agent IOS with specific gains satisfying the cyclic-small-gain condition. Equations (6.39)–(6.40) of the observer are in the same spirit of the reduced-order observer in Section 4.3. Slightly differently, we use ξ_{i1} instead of the unavailable y_i' in (6.39)–(6.40).

With the estimates, a nonlinear local control law is designed as

$$e_{i1} = \xi_{i1}, \tag{6.41}$$

$$e_{ij} = \xi_{ij} - \kappa_{i(j-1)}(e_{i(j-1)}), \quad 2 \le j \le n_i + 1 \tag{6.42}$$

$$v_i = \kappa_{i(n_i+1)}(e_{i(n_i+1)}), \tag{6.43}$$

where $\kappa_{i1}, \ldots, \kappa_{i(n_i+1)}$ are continuously differentiable, odd, strictly decreasing, and radially unbounded functions.

Consider $Z_i = [x'_{i1}, \ldots, x'_{i(n_i+1)}, \xi_{i1}, \ldots, \xi_{i(n_i+1)}]^T$ as the internal state of each controlled agent composed of the transformed agent system (6.30)–(6.32) and the local observer-based controller (6.38)–(6.43). The block diagram of controlled agent i with μ_i as the input and y'_i as the output is shown in Figure 6.3.

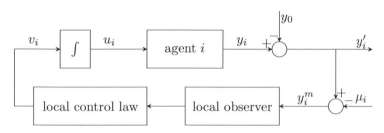

FIGURE 6.3 The block diagram of each controlled agent i.

The following proposition presents the UO and IOS properties of each controlled agent i.

Proposition 6.1 *Each controlled agent i composed of (6.30)–(6.32) and (6.38)–(6.43) has the following UO and IOS properties with μ_i as the input and y'_i as the output: for all $t \ge 0$,*

$$|Z_i(t)| \le \alpha_i^{UO}(|Z_{i0}| + \|\mu_i\|_{[0,t]}) \tag{6.44}$$

$$|y'_i(t)| \le \max\left\{ \beta_i(|Z_{i0}|, t), \chi_i(\|\mu_i\|_{[0,t]}), \gamma_i(\|w_i\|_{[0,t]}) \right\}, \tag{6.45}$$

for any initial state $Z_i(0) = Z_{i0}$ and any μ_i, w_i, where $\beta_i \in \mathcal{KL}$ and $\chi_i, \gamma_i, \alpha_i \in \mathcal{K}_\infty$. Moreover, γ_i can be designed to be arbitrarily small, and for any specified constant $b_i > 1$, χ_i can be designed such that $\chi_i(s) \le b_i s$ for all $s \ge 0$.

The proof of Proposition 6.1 is given in Subsection 6.2.4.

6.2.2 CYCLIC-SMALL-GAIN SYNTHESIS

With the proposed distributed output-feedback controller, the closed-loop multi-agent system has been transformed into a network of IOS subsystems. This subsection presents the main result of output agreement and provides a proof based on the cyclic-small-gain result in digraphs.

Theorem 6.1 *Consider the multi-agent system in the form of* (6.23)–(6.25) *satisfying Assumptions 6.1 and 6.2. If there is at least one informed agent, i.e.,* $\mathcal{L} \neq \emptyset$, *and the communication digraph* $\mathcal{G}^c$ *has a spanning tree with the informed agents as the roots, then we can design distributed observers* (6.38)– (6.40) *and distributed control laws* (6.29), (6.41)–(6.43) *such that all the signals in the closed-loop multi-agent system are bounded, and the output* y_i *of each agent* i *can be steered to within an arbitrarily small neighborhood of the desired agreement value* y_0. *Moreover, if* $w_i = 0$ *for* $i = 1, \ldots, N$, *then each output* y_i *asymptotically converges to* y_0.

Proof. Notice that for any constants $a_1, \ldots, a_n > 0$ satisfying $\sum_{i=1}^{n}(1/a_i) \leq n$, it holds that

$$\sum_{i=1}^{n} d_i = \sum_{i=1}^{n} \frac{1}{a_i} a_i d_i \leq n \max_{1 \leq i \leq n} \{a_i d_i\} \qquad (6.46)$$

for all $d_1, \ldots, d_n \geq 0$.

Recall the definition of μ_i in (6.36) and (6.37). We have

$$|\mu_i| \leq \delta_i \max_{k \in \mathcal{N}_i} \{a_{ik} |y_k'|\}, \qquad (6.47)$$

where $\delta_i = \frac{N_i}{N_i+1}$ if $i \in \mathcal{L}$, $\delta_i = 1$ if $i \notin \mathcal{L}$, and a_{ik} are positive constants satisfying

$$\sum_{k \in \mathcal{N}_i} \frac{1}{a_{ik}} \leq N_i. \qquad (6.48)$$

Then, using the fact that the $\mathcal{N}_i$ in (6.47) is time-invariant, property (6.45) implies

$$|y_i'(t)| \leq \max \left\{ \beta_i(|Z_{i0}|, t), b_i \delta_i \max_{k \in \mathcal{N}_i} \{a_{ik} \|y_k'\|_{[0,t]}\}, \gamma_i(\|w_i\|_{[0,t]}) \right\} \qquad (6.49)$$

for any initial state Z_{i0} and any w_i, for all $t \geq 0$.

It can be observed that the interconnection topology of the controlled agents is in accordance with the information exchange topology, represented by digraph $\mathcal{G}^c$. For $i \in \mathcal{N}$, $k \in \mathcal{N}_i$, we assign the positive value a_{ik} to the edge (k, i) in $\mathcal{G}^c$. Denote $\mathcal{C}$ as the set of all simple cycles in $\mathcal{G}^c$ and $\mathcal{C}_\mathcal{L}$ as the set of all simple cycles through the vertices belonging to $\mathcal{L}$. Denote $A_\mathcal{O}$ as the product of the positive values assigned to the edges of the cycle $\mathcal{O} \in \mathcal{C}$.

Note that b_i can be designed to be arbitrarily close to one. By using the cyclic-small-gain theorem for networks of IOS systems, the closed-loop multi-agent system is IOS if

$$A_\mathcal{O} \frac{N}{N+1} < 1, \quad \mathcal{O} \in \mathcal{C}_\mathcal{L} \qquad (6.50)$$

$$A_\mathcal{O} < 1, \quad \mathcal{O} \in \mathcal{C} \backslash \mathcal{C}_\mathcal{L}. \qquad (6.51)$$

If $\mathcal{G}^c$ has a spanning tree with vertices belonging to $\mathcal{L}$ as the roots, then according to Lemma 6.1, there exist positive constants a_{ik} satisfying (6.48), (6.50) and (6.51). Then, the closed-loop distributed system is UO and IOS with w_i as the inputs and y_i' as the outputs. With Assumption 6.2, the external disturbances w_i are bounded. The boundedness of the signals of the closed-loop distributed system can be directly verified under Assumption 6.2.

By designing the IOS gains γ_i arbitrarily small (this can be done according to Proposition 6.1), the influence of the external disturbances w_i is made arbitrarily small, and y_i' can be driven to within an arbitrarily small neighborhood of the origin. Equivalently, y_i can be driven to within an arbitrarily small neighborhood of y_0. In the case of $w_i = 0$ for $i = 1, \ldots, N$, each output y_i asymptotically converges to y_0. This ends the proof of Theorem 6.1. $\diamond$

6.2.3   ROBUSTNESS TO TIME DELAYS OF INFORMATION EXCHANGE

If there are communication delays, then y_i^m as defined in (6.33) and (6.34) should be modified as

$$y_i^m(t) = \frac{1}{N_i + 1} \left(\sum_{k \in \mathcal{N}_i} (y_i(t) - y_k(t - \tau_{ik}(t))) + (y_i(t) - y_0) \right), \quad i \in \mathcal{L}$$

$$\tag{6.52}$$

$$y_i^m(t) = \frac{1}{N_i} \sum_{k \in \mathcal{N}_i} (y_i(t) - y_k(t - \tau_{ik}(t))), \quad i \in \{1, \ldots, N\} \backslash \mathcal{L}, \tag{6.53}$$

where $\tau_{ik} : \mathbb{R}_+ \to \mathbb{R}_+$ represents non-constant time delays of exchanged information.

In this case, $y_i^m(t)$ can still be written in the form of $y_i^m(t) = y_i'(t) - \mu_i(t)$ with

$$\mu_i(t) = \frac{1}{N_i + 1} \sum_{k \in \mathcal{N}_i} y_k'(t - \tau_{ik}(t)), \quad i \in \mathcal{L} \tag{6.54}$$

$$\mu_i(t) = \frac{1}{N_i} \sum_{k \in \mathcal{N}_i} y_k'(t - \tau_{ik}(t)), \quad i \in \{1, \ldots, N\} \backslash \mathcal{L}. \tag{6.55}$$

We assume that there exists a $\bar{\tau} \geq 0$ such that, for $i = 1, \ldots, N$, $k \in \mathcal{N}_i$, $0 \leq \tau_{ik}(t) \leq \bar{\tau}$ holds for all $t \geq 0$. By considering μ_i and w_i as the external inputs, each controlled agent i composed of (6.30)–(6.32) and (6.38)–(6.43) is still UO and property (6.49) should be modified as

$$|y_i'(t)| \leq \max \left\{ \beta_i(|Z_{i0}|, t), b_i \delta_i \max_{k \in \mathcal{N}_i} \{ a_{ik} \| y_k' \|_{[-\bar{\tau}, \infty)} \}, \gamma_i(\| w_i \|_{[0, \infty)}) \right\} \tag{6.56}$$

for any initial state Z_{i0} and any w_i, for all $t \geq 0$.

By using the time-delay version of the cyclic-small-gain theorem, Theorem 3.3, we can still guarantee the IOS of the closed-loop multi-agent system with y_i' as the outputs and w_i as the inputs, following analysis similar to that for the proof of Theorem 6.1.

6.2.4   PROOF OF UO AND IOS OF EACH CONTROLLED AGENT

This subsection gives the proof of Proposition 6.1.

The Observation Error System

Define $\zeta_{i1} = y_i' - \xi_{i1}$ and $\zeta_{ij} = x_{ij}' - L_{ij}y_i' - \xi_{ij}$ for $2 \le j \le n_i + 1$ as the observation errors. Denote $\bar{\zeta}_{i2} = [\zeta_{i2}, \ldots, \zeta_{i(n_i+1)}]^T$ and $\tilde{\Delta}_{ij}(y_i', y_0, w_i) = \Delta_{ij}(y_i' + y_0, w_i) - \Delta_{ij}(y_0, 0)$ for $1 \le j \le n_i$.

By taking the derivatives of $\zeta_{i1}, \bar{\zeta}_{i2}$, we obtain

$$\dot{\zeta}_{i1} = \rho_{i1}(\zeta_{i1} - \mu_i) + \phi_{i1}(y_0, \zeta_{i1}, \zeta_{i2}, \xi_{i1}, w_i) \tag{6.57}$$

$$\dot{\bar{\zeta}}_{i2} = A_i \bar{\zeta}_{i2} + \bar{\phi}_{i2}(y_0, \zeta_{i1}, \xi_{i1}, w_i), \tag{6.58}$$

where

$$A_i = \begin{bmatrix} -L_{i2} & & \\ \vdots & & I_{n_i-2} \\ -L_{i(n_i-1)} & & \\ -L_{in_i} & 0 & \cdots & 0 \end{bmatrix}, \tag{6.59}$$

$$\bar{\phi}_{i2}(y_0, \zeta_{i1}, \xi_{i1}, w_i) = \begin{bmatrix} \phi_{i2}(y_0, \zeta_{i1}, \xi_{i1}, w_i) \\ \vdots \\ \phi_{i(n_i+1)}(y_0, \zeta_{i1}, \xi_{i1}, w_i) \end{bmatrix}, \tag{6.60}$$

with

$$\phi_{i1} = \tilde{\Delta}_{i1} + \zeta_{i2} + L_{i2}\zeta_{i1} \tag{6.61}$$

$$\phi_{ij} = \tilde{\Delta}_{ij} - L_{ij}\Delta_{i1} + (L_{i(j+1)} - L_{ij}L_{i2})\zeta_{i1}, \quad 2 \le j \le n_i \tag{6.62}$$

$$\phi_{i(n_i+1)} = -L_{i(n_i+1)}\tilde{\Delta}_{i1} - L_{in_i}L_{i2}\zeta_{i1}. \tag{6.63}$$

With Assumption 6.1 satisfied, one can find $\psi_{\phi_{ij}} \in \mathcal{K}_\infty$ such that

$$|\phi_{i1}(y_0, \zeta_{i1}, \zeta_{i2}, \xi_{i1}, w_i)| \le \psi_{\phi_{ij}}(|[\zeta_{i1}, \zeta_{i2}, \xi_{i1}, w_i]^T|), \tag{6.64}$$

$$|\phi_{ij}(y_0, \zeta_{i1}, \xi_{i1}, w_i)| \le \psi_{\phi_{ij}}(|[\zeta_{i1}, \xi_{i1}, w_i]^T|). \tag{6.65}$$

The positive constants $L_{i2}, \ldots, L_{in_i}$ are chosen so that A_i is a Hurwitz matrix, i.e., its eigenvalues have negative real parts.

With Lemma 4.1, we can find a continuously differentiable ρ_{i1} such that for any constants $0 < c_i < 1$, $\ell_{\zeta_{i1}} > 0$ and any $\chi_{\zeta_{i1}}^{\zeta_{i2}}, \chi_{\zeta_{i1}}^{\xi_{i1}}, \chi_{\zeta_{i1}}^{w_i} \in \mathcal{K}_\infty$ being Lipschitz on compact sets, the ζ_{i1}-subsystem is ISS with $V_{\zeta_{i1}}(\zeta_{i1}) = |\zeta_{i1}|$ as an ISS-Lyapunov function, which satisfies

$$V_{\zeta_{i1}}(\zeta_{i1}) \ge \max\left\{ \chi_{\zeta_{i1}}^{\mu_i}(|\mu_i|), \chi_{\zeta_{i1}}^{\zeta_{i2}}(|\zeta_{i2}|), \chi_{\zeta_{i1}}^{\xi_{i1}}(|\xi_{i1}|), \chi_{\zeta_{i1}}^{w_i}(|w_i|) \right\}$$

$$\Rightarrow \nabla V_{\zeta_{i1}}(\zeta_{i1})\dot{\zeta}_{i1} \le -\ell_{\zeta_{i1}} V_{\zeta_{i1}}, \quad \text{a.e.,} \tag{6.66}$$

where $\chi_{\zeta_{i1}}^{\mu_i}(s) = s/c_i$ for $s \in \mathbb{R}_+$.

Noticing that A_i is Hurwitz, there exists a positive definite matrix $P_i = P_i^T \in \mathbb{R}^{(n_i-1) \times (n_i-1)}$ satisfying $P_i A_i + A_i^T P_i = -2I_{n_i-1}$. Define $V_{\bar{\zeta}_{i2}}(\bar{\zeta}_{i2}) = \bar{\zeta}_{i2}^T P_i \bar{\zeta}_{i2}$. Then, there exist $\underline{\alpha}_{\bar{\zeta}_{i2}}, \overline{\alpha}_{\bar{\zeta}_{i2}} \in \mathcal{K}_\infty$ such that $\underline{\alpha}_{\bar{\zeta}_{i2}}(|\bar{\zeta}_{i2}|) \le V_{\bar{\zeta}_{i2}}(\bar{\zeta}_{i2}) \le \overline{\alpha}_{\bar{\zeta}_{i2}}(|\bar{\zeta}_{i2}|)$. With direct calculation, we have

$$
\begin{aligned}
\nabla V_{\bar{\zeta}_{i2}}(\bar{\zeta}_{i2})\dot{\bar{\zeta}}_{i2} = & -2\bar{\zeta}_{i2}^T \bar{\zeta}_{i2} + 2\bar{\zeta}_{i2}^T P_i \bar{\phi}_{i2}(\zeta_{i1}, \xi_{i1}, w_i) \\
\le & -\bar{\zeta}_{i2}^T \bar{\zeta}_{i2} + |P_i|^2 |\bar{\phi}_{i2}(\zeta_{i1}, \xi_{i1}, w_i)|^2 \\
\le & -\frac{1}{\lambda_{\max}(P_i)} V_{\bar{\zeta}_{i2}}(\bar{\zeta}_{i2}) \\
& + |P_i|^2 \left(\psi_{\bar{\phi}_{i2}}^{\zeta_{i1}}(|\zeta_{i1}|) + \psi_{\bar{\phi}_{i2}}^{\xi_{i1}}(|\xi_{i1}|) + \psi_{\bar{\phi}_{i2}}^{w_i}(|w_i|) \right).
\end{aligned}
\tag{6.67}
$$

This means that the $\bar{\zeta}_{i2}$-subsystem is ISS with $V_{\bar{\zeta}_{i2}}$ as an ISS-Lyapunov function. The ISS gains can be chosen as follows. Define $\chi_{\bar{\zeta}_{i2}}^{\zeta_{i1}} = 4\lambda_{\max}(P_i)|P_i^2|\psi_{\bar{\phi}_{i2}}^{\zeta_{i1}}$, $\chi_{\bar{\zeta}_{i2}}^{\xi_{i1}} = 4\lambda_{\max}(P_i)|P_i^2|\psi_{\bar{\phi}_{i2}}^{\xi_{i1}}$ and $\chi_{\bar{\zeta}_{i2}}^{w_i} = 4\lambda_{\max}(P_i)|P_i^2|\psi_{\bar{\phi}_{i2}}^{w_i}$. Then,

$$
\begin{aligned}
& V_{\bar{\zeta}_{i2}}(\bar{\zeta}_{i2}) \ge \max \left\{ \chi_{\bar{\zeta}_{i2}}^{\zeta_{i1}}(|\zeta_{i1}|), \chi_{\bar{\zeta}_{i2}}^{\xi_{i1}}(|\xi_{i1}|), \chi_{\bar{\zeta}_{i2}}^{w_i}(|w_i|) \right\} \\
& \Rightarrow \nabla V_{\bar{\zeta}_{i2}}(\bar{\zeta}_{i2})\dot{\bar{\zeta}}_{i2} \le -\ell_{\bar{\zeta}_{i2}} V_{\bar{\zeta}_{i2}}(\bar{\zeta}_{i2}),
\end{aligned}
\tag{6.68}
$$

where $\ell_{\bar{\zeta}_{i2}} = \frac{1}{4\lambda_{\max}(P_i)}$.

The Control Error System

By taking the derivatives of $e_{i1}, \ldots, e_{i(n_i+1)}$, direct calculation yields:

$$
\dot{e}_{i1} = \kappa_{i1}(e_{i1}) + \bar{\varphi}_{i1}(e_{i1}, e_{i2}, \mu_i, \zeta_{i1}),
\tag{6.69}
$$

$$
\dot{e}_{ij} = \kappa_{ij}(e_{ij}) + \bar{\varphi}_{ij}(e_{i1}, \ldots, e_{i(j+1)}, \mu_i, \zeta_{i1}), \quad 2 \le j \le n_i + 1,
\tag{6.70}
$$

where

$$
\bar{\varphi}_{i1} = e_{i2} + L_{i2}\xi_{i1} + \rho_{i1}(\xi_{i1} - y_i^m)
\tag{6.71}
$$

$$
\begin{aligned}
\bar{\varphi}_{ij} = & e_{i(j+1)} + L_{i(j+1)}\xi_{i1} - L_{ij}(\xi_{i2} + L_{i2}\xi_{i1}) \\
& - \frac{\partial \kappa_{i(j-1)}(e_{i(j-1)})}{\partial e_{i(j-1)}}\dot{e}_{i(j-1)}.
\end{aligned}
\tag{6.72}
$$

By default, $e_{i(n_i+2)} = 0$. Clearly, $\bar{\varphi}_{i1}$ and $\bar{\varphi}_{ij}$ are locally Lipschitz functions. Also, we can find $\psi_{\bar{\varphi}_{ij}} \in \mathcal{K}_\infty$ such that

$$
|\bar{\varphi}_{ij}| \le \psi_{\bar{\varphi}_{ij}}(|[e_{i1}, \ldots, e_{i(j+1)}, \mu_i, \zeta_{i1}]^T|)
\tag{6.73}
$$

for $1 \le j \le n_i + 1$.

With Lemma 4.1, we can find continuously differentiable functions κ_{ij} for $1 \leq j \leq n_i + 1$ such that each e_{ij}-subsystem is ISS with $V_{e_{ij}}(e_{ij}) = |e_{ij}|$ as an ISS-Lyapunov function:

$$V_{e_{ij}}(e_{ij}) \geq \max_{k=1,\ldots,j-1,j+1} \left\{ \chi_{e_{ij}}^{e_{ik}}(|e_{ik}|), \chi_{e_{ij}}^{\mu_i}(|\mu_i|), \chi_{e_{ij}}^{\zeta_{i1}}(|\zeta_{i1}|) \right\}$$

$$\Rightarrow \nabla V_{e_{ij}}(e_{ij}) \dot{e}_{ij} \leq -\ell_{e_{ij}} V_{e_{ij}}(e_{ij}), \quad \text{a.e.,} \tag{6.74}$$

where $\ell_{(\cdot)}$ can be any specified positive constants, $\chi_{e_{i1}}^{e_{i0}}, \chi_{e_{i(n_i+1)}}^{e_{i(n_i+2)}} = 0$ and the other $\chi_{(\cdot)}^{(\cdot)}$'s can be any specified $\mathcal{K}_\infty$ functions that are Lipschitz on compact sets.

UO and IOS

Define $Z_i = [\zeta_{i1}, \bar{\zeta}_{i2}^T, e_{i1}, \ldots, e_{i(n_i+1)}]^T$. Each controlled agent with state Z_i has been transformed into a network of ISS subsystems. With the Lyapunov-based cyclic-small-gain theorem, controlled agent i is ISS with w_i and μ_i as inputs if the composition of the ISS gains along every simple cycle in the system digraph is less than the identity function. The cyclic-small-gain condition can be satisfied by choosing the ISS gains of the $\zeta_{i1}, e_{i1}, \ldots, e_{i(n_i+1)}$-subsystems to be small enough.

With the cyclic-small-gain condition satisfied, to find the IOS gains, we construct an ISS-Lyapunov function in the following form:

$$V_i(Z_i) = \max_{1 \leq j \leq n_i + 1} \left\{ \sigma_{\zeta_{i1}}(V_{\zeta_{i1}}(\zeta_{i1})), \sigma_{\bar{\zeta}_{i2}}(V_{\bar{\zeta}_{i2}}(\bar{\zeta}_{i2})), \sigma_{e_{ij}}(V_{e_{ij}}(e_{ij})) \right\}, \tag{6.75}$$

where $\sigma_{e_{i1}} = \text{Id}$, and the other $\sigma_{(\cdot)}$'s are compositions of $\hat{\chi}_{(\cdot)}$'s which are of class $\mathcal{K}_\infty$, smooth on $(0, \infty)$, slightly larger than the corresponding $\chi_{(\cdot)}^{(\cdot)}$'s, and still satisfy the cyclic-small-gain condition. Thus, $V_i(Z_i)$ is positive definite and radially unbounded with respect to Z_i. Here, it is not necessary to give an explicit representation of the $\sigma_{(\cdot)}$'s.

Accordingly, we define

$$\bar{\chi}_i(|\mu_i|) = \max_{1 \leq j \leq n_i + 1} \left\{ \sigma_{\zeta_{i1}} \circ \chi_{\zeta_{i1}}^{\mu_i}(|\mu_i|), \sigma_{e_{ij}} \circ \chi_{e_{ij}}^{\mu_i}(|\mu_i|) \right\}, \tag{6.76}$$

$$\bar{\gamma}_i(|w_i|) = \max_{1 \leq j \leq n_i + 1} \left\{ \sigma_{\zeta_{i1}} \circ \chi_{\zeta_{i1}}^{w_i}(|w_i|), \sigma_{\bar{\zeta}_{i2}} \circ \chi_{\bar{\zeta}_{i2}}^{w_i}(|w_i|) \right\}. \tag{6.77}$$

By choosing the ISS gains $\chi_{e_{ij}}^{\mu_i}$ and $\chi_{e_{i1}}^{\zeta_{i1}}$ small enough, it can be achieved that

$$\bar{\chi}_i = \sigma_{\zeta_{i1}} \circ \chi_{\zeta_{i1}}^{\mu_i}, \tag{6.78}$$

where $\sigma_{\zeta_{i1}}$ can be designed to be arbitrarily small. Similarly, $\bar{\gamma}_i$ can also be designed to be arbitrarily small.

Then, there exists a continuous and positive definite function α_i such that

$$V_i(Z_i) \geq \max\{\bar{\chi}_i(|\mu_i|), \bar{\gamma}_i(|w_i|)\} \Rightarrow \nabla V_i(Z_i) \dot{Z}_i \leq -\alpha_i(V_i(Z_i)) \tag{6.79}$$

holds almost everywhere. Thus, there exists a $\bar{\beta}_i \in \mathcal{KL}$ such that for all $t \geq 0$,

$$V_i(Z_i(t)) \leq \max\left\{\bar{\beta}_i(V_i(Z_{i0}),t), \bar{\chi}_i(\|\mu_i\|_{[0,t]}), \bar{\gamma}_i(\|w_i\|_{[0,t]})\right\} \qquad (6.80)$$

holds for any initial state Z_{i0}.

Recall the definitions of $\zeta_{i1}, \bar{\zeta}_{i2}, e_{i1}, \ldots, e_{i(n_i+1)}$. One can find $\underline{\alpha}_i, \bar{\alpha}_i \in \mathcal{K}_\infty$ such that $\underline{\alpha}_i(|Z_i|) \leq V_i(Z_i) \leq \bar{\alpha}_i(|Z_i|)$ holds for all Z_i. Based on (6.80), one can find an $\alpha_i^{\text{UO}} \in \mathcal{K}_\infty$ such that the UO property (6.44) holds.

From the definition of $V_i(Z_i)$, using $\sigma_{\zeta_{i1}} = \text{Id}$, we have $|y_i'| \leq |e_{i1}| + |\zeta_{i1}| \leq \sigma_{e_{i1}}^{-1}(V_i(Z_i)) + \sigma_{\zeta_{i1}}^{-1}(V_i(Z_i)) = (\text{Id} + \sigma_{\zeta_{i1}}^{-1})(V_i(Z_i))$. By defining

$$\chi_i(s) = (\text{Id} + \sigma_{\zeta_{i1}}^{-1}) \circ \sigma_{\zeta_{i1}} \circ \chi_{\zeta_{i1}}^{\mu_i} = (\text{Id} + \sigma_{\zeta_{i1}}) \circ \chi_{\zeta_{i1}}^{\mu_i}(s) \qquad (6.81)$$

$$\gamma_i(s) = (\text{Id} + \sigma_{\zeta_{i1}}^{-1}) \circ \bar{\gamma}_i(s) \qquad (6.82)$$

$$\beta_i(s,t) = (\text{Id} + \sigma_{\zeta_{i1}}^{-1}) \circ \bar{\beta}_i(\bar{\alpha}_i(s),t) \qquad (6.83)$$

for $s,t \geq 0$, we can prove (6.45).

Recall that $\chi_{\zeta_{i1}}^{\mu_i}(s) = s/c_i$ for $s \in \mathbb{R}_+$. For any specified constant $b_i > 1$, by choosing c_i to be close enough to one and $\sigma_{\zeta_{i1}}$ to be small enough, χ_i can be designed such that $\chi_i(s) \leq b_i s$ for all $s \geq 0$. With fixed $\sigma_{\zeta_{i1}}$, by choosing $\bar{\gamma}_i$ to be arbitrarily small, γ_i can be designed to be arbitrarily small.

6.3   FORMATION CONTROL OF NONHOLONOMIC MOBILE ROBOTS

Formation control of autonomous mobile agents is aimed at forcing agents to converge toward, and maintain, specific relative positions. Distributed formation control of multi-agent systems based on available local information, e.g., relative position measurements, has attracted tremendous attention from the robotics and controls communities.

Motivated by the cyclic-small-gain design for distributed output-feedback control of nonlinear systems in Section 6.2, this section proposes a class of distributed controllers for leader-following formation control of unicycle robots using the practically available relative position measurements. The kinematics of the unicycle robot are demonstrated by Figure 6.4.

For this purpose, the formation control problem is first transformed into a state agreement problem of double-integrators through dynamic feedback linearization. The nonholonomic constraint causes singularity for dynamic feedback linearization when the linear velocity of the robot is zero. This issue should be well taken into consideration for the validity of the transformed double-integrator models. Then, distributed formation control laws are developed. To avoid the singularity problem caused by the nonholonomic constraint, saturation functions are introduced to the control design to restrict the linear velocities of the robots to be larger than zero. It should be noted that linear analysis methods may not be directly applicable due to the employment of the saturation functions. Then, the closed-loop system is transformed into a dynamic network of IOS systems. The cyclic-small-gain result in digraphs is

used to guarantee the IOS of the dynamic network and thus the achievement
of formation control.

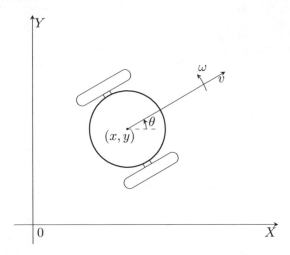

FIGURE 6.4   Kinematics of the unicycle robot, where (x, y) represents the Cartesian
coordinates of the center of mass of the robot, v is the linear velocity, θ is the heading
angle, and ω is the angular velocity.

With the effort mentioned above, the proposed design has three advantages:

1. The proposed distributed formation control law does not use global position
   measurements or assume tree position sensing structures.
2. The formation control objective can be practically achieved in the presence
   of position measurement errors.
3. The linear velocities of the robots can be designed to be less than certain
   desired values, as practically required.

This section considers the formation control problem of a group of $N +$
1 mobile robots. For $i = 0, 1, \ldots, N$, the kinematics of each i-th robot are
described by the unicycle model:

$$\dot{x}_i = v_i \cos \theta_i \tag{6.84}$$

$$\dot{y}_i = v_i \sin \theta_i \tag{6.85}$$

$$\dot{\theta}_i = \omega_i, \tag{6.86}$$

where $[x_i, y_i]^T \in \mathbb{R}^2$ represent the Cartesian coordinates of the center of mass
of the i-th robot, $v_i \in \mathbb{R}$ is the linear velocity, $\theta_i \in \mathbb{R}$ is the heading angle,
and $\omega_i \in \mathbb{R}$ is the angular velocity.

The robot with index 0 is the leader robot, and the robots with indices
$1, \ldots, N$ are follower robots. The linear velocity v_i and the angular velocity
ω_i are considered as the control inputs of the i-th robot for $i = 1, \ldots, N$. For

this system, the formation control objective is to control each i-th follower robot such that

$$\lim_{t \to \infty} (x_i(t) - x_j(t)) = d_{xij} \tag{6.87}$$

$$\lim_{t \to \infty} (y_i(t) - y_j(t)) = d_{yij} \tag{6.88}$$

with d_{xij}, d_{yij} being appropriate constants representing the desired relative positions, and

$$\lim_{t \to \infty} ((\theta_i(t) - \theta_j(t)) \bmod 2\pi) = 0 \tag{6.89}$$

for any $i, j = 0, \ldots, N$, where mod represents the modulo operation. For convenience of notation, let $d_{xii} = d_{yii} = 0$ for any $i = 0, \ldots, N$. We assume that $d_{xij} = d_{xik} - d_{xkj}$ and $d_{yij} = d_{yik} - d_{ykj}$ for any $i, j, k = 0, \ldots, N$.

Assumption 6.3 on v_0 is made throughout this section.

Assumption 6.3 *The linear velocity v_0 of the leader robot is differentiable with bounded derivative, i.e., $\dot{v}_0(t)$ exists and is bounded on $[0, \infty)$, and has upper and lower constant bounds $\bar{v}_0, \underline{v}_0 > 0$ such that $\underline{v}_0 \le v_0(t) \le \bar{v}_0$ for all $t \ge 0$.*

One technical problem of controlling groups of mobile robots is that accurate global positions of the robots are usually not available for feedback, and relative position measurements should be used instead. The hardware for relative position measurements for each follower robot may include a range sensor (e.g., sonar, laser range finder, and light detection and ranging (LIDAR) component) and a gyroscope for orientation measurement [57]. A digraph can be employed to represent the relative position sensing structure between the robots. The position sensing digraph $\mathcal{G}$ has $N + 1$ vertices with indices $0, 1, \ldots, N$ corresponding to the robots. If the relative position between robot i and robot j is available to robot j, then $\mathcal{G}$ has a directed edge from vertex i to vertex j; otherwise $\mathcal{G}$ does not have such an edge.

The goal of this section is to present a class of distributed formation controllers for mobile robots by using local relative position measurements as well as the velocity and acceleration information of the leader. The basic idea of the design is to first transform the unicycle model into the double integrator through dynamic feedback linearization under constraints, and at the same time, to reformulate the formation control problem as a stabilization problem. Then, distributed control laws are designed to make each controlled mobile robot IOS. Finally, the cyclic-small-gain theorem is used to guarantee the achievement of the formation control objective.

6.3.1 DYNAMIC FEEDBACK LINEARIZATION

In this subsection, the distributed formation control problem is reformulated with the dynamic feedback linearization technique. For details of dynamic feedback linearization and its applications to nonholonomic systems, please consult [40, 61].

For each $i = 0, \ldots, N$, introduce a new input $r_i \in \mathbb{R}$ such that

$$\dot{v}_i = r_i. \tag{6.90}$$

Define $v_{xi} = v_i \cos \theta_i$ and $v_{yi} = v_i \sin \theta_i$. Then, $\dot{x}_i = v_{xi}$ and $\dot{y}_i = v_{yi}$. Take the derivatives of v_{xi} and v_{yi}, respectively. Then,

$$\begin{pmatrix} \dot{v}_{xi} \\ \dot{v}_{yi} \end{pmatrix} = \begin{pmatrix} \cos \theta_i & -v_i \sin \theta_i \\ \sin \theta_i & v_i \cos \theta_i \end{pmatrix} \begin{pmatrix} r_i \\ \omega_i \end{pmatrix}. \tag{6.91}$$

In the case of $v_i \neq 0$, by designing

$$\begin{pmatrix} r_i \\ \omega_i \end{pmatrix} = \begin{pmatrix} \cos \theta_i & \sin \theta_i \\ -\frac{\sin \theta_i}{v_i} & \frac{\cos \theta_i}{v_i} \end{pmatrix} \begin{pmatrix} u_{xi} \\ u_{yi} \end{pmatrix}, \tag{6.92}$$

the unicycle model (6.84)–(6.86) can be transformed into two double-integrators with new inputs u_{xi} and u_{yi}:

$$\dot{x}_i = v_{xi}, \quad \dot{v}_{xi} = u_{xi}, \tag{6.93}$$

$$\dot{y}_i = v_{yi}, \quad \dot{v}_{yi} = u_{yi}. \tag{6.94}$$

Define $\tilde{x}_i = x_i - x_0 - d_{xi}$, $\tilde{y}_i = y_i - y_0 - d_{yi}$, $\tilde{v}_{xi} = v_{xi} - v_{x0}$, $\tilde{v}_{yi} = v_{yi} - v_{y0}$, $\tilde{u}_{xi} = u_{xi} - u_{x0}$, and $\tilde{u}_{yi} = u_{yi} - u_{y0}$. Then,

$$\dot{\tilde{x}}_i = \tilde{v}_{xi}, \quad \dot{\tilde{v}}_{xi} = \tilde{u}_{xi}, \tag{6.95}$$

$$\dot{\tilde{y}}_i = \tilde{v}_{yi}, \quad \dot{\tilde{v}}_{yi} = \tilde{u}_{yi}. \tag{6.96}$$

The formation control problem is solvable if we can design control laws for system (6.95)–(6.96) with $\tilde{u}_{xi}$ and $\tilde{u}_{yi}$ as the control inputs, so that $v_i \neq 0$ is guaranteed, and at the same time,

$$\lim_{t \to \infty} \tilde{x}_i(t) = 0, \tag{6.97}$$

$$\lim_{t \to \infty} \tilde{y}_i(t) = 0. \tag{6.98}$$

It should be noted that the validity of (6.93)–(6.94) (and thus (6.95)–(6.96)) for the unicycle model is under the condition that $v_i \neq 0$. Such requirement is basically caused by the nonholonomic constraint of the mobile robot. This leads to the major difference between this problem and the distributed control problem for double-integrators.

To use (6.95)–(6.96) for control design, each follower robot should have access to u_{x0}, u_{y0}, which represent the acceleration of the leader robot. This requirement can be fulfilled if the leader robot can calculate u_{x0}, u_{y0} by using $r_0, \omega_0, \theta_0, v_0$ according to (6.92) and transmit them to the follower robots. Note that ω_0, θ_0, v_0 are usually measurable, and r_0 is normally available as it is the control input of the leader robot.

6.3.2 A CLASS OF IOS CONTROL LAWS

As an ingredient for the distributed control design, this subsection presents a class of nonlinear control laws for the following double-integrator system with an external input, such that the closed-loop system is UO and IOS:

$$\dot{\eta} = \zeta \tag{6.99}$$

$$\dot{\zeta} = \mu \tag{6.100}$$

$$\hat{\eta} = \eta + w, \tag{6.101}$$

where $[\eta, \zeta]^T \in \mathbb{R}^2$ is the state, $\mu \in \mathbb{R}$ is the control input, $w \in \mathbb{R}$ represents an external input, $\hat{\eta}$ can be considered as a measurement of η subject to w, and only $(\hat{\eta}, \zeta)$ are used for feedback. As shown later, each controlled robot can be transformed into the form of (6.99)–(6.101) with w representing the interaction between the robots.

Lemma 6.2 *For system (6.99)–(6.101), consider a control law taking the form:*

$$\mu = -k_\mu(\zeta - \phi(\hat{\eta})). \tag{6.102}$$

For any constant $\overline{\phi} > 0$, one can find an odd, strictly decreasing, continuously differentiable function $\phi : \mathbb{R} \to [-\overline{\phi}, \overline{\phi}]$ and a positive constant k_μ satisfying

$$-\frac{k_\mu}{4} < \frac{d\phi(r)}{dr} < 0 \tag{6.103}$$

for all $r \in \mathbb{R}$, such that the closed-loop system (6.99)–(6.102) is UO with zero offset, and is IOS with the identity function as the gain, i.e., the following properties hold:

$$|\eta(t)| \le \overline{\beta}(|[\eta(0), \zeta(0)]^T|, t) + \|w\|_t \tag{6.104}$$

$$|\zeta(t)| \le |\zeta(0)| + \alpha_{UO}(\|\eta\|_t + \|w\|_t) \tag{6.105}$$

for some $\overline{\beta} \in \mathcal{KL}$, $\alpha_{UO} \in \mathcal{K}_\infty$, and all $t \ge 0$.

It is necessary to note that condition (6.103) is easily checkable for practical implementation of the control law (6.102).

Proof. Denote $\overline{w} = \|w\|_T$ with $T \ge 0$. Recall that $\vec{d}(r, S)$ represents the direct distance from $r \in \mathbb{R}$ to $S \subset \mathbb{R}$. We first introduce the following transformation to state ζ:

$$\tilde{\zeta} = \vec{d}(\zeta, S_\zeta(\eta, \overline{w})) \tag{6.106}$$

with

$$S_\zeta(\eta, \overline{w}) = \{\overline{c}\phi(\eta + w) : |w| \le \overline{w}, \overline{c} \in [c_2, c_1]\}, \tag{6.107}$$

where $0 < c_2 < c_1$ are constants to be defined later. Then, we have $\zeta - \tilde{\zeta} \in S_\zeta(\eta, \overline{w})$.

The $\tilde{\eta}$-subsystem

With $\zeta - \tilde{\zeta} \in S_\zeta(\eta, \overline{w})$, the η-subsystem can be represented by the following differential inclusion:

$$\dot{\eta} \in \left\{ \zeta^* + \tilde{\zeta} : \zeta^* \in S_\zeta(\eta, \overline{w}) \right\}. \tag{6.108}$$

Due to the influence of w, we study the convergence of η to the set $S_\zeta(\overline{w}) = \{w : |w| \leq \overline{w}\}$ when $0 \leq t \leq T$. Define

$$\tilde{\eta} = \vec{d}(\eta, S_\eta(\overline{w})). \tag{6.109}$$

Then, from (6.108), we have

$$\dot{\tilde{\eta}} \in \left\{ \zeta^* + \tilde{\zeta} : \zeta^* \in S_\zeta(\eta, \overline{w}) \right\} := F_{\tilde{\eta}}(\eta, \tilde{\zeta}, \overline{w}). \tag{6.110}$$

The properties of ϕ and the definition of $\tilde{\eta}$ guarantee

$$|\overline{c}\phi(\eta + w)| \geq c_2|\phi(\tilde{\eta})| \tag{6.111}$$
$$\mathrm{sgn}(\overline{c}\phi(\eta + w)) = \mathrm{sgn}(c_2\phi(\tilde{\eta})) = -\mathrm{sgn}(\tilde{\eta}) \tag{6.112}$$

for $\overline{c} \in [c_2, c_1]$ and $|w| \leq \overline{w}$, when $\tilde{\eta} \neq 0$.

We consider the case of $(1 - \delta)c_2|\phi(\tilde{\eta})| \geq |\tilde{\zeta}|$ with $\delta < 1$ and $\tilde{\eta} \neq 0$. In this case, it holds that

$$\begin{aligned}
|\overline{c}\phi(\eta + w) + \tilde{\zeta}| &\geq |\overline{c}\phi(\eta + w)| - |\tilde{\zeta}| \\
&\geq c_2|\phi(\tilde{\eta})| - (1 - \delta)c_2|\phi(\tilde{\eta})| \\
&= \delta c_2|\phi(\tilde{\eta})|
\end{aligned} \tag{6.113}$$

and

$$\mathrm{sgn}(\overline{c}\phi(\eta + w) + \tilde{\zeta}) = \mathrm{sgn}(c_2\phi(\tilde{\eta})) = -\mathrm{sgn}(\tilde{\eta}) \tag{6.114}$$

for $\overline{c} \in [c_2, c_1]$ and $|w| \leq \overline{w}$.

Define $V_{\tilde{\eta}}(\tilde{\eta}) = \tilde{\eta}^2/2$ as a Lyapunov function candidate for the $\tilde{\eta}$-subsystem. Then, in the case of $(1 - \delta)c_2|\phi(\tilde{\eta})| \geq |\tilde{\zeta}|$, direct calculation yields:

$$\max_{f_{\tilde{\eta}} \in F_{\tilde{\eta}}(\eta, \tilde{\zeta}, \overline{w})} \nabla V_{\tilde{\eta}}(\tilde{\eta}) f_{\tilde{\eta}} \leq -\delta c_2|\tilde{\eta}||\phi(\tilde{\eta})| = -\delta c_2\tilde{\eta}\phi(\tilde{\eta}). \tag{6.115}$$

The $\tilde{\zeta}$-subsystem

To simplify the discussions, we only study the case of $\tilde{\zeta} > 0$. The case of $\tilde{\zeta} < 0$ can be studied in the same way. The definition of S_ζ in (6.107) implies

$$\max S_\zeta(\eta, \overline{w}) = \begin{cases} c_2\phi(\eta - \overline{w}), & \text{when } \eta \geq \overline{w}; \\ c_1\phi(\eta - \overline{w}), & \text{when } \eta < \overline{w}. \end{cases} \tag{6.116}$$

Since ϕ is continuously differentiable, $\max S_\zeta(\eta, \overline{w})$ is continuously differentiable almost everywhere. Denote $\frac{d\phi(r)}{dr} = \phi^d(r)$ for $r \in \mathbb{R}$. When $0 \leq t \leq T$, by taking the derivative of $\tilde{\zeta}$ and using the control law (6.102), we can use a differential inclusion to represent the ζ-subsystem:

$$
\begin{aligned}
\dot{\tilde{\zeta}} &\in \{\mu - c\phi^d(\eta - \overline{w})\dot{\eta} : c \in S_c(\eta)\} \\
&= \{-k_\mu(\zeta - \phi(\eta + w)) - c\phi^d(\eta - \overline{w})\zeta : c \in S_c(\eta)\} \\
&\subseteq \left\{-\left(k_\mu + c\phi^d(\eta - \overline{w})\right)\left(\zeta - \frac{k_\mu}{k_\mu + c\phi^d(\eta - \overline{w})}\phi(\eta + w)\right) : \right. \\
&\qquad\qquad \left. c \in [c_2, c_1], |w| \leq \overline{w}\right\} \\
&:= F_{\tilde{\zeta}}(\eta, \zeta, \overline{w}),
\end{aligned}
\tag{6.117}
$$

where $S_c(\eta) = \{c_2\}$ when $\eta > \overline{w}$, $S_c(\eta) = \{c_1\}$ when $\eta < \overline{w}$, and $S_c(\eta) = [c_2, c_1]$ when $\eta = \overline{w}$.

Define $k_\phi = -\inf_{r \in \mathbb{R}}\{\phi^d(r)\} = -\inf_{r \in \mathbb{R}}\{d\phi(r)/dr\}$. Condition (6.103) implies $0 < 4k_\phi < k_\mu$.

Choose $c_1 = k_\mu/2k_\phi$. Then, given $0 < 4k_\phi < k_\mu$, i.e., $(4k_\phi k_\mu - k_\mu^2)/4k_\phi < 0$, we have $k_\phi c_1^2 - k_\mu c_1 + k_\mu < 0$, i.e., $k_\mu/(k_\mu - c_1 k_\phi) < c_1$. Choose $c_2 < 1$. Then, it can be proved that

$$
c_2 < \frac{k_\mu}{k_\mu + c\phi^d(\eta - \overline{w})} < c_1
\tag{6.118}
$$

for $c \in [c_2, c_1]$ and $-k_\phi \leq \phi^d(\eta - \overline{w}) < 0$.

Choose $0 < \overline{k}_\mu \leq \frac{1}{2}k_\mu$. Then,

$$
k_\mu + c\phi^d(\eta - \overline{w}) \geq \overline{k}_\mu
\tag{6.119}
$$

for $c \in [c_2, c_1]$ and $-k_\phi \leq \phi^d(\eta - \overline{w}) < 0$.

Denote

$$
\frac{k_\mu}{k_\mu + c\phi^d(\eta - \overline{w})}\phi(\eta + w) = \Delta(\eta, w, \overline{w}).
\tag{6.120}
$$

When $0 \leq t \leq T$, by using (6.118) and $|w| \leq \overline{w}$, we have

$$
\Delta(\eta, w, \overline{w}) \in \{\overline{c}\phi(\eta + w) : \overline{c} \in [c_2, c_1], |w| \leq \overline{w}\}
\tag{6.121}
$$

for $c \in [c_2, c_1]$ and $-k_\phi \leq \phi^d(\eta - \overline{w}) < 0$, and thus

$$
|\zeta - \Delta(\eta, w, \overline{w})| \geq |\tilde{\zeta}|
\tag{6.122}
$$

$$
\text{sgn}(\zeta - \Delta(\eta, w, \overline{w})) = \text{sgn}(\tilde{\zeta})
\tag{6.123}
$$

when $\tilde{\zeta} \neq 0$.

Define $V_{\tilde{\zeta}}(\tilde{\zeta}) = \frac{1}{2}\tilde{\zeta}^2$ as a Lyapunov function candidate for the $\tilde{\zeta}$-subsystem. Then, by using (6.119)–(6.123), we have

$$\max_{f_{\tilde{\zeta}} \in F_{\tilde{\zeta}}(\eta,\zeta,w)} \nabla V_{\tilde{\zeta}}(\tilde{\zeta})f_{\tilde{\zeta}} \leq -\bar{k}_\mu \tilde{\zeta}^2 = -2\bar{k}_\mu V_{\tilde{\zeta}}(\tilde{\zeta}). \tag{6.124}$$

IOS and UO

From (6.124), we have

$$V_{\tilde{\zeta}}(\tilde{\zeta}(t)) \leq e^{-2\bar{k}_\mu t} V_{\tilde{\zeta}}(\tilde{\zeta}(0)) \tag{6.125}$$

and thus

$$|\tilde{\zeta}(t)| \leq e^{-\bar{k}_\mu t} |\tilde{\zeta}(0)|. \tag{6.126}$$

It can be calculated that it takes $T_0 = \max\{0, T_0'\}$ for $|\tilde{\zeta}(t)|$ to converge to the region $|\tilde{\zeta}| \leq (1-\delta)c_2\bar{\phi}$, where $T_0' = \frac{1}{\bar{k}_\mu}\left(\ln(|\tilde{\zeta}(0)|) - \ln((1-\delta)c_2\bar{\phi})\right)$.

Considering properties (6.111) and (6.112) of set-valued map $S_\zeta(\eta, \overline{w})$, we have

$$\max F_{\tilde{\eta}}(\eta, \tilde{\zeta}, \overline{w}) \leq |\tilde{\zeta}| \tag{6.127}$$

$$\min F_{\tilde{\eta}}(\eta, \tilde{\zeta}, \overline{w}) \geq -|\tilde{\zeta}| \tag{6.128}$$

and thus

$$
\begin{aligned}
|\tilde{\eta}(t)| &\leq |\tilde{\eta}(0)| + \int_0^t |\tilde{\zeta}(\tau)| d\tau \\
&\leq |\tilde{\eta}(0)| + \int_0^t e^{-\bar{k}_\mu \tau} d\tau |\tilde{\zeta}(0)| \\
&\leq |\tilde{\eta}(0)| + \frac{1}{\bar{k}_\mu} |\tilde{\zeta}(0)|,
\end{aligned}
\tag{6.129}
$$

where we used property (6.126) for the second inequality.

When $0 \leq t \leq T$, we first consider the case of $\tilde{\zeta}(t) \geq (1-\delta)c_2\bar{\phi}$. In this case, $t \leq T_0' = T_0$.

Choose $\alpha_{\beta 0}(s) = (2 + \bar{k}_\mu)s^2/(2(1-\delta)c_2\bar{\phi}\bar{k}_\mu)$ for $s \in \mathbb{R}_+$ and $\beta_0(s, t) = e^{-\bar{k}_\mu t}\alpha_{\beta 0}(s)$ for $s, t \in \mathbb{R}_+$. Then, $\beta_0 \in \mathcal{KL}$. It can be proved that

$$
\begin{aligned}
\beta_0(|[\tilde{\eta}(0), \tilde{\zeta}(0)]^T|, T_0') &= \frac{1 + \bar{k}_\mu}{2\bar{k}_\mu |\tilde{\zeta}(0)|}(\tilde{\eta}^2(0) + \tilde{\zeta}^2(0)) \\
&\geq |\tilde{\eta}(0)| + \frac{1}{\bar{k}_\mu}|\tilde{\zeta}(0)|,
\end{aligned}
\tag{6.130}
$$

where Young's inequality—see [153]—is used for the second inequality.

When $0 \le t \le T$, if $\tilde{\zeta}(t) \ge (1 - \delta)c_2\overline{\phi}$, by using (6.129), (6.130), and $t \le T_0'$, we have

$$|\tilde{\eta}(t)| \le \beta_0(|[\tilde{\eta}(0), \tilde{\zeta}(0)]^T|, T_0') \le \beta_0(|[\tilde{\eta}(0), \tilde{\zeta}(0)]^T|, t). \tag{6.131}$$

When $0 \le t \le T$, in the case of $|\tilde{\zeta}(t)| \le (1 - \delta)c_2\overline{\phi}$, we can consider the $(\tilde{\eta}, \tilde{\zeta})$-system as a cascade connection of the input-to-stable $\tilde{\eta}$-subsystem and the asymptotically stable $\tilde{\zeta}$-subsystem, and use the small-gain theorem in [130] to directly prove the existence of $\beta_1 \in \mathcal{KL}$ such that

$$|\tilde{\eta}(t)| \le \beta_1(|[\tilde{\eta}(T_0), \tilde{\zeta}(T_0)]^T|, t - T_0). \tag{6.132}$$

Define $\beta \in \mathcal{KL}$ as

$$\beta(s, t) = \max\{\beta_0(s, t), \beta_1(s, t)\} \tag{6.133}$$

for $s, t \in \mathbb{R}_+$. Then, for any $\tilde{\eta}_0, \tilde{\zeta}_0 \in \mathbb{R}$, with initial condition $\tilde{\eta}(0) = \tilde{\eta}_0$, $\tilde{\zeta}(0) = \tilde{\zeta}_0$, it holds that

$$|\tilde{\eta}(t)| \le \beta(|[\tilde{\eta}_0, \tilde{\zeta}_0]^T|, t) \tag{6.134}$$

for $0 \le t \le T$.

From the definition of $\tilde{\eta}$ in (6.109), we have

$$|\tilde{\eta}(t)| \le |\eta(t)| \le |\tilde{\eta}(t)| + \|w\|_T. \tag{6.135}$$

Recall the definition of $\tilde{\zeta}$ in (6.106). From the fact that

$$\phi(\eta(t)) \in \{\overline{c}\phi(\eta(t) + w) : |w| \le \overline{w}, \overline{c} \in [c_2, c_1]\} \tag{6.136}$$

with $0 < c_2 < 1 < c_1$, we get

$$|\tilde{\zeta}(t)| \le |\zeta(t) - \phi(\eta(t))| \le |\zeta(t)| + |\phi(\eta(t))| \tag{6.137}$$

for $0 \le t \le T$. From the properties of ϕ, we can find $|\phi(r)| \le k_\phi|r|$ for $r \in \mathbb{R}$. Thus, there exists an $\alpha_0 \in \mathcal{K}_\infty$ such that

$$|[\tilde{\eta}(0), \tilde{\zeta}(0)]^T| \le \alpha_0(|[\eta(0), \zeta(0)]^T|). \tag{6.138}$$

Property (6.134) together with (6.135) and (6.138) implies that

$$|\eta(T)| \le \overline{\beta}(|[\eta(0), \zeta(0)]^T|, t) + \|w\|_T, \tag{6.139}$$

where $\overline{\beta}(s, t) := \beta(\alpha_0(s), t)$ for $s, t \in \mathbb{R}_+$.

From (6.126) and (6.137), we have

$$|\tilde{\zeta}(t)| \le |\tilde{\zeta}(0)| \le |\zeta(0)| + |\phi(\eta(0))|. \tag{6.140}$$

Using the definition of $\tilde{\zeta}$, we have

$$|\zeta(t)| \leq |\tilde{\zeta}(t)| + \max_{\zeta^* \in S_\zeta(\eta(t),\overline{w})} |\zeta^*|$$
$$\leq |\zeta(0)| + |\phi(\eta(0))| + c_1|\phi(|\eta(t)| + \|w\|_T)|$$
$$\leq |\zeta(0)| + k_\phi|\eta(0)| + c_1 k_\phi(|\eta(t)| + \|w\|_T)$$
$$\leq |\zeta(0)| + k_\phi|\eta(0)| + c_1 k_\phi(\|\eta\|_T + \|w\|_T) \qquad (6.141)$$

for $0 \leq t \leq T$. Then, one can find an $\alpha_{UO} \in \mathcal{K}_\infty$ such that

$$|\zeta(T)| \leq |\zeta(0)| + \alpha_{UO}(\|\eta\|_T, \|w\|_T). \qquad (6.142)$$

It can be observed that the definitions of $\overline{\beta}$ in (6.139) and α_{UO} in (6.142) do not depend on the signals η, ζ, w and time T, and it can be concluded that for any $\eta_0, \zeta_0 \in \mathbb{R}$, with initial states $\eta(0) = \eta_0$ and $\zeta(0) = \zeta_0$, and input $w : \mathbb{R}_+ \to \mathbb{R}$ being piecewise continuous and bounded, it holds that

$$|\eta(T)| \leq \overline{\beta}(|[\eta_0, \zeta_0]^T|, T) + \|w\|_T \qquad (6.143)$$
$$|\zeta(T)| \leq |\zeta_0| + \alpha_{UO}(\|\eta\|_T + \|w\|_T) \qquad (6.144)$$

for all $T \geq 0$. Note that we used T instead of t in the proof to avoid confusion. This ends the proof of Lemma 6.2. $\diamond$

6.3.3   DISTRIBUTED FORMATION CONTROLLER DESIGN AND SMALL-GAIN ANALYSIS

As discussed in Subsection 6.3.1, for the validity of (6.95)–(6.96) of the formation control design, v_i should be guaranteed to be nonzero. For a specified λ_* satisfying $0 < \lambda_* < \underline{v}_0$, by designing a control law for the i-th robot such that

$$\max\{|\tilde{v}_{xi}|, |\tilde{v}_{yi}|\} \leq \frac{\sqrt{2}}{2}(\underline{v}_0 - \lambda_*) \leq \frac{\sqrt{2}}{2}(v_0 - \lambda_*), \qquad (6.145)$$

it can be guaranteed that $|v_i| = \sqrt{v_{xi}^2 + v_{yi}^2} = \sqrt{(v_{x0} + \tilde{v}_{xi})^2 + (v_{y0} + \tilde{v}_{yi})^2} \geq \lambda_* > 0$ and thus $v_i \neq 0$. In this way, singularity is avoided.

Practically, the linear velocity of each robot is usually required to be less than a desired value. For any given $\lambda^* > \overline{v}_0$, we can also guarantee $|v_i| \leq \lambda^*$ by designing a control law such that

$$\max\{|\tilde{v}_{xi}|, |\tilde{v}_{yi}|\} \leq \frac{\sqrt{2}}{2}(\lambda^* - \overline{v}_0). \qquad (6.146)$$

For specified constants $\lambda_*, \lambda^*, \underline{v}_0, \overline{v}_0$ satisfying $0 < \lambda_* < \underline{v}_0 < \overline{v}_0 < \lambda^*$, we define

$$\lambda = \min\left\{\frac{\sqrt{2}}{2}(\underline{v}_0 - \lambda_*), \frac{\sqrt{2}}{2}(\lambda^* - \overline{v}_0)\right\}. \qquad (6.147)$$

Then, conditions (6.145) and (6.146) can be satisfied if

$$\max\{|\tilde{v}_{xi}|, |\tilde{v}_{yi}|\} \le \lambda. \tag{6.148}$$

The proposed distributed control law is composed of two stages: (a) initialization and (b) formation control. The initialization stage is employed because the formation control stage cannot solely guarantee $v_i \ne 0$ if (6.145) is not satisfied at the beginning of the control procedure. With the initialization stage, the linear velocity and the heading direction of each follower robot can be controlled to satisfy (6.148) within some finite time. Then, the formation control stage is triggered, and thereafter, the satisfaction of (6.148) is guaranteed, and at the same time, the formation control objective is achieved.

Initialization Stage

For this stage, we design the following control law

$$\omega_i = \phi_{\theta i}(\theta_i - \theta_0) + \omega_0 \tag{6.149}$$
$$r_i = \phi_{vi}(v_i - v_0) + \dot{v}_0 \tag{6.150}$$

for each i-th follower robot, where $\phi_{\theta i}, \phi_{vi} : \mathbb{R} \to \mathbb{R}$ are nonlinear functions.

Define $\tilde{\theta}_i = \theta_i - \theta_0$ and $\tilde{v}_i = v_i - v_0$. Taking the derivatives of $\tilde{\theta}_i$ and $\tilde{v}_i$, respectively, and using (6.149) and (6.150), we have

$$\dot{\tilde{\theta}}_i = \phi_{\theta i}(\tilde{\theta}_i), \tag{6.151}$$
$$\dot{\tilde{v}}_i = \phi_{vi}(\tilde{v}_i). \tag{6.152}$$

By designing $\phi_{\theta i}, \phi_{vi}$ such that $-\phi_{\theta i}(s), \phi_{\theta i}(-s), -\phi_{vi}(s), \phi_{vi}(-s)$ are positive definite for $s \in \mathbb{R}_+$, we can guarantee the asymptotic stability of systems (6.151) and (6.152). Moreover, there exist $\beta_{\tilde{\theta}}, \beta_{\tilde{v}} \in \mathcal{KL}$ such that $|\tilde{\theta}(t)| \le \beta_{\tilde{\theta}}(|\tilde{\theta}(0)|, t)$ and $|\tilde{v}(t)| \le \beta_{\tilde{v}}(|\tilde{v}(0)|, t)$.

By directly using the property of continuous functions, there exist $\bar{\delta}_{v0} > 0$ and $\bar{\delta}_{\theta 0} > 0$ such that, for all $v_0 \in [\underline{v}_0, \overline{v}_0]$, $\theta_0 \in \mathbb{R}$, $|\delta v_0| \le \bar{\delta}_{v0}$ and $|\delta_{\theta 0}| \le \bar{\delta}_{\theta 0}$,

$$|(v_0 + \delta_{v0})\cos(\theta_0 + \delta_{\theta 0}) - v_0 \cos\theta_0| \le \lambda, \tag{6.153}$$
$$|(v_0 + \delta_{v0})\sin(\theta_0 + \delta_{\theta 0}) - v_0 \sin\theta_0| \le \lambda. \tag{6.154}$$

Recall that for any $\beta \in \mathcal{KL}$, there exist $\alpha_1, \alpha_2 \in \mathcal{K}_\infty$ such that $\beta(s, t) \le \alpha_1(s)\alpha_2(e^{-t})$ for all $s, t \in \mathbb{R}_+$ according to Lemma 1.1; see also [243, Lemma 8]. With control law (6.149)–(6.150), there exists a finite time T_{Oi} for the i-th robot such that $|\theta_i(T_{Oi}) - \theta_0(T_{Oi})| \le \bar{\delta}_{\theta 0}$ and $|v_i(T_{Oi}) - v_0(T_{Oi})| \le \bar{\delta}_{v0}$, and thus condition (6.148) is satisfied at time T_{Oi}.

It should be noted that if $v_i(0) \le \lambda^*$, then control law (6.150) guarantees $v_i(t) \le \lambda^*$ for $t \in [0, T_{Oi}]$ because of $v_0(t) \le \overline{v}_0 < \lambda^*$.

Formation Control Stage

At time T_{Oi}, the distributed control law for the i-th follower robot is switched to the formation control stage.

In this stage, we design

$$\tilde{u}_{xi} = -k_{xi}(\tilde{v}_{xi} - \phi_{xi}(z_{xi})) \tag{6.155}$$

$$\tilde{u}_{yi} = -k_{yi}(\tilde{v}_{yi} - \phi_{yi}(z_{yi})), \tag{6.156}$$

where $\phi_{xi}, \phi_{yi} : \mathbb{R} \to [-\lambda, \lambda]$ are odd, strictly decreasing, and continuously differentiable functions and k_{xi}, k_{yi} are positive constants satisfying

$$- k_{xi}/4 < d\phi_{xi}(r)/dr < 0 \tag{6.157}$$

$$- k_{yi}/4 < d\phi_{yi}(r)/dr < 0 \tag{6.158}$$

for all $r \in \mathbb{R}$. An example for ϕ_{xi} and ϕ_{yi} is shown in Figure 6.5.

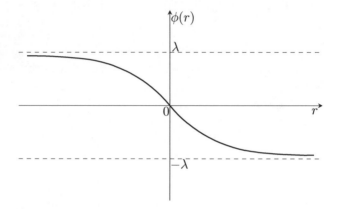

FIGURE 6.5　An example for ϕ_{xi} and ϕ_{yi}.

The variables z_{xi} and z_{yi} are defined as

$$z_{xi} = \frac{1}{N_i} \sum_{j \in \mathcal{N}_i} (x_i - x_j - (d_{xi} - d_{xj})) \tag{6.159}$$

$$z_{yi} = \frac{1}{N_i} \sum_{j \in \mathcal{N}_i} (y_i - y_j - (d_{yi} - d_{yj})), \tag{6.160}$$

where N_i is the size of $\mathcal{N}_i$ with $\mathcal{N}_i$ representing the position sensing structure. If $j \in \mathcal{N}_i$, then the position sensing digraph $\mathcal{G}$ has a directed edge (j, i) from vertex j to vertex i. Note that $d_{xi} - d_{xj}, d_{yi} - d_{yj}$ in (6.159) and (6.160) represent the desired relative position between the i-th robot and the j-th robot. By default, $d_{x0} = d_{y0} = 0$.

In the formation control stage, the control inputs r_i and ω_i are defined as (6.92) with $u_{xi} = \tilde{u}_{xi} + u_{x0}$ and $u_{yi} = \tilde{u}_{yi} + u_{y0}$.

Consider the $(\tilde{v}_{xi}, \tilde{v}_{yi})$-system defined in (6.95)–(6.96). With condition (6.148) satisfied at time T_{Oi}, the boundedness of ϕ_{xi} and ϕ_{yi} together with the control law (6.155)–(6.156) guarantees the satisfaction of (6.148) after T_{Oi}. For the proof of this statement, we can consider $\{(\tilde{v}_{xi}, \tilde{v}_{yi}) : \max\{|\tilde{v}_{xi}|, |\tilde{v}_{yi}|\} \le \lambda\}$ as an invariant set of the $(\tilde{v}_{xi}, \tilde{v}_{yi})$-system.

The main result of distributed formation control is summarized below.

Theorem 6.2 *Consider the multi-robot model* (6.84)–(6.86) *and the distributed control laws defined by* (6.90), (6.92), (6.149), (6.150), (6.155), *and* (6.156) *with parameters* k_{xi}, k_{yi} *satisfying* (6.157)–(6.158). *Under Assumption 6.3, if the position sensing digraph* $\mathcal{G}$ *has a spanning tree with vertex 0 as the root, then for any constants* $d_{xi}, d_{yi} \in \mathbb{R}$ *with* $i = 1, \ldots, N$, *the coordinates* $(x_i(t), y_i(t))$ *and the angle* $\theta_i(t)$ *of each i-th robot asymptotically converge to* $(x_0(t) + d_{xi}, y_0(t) + d_{yi})$ *and* $\theta_0(t) + 2k\pi$ *with* $k \in \mathbb{Z}$, *respectively. Moreover, given any* $\lambda^* > \overline{v}_0$, *if* $v_i(0) \le \lambda^*$ *for* $i = 1, \ldots, N$, *then* $v_i(t) \le \lambda^*$ *for all* $t \ge 0$.

The two-stage distributed control law results in a switching incident of the control signal for each follower robot during the control procedure. The trajectories of such systems can be well defined in the spirit of Rademacher (see e.g., [60]), and the performance of the system can be analyzed by considering the two stages one-by-one.

6.3.4 SMALL-GAIN ANALYSIS AND PROOF OF THEOREM 6.2

Recall the definition of λ in (6.147). With condition (6.148) satisfied after T_{Oi}, we have $v_i \ne 0$ and thus the validity of (6.92) for all $t \ge T_{Oi}$. Under the condition of $v_i(0) \le \lambda^*$, the boundedness of $v_i(t)$, i.e., $v_i(t) \le \lambda^*$, can also be directly proved based on the discussions in Subsection 6.3.3.

Denote $\tilde{x}_0 = 0$ and $\tilde{y}_0 = 0$. We equivalently represent z_{xi} and z_{yi} as

$$
\begin{aligned}
z_{xi} &= \frac{1}{N_i} \sum_{j \in \mathcal{N}_i} (x_i - d_{xi} - x_0 - (x_j - d_{xj} - x_0)) \\
&= \frac{1}{N_i} \sum_{j \in \mathcal{N}_i} (\tilde{x}_i - \tilde{x}_j) = \tilde{x}_i - \frac{1}{N_i} \sum_{j \in \mathcal{N}_i} \tilde{x}_j
\end{aligned}
\tag{6.161}
$$

and similarly,

$$
z_{yi} = \tilde{y}_i - \frac{1}{N_i} \sum_{j \in \mathcal{N}_i} \tilde{y}_j.
\tag{6.162}
$$

Denote

$$\omega_{xi} = \frac{1}{N_i} \sum_{j \in \mathcal{N}_i} \tilde{x}_j, \tag{6.163}$$

$$\omega_{yi} = \frac{1}{N_i} \sum_{j \in \mathcal{N}_i} \tilde{y}_j. \tag{6.164}$$

Then, control laws (6.155) and (6.156) are in the form of (6.102).

In the following proof, we only consider the $(\tilde{x}_i, \tilde{v}_{xi})$-system (6.95). The $(\tilde{y}_i, \tilde{v}_{yi})$-system (6.96) can be studied in the same way.

Define $T_O = \max_{i=1,\ldots,N}\{T_{Oi}\}$. By using Lemma 6.2, for each $i = 1, \ldots, N$, the closed-loop system composed of (6.95) and (6.155) has the following properties: for any $\tilde{x}_{i0}, \tilde{v}_{xi0} \in \mathbb{R}$, with $\tilde{x}_i(T_O) = \tilde{x}_{i0}$ and $\tilde{v}_{xi}(T_O) = \tilde{v}_{xi0}$,

$$|\tilde{x}_i(t)| \le \beta_{xi}(|[\tilde{x}_{i0}, \tilde{v}_{xi0}]^T|, t - T_O) + \|\omega_{xi}\|_{[T_O,t]} \tag{6.165}$$

$$|\tilde{v}_{xi}(t)| \le |\tilde{v}_{xi0}| + \alpha_{xi}(\|\tilde{x}_i\|_{[T_O,t]} + \|\omega_{xi}\|_{[T_O,t]}), \tag{6.166}$$

where $\beta_{xi} \in \mathcal{KL}$ and $\alpha_{xi} \in \mathcal{K}_\infty$.

Notice that for any constants $a_1, \ldots, a_n > 0$ satisfying $\sum_{i=1}^n (1/a_i) \le n$, it holds that $\sum_{i=1}^n d_i = \sum_{i=1}^n (1/a_i) a_i d_i \le n \max_{1 \le i \le n}\{a_i d_i\}$ for all $d_1, \ldots, d_n \ge 0$. We have

$$|\omega_{xi}| \le \delta_i \max_{j \in \overline{\mathcal{N}}_i}\{a_{ij}|\tilde{x}_j|\}, \tag{6.167}$$

where $\delta_i = (N_i - 1)/N_i$, $\overline{\mathcal{N}}_i = \mathcal{N}_i \backslash \{0\}$, and $\sum_{j \in \overline{\mathcal{N}}_i}(1/a_{ij}) \le N_i - 1$ if $0 \in \mathcal{N}_i$; $\delta_i = 1$, $\overline{\mathcal{N}}_i = \mathcal{N}_i$ and $\sum_{j \in \overline{\mathcal{N}}_i}(1/a_{ij}) \le N_i$ if $0 \notin \mathcal{N}_i$.

Then, properties (6.165) and (6.166) imply

$$|\tilde{x}_i(t)| \le \beta_{xi}(|[\tilde{x}_{i0}, \tilde{v}_{xi0}]^T|, t - T_O) + \delta_i \max_{j \in \overline{\mathcal{N}}_i}\{a_{ij}\|\tilde{x}_j\|_{[T_O,t]}\}, \tag{6.168}$$

$$|\tilde{v}_{xi}(t)| \le |\tilde{v}_{xi0}| + \alpha_{xi}(\|\tilde{x}_i\|_{[T_O,t]} + \delta_i \max_{j \in \overline{\mathcal{N}}_i}\{a_{ij}\|\tilde{x}_j\|_{[T_O,t]}\}). \tag{6.169}$$

Define the follower sensing digraph $\mathcal{G}_f$ as a subgraph of $\mathcal{G}$. Digraph $\mathcal{G}_f$ has N vertices with indices $1, \ldots, N$ corresponding to the vertices with indices $1, \ldots, N$ of $\mathcal{G}$ and representing the follower robots. From the definitions of $\overline{\mathcal{N}}_i$ and $\mathcal{G}_f$, it can be observed that, for $i = 1, \ldots, N$, if $j \in \overline{\mathcal{N}}_i$, then there is a directed edge (j, i) from the j-th vertex to the i-th vertex in $\mathcal{G}_f$. Clearly, $\mathcal{G}_f$ represents the interconnection topology of the network composed of the $(\tilde{x}_i, \tilde{v}_{xi})$-systems (6.95).

Define $\mathcal{F}_0 = \{i \in \{1, \ldots, N\} : 0 \in \mathcal{N}_i\}$. Denote $\mathcal{C}_f$ as the set of all simple cycles of $\mathcal{G}_f$, and denote $\mathcal{C}_0 \subseteq \mathcal{C}_f$ as the set of all simple cycles through the vertices with indices belonging to $\mathcal{F}_0$.

For $i = 1, \ldots, N$, $j \in \overline{\mathcal{N}}_i$, we assign the positive value a_{ij} to edge (j, i) in $\mathcal{G}_f$. For a simple cycle $\mathcal{O} \in \mathcal{C}_f$, denote $A_{\mathcal{O}}$ as the product of the positive values assigned to the edges of the cycle.

Consider $\tilde{x}_i$ with $i = 1, \ldots, N$ as the outputs of the network composed of the $(\tilde{x}_i, \tilde{v}_{xi})$-systems (6.95). By using the IOS small-gain theorem for general nonlinear systems in [134, 137], $\tilde{x}_i(t)$ with $i = 1, \ldots, N$ converge to the origin if

$$A_{\mathcal{O}} \frac{N-1}{N} < 1 \text{ for } \mathcal{O} \in \mathcal{C}_0, \tag{6.170}$$

$$A_{\mathcal{O}} < 1 \text{ for } \mathcal{O} \in \mathcal{C}_f \backslash \mathcal{C}_0. \tag{6.171}$$

Note that $A_{\mathcal{O}}(N-1)/N < 1$ is equivalent to $A_{\mathcal{O}} < N/(N-1) = 1 + 1/(N-1)$.

If $\mathcal{G}$ has a spanning tree with vertex 0 as the root, then $\mathcal{G}_f$ has a spanning tree with the indices of the root vertices belonging to $\mathcal{F}_0$. According to Lemma 6.1, there exist positive constants a_{ij} such that both conditions (6.170) and (6.171) are satisfied. For system (6.95), with the convergence of each $\tilde{x}_i$ to the origin and the boundedness of $\tilde{u}_{xi}$, we can guarantee the convergence of $\tilde{v}_{xi}$ to the origin by using Barbalat's lemma [144]. Similarly, we can prove the convergence of $\tilde{v}_{yi}$ to the origin. By using the definitions of $\tilde{v}_{xi}$ and $\tilde{v}_{yi}$, the convergence of θ_i to $\theta_0 + 2k\pi$ with $k \in \mathbb{Z}$ can be concluded. This ends the proof of Theorem 6.2.

6.3.5   ROBUSTNESS TO RELATIVE POSITION MEASUREMENT ERRORS

Measurement errors can decrease the performance of a nonlinear control system. In this section, we discuss the robustness of our distributed formation controller in the presence of relative position measurement errors.

It can be observed that the initialization stage of the distributed control law defined in (6.149)–(6.150) is not affected by the position measurement errors. Also, condition (6.148) still holds for $t \geq T_{Oi}$ for $i = 1, \ldots, N$.

For the formation control stage, in the presence of relative position measurement errors, the z_{xi} and z_{yi} defined for the distributed control law (6.155)–(6.156) should be modified as

$$z_{xi} = \frac{1}{N_i} \sum_{j \in \mathcal{N}_i} \left(x_i - x_j - (d_{xi} - d_{xj}) + \omega_{ij}^x \right) \tag{6.172}$$

$$z_{yi} = \frac{1}{N_i} \sum_{j \in \mathcal{N}_i} \left(y_i - y_j - (d_{yi} - d_{yj}) + \omega_{ij}^y \right), \tag{6.173}$$

where N_i is the size of $\mathcal{N}_i$ and $\omega_{ij}^x, \omega_{ij}^y \in \mathbb{R}$ represent the relative position measurement errors corresponding to $(x_i - x_j)$ and $(y_i - y_j)$, respectively. Due to the boundedness of the designed ϕ_{xi} and ϕ_{yi} in (6.155)–(6.156), condition (6.148) is still satisfied in the existence of position measurement errors, which guarantees the validity of (6.92).

Here, we only consider each $\tilde{x}_i$-subsystem. The $\tilde{y}_i$-subsystems can be studied in the same way. By defining

$$\omega_{xi} = \frac{1}{N_i} \sum_{j \in \mathcal{N}_i} \left(\tilde{x}_j + \omega_{ij}^x \right), \tag{6.174}$$

we have $z_{xi} = \tilde{x}_i - \omega_{xi}$. With such definition, if the measurement errors ω_{ij}^x are piecewise continuous and bounded, then each $\tilde{x}_i$-subsystem still has the IOS and UO properties given by (6.165) and (6.166), respectively.

As in the discussion above (6.167), we have

$$|\omega_{xi}| \leq \max \left\{ \frac{\rho_i}{N_i} \sum_{j \in \mathcal{N}_i} (|\tilde{x}_j|), \frac{\rho_i'}{N_i} \sum_{j \in \mathcal{N}_i} (|\omega_{ij}^x|) \right\}$$

$$:= \max \left\{ \frac{\rho_i}{N_i} \sum_{j \in \mathcal{N}_i} (|\tilde{x}_j|), \omega_{xi}^e \right\}$$

$$\leq \max_{j \in \overline{\mathcal{N}}_i} \left\{ \delta_i a_{ij} |\tilde{x}_j|, \omega_{xi}^e \right\}, \tag{6.175}$$

where $\rho_i, \rho_i' > 0$ satisfying $1/\rho_i + 1/\rho_i' \leq 1$, and $\delta_i = \rho_i(N_i - 1)/N_i$, $\overline{\mathcal{N}}_i = \mathcal{N}_i \backslash \{0\}$ and $\sum_{j \in \overline{\mathcal{N}}_i} (1/a_{ij}) \leq N_i - 1$ if $0 \in \mathcal{N}_i$; $\delta_i = \rho_i$, $\overline{\mathcal{N}}_i = \mathcal{N}_i$ and $\sum_{j \in \overline{\mathcal{N}}_i} (1/a_{ij}) \leq N_i$ if $0 \notin \mathcal{N}_i$.

In the existence of the relative position measurement errors, we can still guarantee the IOS of the closed-loop distributed system by using the cyclic-small-gain theorem. In this case, the cyclic-small-gain condition is as follows:

$$A_{\mathcal{O}} \frac{\rho(N-1)}{N} < 1 \text{ for } \mathcal{O} \in \mathcal{C}_0, \tag{6.176}$$

$$A_{\mathcal{O}} \rho < 1 \text{ for } \mathcal{O} \in \mathcal{C}_f \backslash \mathcal{C}_0, \tag{6.177}$$

where $\rho := \max_{i \in \{1, \dots, N\}} \{\rho_i\}$ is larger than one according to $\frac{1}{\rho_i} + \frac{1}{\rho_i'} \leq 1$, and can be chosen to be very close to one. Lemma 6.1 can guarantee (6.176) and (6.177) if $\mathcal{G}$ has a spanning tree with vertex 0 as the root. Thus, the proposed distributed control law is robust with respect to relative position measurement errors.

6.3.6 A NUMERICAL EXAMPLE

Consider a group of 6 robots with indices $0, 1, \dots, 5$. Notice that the robot with index 0 is the leader. The neighbor sets of the robots are defined as follows: $\mathcal{N}_1 = \{0, 5\}$, $\mathcal{N}_2 = \{1, 3\}$, $\mathcal{N}_3 = \{2, 5\}$, $\mathcal{N}_4 = \{3\}$, $\mathcal{N}_5 = \{4\}$.

By default, the values of all the variables in this simulation are in SI units. For convenience, we omit the units. The desired relative position of the follower robots are defined by $d_{x1} = -\sqrt{3}d/2, d_{x2} = -\sqrt{3}d/2, d_{x3} = 0, d_{x4} = \sqrt{3}d/2, d_{x5} = \sqrt{3}d/2, d_{y1} = -d/2, d_{y2} = -3d/2, d_{y3} = -2d, d_{y4} = $

$-3d/2$, $d_{y5} = -d/2$ with $d = 30$. Figure 6.6 shows the position sensing graph of the formation control system. Clearly, the position sensing graph has a spanning tree with vertex 0 as the root.

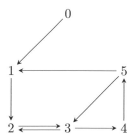

FIGURE 6.6 The position sensing graph of the formation control system.

It should be noted that the control law for each follower robot also uses the velocity and acceleration information of the leader robot, the communication topology of which is not shown in Figure 6.6.

The control inputs of the leader robot are $r_0(t) = 0.1\sin(0.4t)$ and $\omega_0(t) = 0.1\cos(0.2t)$. With such control inputs, the linear velocity v_0 with $v_0(0) = 3$ satisfies $\underline{v}_0 \leq v_0(t) \leq \overline{v}_0$ with $\underline{v}_0 = 3$ and $\overline{v}_0 = 3.5$.

Choose $\lambda_* = 0.45$ and $\lambda^* = 6.05$. The distributed control laws for the initialization stage are in the form of (6.149)–(6.150) with $\phi_{\theta i}(r) = \phi_{vi}(r) = -0.5(1-\exp(-0.5r))/(1+\exp(-0.5r))$ for $i = 1, \ldots, 5$. The distributed control laws for the formation control stage are in the form of (6.155)–(6.156) with $k_{xi} = k_{yi} = 2$ and $\phi_{xi}(r) = \phi_{yi}(r) = -1.8(1 - \exp(-0.5r))/(1 + \exp(-0.5r))$ for $i = 1, \ldots, 5$. With direct calculation, it can be verified that the designed $k_{xi}, k_{yi}, \phi_{xi}, \phi_{yi}$ satisfy (6.157) and (6.158). Also, $\phi_{xi}(r), \phi_{yi}(r) \in [-1.8, 1.8]$ for all $r \in \mathbb{R}$. With $\underline{v}_0 = 3$ and $\overline{v}_0 = 3.5$, the control laws can restrict the linear velocities of the follower robots to be in the range of $[3 - 1.8\sqrt{2}, 3.5 + 1.8\sqrt{2}] = [0.454, 6.046] \subset [\lambda_*, \lambda^*]$.

The initial states of the robots are chosen as

i	$(x_i(0), y_i(0))$	$v_i(0)$	$\theta_i(0)$
0	$(0, 0)$	3	$\pi/6$
1	$(-40, 10)$	4	π
2	$(-20, -40)$	3.5	$5\pi/6$
3	$(5, -40)$	2.5	0
4	$(50, -10)$	2	$-2\pi/3$
5	$(50, 10)$	3	0

The measurement errors are: $\omega_{ij}^x(t) = 0.3(\cos(t + i\pi/6) + \cos(t/3 + i\pi/6) + \cos(t/5 + i\pi/6) + \cos(t/7 + i\pi/6))$ and $\omega_{ij}^y(t) = 0.3(\sin(t + i\pi/6) + \sin(t/3 + i\pi/6) + \sin(t/5 + i\pi/6) + \sin(t/7 + i\pi/6))$ for $i = 1, \ldots, N$, $j \in \mathcal{N}_i$.

The linear velocities and the angular velocities of the robots are shown in Figure 6.7. The stage changes of the distributed controllers are shown in Figure

6.8 with "0" representing initialization stage and "1" representing formation control stage. Figure 6.9 shows the trajectories of the robots. The simulation verifies the theoretical results.

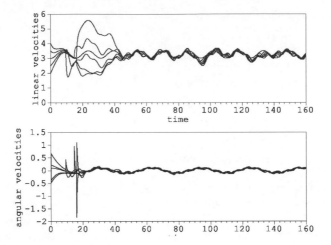

FIGURE 6.7 The linear velocities and the angular velocities of the robots.

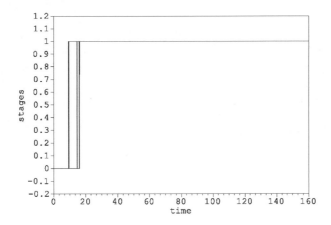

FIGURE 6.8 The stages of the distributed controllers.

6.4 DISTRIBUTED CONTROL WITH FLEXIBLE TOPOLOGIES

In Section 6.3, the formation control problem of the nonholonomic mobile robots is transformed into the distributed nonlinear control problem of double-

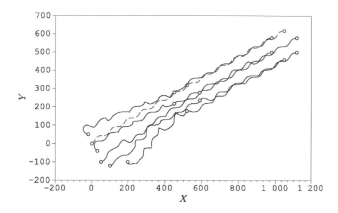

FIGURE 6.9 The trajectories of the robots. The dashed curve represents the trajectory of the leader.

integrators. With fixed information exchange topologies under a connectivity condition, the control objective can be achieved. This section takes a step forward toward solving the distributed nonlinear control problem for multi-agent systems modeled by double integrators under switching information exchange topologies. The goal is to develop a new class of distributed nonlinear control laws to solve a strong output agreement problem, that is, the outputs of the agents converge to each other and the internal states converge to the origin. From a practical point of view, it is assumed that the double-integrators interact with each other through output interconnections (more specifically, the differences of the outputs), for coordination. In this section, an invariant set method is developed such that the strong output agreement problem is solvable if the information exchange graph satisfies a mild joint connectivity condition. Moreover, the proposed design is also valid for systems under physical constraints such as velocity limitation. The result proposed in this section is also used for distributed formation control of nonholonomic mobile robots.

Strong Output Agreement Problem

Consider a group of N double-integrator agents with switching topologies by distributed control:

$$\dot{\eta}_i = \zeta_i \tag{6.178}$$

$$\dot{\zeta}_i = \mu_i, \tag{6.179}$$

where $[\eta_i, \zeta_i]^T \in \mathbb{R}^2$ is the state and $\mu_i \in \mathbb{R}$ is the control input.

The basic idea of the strong output agreement problem is to design a class
of distributed nonlinear control laws in the form of

$$\mu_i = \overline{\varphi}_i(\zeta_i, \xi_i) \tag{6.180}$$

$$\xi_i = \overline{\phi}_i^{\sigma(t)}(\eta_1, \ldots, \eta_N), \tag{6.181}$$

where $\sigma : [0, +\infty) \to \mathcal{P}$ is a piecewise constant signal representing switching
information exchange topology with $\mathcal{P} \subset \mathbb{N}$ being a finite set representing all
the possible information exchange topologies, $\overline{\varphi}_i : \mathbb{R}^2 \to \mathbb{R}$ and $\overline{\phi}_i^p : \mathbb{R}^N \to \mathbb{R}$
for each $p \in \mathcal{P}$, such that the following properties hold:

$$\lim_{t \to \infty} (\eta_i(t) - \eta_j(t)) = 0, \quad \text{for any } i, j = 1, \ldots, N, \tag{6.182}$$

$$\lim_{t \to \infty} \zeta_i(t) = 0, \quad \text{for } i = 1, \ldots, N. \tag{6.183}$$

Notice that the strong output agreement as defined above is a stronger property of state agreement. For state agreement, the internal states ξ_i are only
required to converge to each other.

6.4.1 PROPERTIES OF A CLASS OF NONLINEAR SYSTEMS

Our strong output agreement result is developed based on several properties
of the following second-order nonlinear system:

$$\dot{\eta} = \zeta \tag{6.184}$$

$$\dot{\zeta} = \varphi(\zeta - \phi(\eta - \omega)), \tag{6.185}$$

where $[\eta, \zeta]^T \in \mathbb{R}^2$ is the state, $\omega \in \mathbb{R}$ is an external disturbance input, and
$\varphi, \phi : \mathbb{R} \to \mathbb{R}$ are nonincreasing and locally Lipschitz functions.

For convenience of notation, we define two new classes of functions. A
function $\beta : \mathbb{R}_+ \times \mathbb{R}_+ \to \mathbb{R}_+$ is called an $\mathcal{I}^+\mathcal{L}$ function, denoted by $\beta \in \mathcal{I}^+\mathcal{L}$,
if $\beta \in \mathcal{KL}$, $\beta(s, 0) = s$ for $s \in \mathbb{R}_+$, and for any specified $T > 0$, there exist
continuous, positive definite, and nondecreasing $\alpha_1, \alpha_2 < \text{Id}$ such that for all
$s \in \mathbb{R}_+$, $\beta(s, t) \geq \alpha_1(s)$ for $t \in [0, T]$ and $\beta(s, t) \leq \alpha_2(s)$ for $t \in [T, \infty)$. A
function $\beta : \mathbb{R} \times \mathbb{R}_+ \to \mathbb{R}$ is called an $\mathcal{IL}$ function, denoted by $\beta \in \mathcal{IL}$, if
there exist $\beta', \beta'' \in \mathcal{I}^+\mathcal{L}$ such that for $t \geq 0$, $\beta(r, t) = \beta'(r, t)$ for $r \geq 0$, and
$\beta(r, t) = -\beta''(-r, t)$ for $r < 0$. The new notations are necessary when we
want to avoid the finite-time convergence in a network with a time-variable
topology, which may lead to oscillation. A similar problem arises in the state
agreement problem of coupled nonlinear systems. See [172, Lemma 5.2] for
some details.

Proposition 6.2 *If $\omega \in [\underline{\omega}, \overline{\omega}]$ with $\underline{\omega} \leq \overline{\omega}$ being constants, and if functions*

φ and ϕ satisfy

$$\varphi(0) = \phi(0) = 0, \tag{6.186}$$

$$\varphi(r)r < 0, \ \phi(r)r < 0 \ for \ r \neq 0, \tag{6.187}$$

$$\sup_{r \in \mathbb{R}}\{\max \partial\varphi(r)\} < 4 \inf_{r \in \mathbb{R}}\{\min \partial\phi(r)\}, \tag{6.188}$$

then system (6.184)–(6.185) has the following properties:

1. *There exist strictly decreasing and locally Lipschitz functions $\underline{\psi}, \overline{\psi} : \mathbb{R} \to \mathbb{R}$ satisfying $\underline{\psi}(0) = \overline{\psi}(0) = 0$ such that*

$$S(\underline{\omega}, \overline{\omega}) = \{(\eta, \zeta) : \underline{\psi}(\eta - \underline{\omega}) \leq \zeta \leq \overline{\psi}(\eta - \overline{\omega})\} \tag{6.189}$$

   is an invariant set of system (6.184)–(6.185).
2. *For any specified initial state $(\eta(0), \zeta(0))$, there exist a finite time t_1 and constants $\underline{\mu}, \overline{\mu} \in \mathbb{R}$ such that*

$$\underline{\psi}(\eta(t_1) - \underline{\mu}) \leq \zeta(t_1) \leq \overline{\psi}(\eta(t_1) - \overline{\mu}). \tag{6.190}$$

3. *For any specified $\underline{\sigma}, \overline{\sigma} \in \mathbb{R}$, if $(\eta(t), \zeta(t)) \in S(\underline{\sigma}, \overline{\sigma})$ for $t \in [0, T]$, then there exist $\underline{\beta}_1, \overline{\beta}_1 \in \mathcal{IL}$ such that*

$$-\underline{\beta}_1(\underline{\sigma} - \eta(0), t) + \underline{\sigma} \leq \eta(t) \leq \overline{\beta}_1(\eta(0) - \underline{\sigma}, t) + \overline{\sigma} \tag{6.191}$$

   for all $t \in [0, T]$.
4. *For any specified compact $M \subset \mathbb{R}$, there exist $\underline{\beta}_2, \overline{\beta}_2 \in \mathcal{IL}$ such that if $(\eta(0), \zeta(0)) \in S(\underline{\mu}_0, \overline{\mu}_0)$ with $\underline{\mu}_0 \leq \overline{\mu}_0$ belonging to M, then one can find $\underline{\mu}(t), \overline{\mu}(t)$ satisfying*

$$-\underline{\beta}_2(\underline{\omega} - \underline{\mu}_0, t) + \underline{\omega} \leq \underline{\mu}(t) \leq \overline{\mu}(t) \leq \overline{\beta}_2(\overline{\mu}_0 - \overline{\omega}, t) + \overline{\omega} \tag{6.192}$$

   such that

$$(\eta(t), \zeta(t)) \in S(\underline{\mu}(t), \overline{\mu}(t)) \tag{6.193}$$

   for all $t \geq 0$.

The proof of Proposition 6.2 is given in Subsection 6.4.2.

Figure 6.10 shows Property 1 of system (6.184)–(6.185) given in Proposition 6.2. The region between $\zeta = \underline{\psi}(\eta - \underline{\omega})$ and $\zeta = \overline{\psi}(\eta - \overline{\omega})$ forms an invariant set of system (6.184)–(6.185). Properties 2, 3, and 4 are based on the definitions of $\underline{\psi}$ and $\overline{\psi}$.

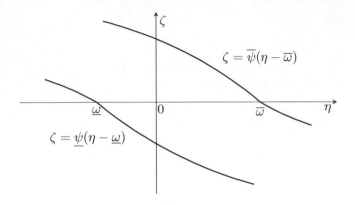

FIGURE 6.10 The boundaries of the invariant set in Property 1 of Proposition 6.2.

6.4.2 PROOF OF PROPOSITION 6.2

We first present a technical lemma on a class of first-order nonlinear systems.

Lemma 6.3 *Consider the following first-order system*

$$\dot{\varsigma} = \alpha(\varsigma), \tag{6.194}$$

where $\varsigma \in \mathbb{R}$ is the state and α is a nonincreasing and locally Lipschitz function satisfying $\alpha(0) = 0$ and $r\alpha(r) < 0$ for all $r \neq 0$. There exists $\beta \in \mathcal{IL}$ such that for any $\varsigma_0 \in \mathbb{R}$, with initial condition $\varsigma(0) = \varsigma_0$, it holds that

$$\varsigma(t) \leq \beta(\varsigma_0, t) \tag{6.195}$$

for all $t \geq 0$.

Proof. Denote $\varsigma^*(\varsigma_0, t)$ as the solution of system (6.194) with initial condition $\varsigma(0) = \varsigma_0$. Define $\beta'(s, t) = \varsigma^*(s, t)$ and $\beta''(s, t) = -\varsigma^*(-s, t)$ for $s, t \in \mathbb{R}_+$.

Consider the case of $\varsigma(0) \geq 0$.

Since α is locally Lipschitz, for any specified $\bar{\varsigma} > 0$, there exists a constant $k_\alpha > 0$ such that $\alpha(s) \geq -k_\alpha s$ for $s \leq \bar{\varsigma}$ and thus

$$\dot{\varsigma}(t) \geq -k_\alpha \varsigma(t) \tag{6.196}$$

for $0 \leq \varsigma(t) \leq \bar{\varsigma}$.

For any specified $T > 0$, define

$$\alpha_1(s) = e^{-k_\alpha T} \min \{s, \bar{\varsigma}\} \tag{6.197}$$

for $s \in \mathbb{R}_+$. Then, α_1 is continuous and positive definite.

If $\varsigma(0) \leq \bar{\varsigma}$, then

$$\varsigma(T) \geq \varsigma(0)e^{-k_\alpha T} \geq \alpha_1(\varsigma(0)). \tag{6.198}$$

Consider the case of $\varsigma(0) > \bar{\varsigma}$. If there is a time $0 < t' \leq T$ such that $\varsigma(t') = \bar{\varsigma}$, then

$$\varsigma(T) \geq \varsigma(t')e^{-k_\alpha(T-t')} \geq \bar{\varsigma}e^{-k_\alpha T} > \alpha_1(\varsigma(0)); \qquad (6.199)$$

otherwise,

$$\varsigma(T) > \bar{\varsigma} > \bar{\varsigma}e^{-k_\alpha T} > \alpha_1(\varsigma(0)). \qquad (6.200)$$

According to the definition of β', for the specified $T > 0$, it holds that $\beta'(s,T) \geq \alpha_1(s)$. Because of the nonincreasing property of $\varsigma(t)$ with $\varsigma(0) \geq 0$, we have

$$\beta'(s,t) \geq \alpha_1(s) \qquad (6.201)$$

for $t \in [0,T]$.

For any specified $T > 0$, define

$$\alpha_2(s) = \max\left\{\frac{1}{2}s, s + T \max_{\frac{1}{2}s \leq \tau \leq s} \alpha(\tau)\right\} \qquad (6.202)$$

for $s \in \mathbb{R}_+$. It can be verified that α_2 is continuous, positive definite, and less than Id.

If $\varsigma(T) \geq \frac{1}{2}\varsigma(0)$, then $\frac{1}{2}\varsigma(0) \leq \varsigma(t) \leq \varsigma(0)$ for $0 \leq t \leq T$, and

$$\dot{\varsigma}(t) \leq \max_{\frac{1}{2}\varsigma(0) \leq \tau \leq \varsigma(0)} \alpha(\tau). \qquad (6.203)$$

Then, we have

$$\varsigma(T) \leq \varsigma(0) + \int_0^T \max_{\frac{1}{2}\varsigma(0) \leq \tau \leq \varsigma(0)} \alpha(\tau)dt$$
$$= \varsigma(0) + T \max_{\frac{1}{2}\varsigma(0) \leq \tau \leq \varsigma(0)} \alpha(\tau)$$
$$\leq \alpha_2(\varsigma(0)). \qquad (6.204)$$

If $\varsigma(T) < \varsigma(0)/2$, then $\varsigma(T) < \alpha_2(\varsigma(0))$ automatically. According to the definition of β', for the specified $T > 0$, it holds that $\beta'(s,T) \leq \alpha_2(s)$. Because of the nonincreasing property of $\varsigma(t)$ with $\varsigma(0) \geq 0$, we have

$$\beta'(s,t) \leq \alpha_2(s) \qquad (6.205)$$

for $t \in [T,\infty)$.

It can be directly verified that $\beta' \in \mathcal{KL}$ and $\beta'(s,0) = s$ for $s \geq 0$. By also using (6.201) and (6.205), we can prove $\beta' \in \mathcal{I}^+\mathcal{L}$. Due to symmetry, we can also prove $\beta'' \in \mathcal{I}^+\mathcal{L}$. Thus, $\beta \in \mathcal{IL}$.

Define $\beta(r,t) = \beta'(r,t)$ for $r \geq 0$, $t \geq 0$ and $\beta(r,t) = -\beta''(-r,t)$ for $r \leq 0$, $t \geq 0$. Then, (6.195) holds and $\beta \in \mathcal{IL}$. This ends the proof. $\diamondsuit$

Property 1

Denote $S^a(\overline{\omega}) = \{(\eta, \zeta) : \zeta \leq \overline{\psi}(\eta - \overline{\omega})\}$ and $S^b(\underline{\omega}) = \{(\eta, \zeta) : \zeta \geq \underline{\psi}(\eta - \overline{\omega})\}$. Then, $S(\underline{\omega}, \overline{\omega}) = S^a(\overline{\omega}) \cap S^b(\underline{\omega})$. If both $S^a(\overline{\omega})$ and $S^b(\underline{\omega})$ are invariant sets of system (6.184)–(6.185) and $\underline{\psi}(r) \leq \overline{\psi}(r)$ for all $r \in \mathbb{R}$, then $S(\underline{\omega}, \overline{\omega})$ is an invariant set. In the following procedure, we find appropriate $\overline{\psi}$ such that $S^a(\overline{\omega})$ is an invariant set. Function $\underline{\psi}$ can be found in the same way.

For a nonincreasing and locally Lipschitz function ϕ satisfying (6.186) and (6.187), there exists a function $\overline{\phi} : \mathbb{R} \to \mathbb{R}$ which is strictly decreasing and continuously differentiable on $(-\infty, 0) \cup (0, \infty)$ such that $\overline{\phi}(0) = 0$, $\overline{\phi}(r) \geq \phi(r)$ for $r \in \mathbb{R}$ and

$$\inf_{r \in \mathbb{R}} \{\min \partial \overline{\phi}(r)\} \geq \inf_{r \in \mathbb{R}} \{\min \partial \phi(r)\} - \epsilon \tag{6.206}$$

for any specified arbitrarily small $\epsilon > 0$.

Define

$$\overline{\psi}(r) = \max \{c\overline{\phi}(r) : c \in [c_1, c_2]\}, \tag{6.207}$$

where c_1 and c_2 are constants satisfying $0 < c_2 < 1 < c_1$ to be determined later. Then, $\overline{\psi}$ is strictly decreasing and continuously differentiable on $(-\infty, 0) \cup (0, \infty)$ and $\overline{\psi}(0) = 0$.

Define $\tilde{\zeta} = \zeta - \overline{\psi}(\eta - \overline{\omega})$. When $\zeta \geq \overline{\psi}(\eta - \overline{\omega})$, directly taking the derivative of $\tilde{\zeta}$ yields:

$$\dot{\tilde{\zeta}} \in \left\{\dot{\zeta} - \overline{\psi}^d \dot{\eta} : \overline{\psi}^d \in \partial \overline{\psi}(\eta - \overline{\omega})\right\}$$

$$= \left\{\varphi(\zeta - \phi(\eta - \omega)) - \overline{\psi}^d \dot{\zeta} : \overline{\psi}^d \in \partial \overline{\psi}(\eta - \overline{\omega})\right\}$$

$$\subseteq \left\{\varphi(\zeta - \phi(\eta - \omega)) - \overline{\psi}^d \dot{\zeta} : \overline{\psi}^d \in \partial \overline{\psi}(\eta - \overline{\omega}), \underline{\omega} \leq \omega \leq \overline{\omega}\right\}$$

$$= \left\{-(k_\varphi + \overline{\psi}^d)\left(\zeta - \frac{k_\varphi \phi(\eta - \omega)}{k_\varphi + \overline{\psi}^d}\right) + \tilde{\varphi}(\zeta - \phi(\eta - \omega)) : \right.$$

$$\left. \overline{\psi}^d \in \partial \overline{\psi}(\eta - \overline{\omega}), \underline{\omega} \leq \omega \leq \overline{\omega}\right\}$$

$$:= F_{\tilde{\zeta}}(\eta, \zeta, \underline{\omega}, \overline{\omega}), \tag{6.208}$$

where $k_\varphi := -\sup\{\partial \varphi(r) : r \in \mathbb{R}\}$ and $\tilde{\varphi}(r) := \varphi(r) + k_\varphi(r)$ for $r \in \mathbb{R}$. Clearly,

$$\tilde{\varphi}(r)r \leq 0 \tag{6.209}$$

for $r \in \mathbb{R}$.

Denote $k_{\overline{\phi}} = -\inf_{r \in \mathbb{R}}\{\min \partial \overline{\phi}(r)\}$. With condition (6.188) and (6.206) satisfied, we can choose $\overline{\phi}$ such that $0 < 4k_{\overline{\phi}} \leq k_\varphi$. Choose $c_1 = k_\varphi/2k_{\overline{\phi}}$. Then, $k_{\overline{\phi}}c_1^2 - k_\varphi c_1 + k_\varphi \leq 0$, i.e., $k_\varphi/(k_\varphi - c_1 k_{\overline{\phi}}) \leq c_1$. Clearly, $c_1 \geq 2$. Choose $c_2 \leq 1$.

The definition of $\overline{\psi}$ in (6.207) implies $\partial\overline{\psi}(r) \subseteq \left\{c\overline{\phi}^d : c \in [c_2, c_1], \overline{\phi}^d \in \partial\overline{\phi}(r)\right\}$, and thus $\inf_{r\in\mathbb{R}}\{\min \partial\overline{\psi}(r)\} \geq -c_1 k_{\overline{\phi}}$. By also using $\sup_{r\in\mathbb{R}}\{\max \partial\overline{\psi}(r)\} \leq 0$ (due to the strict decreasing of $\overline{\psi}$), we can prove

$$c_2 \leq \frac{k_\varphi}{k_\varphi + \overline{\psi}^d} \leq c_1 \tag{6.210}$$

for $\overline{\psi}^d \in \partial\overline{\psi}(\eta - \overline{\omega})$.

Using $\omega \leq \overline{\omega}$ and the nonincreasing of ϕ and $\overline{\phi}$, from (6.210) we have

$$\frac{k_\varphi \phi(\eta - \omega)}{k_\varphi + \overline{\psi}^d} \leq \max\left\{c\overline{\phi}(\eta - \overline{\omega}) : c \in [c_1, c_2]\right\} = \overline{\psi}(\eta - \overline{\omega}), \tag{6.211}$$

which implies

$$\zeta - \frac{k_\varphi \phi(\eta - \omega)}{k_\varphi + \overline{\psi}^d} \geq \tilde{\zeta} \tag{6.212}$$

for $\overline{\psi}^d \in \partial\overline{\psi}(\eta - \overline{\omega})$.

Based on the definitions of $\overline{\psi}^d$ and c_1, we also have

$$k_\varphi + \overline{\psi}^d \geq k_\varphi + \inf_{r\in\mathbb{R}}\{\min \partial\overline{\psi}(r)\} = k_\varphi - c_1 k_{\overline{\phi}} = \frac{1}{2}k_\varphi \tag{6.213}$$

for $\overline{\psi}^d \in \partial\overline{\psi}(\eta - \overline{\omega})$.

Based on (6.208), (6.209), (6.212), and (6.213), it can be proved that

$$\max_{f_{\tilde{\zeta}} \in F_{\tilde{\zeta}}(\eta, \zeta, \underline{\omega}, \overline{\omega})} f_{\tilde{\zeta}} \leq -\frac{1}{2}k_\varphi \tilde{\zeta} \tag{6.214}$$

when $\zeta \geq \overline{\psi}(\eta - \overline{\omega})$, i.e., $\tilde{\zeta} \geq 0$. This guarantees the invariance of set $S^a(\overline{\omega})$.

Following a similar approach, we can also find $\underline{\psi} : \mathbb{R} \to \mathbb{R}$ such that it is strictly decreasing and continuously differentiable on $(-\infty, 0) \cup (0, \infty)$ and satisfies $\underline{\psi}(0) = 0$ and $\underline{\psi} \leq \overline{\psi}(r)$ for all $r \in \mathbb{R}$, and prove that $S^b(\underline{\omega})$ is an invariant set.

Property 2

We present only the proof of the second inequality in (6.190). The first inequality in (6.190) can be proved in the same way.

We first consider the case in which $\phi(r) \to \infty$ as $r \to -\infty$. In this case, $\overline{\psi}(r) \to \infty$ as $r \to -\infty$ according to the definition of $\overline{\psi}$ in (6.207). In this case, for any $(\eta(0), \zeta(0))$, one can always find a $\overline{\mu}$ such that the second inequality in (6.190) holds with $t_1 = 0$.

If the condition for the first case is not satisfied, then there exist constants $\phi^u > 0$ and $2/3 < \phi^\delta < 1$ such that $\phi(r) \le \phi^u$ for all $r \in \mathbb{R}$ and one can find an r^* satisfying $\phi(r^*) \ge \phi^\delta \phi^u$. According to the definition of $\overline{\psi}$ in (6.207), it holds that $\overline{\psi}(r^*) \ge c_1 \phi^\delta \phi^u$, where $c_1 \ge 2$, and thus $\overline{\psi}(r^*) \ge 4\phi^u/3$.

Define $\tilde{\zeta} = \zeta - \phi^u$. When $\zeta \ge \phi^u$, taking the derivative of $\tilde{\zeta}$ yields

$$\dot{\tilde{\zeta}} = \dot{\zeta} = \varphi(\zeta - \phi(\eta - \omega)) \le \phi(\zeta - \phi^u) = \varphi(\tilde{\zeta}), \tag{6.215}$$

where φ satisfies (6.186), (6.187), and (6.188). Then, there exists a $\beta \in \mathcal{KL}$ such that for any $\zeta(0) \ge \phi^u$,

$$\tilde{\zeta}(t) \le \beta(\tilde{\zeta}(0), t) \tag{6.216}$$

for all $t \ge 0$. According to [243, Lemma 8], there exist $\alpha_{\beta 1}, \alpha_{\beta 2} \in \mathcal{K}_\infty$ such that $\beta(s, t) \le \alpha_{\beta 1}(s)\alpha_{\beta 2}(e^{-t})$ for all $s, t \in \mathbb{R}_+$, and thus there exists a finite time t_1 such that $\beta(\tilde{\zeta}(0), t_1) \le \phi^u/3$, which guarantees $\tilde{\zeta}(t_1) \le \phi^u/3$, i.e., $\zeta(t_1) \le 4\phi^u/3$. During finite time interval $[0, t_1]$, the boundedness of $\zeta(t)$ implies the boundedness of $\eta(t)$. Then, one can find a $\overline{\mu}$ such that the second inequality in (6.190) holds.

Property 3

For $t \in [0, T]$, it holds that

$$\dot{\eta}(t) = \zeta(t) \le \overline{\psi}(\eta(t) - \overline{\sigma}). \tag{6.217}$$

Define $\varsigma(t)$ as the solution of the initial-value problem

$$\dot{\varsigma}(t) = \overline{\psi}(\varsigma(t) - \overline{\sigma}) \tag{6.218}$$

with $\varsigma(0) = \eta(0)$. By using the comparison principle (see e.g., [144]), it can be proved that

$$\eta(t) \le \varsigma(t) \tag{6.219}$$

for $t \in [0, T]$.

Note that $\overline{\psi}$ is locally Lipschitz and satisfies $r\overline{\psi}(r) < 0$ for $r \ne 0$. Define $\tilde{\varsigma} = \varsigma - \overline{\sigma}$. Then, (6.218) implies $\dot{\tilde{\varsigma}}(t) = \overline{\psi}(\tilde{\varsigma}(t))$. By using Lemma 6.3, there exists a $\overline{\beta}_1 \in \mathcal{IL}$ such that $\tilde{\varsigma}(t) \le \overline{\beta}_1(\tilde{\varsigma}(0), t)$ for $t \in [0, T]$, i.e.,

$$\varsigma(t) \le \overline{\beta}_1(\varsigma(0) - \overline{\sigma}, t) + \overline{\sigma}. \tag{6.220}$$

The second inequality in (6.191) is proved by using $\varsigma(0) = \eta(0)$ and $\eta(t) \le \varsigma(t)$ for $t \in [0, T]$. The first inequality in (6.191) can be proved in the same way.

Property 4

Define $\underline{\omega}^* = \min\{\underline{\mu}_0, \eta(0), \underline{\omega}\}$ and $\overline{\omega}^* = \max\{\overline{\mu}_0, \eta(0), \overline{\omega}\}$. Then, $(\eta(0), \zeta(0)) \in S(\underline{\omega}^*, \overline{\omega}^*) \cap \{(\eta, \zeta) : \underline{\omega}^* \leq \eta \leq \overline{\omega}^*\} := \breve{S}(\underline{\omega}^*, \overline{\omega}^*)$.

Because $\omega \in [\underline{\omega}, \overline{\omega}] \subseteq [\underline{\omega}^*, \overline{\omega}^*]$, $S(\underline{\omega}^*, \overline{\omega}^*)$ is an invariant set, and $\breve{S}(\underline{\omega}^*, \overline{\omega}^*)$ is also an invariant set. Given $(\eta(0), \zeta(0)) \in \breve{S}(\underline{\omega}^*, \overline{\omega}^*)$, it holds that $(\eta(t), \zeta(t)) \in \breve{S}(\underline{\omega}^*, \overline{\omega}^*)$ for all $t \geq 0$.

In the following proof, we adopt some idea from kinematics of the plane translational motion of a rigid body; see e.g., [96]. Define

$$\eta^d = \zeta, \tag{6.221}$$

$$\zeta^d = \varphi(\zeta - \phi(\eta - \omega)), \tag{6.222}$$

$$v = [\eta^d, \zeta^d]^T, \tag{6.223}$$

$$v_1 = \left[\min_{\overline{\psi}^d \in \partial\overline{\psi}(\eta)} \frac{\zeta^d}{\overline{\psi}^d}, \zeta^d\right]^T, \tag{6.224}$$

$$v_2 = \left[\eta^d - \min_{\overline{\psi}^d \in \partial\overline{\psi}(\eta)} \frac{\zeta^d}{\overline{\psi}^d}, 0\right]^T. \tag{6.225}$$

Clearly, $v_2(t) = v(t) - v_1(t)$.

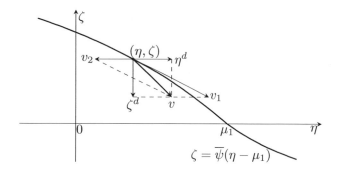

FIGURE 6.11 The motion of the point (η, ζ) and the rigid body $\zeta = \overline{\psi}(\eta - \mu)$: v is the velocity of the point, which is the composition of η^d and ζ^d and also the composition of v_1 and v_2; v_1 represents the relative velocity of the point along the rigid body and v_2 represents the translational motion velocity of the rigid body.

For any specified $\overline{\mu}$, if $\omega \in [\underline{\omega}^*, \overline{\mu}]$, then $S(\underline{\omega}^*, \overline{\mu})$ is an invariant set. From (6.214), for any (η, ζ) satisfying $\zeta = \overline{\psi}(\eta - \mu)$, it holds that

$$\eta^d - \min_{\overline{\psi}^d \in \partial\overline{\psi}(\eta)} \frac{\zeta^d}{\overline{\psi}^d} \leq 0, \tag{6.226}$$

i.e.,

$$\zeta - \min_{\overline{\psi}^d \in \partial \overline{\psi}(\eta)} \frac{\varphi(\zeta - \phi(\eta - \omega))}{\overline{\psi}^d} \leq 0 \qquad (6.227)$$

for all $\omega \in [\underline{\omega}^*, \overline{\mu}]$. Thus, it can be concluded that

$$\zeta - \min_{\overline{\psi}^d \in \partial \overline{\psi}(\eta)} \frac{\varphi(\zeta - \phi(\eta - \overline{\mu}))}{\overline{\psi}^d} \leq 0 \qquad (6.228)$$

for any (η, ζ) satisfying $\zeta = \overline{\psi}(\eta - \overline{\mu})$.

Then, for any (η, ζ) satisfying $\zeta = \overline{\psi}(\eta - \overline{\mu})$, it holds that

$$\eta^d - \min_{\overline{\psi}^d \in \partial \overline{\psi}(\eta)} \frac{\zeta^d}{\overline{\psi}^d}$$

$$= \zeta - \min_{\overline{\psi}^d \in \partial \overline{\psi}(\eta)} \frac{\varphi(\zeta - \phi(\eta - \overline{\omega}))}{\overline{\psi}^d}$$

$$= \zeta - \min_{\overline{\psi}^d \in \partial \overline{\psi}(\eta)} \frac{\varphi(\zeta - \phi(\eta - \overline{\mu}))}{\overline{\psi}^d}$$

$$\quad - \min_{\overline{\psi}^d \in \partial \overline{\psi}(\eta)} \frac{\varphi(\zeta - \phi(\eta - \overline{\omega})) - \varphi(\zeta - \phi(\eta - \overline{\mu}))}{\overline{\psi}^d}$$

$$\leq - \min_{\overline{\psi}^d \in \partial \overline{\psi}(\eta)} \frac{\varphi(\zeta - \phi(\eta - \overline{\omega})) - \varphi(\zeta - \phi(\eta - \overline{\mu}))}{\overline{\psi}^d}, \qquad (6.229)$$

where we used (6.228) for the inequality.

In the case of $\overline{\mu} \geq \overline{\omega}$, by using the continuous and strictly decreasing properties of φ and ϕ, one can always find a positive definite, nondecreasing, locally Lipschitz $\overline{\alpha}_2^a$ such that

$$\min_{\overline{\psi}^d \in \partial \overline{\psi}(\eta)} \frac{\varphi(\zeta - \phi(\eta - \overline{\omega})) - \varphi(\zeta - \phi(\eta - \overline{\mu}))}{\overline{\psi}^d} \geq \overline{\alpha}_2^a(\overline{\mu} - \overline{\omega}) \qquad (6.230)$$

for $(\eta, \zeta) \in \check{S}(\underline{\omega}^*, \overline{\omega}^*)$ and $\omega \in [\underline{\omega}, \overline{\omega}]$.

In the case of $\overline{\mu} < \overline{\omega}$, because φ and ϕ are locally Lipschitz and strictly decreasing, one can find a positive definite, nondecreasing, locally Lipschitz $\overline{\alpha}_2^b$ such that

$$\min_{\overline{\psi}^d \in \partial \overline{\psi}(\eta)} \frac{\varphi(\zeta - \phi(\eta - \overline{\omega})) - \varphi(\zeta - \phi(\eta - \overline{\mu}))}{\overline{\psi}^d} \geq -\overline{\alpha}_2^b(\overline{\omega} - \overline{\mu}) \qquad (6.231)$$

for $(\eta, \zeta) \in \check{S}(\underline{\omega}^*, \overline{\omega}^*)$ and $\omega \in [\underline{\omega}, \overline{\omega}]$.

Define

$$\overline{\alpha}_2(r) = \begin{cases} -\overline{\alpha}_2^a(r) & \text{for } r \geq 0; \\ \overline{\alpha}_2^b(-r) & \text{for } r < 0. \end{cases} \qquad (6.232)$$

Then, $\overline{\alpha}_2(0) = 0$, $r\overline{\alpha}_2(r) < 0$ for all $r \neq 0$, and $\overline{\alpha}_2$ is nonincreasing and locally Lipschitz.

Define $\overline{\varsigma}(t)$ as the solution of the initial-value problem

$$\dot{\overline{\varsigma}}(t) = \overline{\alpha}_2(\overline{\varsigma}(t) - \overline{\varpi}) \tag{6.233}$$

with initial condition $\overline{\varsigma}(0) = \overline{\mu}_0$. Then, $\zeta(t) \leq \overline{\psi}(\eta(t) - \overline{\varsigma}(t))$ for $t \geq 0$. If $\overline{\mu}(t) \geq \overline{\sigma}(t)$ for $t \geq 0$, then $\zeta(t) \leq \overline{\psi}(\eta(t) - \overline{\mu}(t))$ for $t \geq 0$.

Similarly, one can find a nonincreasing, locally Lipschitz $\underline{\alpha}_2$ which satisfies $\underline{\alpha}_2(0) = 0$, $r\underline{\alpha}_2(r) < 0$ for all $r \neq 0$, such that $\zeta(t) \geq \underline{\psi}(\eta(t) - \underline{\mu}(t))$ for $t \geq 0$, if $\underline{\mu}(t) \leq \underline{\sigma}(t)$ for $t \geq 0$, where $\underline{\varsigma}(t)$ is the solution of the initial-value problem

$$\dot{\underline{\varsigma}}(t) = \underline{\alpha}_2(\underline{\varsigma}(t) - \underline{\omega}) \tag{6.234}$$

with initial condition $\underline{\varsigma}(0) = \underline{\mu}_0$.

Define $\overline{\alpha}_2' = \max\{\overline{\alpha}_2(r), \underline{\alpha}_2(r)\}$ and $\underline{\alpha}_2' = \min\{\overline{\alpha}_2(r), \underline{\alpha}_2(r)\}$ for $r \in \mathbb{R}$. Define $\overline{\mu}(t)$ and $\underline{\mu}(t)$ as the solutions of the initial-value problems

$$\dot{\overline{\mu}}(t) = \overline{\alpha}_2'(\overline{\mu}(t) - \overline{\varpi}) \tag{6.235}$$

$$\dot{\underline{\mu}}(t) = \underline{\alpha}_2'(\underline{\mu}(t) - \underline{\omega}) \tag{6.236}$$

with initial conditions $\overline{\mu}(0) = \overline{\mu}_0$ and $\underline{\mu}(0) = \underline{\mu}_0$. Then, the comparison principle can guarantee $\overline{\mu}(t) \leq \overline{\sigma}(t)$ and $\underline{\mu}(t) \leq \underline{\sigma}(t)$. Moreover, $\overline{\mu}(t) \geq \underline{\mu}(t)$ for $t \geq 0$.

By using Lemma 6.3, one can find $\underline{\beta}_2, \overline{\beta}_2 \in \mathcal{IL}$ such that (6.192) holds.

6.4.3   MAIN RESULTS OF STRONG OUTPUT AGREEMENT WITH FLEXIBLE TOPOLOGIES

Consider the multi-agent system (6.178)–(6.179). We propose a class of distributed control laws in the following form:

$$\mu_i = \varphi_i(\zeta_i - \phi_i(\xi_i)) \tag{6.237}$$

$$\xi_i = \frac{1}{\sum_{j \in \mathcal{N}_i(\sigma(t))} a_{ij}} \sum_{j \in \mathcal{N}_i(\sigma(t))} a_{ij}(\eta_i - \eta_j), \tag{6.238}$$

where $\sigma : [0, \infty) \to \mathcal{P}$ is a piecewise constant signal, which describes the information exchange between the systems, with $\mathcal{P} \subset \mathbb{N}$ being a finite set representing all the possible information exchange topologies, and $\mathcal{N}_i(p) \subseteq \{1, \ldots, N\}$ denotes the neighbor set of agent i for each $i = 1, \ldots, N$ and each $p \in \mathcal{P}$. In (6.237)–(6.238), constant $a_{ij} > 0$ if $i \neq j$ and $a_{ij} \geq 0$ if $i = j$. The functions $\varphi_i, \phi_i : \mathbb{R} \to \mathbb{R}$ are nonincreasing, locally Lipschitz, and satisfy

$$\varphi_i(0) = \phi_i(0) = 0, \tag{6.239}$$

$$\varphi_i(r)r < 0, \quad \phi_i(r)r < 0 \quad \text{for} \quad r \neq 0, \tag{6.240}$$

$$\sup_{r \in \mathbb{R}}\{\max \partial \varphi_i(r) : r \in \mathbb{R}\} < 4 \inf_{r \in \mathbb{R}}\{\min \partial \phi_i(r)\}, \tag{6.241}$$

for $i = 1, \ldots, N$.

By defining $\omega_i = \sum_{j \in \mathcal{N}_i(\sigma(t))} a_{ij} \eta_j / \sum_{j \in \mathcal{N}_i(\sigma(t))} a_{ij}$, it can be observed that each controlled agent (6.178)–(6.179) with control law (6.237)–(6.238) is in the form of (6.184)–(6.185), and conditions (6.239)–(6.241) are in accordance with conditions (6.186)–(6.188).

Before proposing our main result on strong output agreement, we first use a switching digraph $\mathcal{G}(\sigma(t)) = (\mathcal{N}, \mathcal{E}(\sigma(t)))$ to represent the information exchange topology between the agents, where $\mathcal{N}$ is the set of N vertices corresponding to the agents, and for each $p \in \mathcal{P}$, if $j \in \mathcal{N}_i(p)$, then there is a directed edge (j, i) belonging to $\mathcal{E}(p)$. By default, (i, i) for $i \in \mathcal{N}$ belong to $\mathcal{E}(p)$ for all $p \in \mathcal{P}$.

A digraph $\mathcal{G} = (\mathcal{N}, \mathcal{E})$ is quasi-strongly connected (QSC) if there exists a $c \in \mathcal{N}$ such that there is a directed path from c to i for each $i \in \mathcal{N}$; vertex c is called the center of $\mathcal{G}$. For a switching digraph $\mathcal{G}(\sigma(t)) = (\mathcal{N}, \mathcal{E}(\sigma(t)))$, we define the union digraph over $[t_1, t_2]$ as $\mathcal{G}(\sigma([t_1, t_2])) = (\mathcal{N}, \bigcup_{t \in [t_1, t_2]} \mathcal{E}(\sigma(t)))$. A switching digraph $\mathcal{G}(\sigma(t))$ with $\sigma : [0, \infty) \to \mathcal{P}$ is said to be uniformly quasi-strongly connected (UQSC) with time constant $T > 0$ if $\mathcal{G}(\sigma([t, t+T]))$ is QSC for all $t \geq 0$. A switching digraph $\mathcal{G}(\sigma(t))$ with $\sigma : [0, \infty) \to \mathcal{P}$ has an edge dwell time $\tau_D > 0$ if for each $t \in [0, \infty)$, for any directed edge $(i_1, i_2) \in \mathcal{E}(\sigma(t))$, there exists a $t^* \geq 0$ such that $t \in [t^*, t^* + \tau_D]$ and $(i_1, i_2) \in \mathcal{E}(\sigma(\tau))$ for $\tau \in [t^*, t^* + \tau_D]$.

Lemma 6.4 *Consider a switching digraph* $\mathcal{G}(\sigma(t)) = (\mathcal{N}, \mathcal{E}(\sigma(t)))$ *with* $\sigma : [0, \infty) \to \mathcal{P}$, *which is UQSC with time constant* $T > 0$ *and has an edge dwell time* $\tau_D > 0$. *If* $c \in \mathcal{N}$ *is a center of* $\mathcal{G}(\sigma([t, t + T]))$, *then for any* $\mathcal{N}_1$ *such that* $c \in \mathcal{N}_1$, *there exist* $i_1 \in \mathcal{N}_1$, $i_2 \in \mathcal{N} \backslash \mathcal{N}_1$, *and* $t' \in [t - \tau_D, t + T]$ *such that* $(i_1, i_2) \in \mathcal{E}(\sigma(\tau))$ *for* $\tau \in [t', t' + \tau_D]$.

Lemma 6.4 can be proved by directly using the definitions of UQSC and edge dwell-time.

The following theorem presents our main result on the strong output agreement problem.

Theorem 6.3 *Consider the double-integrators* (6.178)–(6.179) *with control laws in the form of* (6.237)–(6.238). *Assume conditions* (6.239)–(6.241) *are satisfied. If* $\mathcal{G}(\sigma(t)) = (\mathcal{N}, \mathcal{E}(\sigma(t)))$ *with* $\sigma : [0, \infty) \to \mathcal{P}$ *is UQSC and has an edge dwell-time* $\tau_D > 0$, *then the strong output agreement problem is solvable.*

The proof of Theorem 6.3, which is based on Proposition 6.2, is in Section 6.4.4.

In previously published papers on coordinated control of continuous-time systems with switching topologies, e.g., [111, 172, 236], it is usually assumed that the switching graph has a dwell-time τ_D', which means each specific topology remains unchanged for a period larger than τ_D'. As edge dwell-time is directly used in our analysis, we assume edge dwell-time instead of the customary graph dwell-time. It should be noted that the assumption of edge dwell-time does not cause restrictions to the main results. In fact, the edge dwell-time is normally larger than the graph dwell-time for a specific switching graph.

In a leader–follower structure, the motion of the leader does not depend on the outputs of the followers and the output of the leader is accessible to some of the followers. For a group of systems with a leader i^*, to achieve strong output agreement, it is required that there exists a finite constant $T > 0$ such that for all $t \geq 0$, the union digraph $\mathcal{G}(\sigma([t, t+T]))$ is QSC with i^* as a center, according to Theorem 6.3.

Conditions (6.239)–(6.241) allow us to choose bounded and nonsmooth ϕ_i for $i = 1, \ldots, N$. One example of ϕ_i is

$$
\phi_i(r) = \begin{cases} -1, & \text{when } r > 1; \\ -r, & \text{when } -1 \leq r \leq 1; \\ 1, & \text{when } r < -1. \end{cases} \tag{6.242}
$$

Correspondingly, we can choose $\varphi_i(r) = -kr$ with constant $k > 4$. This could be of practical interest when the velocities ζ_i are required to be bounded in the process of controlling the positions η_i to achieve agreement. Consider the ζ_i-system (6.179) with μ_i defined by (6.237). With bounded ϕ_i, the velocity ζ_i can be restricted to within a specific bounded range depending on the initial state $\zeta_i(0)$ and the bounds of ϕ_i. With bounded velocities, we can also guarantee the boundedness of the control signals μ_i, which may be required to be bounded due to actuator saturation.

6.4.4 PROOF OF THEOREM 6.3

Define

$$
\omega_i = \frac{\sum_{j \in \mathcal{N}_i(\sigma(t))} a_{ij} \eta_j}{\sum_{j \in \mathcal{N}_i(\sigma(t))} a_{ij}}. \tag{6.243}
$$

Then, $\xi_i = \eta_i - \omega_i$, and each (η_i, ζ_i)-system (6.178)–(6.179) with μ_i defined in (6.237)–(6.238) is in the form of (6.184)–(6.185). With conditions (6.239)–(6.241) satisfied, each closed-loop (η_i, ζ_i)-system has the properties given in Proposition 6.2.

According to Property 1 in Proposition 6.2, for $i = 1, \ldots, N$, one can find $\underline{\psi}_i, \overline{\psi}_i$ such that

$$
S_i(\underline{\omega}_i, \overline{\omega}_i) = \left\{ (\eta_i, \zeta_i) : \underline{\psi}_i(\eta_i - \omega_i) \leq \zeta_i \leq \overline{\psi}_i(\eta_i - \overline{\omega}) \right\} \tag{6.244}
$$

is an invariant set of the (η_i, ζ_i)-system if $\omega_i \in [\underline{\omega}_i, \overline{\omega}_i]$.

Given the chosen $\underline{\psi}_i, \overline{\psi}_i$, we suppose that there exist $\underline{\mu}_i(0), \overline{\mu}_i(0)$ such that

$$(\eta_i(0), \zeta_i(0)) \in S_i(\underline{\mu}_i(0), \overline{\mu}_i(0)) \tag{6.245}$$

$$\underline{\mu}(0) \leq \eta_i(0) \leq \overline{\mu}_i(0) \tag{6.246}$$

for $i \in \mathcal{N}$. Otherwise, there exists a finite time t^*, at which property (6.245) holds, according to Property 2 in Proposition 6.2.

Denote $\eta = [\eta_1, \ldots, \eta_N]^T$ and $\zeta = [\zeta_1, \ldots, \zeta_N]^T$. Then, $(\eta(0), \zeta(0)) \in S(\underline{\mu}(0), \overline{\mu}(0))$ with

$$S(\underline{\mu}(0), \overline{\mu}(0)) = \{(\eta, \zeta) : (\eta_i, \zeta_i) \in S_i(\underline{\mu}(0), \overline{\mu}(0)) \text{ for } i \in \mathcal{N}\}, \tag{6.247}$$

where $\underline{\mu}(0) = \min_{i \in \mathcal{N}} \underline{\mu}_i(0)$ and $\overline{\mu}(0) = \max_{i \in \mathcal{N}} \overline{\mu}_i(0)$ with $\underline{\mu}_i(0), \overline{\mu}_i(0)$ satisfying (6.245) and (6.246).

For each $i \in \mathcal{N}$, (6.243) implies $\underline{\mu}(0) \leq \omega_i \leq \overline{\mu}(0)$ if $\underline{\mu}(0) \leq \eta_i \leq \overline{\mu}(0)$. Based on Proposition 6.2, it can be proved that $S(\underline{\mu}(0), \overline{\mu}(0))$ is an invariant set of the interconnected system with (η, ζ) as the state. Thus, $(\eta(t), \zeta(t)) \in S(\underline{\mu}(0), \overline{\mu}(0))$ and $\underline{\mu}(0) \leq \omega_i(t) \leq \overline{\mu}(0)$ for all $t \geq 0$.

The basic idea of the proof is to find appropriate $\underline{\mu}_i(t), \overline{\mu}_i(t)$ such that

$$\underline{\psi}_i(\eta_i(t) - \overline{\mu}_i(t)) \leq \zeta_i(t) \leq \overline{\psi}_i(\eta_i(t) - \overline{\mu}_i(t)). \tag{6.248}$$

We define two sets, $\mathcal{Q}_1$ and $\mathcal{Q}_2$, which satisfy $\mathcal{Q}_1 \cup \mathcal{Q}_2 = \mathcal{N}$ and have the following properties: if $i \in \mathcal{Q}_1$, then the $\underline{\mu}_i(0)$ defined in (6.245) satisfies

$$\underline{\mu}_i(0) \geq (\underline{\mu}(0) + \overline{\mu}(0))/2; \tag{6.249}$$

if $i \in \mathcal{Q}_2$, then the $\overline{\mu}_i(0)$ defined in (6.245) satisfies

$$\overline{\mu}_i(0) \leq (\underline{\mu}(0) + \overline{\mu}(0))/2. \tag{6.250}$$

Note that either $\mathcal{Q}_1$ or $\mathcal{Q}_2$ can be an empty set. Also, the existence of the pair $(\mathcal{Q}_1, \mathcal{Q}_2)$ may not be unique.

Define $T^* = \Delta_T + N(T + 2\tau_D + \Delta_T)$ with $\Delta_T > 0$. For each $i \in \mathcal{Q}_1$, by using Property 4 in Proposition 6.2, there exists a $\underline{\mu}_i(t)$ satisfying

$$\underline{\mu}_i(t) - \underline{\mu}(0) \geq \alpha_i(\underline{\mu}_i(0) - \underline{\mu}(0)) \tag{6.251}$$

such that the first inequality of (6.248) holds for $0 \leq t \leq T^*$, where α_i is continuous, positive definite, and less than Id.

For each $i \in \mathcal{Q}_1$, by also using (6.249), we have

$$\begin{aligned}
&\underline{\mu}_i(t) - \underline{\mu}(0) \\
&\geq \min\{\alpha_i(r - \underline{\mu}(0)) : (\underline{\mu}(0) + \overline{\mu}(0))/2 < r \leq \overline{\mu}(0)\} \\
&= \min\{\alpha_i(r') : (\overline{\mu}(0) - \underline{\mu}(0))/2 < r' \leq \overline{\mu}(0) - \underline{\mu}(0)\} \\
&= \breve{\alpha}_i^l(\overline{\mu}(0) - \underline{\mu}(0))
\end{aligned} \tag{6.252}$$

for $0 \leq t \leq T^*$, where $\breve{\alpha}_i^l(s) := \min\{\alpha_i(s') : s/2 \leq s' \leq s\}$ for $s \in \mathbb{R}_+$. Clearly, $\breve{\alpha}_i^l$ is continuous, positive definite, and less than Id. For each $i \in \mathcal{Q}_1$, using Property 3 in Proposition 6.2, for specific $\Delta_T > 0$, one can find a continuous and positive definite function $\hat{\alpha}_i^{l0} < \text{Id}$ such that

$$
\begin{aligned}
\eta_i(t) - \underline{\mu}(0) &\geq \hat{\alpha}_i^{l0}\big(\min_{0 \leq t \leq T^*} \underline{\mu}_i(t) - \eta_i(0)\big) \\
&\geq \hat{\alpha}_i^{l0}\big(\min_{0 \leq t \leq T^*} \underline{\mu}_i(t) - \underline{\mu}(0)\big) \\
&\geq \hat{\alpha}_i^{l0} \circ \breve{\alpha}_i^l(\overline{\mu}(0) - \underline{\mu}(0)) \\
&:= \hat{\alpha}_i^l(\overline{\mu}(0) - \underline{\mu}(0))
\end{aligned}
\tag{6.253}
$$

for $t \in [\Delta_T, T^*]$. Clearly, $\hat{\alpha}_i^l$ is continuous, positive definite, and less than Id.

Similarly, for each $i \in \mathcal{Q}_2$, one can find $\underline{\mu}_i(t)$ satisfying

$$
\overline{\mu}(0) - \overline{\mu}_i(t) \geq \breve{\alpha}_i^u(\overline{\mu}(0) - \underline{\mu}(0))
\tag{6.254}
$$

such that the second inequality of (6.248) holds for $0 \leq t \leq T^*$, where $\breve{\alpha}_i^u$ is continuous, positive definite, and less than Id. Also, one can find continuous and positive definite $\hat{\alpha}_i^u < \text{Id}$ such that

$$
\overline{\mu}(0) - \eta_i(t) \geq \hat{\alpha}_i^u(\overline{\mu}(0) - \underline{\mu}(0))
\tag{6.255}
$$

for $t \in [\Delta_T, T^*]$.

Initial Step: Because $\mathcal{G}(\sigma(t))$ is UQSC, $\mathcal{G}(\sigma([\Delta_T + \tau_D, \Delta_T + T + \tau_D]))$ has a center, denoted by l_1. Suppose $l_1 \in \mathcal{Q}_2$. (If $l_1 \in \mathcal{Q}_1$, then the theorem can be proved in the same way.) Then, according to Lemma 6.4, there exist $l_1' \in \mathcal{Q}_2$, $f_1 \in \mathcal{Q}_1$ and $t' \in [\Delta_T, \Delta_T + T + \tau_D]$ such that $(l_1', f_1) \in \mathcal{E}(\sigma(t))$ for $t \in [t', t' + \tau_D] \subseteq [\Delta_T, \Delta_T + T + 2\tau_D]$.

By using the fact that $\overline{\mu}(0) - \eta_{l_1'}(t) \geq \hat{\alpha}_{l_1'}^u(\overline{\mu}(0) - \underline{\mu}(0))$, we have

$$
\begin{aligned}
\omega_{f_1}(t) &= \frac{\sum_{j \in \mathcal{N}_i(\sigma(t))} a_{ij} \eta_j(t)}{\sum_{j \in \mathcal{N}_i(\sigma(t))} a_{ij}} \\
&\leq \overline{\mu}(0) - \frac{a_{f_1 l_1'} \hat{\alpha}_{l_1'}^u(\overline{\mu}(0) - \underline{\mu}(0))}{\sum_{j \in \mathcal{N}_i(\sigma(t))} a_{ij}}.
\end{aligned}
\tag{6.256}
$$

Also, we have $\omega_{f_1}(t) \geq \underline{\mu}(0)$ for $t \in [t', t' + \tau_D]$. Property 4 in Proposition 6.2 guarantees that one can find $\underline{\mu}_{f_1}(t)$ and $\overline{\mu}_{f_1}(t)$ satisfying $\underline{\mu}_{f_1}(t) \geq \underline{\mu}(0)$ and

$$
\begin{aligned}
\overline{\mu}_{f_1}(t) &\leq \overline{\mu}(0) - \breve{\alpha}_{f_1}^{u0}\big(\overline{\mu}(0) - \max_{t \in [t', t' + \tau_D]} \omega_{f_1}(t)\big) \\
&\leq \overline{\mu}(0) - \breve{\alpha}_{f_1}^{u0}\left(\frac{a_{f_1 l_1'} \hat{\alpha}_{l_1'}^u(\overline{\mu}(0) - \underline{\mu}(0))}{\sum_{j \in \mathcal{N}_i(\sigma(t))} a_{ij}}\right)
\end{aligned}
\tag{6.257}
$$

such that (6.248) with $i = f_1$ holds at $t = t' + \tau_D$, where $\breve{\alpha}_{f_1}^{u0}$ is continuous, positive definite, nondecreasing, and less than Id.

For $t \in [t' + \tau_D, T^*]$, it can be guaranteed that $\underline{\mu}(0) \leq \omega_{f_1}(t) \leq \overline{\mu}(0)$. By using Property 4 in Proposition 6.2 again, one can find $\underline{\mu}_{f_1}(t) \geq \underline{\mu}(0)$ and

$$\overline{\mu}_{f_1}(t) \leq \overline{\mu}(0) - \breve{\alpha}_{f_1}^u(\overline{\mu}(0) - \underline{\mu}(0)) \tag{6.258}$$

such that (6.248) with $i = f_1$ holds for $t \in [\Delta_T + T + 2\tau_D, T^*] \subseteq [t' + \tau_D, T^*]$, where $\breve{\alpha}_{f_1}^u < \mathrm{Id}$ is continuous, positive definite, and less than Id.

By using Property 3 in Proposition 6.2, for specific $\Delta_T > 0$, there exists a continuous and positive definite $\hat{\alpha}_{f_1}^{u0} < \mathrm{Id}$ such that

$$\begin{aligned}
\overline{\mu}(0) - \eta_{f_1}(t) &\geq \hat{\alpha}_{f_1}^{u0}(\overline{\mu}(0) - \max_{0 \leq t \leq T^*} \overline{\mu}_{f_1}(t)) \\
&\geq \hat{\alpha}_{f_1}^{u0} \circ \breve{\alpha}_{f_1}^u(\overline{\mu}(0) - \underline{\mu}(0)) \\
&:= \hat{\alpha}_{f_1}^u(\overline{\mu}(0) - \underline{\mu}(0)),
\end{aligned} \tag{6.259}$$

i.e.,

$$\eta_{f_1}(t) \leq \overline{\mu}(0) - \hat{\alpha}_{f_1}^u(\overline{\mu}(0) - \underline{\mu}(0)) \tag{6.260}$$

for $t \in [\Delta_T + (T + 2\tau_D + \Delta_T), T^*]$. According to the definitions, $\hat{\alpha}_{f_1}^u$ is continuous, positive definite, and less than Id. Since $f_1 \in \mathcal{Q}_1$, according to (6.253), there also exists a continuous and positive definite $\hat{\alpha}_{f_1}^l < \mathrm{Id}$ such that

$$\eta_{f_1}(t) \geq \underline{\mu}(0) + \hat{\alpha}_{f_1}^l(\overline{\mu}(0) - \underline{\mu}(0)) \tag{6.261}$$

for $t \in [\Delta_T + (T + 2\tau_D + \Delta_T), T^*]$.

Recursive Step: Denote $\mathcal{F}_k = \{f_1, \ldots, f_k\} \subset \mathcal{N}$. Suppose that for each $i \in \mathcal{F}_k$, there exist continuous and positive definite functions $\hat{\alpha}_i^l, \hat{\alpha}_i^u < \mathrm{Id}$ such that

$$\eta_i(t) \geq \underline{\mu}(0) + \hat{\alpha}_i^l(\overline{\mu}(0) - \underline{\mu}(0)), \tag{6.262}$$

$$\eta_i(t) \leq \overline{\mu}(0) - \hat{\alpha}_i^u(\overline{\mu}(0) - \underline{\mu}(0)) \tag{6.263}$$

for $t \in [\Delta_T + k(T + 2\tau_D + \Delta_T), T^*]$.

Note that, according to (6.253) and (6.255), for each $i \in \mathcal{Q}_1 \cup \mathcal{F}_k \backslash \mathcal{F}_k$, there exists a continuous and positive definite $\hat{\alpha}_i^l < \mathrm{Id}$ such that (6.262) holds for all $t \in [\Delta_T + k(T + 2\tau_D + \Delta_T), T^*]$, and for each $i \in \mathcal{Q}_2 \cup \mathcal{F}_k \backslash \mathcal{F}_k$, there exists a continuous and positive definite $\hat{\alpha}_i^u < \mathrm{Id}$ such that (6.263) holds for $t \in [\Delta_T + k(T + 2\tau_D + \Delta_T), T^*]$.

Due to the UQSC of $\mathcal{G}(\sigma(t))$, the union digraph $\mathcal{G}(\sigma([\Delta_T + k(T + 2\tau_D + \Delta_T) + \tau_D, \Delta_T + k(T + 2\tau_D + \Delta_T) + T + \tau_D))$ has a center, denoted by l_{k+1}. There are two possible cases: $l_{k+1} \in \mathcal{Q}_1$ and $l_{k+1} \in \mathcal{Q}_2$. We only consider the first case, while the second case can be studied following a quite similar approach.

If $l_{k+1} \in \mathcal{Q}_1$, then $l_{k+1} \in \mathcal{Q}_1 \cup \mathcal{F}_k$ and according to Lemma 6.4, there exist $l'_{k+1} \in \mathcal{Q}_1 \cup \mathcal{F}_k$, $f_{k+1} \in \mathcal{Q}_2 \cup \mathcal{F}_k \backslash \mathcal{F}_k$, and $t' \in [\Delta_T + k(T + 2\tau_D +$

$\Delta_T), \Delta_T + k(T + 2\tau_D + \Delta_T) + T + \tau_D]$ such that $(l'_{k+1}, f_{k+1}) \in \mathcal{E}(\sigma(t))$ for $t \in [t', t' + \tau_D] \subseteq [\Delta_T + k(T + 2\tau_D + \Delta_T), \Delta_T + k(T + 2\tau_D + \Delta_T) + T + 2\tau_D]$.

By using (6.262) with $i = l'_{k+1}$, we have

$$
\begin{aligned}
\omega_{f_{k+1}}(t) &= \frac{\sum_{j \in \mathcal{N}_{f_{k+1}}(\sigma(t))} a_{f_{k+1}j} \eta_j(t)}{\sum_{j \in \mathcal{N}_{f_{k+1}}(\sigma(t))} a_{f_{k+1}j}} \\
&\geq \underline{\mu}(0) + \frac{a_{f_{k+1}l'_{k+1}} \hat{\alpha}^l_{l'_{k+1}} (\overline{\mu}(0) - \underline{\mu}(0))}{\sum_{j \in \mathcal{N}_{f_{k+1}}(\sigma(t))} a_{f_{k+1}j}}
\end{aligned}
\tag{6.264}
$$

for $t \in [t', t' + \tau_D]$. Also, we have $\omega_{f_{k+1}}(t) \leq \overline{\mu}(0)$ for $t \in [t', t' + \tau_D]$. Property 4 in Proposition 6.2 guarantees that one can find $\overline{\mu}_{f_{k+1}}(t)$ and $\overline{\mu}_{f_{k+1}}(t)$ such that $\overline{\mu}_{f_{k+1}}(t) \leq \overline{\mu}(0)$ and

$$
\begin{aligned}
\underline{\mu}_{f_{k+1}}(t) &\geq \underline{\mu}(0) + \check{\alpha}^{l0}_{f_{k+1}} \left(\min_{t \in [t', t' + \tau_D]} \omega_{f_{k+1}}(t) - \underline{\mu}(0) \right) \\
&\geq \underline{\mu}(0) + \check{\alpha}^{l0}_{f_{k+1}} \left(\frac{a_{f_{k+1}l'_{k+1}} \hat{\alpha}^l_{l'_{k+1}} (\overline{\mu}(0) - \underline{\mu}(0))}{\sum_{j \in \mathcal{N}_{f_{k+1}}(\sigma(t))} a_{f_{k+1}j}} \right)
\end{aligned}
\tag{6.265}
$$

such that (6.248) with $i = f_{k+1}$ holds at $t = t' + \tau_D$, where $\check{\alpha}^{l0}_{f_{k+1}}$ is continuous, positive definite, nondecreasing, and less than Id.

For $t \in [t' + \tau_D, T^*]$, we have $\underline{\mu}(0) \leq \omega_{f_{k+1}}(t) \leq \overline{\mu}(0)$. By using Property 4 in Proposition 6.2 again, we can find $\overline{\mu}_{f_{k+1}}(t) \leq \overline{\mu}(0)$ and

$$
\underline{\mu}_{f_{k+1}}(t) \geq \underline{\mu}(0) + \check{\alpha}^l_{f_{k+1}} (\overline{\mu}(0) - \underline{\mu}(0))
\tag{6.266}
$$

such that (6.248) with $i = f_{k+1}$ holds for $t \in [\Delta_T + k(T + 2\tau_D + \Delta_T) + T + 2\tau_D, T^*]$, where $\check{\alpha}^l_{f_{k+1}}$ is continuous, positive definite, and less than Id.

By using Property 3 in Proposition 6.2, for specific $\Delta_T > 0$, there exists a continuous, positive definite $\hat{\alpha}^{l0}_{f_{k+1}} < \text{Id}$ such that

$$
\begin{aligned}
\eta_{f_{k+1}}(t) - \underline{\mu}(0) &\geq \hat{\alpha}^{l0}_{f_{k+1}} \left(\min_{t \in \mathcal{T}} \underline{\mu}_{f_{k+1}}(t) - \underline{\mu}(0) \right) \\
&\geq \hat{\alpha}^{l0}_{f_{k+1}} \circ \check{\alpha}^l_{f_{k+1}} (\overline{\mu}(0) - \underline{\mu}(0)) \\
&:= \hat{\alpha}^l_{f_{k+1}} (\overline{\mu}(0) - \underline{\mu}(0)),
\end{aligned}
\tag{6.267}
$$

i.e.,

$$
\eta_{f_{k+1}}(t) \geq \underline{\mu}(0) + \hat{\alpha}^l_{f_{k+1}} (\overline{\mu}(0) - \underline{\mu}(0))
\tag{6.268}
$$

for $t \in [\Delta_T + (k + 1)(T + 2\tau_D + \Delta_T), T^*]$, where $\mathcal{T} = [\Delta_T + k(T + 2\tau_D + \Delta_T) + T + 2\tau_D, T^*]$. Since $f_{k+1} \in \mathcal{Q}_2$, according to (6.255), there also exists a continuous and positive definite $\hat{\alpha}^u_{f_{k+1}} < \text{Id}$ such that

$$
\eta_{f_{k+1}}(t) \leq \overline{\mu}(0) - \hat{\alpha}^u_{f_{k+1}} (\overline{\mu}(0) - \underline{\mu}(0))
\tag{6.269}
$$

for $t \in [\Delta_T + (k+1)(T + 2\tau_D + \Delta_T), T^*]$.

Denote $\mathcal{F}_{k+1} = \{f_1, \ldots, f_{k+1}\}$. For each $i \in \mathcal{F}_{k+1}$, there exist continuous and positive definite functions $\hat{\alpha}_i^l, \hat{\alpha}_i^u < \mathrm{Id}$ such that (6.262) and (6.263) hold for all $t \in [\Delta_T + (k+1)(T + 2\tau_D + \Delta_T), T^*]$.

Final Step: Repeat the procedure in Step $k+1$ until $k+1 = k^*$ such that

$$\mathcal{Q}_1 \subseteq \mathcal{F}_{k^*}, \tag{6.270}$$

or

$$\mathcal{Q}_2 \subseteq \mathcal{F}_{k^*}. \tag{6.271}$$

Note that for $i = 1, \ldots, k^* - 1$, $f_{i+1} \in \mathcal{Q}_1 \cup \mathcal{F}_i \backslash \mathcal{F}_i$ or $f_{i+1} \in \mathcal{Q}_2 \cup \mathcal{F}_i \backslash \mathcal{F}_i$. It can be concluded that $f_{i+1} \notin \mathcal{F}_i$ and thus $k^* \leq N$. Otherwise, the size of $\mathcal{F}_{k^*}$ is larger than the size of $\mathcal{N}$.

Also note that it is a special case that $\mathcal{Q}_1 = \emptyset$ or $\mathcal{Q}_2 = \emptyset$.

Convergence: Recall that $\mathcal{Q}_1 \cup \mathcal{Q}_2 = \mathcal{N}$. Condition (6.270) implies $\mathcal{F}_{k^*} \cup \mathcal{Q}_2 = \mathcal{N}$; condition (6.271) implies $\mathcal{F}_{k^*} \cup \mathcal{Q}_1 = \mathcal{N}$.

If (6.270) holds, then for $i \in \mathcal{F}_{k^*} \cup \mathcal{Q}_2 = \mathcal{N}$, define

$$\overline{\mu}_i(T^*) = \overline{\mu}(0) - \breve{\alpha}_i^u(\overline{\mu}(0) - \underline{\mu}(0)) \tag{6.272}$$

and define

$$\overline{\mu}(T^*) = \max_{i \in \mathcal{N}} \overline{\mu}_i(T^*). \tag{6.273}$$

Then, there exists a continuous and positive definite $\breve{\alpha}^u < \mathrm{Id}$ such that $\overline{\mu}(T^*) \leq \overline{\mu}(0) - \breve{\alpha}^u(\overline{\mu}(0) - \underline{\mu}(0))$. By defining $\underline{\mu}(T^*) = \underline{\mu}(0)$ and $\tilde{\mu} = \overline{\mu} - \underline{\mu}$, we can achieve

$$\begin{aligned}
\tilde{\mu}(T^*) &= \overline{\mu}(T^*) - \underline{\mu}(T^*) \\
&\leq \overline{\mu}(0) - \breve{\alpha}^u(\overline{\mu}(0) - \underline{\mu}(0)) - \underline{\mu}(0) \\
&= \tilde{\mu}(0) - \breve{\alpha}^u(\tilde{\mu}(0)).
\end{aligned} \tag{6.274}$$

If (6.271) holds, then for $i \in \mathcal{F}_{k^*} \cup \mathcal{Q}_1 = \mathcal{N}$, define

$$\underline{\mu}_i(T^*) = \underline{\mu}(0) + \breve{\alpha}_i^l(\overline{\mu}(0) - \underline{\mu}(0)) \tag{6.275}$$

and define

$$\underline{\mu}(T^*) = \min_{i \in \mathcal{N}} \underline{\mu}_i(T^*). \tag{6.276}$$

By defining $\overline{\mu}(T^*) = \overline{\mu}(0)$ and $\tilde{\mu} = \overline{\mu} - \underline{\mu}$, we can achieve

$$\tilde{\mu}(T^*) \leq \tilde{\mu}(0) - \breve{\alpha}^l(\tilde{\mu}(0)), \tag{6.277}$$

where $\breve{\alpha}^l$ is continuous, positive definite, and less than Id.

Define $\check{\alpha}(s) = \min\{\check{\alpha}^l(s), \check{\alpha}^u(s)\}$ for $s \in \mathbb{R}_+$. Then, $\check{\alpha}$ is continuous, positive definite, and less than Id. By recursively analyzing the system, we can achieve

$$\tilde{\mu}((k+1)T^*) \leq \tilde{\mu}(kT^*) - \check{\alpha}(\tilde{\mu}(kT^*)) \tag{6.278}$$

for $k \in \mathbb{Z}_+$. By using the asymptotic stability result for discrete-time nonlinear systems in [132], we can conclude that $\tilde{\mu}(kT^*) \to 0$ as $k \to \infty$.

Define $\overline{\mu}(t) = \overline{\mu}(kT^*)$ and $\underline{\mu}(t) = \underline{\mu}(kT^*)$ if $t \in [kT^*, (k+1)T^*)$. According to the analysis above, during the control procedure, it always holds that

$$(\eta_i(t), \zeta_i(t)) \in S_i(\underline{\mu}(t), \overline{\mu}(t)) \tag{6.279}$$

and

$$\underline{\mu}(t) \leq \eta_i(t) \leq \overline{\mu}(t). \tag{6.280}$$

Properties (6.182) and (6.183) can be proved as $\tilde{\mu} = \overline{\mu} - \underline{\mu}$ asymptotically converges to the origin. This ends the proof of Theorem 6.3.

6.4.5 DISTRIBUTED FORMATION CONTROL OF MOBILE ROBOTS

As a practical engineering application, we study the distributed formation control of a group of $N+1$ nonholonomic mobile robots under switching position measurement topology. For $i = 0, \ldots, N$, each i-th robot is modeled by the unicycle model (6.84)–(6.86).

The robot with index 0 is the leader robot, and the robots with indices $1, \ldots, N$ are follower robots. We consider v_i and ω_i as the control inputs of the i-th robot for $i = 1, \ldots, N$. For system (6.84)–(6.86), the position of the leader robot is assumed to be accessible to (some of) the follower robots, and the control objective is still to achieve (6.87)–(6.89). That is, $(x_i(t), y_i(t))$ converges to $(x_0(t) + d_{xi}, y_0(t) + d_{yi})$ with d_{xi}, d_{yi} being constants and $\theta_i(t)$ converges to $\theta_0(t) + 2k\pi$ with $k \in \mathbb{Z}$.

The following assumption is made throughout this section.

Assumption 6.4 *The linear velocity v_0 of the leader robot is differentiable with bounded derivative, i.e., $\dot{v}_0(t)$ exists and is bounded on $[0, \infty)$, and there exists constants $\overline{v}_0 > \underline{v}_0 > 0$ such that $\underline{v}_0 \leq v_0(t) \leq \overline{v}_0$ for all $t \geq 0$.*

Global position measurements of the robots are usually unavailable and the sensing topology may be switching in practical formation control systems. We employ the strong output agreement result proposed in the previous section to develop a new class of coordinated controllers, which are capable of overcoming the problems caused by the nonholonomic constraint and achieving the formation control objective by using local relative position measurements under switching position sensing topologies as well as the velocity information of the leader. With our new coordinated controller, the velocity v_i of each i-th follower robot can also be guaranteed to be upper bounded by a constant $\lambda_i^* > \bar{v}_0$ if required.

The distributed controller for each follower robot is composed of two stages: (a) initialization and (b) formation control. With the initialization stage, the heading direction of each follower robot can be controlled to converge to within desired ranges in some finite time. Then, the formation control stage is triggered and the formation control objective is achieved during the formation control stage.

Initialization Stage

The objective of the initialization stage is to control the angles $\theta_i(t)$ for $i = 1, \ldots, N$ to within a specific small neighborhood of $\theta_0(t)$.

For each i-th mobile robot (6.84)–(6.86), we propose the following initialization control law:

$$\omega_i = \phi_{\theta i}(\theta_i - \theta_0) + \omega_0 \tag{6.281}$$

$$v_i = v_0, \tag{6.282}$$

where $\phi_{\theta i} : \mathbb{R} \to \mathbb{R}$ is a nonlinear function such that $\phi_{\theta i}(r)r < 0$ for $r \neq 0$ and $\phi_{\theta i}(0) = 3$.

Define $\tilde{\theta}_i = \theta_i - \theta_0$. By taking the derivative of $\tilde{\theta}_i$ and using (6.281) and (6.86), we have

$$\dot{\tilde{\theta}}_i = \phi_{\theta i}(\tilde{\theta}_i). \tag{6.283}$$

With the appropriately designed $\phi_{\theta i}$, we can guarantee the asymptotic stability of system (6.283). Moreover, there exists $\beta_{\tilde{\theta}} \in \mathcal{KL}$ such that $|\tilde{\theta}(t)| \leq \beta_{\tilde{\theta}}(|\tilde{\theta}(0)|, t)$ for $t \geq 0$.

For specified $0 < \lambda_* < \underline{v}_0 < \bar{v}_0 < \lambda^*$, define

$$\lambda = \min\left\{ \frac{\sqrt{2}}{2}(\underline{v}_0 - \lambda_*), \frac{\sqrt{2}}{2}(\lambda^* - \bar{v}_0) \right\}. \tag{6.284}$$

By directly using the property of continuous functions, there exists a $\bar{\delta}_{\theta 0} > 0$ such that

$$|v_0 \cos(\theta_0 + \delta_{\theta 0}) - v_0 \cos \theta_0| \leq \lambda, \tag{6.285}$$

$$|v_0 \sin(\theta_0 + \delta_{\theta 0}) - v_0 \sin \theta_0| \leq \lambda \tag{6.286}$$

for all $v_0 \in [\underline{v}_0, \bar{v}_0]$, $\theta_0 \in \mathbb{R}$ and $|\delta_{\theta 0}| \leq \bar{\delta}_{\theta 0}$. Recall that for any $\beta \in \mathcal{KL}$, there exist $\alpha_1, \alpha_2 \in \mathcal{K}_\infty$ such that $\beta(s,t) \leq \alpha_1(s)\alpha_2(e^{-t})$ for all $s, t \in \mathbb{R}_+$ [243, Lemma 8]. With control law (6.281)–(6.282), there exists a finite time T_{Oi} for the i-th robot such that $|\theta_i(T_{Oi}) - \theta_0(T_{Oi})| \leq \bar{\delta}_{\theta 0}$, and thus,

$$|v_i(T_{Oi}) \cos \theta_i(T_{Oi}) - v_0(T_{Oi}) \cos \theta_0(T_{Oi})| \leq \lambda, \qquad (6.287)$$

$$|v_i(T_{Oi}) \sin \theta_i(T_{Oi}) - v_0(T_{Oi}) \sin \theta_0(T_{Oi})| \leq \lambda. \qquad (6.288)$$

At time T_{Oi}, the distributed control law for the i-th follower robot is switched to the formation control stage.

Formation Control Stage

With the dynamic feedback linearization technique recalled in Subsection 6.3.1, the unicycle model (6.84)–(6.86) can be transformed into two double-integrators in the form of (6.95)–(6.96) by introducing a new input r_i for (6.90), if $v_i \neq 0$ is satisfied. The formation control objective is achieved if control laws can be designed for the robots to guarantee $v_i \neq 0$, and at the same time, stabilize (6.95)–(6.96) at the origin.

In this way, the formation control problem is transformed into the issue of designing control laws for system (6.95)–(6.96) with $\tilde{u}_{xi}$ and $\tilde{u}_{yi}$ as the control inputs, so that $v_i \neq 0$ is guaranteed, and at the same time, the formation control objective is achieved.

The condition $v_i \neq 0$ for the validity of (6.93)–(6.94) can be equivalently represented by $\sqrt{v_{xi}^2 + v_{yi}^2} > 0$ based on the definition of v_{xi} and v_{yi}. To implement the transformation in (6.92), we need to design the control law for the i-th robot such that

$$\max\{|\tilde{v}_{xi}|, |\tilde{v}_{yi}|\} \leq \frac{\sqrt{2}}{2}(\underline{v}_0 - \lambda_*) \qquad (6.289)$$

for a specified $0 < \lambda_* < \underline{v}_0$. In doing so, we can guarantee that $|v_i| = \sqrt{v_{xi}^2 + v_{yi}^2} = \sqrt{(v_{x0} + \tilde{v}_{xi})^2 + (v_{y0} + \tilde{v}_{yi})^2} \geq \lambda_* > 0$ and thus $v_i \neq 0$.

Similarly, to guarantee that $|v_i| \leq \lambda^*$ for any given $\lambda^* > \bar{v}_0$, we design a control law such that

$$\max\{|\tilde{v}_{xi}|, |\tilde{v}_{yi}|\} \leq \frac{\sqrt{2}}{2}(\lambda^* - \bar{v}_0). \qquad (6.290)$$

The formation control problem is now transformed into the issue of designing control laws for system (6.95)–(6.96) with $\tilde{u}_{xi}$ and $\tilde{u}_{yi}$ as the control inputs, so that (6.289) and (6.290) are guaranteed during the control procedure, and at the same time, the formation control objective is achieved.

Recall the definition of λ in (6.284). Both (6.289) and (6.290) can be satisfied if

$$\max\{|\tilde{v}_{xi}|, |\tilde{v}_{yi}|\} \leq \lambda. \qquad (6.291)$$

After the initialization stage, at time T_{Oi}, the satisfaction of (6.287) and (6.288) implies that (6.291) is satisfied at time T_{Oi}.

Considering the requirement of relative position measurement, we propose a distributed control law in the following form:

$$\tilde{u}_{xi} = -\varphi_{xi}(\tilde{v}_{xi} - \phi_{xi}(z_{xi})) \tag{6.292}$$

$$\tilde{u}_{yi} = -\varphi_{yi}(\tilde{v}_{yi} - \phi_{yi}(z_{yi})), \tag{6.293}$$

where $\varphi_{xi}, \varphi_{yi}, \phi_{xi}, \phi_{yi}$ are strictly decreasing and locally Lipschitz, and satisfy $\varphi_{xi}(0) = \varphi_{yi}(0) = \phi_{xi}(0) = \phi_{yi}(0) = 0$, $\varphi_{xi}(r)r < 0$, $\varphi_{yi}(r)r < 0$, $\phi_{xi}(r)r < 0$ and $\phi_{yi}(r)r < 0$ for $r \neq 0$, and

$$\sup_{r \in \mathbb{R}}\{\max \partial \varphi_{xi}(r) : r \in \mathbb{R}\} < 4 \inf_{r \in \mathbb{R}}\{\min \partial \phi_{xi}(r)\}, \tag{6.294}$$

$$\sup_{r \in \mathbb{R}}\{\max \partial \varphi_{yi}(r) : r \in \mathbb{R}\} < 4 \inf_{r \in \mathbb{R}}\{\min \partial \phi_{yi}(r)\}, \tag{6.295}$$

for $i = 1, \ldots, N$. The functions ϕ_{xi}, ϕ_{yi} are also designed to satisfy

$$-\lambda \leq \phi_{xi}(r), \phi_{yi}(r) \leq \lambda \tag{6.296}$$

for $r \in \mathbb{R}$.

The variables z_{xi} and z_{yi} are defined as

$$z_{xi} = \frac{\sum_{j \in \mathcal{N}_i(\sigma(t))} a_{ij}(x_i - x_j - d_{xij})}{\sum_{j \in \mathcal{N}_i(\sigma(t))} a_{ij}} \tag{6.297}$$

$$z_{yi} = \frac{\sum_{j \in \mathcal{N}_i(\sigma(t))} b_{ij}(y_i - y_j - d_{yij})}{\sum_{j \in \mathcal{N}_i(\sigma(t))} b_{ij}}, \tag{6.298}$$

where $\sigma : [0, \infty) \to \mathcal{P}$ is a piecewise constant switching signal describing the position sensing topology with $\mathcal{P}$ being the set of all the possible position sensing topologies, $\mathcal{N}_i(p) \subseteq \{0, \ldots, N\}$ for each $i = 1, \ldots, N$ and each $p \in \mathcal{P}$, constant $a_{ij} > 0$ if $i \neq j$ and $a_{ij} \geq 0$ if $i = j$. Note that d_{xij}, d_{yij} in (6.297) and (6.298) represent the desired relative position between the i-th robot and the j-th robot. By default, $d_{xii} = d_{yii} = 0$.

Consider the $(\tilde{v}_{xi}, \tilde{v}_{yi})$-system defined in (6.93)–(6.94). With (6.287) and (6.288) achieved, the boundedness of ϕ_{xi} and ϕ_{yi} in (6.296) together with the control law (6.292)–(6.293) guarantees that

$$\max \{|\tilde{v}_{xi}(t)|, |\tilde{v}_{yi}(t)|\} \leq \lambda \tag{6.299}$$

for $t \geq T_{Oi}$. By considering $\{(\tilde{v}_{xi}, \tilde{v}_{yi}) : \max\{|\tilde{v}_{xi}|, |\tilde{v}_{yi}|\} \leq \lambda\}$ as an invariant set of the $(\tilde{v}_{xi}, \tilde{v}_{yi})$-system, (6.299) can be proved. With (6.299) achieved, we have $\max_{i=1,\ldots,N} \{|\tilde{v}_{xi}(t)|, |\tilde{v}_{yi}(t)|\} \leq \lambda$ for all $t \geq T_O$ with $T_O := \max_{i=1,\ldots,N}\{T_{Oi}\}$. This guarantees the validity of the transformed model (6.93)–(6.94).

Consider the multi-robot model (6.84)–(6.86) and the distributed control laws defined by (6.281), (6.282), (6.90), (6.92), (6.292), and (6.293) with nonlinear functions $\varphi_{xi}, \varphi_{yi}, \phi_{xi}, \phi_{yi}$ satisfying (6.294), (6.295), and (6.296).

We represent the switching position sensing topology by a switching digraph $\mathcal{G}(\sigma(t)) = (\mathcal{N}, \mathcal{E}(\sigma(t)))$, where $\mathcal{N} = \{0, \ldots, N\}$ and $\mathcal{E}(\sigma(t))$ is defined based on $\mathcal{N}_i(\sigma(t))$ given in (6.297) and (6.298) for $i = 1, \ldots, N$ and $\mathcal{N}_0(\sigma(t)) \equiv \{0\}$. Theorem 6.4 presents our main result on formation control of unicycle mobile robots.

Theorem 6.4 *Under Assumption 6.4, if* $\mathcal{G}(\sigma(t)) = (\mathcal{N}, \mathcal{E}(\sigma(t)))$ *with* $\sigma :$ $[0, \infty) \to \mathcal{P}$ *is UQSC and has an edge dwell-time* $\tau_D > 0$*, then* (6.87)–(6.89) *can be achieved for any* $i, j = 0, \ldots, N$*. Moreover, given any* $\lambda^* > \bar{v}_0$*, if* $v_i(0) \leq \lambda^*$ *for* $i = 1, \ldots, N$*, then* $v_i(t) \leq \lambda^*$ *for all* $t \geq 0$*.*

Proof. The states of the mobile robots remain bounded during the finite time interval $[0, T_O]$. We study the motion of the robots during $[T_O, \infty)$. Note that the model (6.95)–(6.96) is valid during $[T_O, \infty)$.

Note that $\tilde{v}_{x0} = \tilde{v}_{y0} = \tilde{u}_{x0} = \tilde{u}_{y0} = 0$. One can find appropriate $\varphi_{x0}, \varphi_{y0}, \phi_{x0}, \phi_{y0}, a_{00}, b_{00}$ to represent $\tilde{u}_{x0}$ and $\tilde{u}_{y0}$ by (6.292) and (6.293) with z_{x0} and x_{y0} in the form of (6.297) and (6.298), respectively.

For $i = 0, \ldots, N$, rewrite

$$z_{xi} = \frac{\sum_{j \in \mathcal{N}_i(\sigma(t))} a_{ij}(\tilde{x}_i - \tilde{x}_j)}{\sum_{j \in \mathcal{N}_i(\sigma(t))} a_{ij}}, \tag{6.300}$$

$$z_{yi} = \frac{\sum_{j \in \mathcal{N}_i(\sigma(t))} b_{ij}(\tilde{y}_i - \tilde{y}_j)}{\sum_{j \in \mathcal{N}_i(\sigma(t))} b_{ij}}. \tag{6.301}$$

For $i = 0, \ldots, N$, all the $\tilde{u}_{xi}$ and $\tilde{u}_{yi}$ are in the form of μ_i defined in (6.237) and all the z_{xi} and z_{yi} are in the form of ξ_i defined in (6.238).

Theorem 6.3 guarantees that

$$\lim_{t \to \infty} (\tilde{x}_i(t) - \tilde{x}_j(t)) = 0, \tag{6.302}$$

$$\lim_{t \to \infty} (\tilde{y}_i(t) - \tilde{y}_j(t)) = 0, \tag{6.303}$$

for any $i, j = 0, \ldots, N$. Then, we can prove (6.87) and (6.88) by using the definitions of $\tilde{x}_i$ and $\tilde{y}_i$ and the fact that $\tilde{x}_0(t) = \tilde{y}_0(t) \equiv 0$. The result of $v_i(t) \leq \lambda^*$ can be proved based on the discussions below (6.299).

By using Theorem 6.3, we can also prove the convergence of $\tilde{v}_{xi}, \tilde{v}_{yi}$ to the origin. By using the definitions of $\tilde{v}_{xi}, \tilde{v}_{yi}$, the convergence such that (6.89) can be proved. This ends the proof. $\diamond$

If there is no leader in the mobile robot system, the velocities of the robots are usually hard to control. To overcome this problem, we may employ a virtual leader, which generates a reference velocity for the follower robots. Note that the global and relative positions of the virtual leader are usually not available. In this case, our proposed formation control strategy is still valid such that the follower robots are controlled to converge to specific relative positions of each other, i.e., (6.87) and (6.88) are achieved for any $i, j = 1, \ldots, N$.

We represent the switching position sensing topology of the follower robots with a switching digraph $\mathcal{G}^f(\sigma(t)) = (\mathcal{N}^f, \mathcal{E}^f(\sigma(t)))$, where $\mathcal{N}^f = \{1, \ldots, N\}$ and $\mathcal{E}^f(\sigma(t))$ is defined based on $\mathcal{N}_i(\sigma(t))$ given in (6.297) and (6.298) for $i = 1, \ldots, N$.

Theorem 6.5 presents such an extension of our main formation control result.

Theorem 6.5 *Under Assumption 6.4, if $\mathcal{G}^f(\sigma(t)) = (\mathcal{N}^f, \mathcal{E}^f(\sigma(t)))$ with $\sigma : [0, \infty) \to \mathcal{P}$ is UQSC and has an edge dwell-time $\tau_D^f > 0$, then (6.87)–(6.89) can be achieved for any $i, j = 1, \ldots, N$. Moreover, given any $\lambda^* > \overline{v}_0$, if $v_i(0) \le \lambda^*$ for $i = 1, \ldots, N$, then $v_i(t) \le \lambda^*$ for all $t \ge 0$.*

Proof. The proof of Theorem 6.5 is still based on Theorem 6.3 and quite similar as the proof of Theorem 6.4. In the case of Theorem 6.5, we need not consider $\tilde{x}_0$ and $\tilde{y}_0$, and properties (6.302) and (6.302) can be achieved for any $i, j = 1, \ldots, N$. Given the definitions of $\tilde{x}_i$ and $\tilde{y}_i$ for $i = 1, \ldots, N$, we can prove (6.87) and (6.88) for any $i, j = 1, \ldots, N$. $\diamond$

6.4.6 SIMULATION RESULTS

Consider a group of 6 robots with indices $0, 1, \ldots, 5$ with robot 0 being the leader. By default, the values of all the variables in this simulation are in SI units.

The control inputs of the leader robot are $r_0(t) = 0.1 \sin(0.4t)$ and $\omega_0(t) = 0.1 \cos(0.2t)$. With such control inputs, the linear velocity v_0 satisfies $\underline{v}_0 \le v_0(t) \le \overline{v}_0$ with $\underline{v}_0 = 3$ and $\overline{v}_0 = 3.5$, given $v_0(0) = 0$. The desired relative position of the follower robots are defined by $d_{x01} = -\sqrt{3}d/2, d_{x02} = -\sqrt{3}d/2, d_{x03} = 0, d_{x04} = \sqrt{3}d/2, d_{x05} = \sqrt{3}d/2, d_{y01} = -d/2, d_{y02} = -3d/2, d_{y03} = -2d, d_{y04} = -3d/2, d_{y05} = -d/2$ with $d = 80$.

Choose $\lambda_* = 0.45$ and $\lambda^* = 6.05$. The distributed control laws for the initialization stage are designed in the form of (6.281)–(6.282) with

$$\phi_{\theta i}(r) = \phi_{vi}(r) = -0.5(1 - \exp(-0.5r))/(1 + \exp(-0.5r)) \qquad (6.304)$$

for $i = 1, \ldots, 5$. The distributed control laws for the formation control stage are in the form of (6.292)–(6.293) with $k_{xi} = k_{yi} = 6$ and

$$\phi_{xi}(r) = \phi_{yi}(r) = -1.8(1 - \exp(-r))/(1 + \exp(-r)) \qquad (6.305)$$

for $i = 1, \ldots, 5$. It can be verified that $k_{xi}, k_{yi}, \phi_{xi}, \phi_{yi}$ satisfy (6.294) and (6.295). Also, $\phi_{xi}(r), \phi_{yi}(r) \in [-1.8, 1.8]$ for all $r \in \mathbb{R}$. With $\underline{v}_0 = 3$ and $\overline{v}_0 = 3.5$, the linear velocities of the robots are restricted to be within the range of $[3 - 1.8\sqrt{2}, 3.5 + 1.8\sqrt{2}] = [0.454, 6.046] \subset [\lambda_*, \lambda^*]$.

The initial states of the robots are chosen as

i	$(x_i(0), y_i(0))$	$v_i(0)$	$\theta_i(0)$
0	$(0,0)$	3	$\pi/6$
1	$(-20, 50)$	4	π
2	$(30, -40)$	3.5	$5\pi/6$
3	$(50, -100)$	2.5	0
4	$(200, -100)$	2	$-2\pi/3$
5	$(100, -120)$	3	0

The information exchange topology switches between the digraphs in Figure 6.12, and the switching sequence is shown in Figure 6.13.

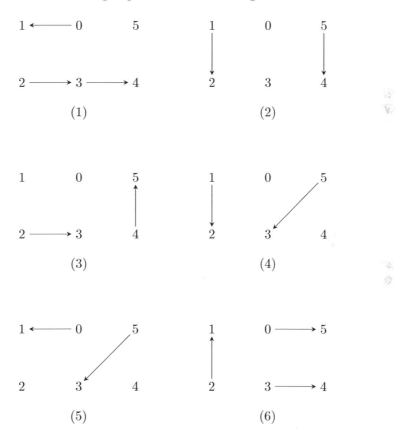

FIGURE 6.12 Digraphs representing the switching information exchange topology.

The linear velocities and the angular velocities of the robots are shown in Figure 6.14. The stage changes of the distributed controllers are shown in Figure 6.15 with "0" representing the initialization stage and "1" representing the formation control stage. Figure 6.16 shows the trajectories of the robots.

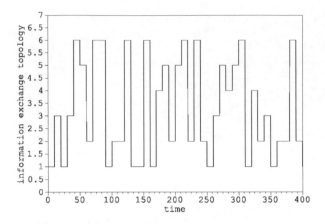

FIGURE 6.13 The switching sequence of the information exchange topology.

The simulation verifies the theoretical result of the paper.

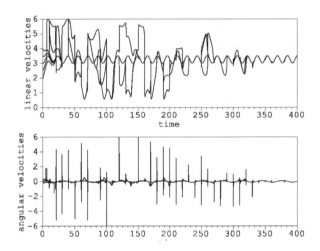

FIGURE 6.14 The linear velocities and angular velocities of the robots.

6.5 NOTES

Distributed control of multi-agent systems for group coordination has recently attracted significant attention within the controls and robotics communities: see, for example, [212, 214, 172, 236, 176] based on Lyapunov methods; [8, 17] using passivity methods; [58, 215, 39, 240, 225, 253, 162] based on linear al-

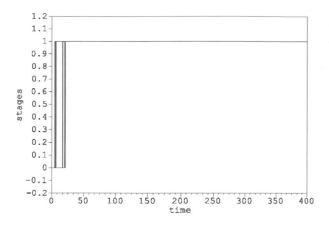

FIGURE 6.15 The stages of the distributed controllers.

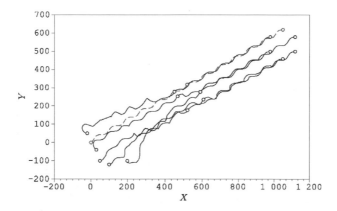

FIGURE 6.16 The trajectories of the robots. The trajectory of the leader robot is represented by the dashed curve.

gebra and graph theory; and [271, 275, 254] using output regulation theory. The main objective of distributed control is to achieve some desired group behavior for multi-agent systems by taking advantage of local system information and information exchanges among neighboring systems. Distributed control may find applications in sensor networks [213], vehicle coordination and formation [256, 229, 111, 68, 282], and smart power grids [280]. Reference [192] studies the synchronization problem of multi-agent systems without connectivity assumptions. In [287], the synchronization of dynamic networks with nonidentical nodes is achieved by reorganizing the connection topologies. One group behavior of wide interest is the agreement property, for which the related variables of multi-agent systems are steered to a common value. It should be noted that most of the previously published papers focus on linear models.

This chapter has presented cyclic-small-gain tools for distributed control of *nonlinear* multi-agent systems. In Section 6.2, with the proposed distributed observers and control laws, the outputs of the agents can be steered to within an arbitrarily small neighborhood of the desired agreement value under external disturbances. Asymptotic output agreement can be achieved if the system is disturbance-free. The robustness to bounded time delays of exchanged information can also be guaranteed. In the problem setting, each agent can use the output of itself and the outputs of its neighbors, while only the informed agents have access to the desired agreement value. This makes the distributed control problem considered in this chapter significantly different from the decentralized control problem, in which each decentralized controller often assumes the *a priori* knowledge of the reference signal and does not take advantage of the available information of neighboring agents; see e.g., [239, 129]. It should be noted that the term of distributed control is also used for decentralized control under arbitrary information structure constraints in some recent works [285, 286].

Section 6.2 only considers the case with time-invariant agreement value y_0. It is practically interesting to further study the distributed nonlinear control for agreement with a time-varying agreement value. Recent developments on the output-feedback tracking control of nonlinear systems (see, e.g., [116]) should be helpful for the research in this direction.

Section 6.3 studies the formation control of nonholonomic mobile robots. For the formation control of mobile robots, assuming a tree sensing structure generally leads to cascade interconnection structures of the closed-loop systems [47]. Along this line of research, [258] employs nonlinear gains to estimate the influence of leader behavior on the formation by using the concept of ISS. Several researchers have attempted to relax the assumption of tree sensing structures, at the price of using global position measurements. In [212], a control Lyapunov function approach was introduced to multi-agent coordination. The authors of [50] proposed a constructive design method for the formation tracking control and collision avoidance of unicycle robots. Reference [51] transforms the cooperative control problem into a decentralized backstepping design problem. In [161], the artificial potential function-based approach was studied for the flocking control of nonholonomic mobile robots. The wiggling controller proposed in [171] does not use global position measurements and is capable of controlling unicycles to stationary points, but the controller seems unable to overcome the nonholonomic constraint of the unicycle to achieve moving formations. It should be noted that if each robot has access to its desired path, as in the coordinated path following problems studied in [2, 103, 157], then the requirement of global position measurements can be relaxed.

The distributed formation control law proposed in Section 6.3 uses relative position measurements without assuming a tree structure. For this purpose, the formation control problem is first transformed into a state agreement problem of double-integrators with dynamic feedback linearization [40]. The nonholonomic constraint leads to singularity for dynamic feedback linearization when the linear velocity of the robot is zero. Then, a class of distributed nonlinear control laws is designed. Saturation functions are employed to restrict the linear velocities of the robots to be larger than zero to avoid the singularity. With the proposed distributed nonlinear control law, the closed-loop system can be transformed into a network of IOS systems, and the achievement of the formation control objective can be guaranteed by using the cyclic-small-gain theorem. The special case in which there are only two robots and the desired relative positions are zero has been studied extensively in the past literature; see [127, 49] and the references therein.

To further relax the requirement on the fixed topologies, Section 6.4 has presented a modified distributed nonlinear control design for strong output agreement of multi-agent systems modeled by double-integrators. With the new design, the information exchange topology of the large-scale interconnected multi-agent system is allowed to be directed and switching, as long as a mild connectivity condition is satisfied. By appropriately designing the distributed control law, the internal states of the agents can be restricted to be within an arbitrarily small neighborhood of the origin, which is of practical interest. As an application, a distributed formation control algorithm has also been developed for groups of unicycle mobile robots with flexible topologies and relative position measurements. The singularity problem caused by the nonholonomic constraint is solved by properly designing the distributed control law.

By showing that a distributed control problem can be transformed into the stability/convergence problem of a dynamic network composed of IOS subsystems, Sections 6.1–6.3 provide some partial answers to the question asked by Open Problem #5 in [137]: "Application of small-gain results for distributed feedback design of large-scale nonlinear systems." More discussions on the application of the cyclic-small-gain theorem to distributed control can be found in [184, 185, 180, 183].

7 Conclusions and Future Challenges

This book has presented recent results on the stability and control of interconnected systems which are modeled as dynamic networks. Such networks arise naturally in engineering applications, such as robotic networks and power systems, as well as in other areas such as biology, physics, and economics. The treatment is all based on the concept of input-to-state stability (ISS) and the idea of small-gain in loops of the network as a means to achieve stability when systems are interconnected. The ISS property for each subsystem includes nonlinear gain functions and corresponds to the existence of an ISS-Lyapunov function. The stability conditions are intuitively and conveniently expressed in terms of compositions of the gains associated with cycles in the system graph being less than one, generalizing the well-known small-gain theorem for feedback systems. After establishing the small-gain theorems for classes of dynamic networks (continuous-time, discrete-time, and hybrid), the book proposed a set of tools for input-to-state stabilization and robust control of complex nonlinear systems from the viewpoint of dynamic networks. Among these tools and applications are:

- Lyapunov-based cyclic-small-gain theorems for continuous-time, discrete-time, and hybrid dynamic networks composed of multiple ISS subsystems;
- novel small-gain-based static feedback and dynamic feedback designs for robust control of nonlinear uncertain systems with disturbed measurements;
- quantized stabilization designs for nonlinear uncertain systems with static quantization and dynamic quantization; and
- distributed coordination control of nonlinear multi-agent systems under information exchange constraints.

The idea of small-gain is about 40 years old, and thinking of the Nyquist criterion interpretation, is one of two fundamental ideas for preserving stability in a feedback loop. Nevertheless, after all the work of numerous researchers, taking this idea to ever more complex systems, the topics considered in this book show that many interesting research problems, in particular at the level of control synthesis, remain to be considered. These new challenges relate to theoretical advances as well as further application of the results. Some suggestions for such work will now be discussed.

The hybrid dynamic network considered in Chapter 3 involves only stable dynamics. However, it is well known that appropriately switching between unstable dynamics may still lead to stability. In fact, this can be a way to achieve higher performance. This behavior is often formulated with the "dwell-time" condition. There are two major difficulties in introducing this approach to hybrid dynamic networks: (a) The subsystems may have different impulsive time instants, which may lead to restrictiveness if we consider the dynamic network as one single system and directly apply the "dwell-time" method; (b) Some necessary transformation is often needed to transform the decreasing/increasing rates of the subsystems from non-exponential to exponential, without influencing the validity of the cyclic-small-gain condition. Recently, researchers have started to consider the possibility of unstable subsystems. One recent contribution in this direction provides revised small-gain theorems for hybrid feedback systems [169]. ISS stability criteria for hybrid systems which include unstable subsystems (discrete-time or continuous-time) have been given in [174, 169]. Generally, there seem to be possibilities to advance the complexity to include switching, impulses, delays, and interconnections [173].

A major problem in analyzing complex systems and networks can be to find ways to manage the computations involved as the system scale becomes larger. This can arise from the complexity of the dynamics and/or the network structure, and so affect finding ISS-Lyapunov functions and testing the small-gain criteria. The latter involves functional tests which grow with the diameter and connectivity of the network. It would appear that much work can yet be done to find computational tools and results which deal with special classes of systems. One such tool is the sum-of-squares technique, which has recently been applied to networked systems [79]. These issues of scale have been around since the effort on so-called large-scale systems in the 1970s. For stability analysis, the paradigm was generally to establish strongly stable subsystems and weak interconnections, which reduced the stability test to a matrix condition [205, 207]. In its most accessible form, this expressed the dominance of the subsystem stability over the strength of the interconnections. It has been of interest ever since to find stability results which allow more diverse interplay between subsystem and interconnection properties. The synchronization theory for dynamical networks demonstrated stability criteria with strong coupling. However, again, as systems get more complicated, there is motivation to find ways to "divide and conquer" the system scale. One idea is to think in terms of stability certificates where a sequence of stability tests are applied to sub-regions of the network, perhaps each subsystem and its nearest neighbors [216]. It follows that there is no need for an overall stability test. Another idea exploits detailed system structure [226]. These ideas are in their infancy and it remains to see how far they can be taken. Computational aspects need to be sensitive to the properties of solutions to dissipation inequalities [112].

The control designs proposed in this book are restricted to systems in specific forms: strict-feedback form and output-feedback form. While popular in research to demonstrate designs with supporting theory, they are based more in mathematical convenience than any physical forms. However, in given practical case studies, such results are a useful guide if not directly applicable. Nevertheless, despite the popularity of these forms, extensions of the designs to more general nonlinear systems would contribute a lot to nonlinear control theory. Again, these structural issues have an earlier form in discussions of how to extend control designs based on backstepping and forwarding to more general structures [235].

In modern networked control systems, data transmission through communication channels inevitably causes time delays and thus late response of the control system to control commands. Similar problems also arise from data sampling. Reference [261] discovered the connections between ISS small-gain theorems and the Razumikhin theorem, the latter being dedicated to time-delay nonlinear systems. Recently, [136] presented a sufficient and necessary condition for stabilizability of nonlinear, time-varying systems with delayed state measurements. However, robust control designs for nonlinear systems (in the popular forms) with both uncertain dynamics and time delay are still to be developed. Related problems have also been noted in [137]. It should be mentioned that the recent work of Krstić [149, 150] has proposed new solutions to stabilization with input delays.

Uncertain actuator dynamics may cause the performance of a nonlinear control system to deteriorate. Quite a few references (see, e.g., [123, 119] and the references therein) studied the controller design problem for nonlinear systems with uncertain actuator dynamics. Another source of destabilizing factors is the uncertain measurement dynamics (due to inertia of the sensor, for example). However, there has not been much research on robust design of nonlinear control systems in the presence of uncertain measurement dynamics. The tools developed in this book may provide some potential solution to this problem. As we can design input-to-state stabilizing controllers for nonlinear systems with measurement errors, one possible solution to the problem is to find some way to formulate the uncertain measurement dynamics with ISS and use the ISS cyclic-small-gain theorem to guarantee stabilization.

An alternative approach to the control of uncertain systems is adaptive control, which is extremely useful for systems with "large" parameterized uncertainties. Passivity has played a central role in adaptive control designs since its earliest days [12]. Robust adaptive control methods have also been developed for systems with both parameterized uncertainties and disturbances. Design techniques for nonlinear systems have been obtained via backstepping and passivity in [153]. However, the influence of sensor noise on adaptive control performance of nonlinear uncertain systems has not been systematically investigated, and refined robust adaptive control methods remain to be achieved. Considering the usefulness of passivity to adaptive control and the importance of ISS and small-gain theorems to robust control (and their more minor roles, each in the other area), one may think about the combination of the methods. Preliminary results were obtained in [115]. The notion of dissipativity [277] has also been used to realize such a combination for interconnected systems, and apply these to giving general stability criteria which combine passivity and finite gain aspects in [90, 207, 93]. Another approach has suggested using the passivity and finite gain properties side by side [72]. In these results, the supply functions are in the quadratic form. To deal with complex nonlinear dynamics, more efforts are needed for the combination of passivity and ISS small-gain methods for systems with more general supply functions. With the expected generalization, the problem mentioned above and more general robust adaptive control problems may hopefully be solved for nonlinear uncertain systems. Similar problems also arise from the distributed adaptive control.

Another promising research area, still in its infancy, is the quantized nonlinear control which, as shown in this book, is strongly connected to the robust nonlinear control. Advanced robust control design methods, such as the ISS small-gain approach, are powerful in handling the new problems caused by quantization in nonlinear control. Expectedly, the preliminary results presented in the book can be further generalized in several directions, in view of the rich literature of nonlinear control over the last three decades. It should be mentioned that there are more open problems in this field than the available results. Some open problems of great interest are stated here:

- *Geometric nonlinear control with quantized signals.* The classical yet important topic of controllability and observability for nonlinear systems needs to be revisited [105, 244], when only quantized signals are allowed. In addition, the relationships between controllability and stabilizability [37], and feedback linearization theory [105], need to be revisited as well in the context of quantized feedback control.

- *Tracking via quantized feedback.* While this book focuses on quantized stabilization, the problem of quantized feedback tracking is of more practical interest and covers stabilization as a special case. Instead of forcing the state or the output to the origin or a set point of interest, the quantized feedback tracking problem seeks a quantized feedback controller so that the output follows a desirable reference signal or the state follows the desired state of a reference model. This problem has received practically no attention in the present literature. Closely related to this problem is the output regulation theory [99, 21] that consists of searching for (unquantized) feedback control laws to achieve asymptotic tracking with disturbance rejection, when the disturbance and reference signals are generated by an exo-system. The well-known *internal model principle* serves as a bridge to convert the output regulation problem into a stabilization problem for a transformed system. To what extent will the internal model principle remain valid and applicable when only quantized signals are used?

- *Decentralized quantized control.* In the decentralized control setting, local controllers are used to control the subsystems of a large-scale system [239]. Among the main characteristics of decentralized control are the dramatic reduction of computational complexity and the enhancement of robustness against uncertain interactions or loss of interaction. The ISS small-gain designs for decentralized control [129, 117] makes it possible to further take into account the effect of quantization; see also the survey [118]. The decentralized measurement feedback control problem, which is closely related to decentralized quantized control, has been studied in this book. For dynamic quantization, the zooming variables of the quantizers of different subsystems should coordinate with each other. This is still the case when decentralized control is reduced to centralized control, as shown in Section 5.2. Another problem with decentralized dynamic quantization is that the updates of the zooming variables of the quantizers may not be synchronized. For such problem, the small-gain results for hybrid systems [137, 138, 188] should be helpful.

- *Quantized adaptive control.* Controllers are expected to possess adaptive capabilities to cope with "large" uncertainties. A further extension of the previously developed methodology to quantized adaptive control is of practical interest for engineering applications. The recent achievements [81, 82] provide a basis for future research in this direction. Reference [81] proposed a Lyapunov-based framework for adaptive quantized control of linear uncertain systems modeled in discrete-time. In [82], a direct adaptive control strategy was developed for nonlinear uncertain systems with input quantizers under the assumption that the system is robustly stabilizable with respect to sector bounded uncertainties.

- *Quantized control systems with time delays.* As discussed above, data transmission through communication channels inevitably results in time delays, a severe cause of poor performance and even instability of the system in question. Recently, a necessary and sufficient condition for robust stabilizability of nonlinear, time-varying systems with delayed state measurements was presented in [136]. A new framework capable of dealing with quantization and time delay at the same time is of paramount importance for the transition of advanced nonlinear control theory to practice.

The potential for application to physical systems certainly needs further exploration. Power systems are the most complex nonlinear dynamic networks within engineering systems. They connect generation and load devices across whole continents with dynamics on time scales from microseconds to hours, and with such distances the effect of time delays has to be considered. In terms of power flows, the device dynamics and interconnections are highly nonlinear. Further, the control loops routinely involve switching leading to large hybrid networks. There are many sources of modeling uncertainty. So for stability analysis and control, techniques associated with the whole theoretical toolkit, especially nonlinearity, uncertainty, hybrid models, large scale, robustness, discrete events, and adaptivity are all potentially useful. Certainly, the major stability and control questions can be formulated as nonlinear and/or network control problems. Trends in the industry, including new approaches to reliability (away from worst-case scenarios) and increasing use of renewables mean there will be much more uncertainty and systems will be driven harder. This translates into a need for nonlinear systems and methods, which reduce the effects of uncertainty. There has been a lot of work on stability analysis [34] and nonlinear control [27, 194] using modern nonlinear systems techniques, including gain-related ideas to handle disturbances [75]. The new concern is how stability and control become robust to major changes such as placement, size, and variability of renewable generation [29]. There appears to be considerable work to be done to use methods based on ISS to explore such issues. The recently developed robust adaptive dynamic programming methods [113, 114] provide new solutions; see also [120] for a recent review.

Similarly, in other areas involving interconnections of systems in network structures, there are ideas to pursue. The distributed control problems of robotic networks have been partially studied in this book. Another area where small-gain and dissipativity concepts appear useful, which is outside the experience of the authors, is biology. A recent book has studied biomolecular systems using ISS and related concepts [145], including contractive systems [193] which are closely related to the small-gain idea. Also see the recent work of Sontag and coworkers in systems biology [246, 45, 56]. The recent review article [120] gives an application of the ISS small-gain theorem to biological motor control, an important problem in systems neuroscience.

These topics—and no doubt there are more that can be seen in the work of others—promise that the fundamental concepts of ISS and small-gain can be developed even further to provide important tools for stability analysis and control of nonlinear interconnected systems.

A Related Notions in Graph Theory

This section gives the standard definitions of the notions in graph theory that are used in this book.

Definition A.1 *A graph $\mathcal{G}$ is a collection of points $\mathcal{V}_1, \ldots, \mathcal{V}_n$ and a collection of lines $a_1, \ldots, a_m$ joining all or some of these points. The points are called vertices, and the lines, denoted by the pairs of points they connect, are called links.*

Definition A.2 *A graph $\mathcal{G}$ is called a directed graph or simply a digraph if the lines in it have a direction. The lines are called directed links or arcs.*

Definition A.3 *A path in a digraph $\mathcal{G}$ is any sequence of arcs where the final vertex of one is the initial vertex of the next one, denoted as the sequence of the vertices it contains. If a path has no repeated vertices, then it is called a simple path.*

Definition A.4 *In a digraph $\mathcal{G}$, if there exists a path leading from vertex i to vertex j, then vertex j is reachable from vertex i. Specifically, any vertex i is reachable from itself.*

Definition A.5 *In a digraph $\mathcal{G}$, the reaching set of a vertex j, denoted by $\mathcal{RS}(j)$, is the set of the vertices from which vertex j is reachable.*

Definition A.6 *In a digraph $\mathcal{G}$, a path such that the starting vertex and the ending vertex are the same is called a cycle. If a cycle has no repeated vertices other than the starting and ending vertices, then it is called simple cycle.*

Definition A.7 *A directed tree is a digraph which has no cycle and there exists a vertex from which all the other vertices are reachable.*

Definition A.8 *A spanning tree $\mathcal{T}$ of a digraph $\mathcal{G}$ is a directed tree formed by all the vertices and some or all of the edges of $\mathcal{G}$.*

B Systems with Discontinuous Dynamics

This appendix provides some basic concepts and preliminary results of systems with discontinuous dynamics. Detailed studies on this topic can be found in [36, 60, 13, 84].

B.1 BASIC DEFINITIONS

Definition B.1 *For a set $\mathcal{M} \in \mathbb{R}^n$, a point $x \in \mathbb{R}^n$ is an interior point of $\mathcal{M}$ if there exists an open ball centered at x which is contained in $\mathcal{M}$. The interior of $\mathcal{M}$, denoted by $\mathrm{int}(\mathcal{M})$, contains all the interior points of $\mathcal{M}$.*

Definition B.2 *A set $\mathcal{M} \subseteq \mathbb{R}^n$ is called convex if for every $x, y \in \mathcal{M}$ and every $\lambda \in [0,1]$, $\lambda x + (1-\lambda)y \in \mathcal{M}$.*

Definition B.3 *The convex hull of a set $\mathcal{M} \subseteq \mathbb{R}^n$, denoted by $\mathrm{co}(\mathcal{M})$, is the intersection of all the convex sets containing $\mathcal{M}$.*

Definition B.4 *The closed convex hull of a set $\mathcal{M} \subseteq \mathbb{R}^n$, denoted by $\overline{\mathrm{co}}(\mathcal{M})$, is the intersection of all the closed convex sets containing $\mathcal{M}$.*

Let $\mathcal{X}$ and $\mathcal{Y}$ be two sets in Euclidian spaces.

Definition B.5 *A set-valued map $F : \mathcal{X} \rightsquigarrow \mathcal{Y}$ is a map that associates with any $x \in \mathcal{X}$ a subset $F(x)$ of $\mathcal{Y}$, and the subsets $F(x)$ are called the images of F.*

Definition B.6 *The domain of a set-valued map $F : \mathcal{X} \rightsquigarrow \mathcal{Y}$ is defined as*

$$\mathrm{dom}(F) \stackrel{\mathrm{def}}{=} \{x \in \mathcal{X} : F(x) \neq \emptyset\}. \tag{B.1}$$

Definition B.7 *A set-valued map $F : \mathcal{X} \rightsquigarrow \mathcal{Y}$ is called strict if $\mathrm{dom}(F) = \mathcal{X}$.*

Definition B.8 *A function $f : \mathcal{X} \rightarrow \mathcal{Y}$ is a selection of a strict set-valued map $F : \mathcal{X} \rightsquigarrow \mathcal{Y}$ if $f(x) \in F(x)$ for all $x \in \mathcal{X}$.*

Definition B.9 *The graph of a set-valued map $F : \mathcal{X} \rightsquigarrow \mathcal{Y}$ is the subset of pairs (x, y) where $y \in F(x)$, that is,*

$$\mathrm{graph}(F) \stackrel{\mathrm{def}}{=} \{(x, y) \in \mathcal{X} \times \mathcal{Y} : y \in F(x)\}. \tag{B.2}$$

Definition B.10 *The range of a set-valued map $F : \mathcal{X} \rightsquigarrow \mathcal{Y}$, denoted by* range($F$), *is defined as*

$$\text{range}(F) \stackrel{\text{def}}{=} \bigcup_{x \in \mathcal{X}} F(x). \tag{B.3}$$

Definition B.11 *A set-valued map $F : \mathcal{X} \rightsquigarrow \mathcal{Y}$ is upper semi-continuous (USC) at $x_0 \in \mathcal{X}$ if for any open $\mathcal{N}$ containing $F(x_0)$ there exists a neighborhood $\mathcal{M}$ of x_0 such that $F(\mathcal{M}) \subset \mathcal{N}$.*

Let $\mathcal{B}(y, d)$ with $y \in \mathbb{R}^m$ and $d \in \mathbb{R}_+$ represent the unit ball with center y and radius d in $\mathbb{R}^m$. With $F : \mathcal{X} \rightsquigarrow \mathcal{Y}$, define

$$\bar{\mathcal{B}}(F(x), d) = \{y : \mathcal{B}(y, d) \cap F(x) \neq \emptyset\} \tag{B.4}$$

for $x \in \mathcal{X}$ and $d \in \mathbb{R}_+$.

Definition B.12 *A set-valued map $F : \mathcal{X} \rightsquigarrow \mathcal{Y}$ is locally Lipschitz if for any $x_0 \in \mathcal{X}$, there exist a neighborhood $\mathcal{N}(x_0) \subset \mathcal{X}$ and a constant $L \geq 0$ such that for any $x, x' \in \mathcal{N}(x_0)$,*

$$F(x) \subseteq \bar{\mathcal{B}}(F(x'), L|x - x'|). \tag{B.5}$$

Definition B.13 *A set-valued map $F : \mathcal{X} \rightsquigarrow \mathcal{Y}$ is Lipschitz if there exists a constant $L \geq 0$ such that for any $x, x' \in \mathcal{X}$,*

$$F(x) \subseteq \bar{\mathcal{B}}(F(x'), L|x - x'|). \tag{B.6}$$

B.2 EXTENDED FILIPPOV SOLUTION

Consider the continuous-time system

$$\dot{x} = f(x, u), \tag{B.7}$$

where $x \in \mathbb{R}^n$ is the state and $u \in \mathbb{R}^m$ is the control input.

The vector field f is assumed to be a piecewise continuous function from $\mathbb{R}^n \times \mathbb{R}^m$ to $\mathbb{R}^n$ in the sense that

$$f(x, u) = f^i(x, u) \tag{B.8}$$

when $[x^T, u^T]^T \in \Omega^i$, $i \in M$, where $\bar{M} = \{1, \ldots, M\}$, $\Omega^1, \ldots, \Omega^M$ are closed subsets of $\mathbb{R}^n \times \mathbb{R}^m$ satisfying $\bigcup_{i \in \bar{M}} \Omega^i = \mathbb{R}^n \times \mathbb{R}^m$ and $\text{int}(\Omega^i) \cap \text{int}\Omega^j = \emptyset$ for $i \neq j$. The dynamics of system (B.7) are discontinuous and the system is called a discontinuous system for convenience.

It is assumed that $f^i : \Omega^i \rightarrow \mathbb{R}^n$ is locally Lipschitz on the domain Ω^i, and $\text{cl}(\text{int}(\Omega^i)) = \Omega^i$ for all $i \in \bar{M}$.

The following differential inclusion is used to define the extended Filippov solution:

$$\dot{x} \in F(x, u) \tag{B.9}$$

with

$$F(x, u) := \overline{\text{co}}\{f^i(x, u) : i \in I(x, u)\}, \tag{B.10}$$

$$I(x, u) := \{i \in \bar{M} : [x^T, u^T]^T \in \Omega^i\}. \tag{B.11}$$

The definition of F implies that

1. F is strict, that is, $\text{Dom}(F) = \mathbb{R}^n \times \mathbb{R}^m$;
2. $F(x, u)$ is a compact convex subset of $\mathbb{R}^n$ for every pair (x, u) in $\mathbb{R}^n \times \mathbb{R}^m$;
3. F is upper semi-continuous.

Definition B.14 *A function* $x : [a, b] \to \mathbb{R}^n$ *is an extended Filippov solution to the discontinuous system* (B.7) *with* $u : \mathbb{R}_+ \to \mathbb{R}^m$ *measurable and essentially ultimately bounded, if* x *is locally absolutely continuous and satisfies*

$$\dot{x}(t) \in F(x(t), u(t)) \tag{B.12}$$

for almost all $t \in [a, b]$.

For a general discontinuous $f(x, u)$, the set-valued map $F(x, u)$ can be defined as

$$F(x, u) = \bigcap_{\epsilon > 0} \bigcap_{\mu(\tilde{\mathcal{M}}=0)} \overline{\text{co}} f(\mathcal{B}_\epsilon(x, u) \backslash \tilde{\mathcal{M}}), \tag{B.13}$$

where $\mathcal{B}_\epsilon(x, u)$ is an open ball of radius ϵ around (x, u), and $\tilde{\mathcal{M}}$ represents all sets of zero measure (i.e., $\mu(\tilde{\mathcal{M}}) = 0$).

Compared with the standard Filippov solution [60], the definition of the extended Filippov solution takes into account both state x and input u. This treatment is helpful for the study of interconnected discontinuous systems. It should be noted that if the system does not have external input, then the definition of extended Filippov solution is reduced to the definition of Filippov solution. See also books [281, 33] and the tutorial [38] for more basic concepts of discontinuous systems. For related concepts in set-valued maps, see [13].

B.3 INPUT-TO-STATE STABILITY

Definition B.15 *System* (B.7) *is said to be ISS if there exist* $\beta \in \mathcal{KL}$ *and* $\chi \in \mathcal{K}$ *such that for any initial state* $x(0) = x_0$ *and any measurable and locally essentially bounded* u, *the extended Filippov solution exits and satisfies*

$$|x(t)| \leq \max\{\beta(|x_0|, t), \chi(\|u\|)\} \tag{B.14}$$

for $t \geq 0$.

An ISS-Lyapunov function candidate of a discontinuous system can be piecewisely defined as:

$$V(x) = V^i(x) \qquad (B.15)$$

when $x \in \Gamma^i$ for $i \in \bar{M}_V$, where $\bar{M}_V = \{1, \ldots, M_V\}$, and $\Gamma^1, \ldots, \Gamma^{M_V}$ are closed subsets of $\mathbb{R}^n$ satisfying $\bigcup_{i \in \bar{M}_V} \Gamma^i = \mathbb{R}^n$ and $\text{int}(\Gamma^i) \cap \text{int}(\Gamma^j) = \emptyset$ for $i \neq j$.

Each V_i is supposed to be continuous differentiable on some open domain containing Γ^i for $i \in \bar{M}_V$. Moreover, it is assumed that

$$V_i(x) = V_j(x) \qquad (B.16)$$

when $x \in \Gamma^i \cap \Gamma^j \neq \emptyset$. Define

$$J(x) = \{j \in \bar{M}_V : x \in \Gamma_j\}. \qquad (B.17)$$

Definition B.16 *The ISS-Lyapunov function candidate V of form (B.15) is said to be an ISS-Lyapunov function for system (B.7) if*

1. *V is locally Lipschitz;*
2. *V is positive definite and radially unbounded, that is, there exist $\underline{\alpha}, \bar{\alpha} \in \mathcal{K}_\infty$ such that for all $x \in \mathbb{R}^n$,*

$$\underline{\alpha}(|x|) \leq V(x) \leq \bar{\alpha}(|x|), \quad \forall x; \qquad (B.18)$$

3. *there exist a $\gamma \in \mathcal{K}$ and a continuous, positive definite α such that for almost all x, u,*

$$V(x) \geq \gamma(|u|) \Rightarrow \nabla V(x) f^i(x, u) \leq -\alpha(V(x)), \quad \forall i \in I(x, u). \qquad (B.19)$$

Theorem B.1 *System (B.7) is ISS if it admits an ISS-Lyapunov function.*

The proof of Theorem B.1 can be found in [84].

For a discontinuous system (B.7) with its Filippov solution defined by differential inclusion (B.9), condition (B.19) can be replaced with

$$V(x) \geq \gamma(|u|) \Rightarrow \max_{f \in F(x,u)} \nabla V(x) f \leq -\alpha(V(x)). \qquad (B.20)$$

B.4 LARGE-SCALE DYNAMIC NETWORKS OF DISCONTINUOUS SUBSYSTEMS

Consider the network of discontinuous subsystems

$$\dot{x}_i = f_i(x, u_i) = f_i^{j_i}(x, u_i), \quad i = 1, \ldots, N \qquad (B.21)$$

when $[x^T, u_i^T]^T \in \Omega_i^{j_i}$ for $j_i \in \bar{M}_i$, where $x_i \in \mathbb{R}^{n_i}$ is the state of the i-th subsystem, $x = [x_1^T, \ldots, x_N^T]^T$ is the state of the dynamic network, $u_i \in \mathbb{R}^{m_i}$ is the input of the x_i-subsystem. By considering u_i as the input of the i-th subsystem, assume that u_i is measurable and locally essentially bounded.

The dynamic network with state $x = [x_1^T, \ldots, x_N^T]^T$ and external input $u = [u_1^T, \ldots, u_N^T]^T$ can be rewritten as

$$\dot{x} = [f_1^{j_1 T}(x, u_1), \ldots, f_N^{j_N T}(x, u_N)]^T$$
$$:= f(x, u) \tag{B.22}$$

when $[x^T, u^T]^T \in \Omega^{(j_1, \ldots, j_N)}$ with

$$\Omega^{(j_1, \ldots, j_N)} := \left\{ [x^T, u^T]^T : [x^T, u_i^T]^T \in \Omega_i^{j_i}, i = 1, \ldots, N \right\} \tag{B.23}$$

for each combination of $(j_1, \ldots, j_N) \in \bar{M}^1 \times \cdots \bar{M}^N$.

Define

$$F_i(x, u_i) = \overline{\text{co}}\{ f_i^{j_i} : j_i \in I_i(x, u_i) \} \tag{B.24}$$
$$I_i(x, u_i) = \{ j_i \in \bar{M}^i : [x^T, u_i^T]^T \in \Omega_i^{j_i} \} \tag{B.25}$$
$$F(x, u) = F_1(x, u_1) \times \cdots \times F_N(x, u_N). \tag{B.26}$$

Then, it can be proved that

1. F is strict, that is, $\text{dom}(F) = \mathbb{R}^n \times \mathbb{R}^{n_u}$;
2. $F(x, u)$ is a compact convex subset of $\mathbb{R}^n$ for every pair (x, u) in $\mathbb{R}^n \times \mathbb{R}^{n_u}$;
3. F is upper semi-continuous.

If $\text{cl}(\text{int}(\Omega_i^{j_i})) = \Omega_i^{j_i}$, then the extended Filippov solution $x : [a, b] \rightarrow \mathbb{R}^{\sum_{i=1, \ldots, N} n_i}$ of the discontinuous dynamic network can be defined by differential inclusion

$$\dot{x}(t) \in F(x, u) \tag{B.27}$$

for almost all $t \in [a, b]$.

C Technical Lemmas Related to Comparison Functions

Lemma C.1 *Consider $\chi \in \mathcal{K}$ and $\chi_i \in \mathcal{K} \cup \{0\}$ for $i = 1, \ldots, n$. If $\chi \circ \chi_i < \mathrm{Id}$ for $i = 1, \ldots, n$, then there exists a $\hat{\chi} \in \mathcal{K}_\infty$ such that $\hat{\chi} > \chi$, $\hat{\chi}$ is continuously differentiable on $(0, \infty)$, and $\hat{\chi} \circ \chi_i < \mathrm{Id}$ for $i = 1, \ldots, n$.*

Proof. Define $\bar{\chi}(s) = \max_{i=1,\ldots,n} \{\chi_i(s)\}$ for all $s \geq 0$. Then, $\bar{\chi} \in \mathcal{K} \cup \{0\}$ and $\chi \circ \bar{\chi} < \mathrm{Id}$. Following the proofs of Theorem 3.1 and Lemma A.1 in [126], one can find a $\hat{\chi} \in \mathcal{K}_\infty$ such that $\hat{\chi} > \chi$, $\hat{\chi}$ is continuously differentiable on $(0, \infty)$ and $\hat{\chi} \circ \bar{\chi} < \mathrm{Id}$. It is easy to verify that $\hat{\chi} \circ \chi_i < \mathrm{Id}$ for $i = 1, \ldots, n$. $\Diamond$

Lemma C.2 *Consider $\chi_{i1}, \chi_{i2} \in \mathcal{K} \cup \{0\}$ for $i = 1, \ldots, n$. If $\chi_{i1} \circ \chi_{i2} < \mathrm{Id}$ for $i = 1, \ldots, n$, then there exists a positive definite function η such that $(\mathrm{Id} - \eta) \in \mathcal{K}_\infty$ and $\chi_{i1} \circ (\mathrm{Id} - \eta)^{-1} \circ \chi_{i2} < \mathrm{Id}$ for $i = 1, \ldots, n$.*

Proof. Recall the fact that for any $\chi_1, \chi_2 \in \mathcal{K} \cup \{0\}$, $\chi_1 \circ \chi_2 < \mathrm{Id} \Leftrightarrow \chi_2 \circ \chi_1 < \mathrm{Id}$. Property $\chi_{i1} \circ (\mathrm{Id} - \eta)^{-1} \circ \chi_{i2} < \mathrm{Id}$ is equivalent to $(\mathrm{Id} - \eta)^{-1} \circ \chi_{i2} \circ \chi_{i1} < \mathrm{Id}$.

Define $\chi_0(s) = \min\{\frac{1}{2}(\chi_{i1}^{-1} \circ \chi_{i2}^{-1}(s) + s)\}$ for $s \geq 0$. Obviously, $\chi_0 \in \mathcal{K}_\infty$. For all $i = 1, \ldots, n$, since $\chi_{i2} \circ \chi_{i1} < \mathrm{Id}$, we have $\chi_{i1}^{-1} \circ \chi_{i2}^{-1} > \mathrm{Id}$. Thus, $\chi_0 > \mathrm{Id}$. We also have $\chi_0 \circ \chi_{i2} \circ \chi_{i1} \leq \frac{1}{2}(\mathrm{Id} + \chi_{i2} \circ \chi_{i1}) < \mathrm{Id}$ for all $i = 1, \ldots, n$. Define $\bar{\eta} = \chi_0 - \mathrm{Id}$. Then, $\bar{\eta}$ is positive definite, $(\mathrm{Id} + \bar{\eta}) \in \mathcal{K}_\infty$, and $(\mathrm{Id} + \bar{\eta}) \circ \chi_{i2} \circ \chi_{i1} < \mathrm{Id}$ for $i = 1, \ldots, n$. The proof follows readily by defining $\eta = \mathrm{Id} - (\mathrm{Id} + \bar{\eta})^{-1}$, or equivalently $\eta = \bar{\eta} \circ (\mathrm{Id} + \bar{\eta})^{-1}$. $\Diamond$

Lemma C.3 *For any positive definite function α, and any class $\mathcal{K}_\infty$ function χ, there exists a positive definite function $\tilde{\alpha}$ such that $\chi(s') - \chi(s) \geq \tilde{\alpha}(s')$ for any pair of nonnegative numbers (s, s') satisfying $s' - s \geq \alpha(s')$.*

Proof. $s' - s \geq \alpha(s')$ can be written as $(\mathrm{Id} - \alpha)(s') \geq s$. Assume $(\mathrm{Id} - \alpha) \in \mathcal{K}$. (Otherwise, one can find an smaller α' to replace α such that $(\mathrm{Id} - \alpha') \in \mathcal{K}$.)

Note that $\chi^{-1} \circ \chi \circ (\mathrm{Id} - \alpha) = \mathrm{Id} - \alpha < \mathrm{Id}$ implies $\chi \circ (\mathrm{Id} - \alpha) \circ \chi^{-1} < \mathrm{Id}$. With Lemma C.2, we can find a positive definite function $\bar{\alpha}$ satisfying $(\mathrm{Id} - \bar{\alpha}) \in \mathcal{K}_\infty$, such that

$$(\mathrm{Id} - \bar{\alpha})^{-1} \circ \chi \circ (\mathrm{Id} - \alpha) \circ \chi^{-1} < \mathrm{Id}. \tag{C.1}$$

Consequently,

$$\chi \circ (\mathrm{Id} - \alpha) < (\mathrm{Id} - \bar{\alpha}) \circ \chi. \tag{C.2}$$

Define $\alpha_0 = \bar{\alpha} \circ \chi$. Then, α_0 is positive definite and for any one positive definite function $\alpha' \leq \alpha_0$,

$$(\chi - \alpha')(s') = (\mathrm{Id} - \bar{\alpha}) \circ \chi(s') \geq \chi \circ (\mathrm{Id} - \alpha)(s') \geq \chi(s) \qquad \text{(C.3)}$$

holds for any pair of nonnegative numbers (s, s') satisfying $s' - s \geq \alpha(s')$. $\Diamond$

Lemma C.4 *For any $\mathcal{K}_\infty$ function χ and any continuous, positive definite function ε satisfying $(\mathrm{Id} - \varepsilon) \in \mathcal{K}_\infty$, there exists a continuous, positive definite function μ satisfying $(\mathrm{Id} - \mu) \in \mathcal{K}_\infty$ such that $\chi \circ (\mathrm{Id} - \mu) = (\mathrm{Id} - \varepsilon) \circ \chi$.*

Proof. $(\mathrm{Id} - \varepsilon) \circ \chi \circ \chi^{-1} = \mathrm{Id} - \varepsilon < \mathrm{Id}$ implies $\chi^{-1} \circ (\mathrm{Id} - \varepsilon) \circ \chi < \mathrm{Id}$. The result is proved by defining $\mu = \mathrm{Id} - \chi^{-1} \circ (\mathrm{Id} - \varepsilon) \circ \chi$. $\Diamond$

Lemma C.5 *For any $\hat{\chi}_i \in \mathcal{K}_\infty$ and $\chi_i \in \mathcal{K}$ satisfying $\hat{\chi}_i > \chi_i$ for $i = 1, \ldots, n$, there exist continuous and positive definite κ satisfying $(\mathrm{Id} - \kappa) \in \mathcal{K}_\infty$ and continuous and positive definite κ' satisfying $(\mathrm{Id} - \kappa') \in \mathcal{K}_\infty$ such that $\hat{\chi}_i \circ (\mathrm{Id} - \kappa) > \chi_i$ and $(\mathrm{Id} - \kappa') \circ \hat{\chi}_i > \chi_i$ for $i = 1, \ldots, n$.*

Proof. From the proof of the Lemma A.1 in [126], there exists a $\mathcal{K}_\infty$ function $\bar{\chi}_i$ such that $\hat{\chi}_i > \bar{\chi}_i > \chi_i$. The proof is concluded by defining $\kappa(s) = \min_{i=1,\ldots,n} \{s - \hat{\chi}_i^{-1} \circ \bar{\chi}_i(s)\}$ and $\kappa'(s) = \min_{i=1,\ldots,n} \{s - \bar{\chi}_i \circ \hat{\chi}_i^{-1}(s)\}$ for $s \geq 0$. $\Diamond$

Lemma C.6 *For any $a, b \in \mathbb{R}$, if there exists a $\theta \in \mathcal{K}$ and a constant $c \geq 0$ such that*

$$|a - b| \leq \max\{\theta \circ (\mathrm{Id} + \theta)^{-1}(|a|), c\}, \qquad \text{(C.4)}$$

then

$$|a - b| \leq \max\{\theta(|b|), c\}. \qquad \text{(C.5)}$$

Proof. We first consider the case of $\theta \circ (\mathrm{Id} + \theta)^{-1}(|a|) \geq c$, which together with (C.4) implies

$$|a - b| \leq \theta \circ (\mathrm{Id} + \theta)^{-1}(|a|). \qquad \text{(C.6)}$$

In this case,

$$|a| - |b| \leq \theta \circ (\mathrm{Id} + \theta)^{-1}(|a|), \qquad \text{(C.7)}$$

and thus,

$$(\mathrm{Id} - \theta \circ (\mathrm{Id} + \theta)^{-1})(|a|) \leq |b|. \qquad \text{(C.8)}$$

Notice that $\text{Id}-\theta\circ(\text{Id}+\theta)^{-1} = (\text{Id}+\theta)\circ(\text{Id}+\theta)^{-1}-\theta\circ(\text{Id}+\theta)^{-1} = (\text{Id}+\theta)^{-1}$. Then, we have

$$|a| \leq (\text{Id} + \theta)(|b|). \tag{C.9}$$

By using (C.6) again, it can be achieved that

$$|a - b| \leq \theta(|b|). \tag{C.10}$$

Property (C.5) is then proved by also considering the case of $\theta \circ (\text{Id} + \theta)^{-1}(|a|) < c$, i.e., $|a - b| \leq c$. $\diamond$

Lemma C.7 *Consider a signal $\mu : [t_0, \infty) \to \mathbb{R}_+$ which is right-continuous and differentiable almost everywhere on $[t_0, \infty)\backslash\varpi$ with $\varpi = \{\tau_k : k \in \mathbb{Z}_+\} \subset [t_0, \infty)$ being a strictly increasing sequence. Suppose that there exists a constant $\omega \geq 0$ such that*

$$\mu(t) \geq \omega \Rightarrow \dot{\mu}(t) \leq -\varphi(\mu(t)) \tag{C.11}$$

for almost all $t \in [t_0, \infty)$, with φ being positive semi-definite and locally Lipschitz, and

$$\mu(t) \leq \max\{\mu(t^-), \omega\} \tag{C.12}$$

when $t \in \varpi$. Then,

$$\mu(t) \leq \max\{\eta(t), \omega\} \tag{C.13}$$

for all $t \in [t_0, \infty)$, where $\eta(t)$ is the unique solution of $\dot{\eta} = -\varphi(\eta)$ with $\eta(t_0) \geq \mu(t_0)$.

Proof. If $\mu(t_1) \leq \omega$ for some $t_1 \geq t_0$, then $\mu(t) \leq \omega$ for all $t \in [t_1, \infty)$. If $\mu(t_2) > \omega$ for some $t_2 \geq t_0$, then there exists a $t_3 > t_2$ such that $\mu(t) > \omega$ for all $t \in [t_0, t_3)$.

We study the case where $\mu(t_0) > \omega$. For any t_3 such that $\mu(t) > \omega$ for all $t \in [t_0, t_3)$, it holds that

$$\mu(t) \leq \max\{\mu(t^-), \omega\} \leq \mu(t^-) \tag{C.14}$$

for all $t \in [t_0, t_3) \cap \varpi$.

By using [155, Theorem 1.10.2] or [264, Lemma 1], we have

$$\mu(t) \leq \eta(t) \tag{C.15}$$

for all $t \in [t_0, t_3) \cap [\tau_k, \tau_{k+1})$ for any $k \in \mathbb{Z}_+$ satisfying $[t_0, t_3) \cap [\tau_k, \tau_{k+1}) \neq \emptyset$.

Inequalities (C.14) and (C.15) together imply $\mu(t) \leq \eta(t)$ for all $t \in [t_0, t_3)$ with any t_3 satisfying $\mu(t) > \omega$ for all $t \in [t_0, t_3)$.

If there exists a t_μ^* such that $\mu(t) > \omega$ for all $t \in [t_0, t_\mu^*)$ and $\mu(t) \leq \omega$ for all $t \in [t_\mu^*, \infty)$, then $\mu(t) \leq \eta(t)$ for all $t \in [t_0, t_\mu^*)$ and $\mu(t) \leq \omega$ for all $t \in [t_\mu^*, \infty)$. Thus, $\mu(t) \leq \max\{\eta(t), \omega\}$ for $t \in [t_0, \infty)$.

If $\mu(t) > \omega$ for all $t \in [t_0, \infty)$, then $\mu(t) \leq \eta(t)$ for all $t \in [t_0, \infty)$. $\diamond$

Lemma C.8 *For any function $\gamma \in \mathcal{K}$ and any $\delta > 0$, there exist constants $k, \delta' > 0$ and a continuously differentiable $\bar{\gamma} \in \mathcal{K}_\infty$ such that*

$$\gamma(s) \leq \bar{\gamma}(s) \ for \ s \geq \delta, \tag{C.16}$$

$$\bar{\gamma}(s) = ks \ for \ s \in [0, \delta'). \tag{C.17}$$

Moreover, if γ is linearly bounded near zero, then (C.16) can be satisfied with $\delta = 0$; if γ is globally bounded by a linear function, then (C.16) and (C.17) hold with $\delta = 0$ and $\delta' = \infty$. In particular, such functions can be taken convex.

See [123, Lemma 1] for the proof of Lemma C.8.

D Proofs of the Small-Gain Theorems 2.1, 3.2 and 3.6

D.1 A USEFUL TECHNICAL LEMMA

Lemma D.1 is used for the proofs of the trajectory-based small-gain theorems. See [265, Lemma 5.4] for the original version.

Lemma D.1 *Let $\beta \in \mathcal{KL}$, let $\rho \in \mathcal{K}$ such that $\rho < \mathrm{Id}$, and let μ be a real number in $(0,1]$. There exists a $\hat{\beta} \in \mathcal{KL}$ such that for any nonnegative real numbers s and d, and any nonnegative real function z defined on $[0,\infty)$ and satisfying*

$$z(t) \leq \max\{\beta(s,t), \rho(\|z\|_{[\mu t,\infty)}), d\} \tag{D.1}$$

for all $t \in [0,\infty)$, it holds that

$$z(t) \leq \max\{\hat{\beta}(s,t), d\} \tag{D.2}$$

for all $t \in [0,\infty)$.

The employment of this kind of technical lemma is motivated by the original small-gain result developed by [130]; see [130, Lemma A.1]. It should be noted that [130] mainly considers "plus"-type interconnections, while Lemma D.1 is used for the systems with "max"-type interconnections in this book. The major difference is that the signal $z(t)$ in [130, Lemma A.1] satisfies $z(t) \leq \beta(s,t) + \rho(\|z\|_{[\mu t,\infty)}) + d$ instead of (D.1), and the corresponding result is in the form of $z(t) \leq \hat{\beta}(s,t) + d'$ instead of (D.2).

D.2 PROOF OF THEOREM 2.1: THE ASYMPTOTIC GAIN APPROACH

Consider the interconnected system composed of two subsystems in the form of (2.14)–(2.15) satisfying (2.16). Assume that the small-gain condition (2.17) is satisfied.

Consider any specific initial state $x(0)$ and any piecewise continuous, bounded input u. Denote $x_i^* = \max\{\bar{\sigma}_{i1}(|x(0)|), \bar{\sigma}_{i2}(\|u\|_\infty)\}$ for $i = 1,2$.

Due to time invariance and causality properties, (2.16) implies

$$|x_i(t)| \leq \max\{\beta_i(|x_i(t_0)|, t - t_0), \gamma_{i(2-i)}(\|x_{2-i}\|_{[t_0,t]}), \gamma_i^u(\|u_i\|_\infty)\}. \tag{D.3}$$

Then, direct calculation yields:

$$\varlimsup_{t\to\infty} \|x_i\|_{[4t,8t]} = \varlimsup_{t\to\infty} \max_{0\le\tau\le 4t} |x(4t+\tau)|$$

$$\le \varlimsup_{t\to\infty} \max_{0\le\tau\le 4t} \{\beta_i(|x_i(2t)|, 2t+\tau),$$

$$\gamma_{i(2-i)}(\|x_{2-i}\|_{[2t,4t+\tau]}), \gamma_i^u(\|u_i\|_\infty)\}$$

$$\le \varlimsup_{t\to\infty} \max\{\beta_i(x_i^*, 2t), \gamma_{i(2-i)}(\|x_{2-i}\|_{[2t,8t]}), \gamma_i^u(\|u_i\|_\infty)\},$$

$$\text{(D.4)}$$

where

$$\|x_{2-i}\|_{[2t,8t]} = \max_{0\le\tau'\le 6t} |x(2t+\tau')|$$

$$\le \max_{0\le\tau'\le 6t} \{\beta_{2-i}(|x_{2-i}(t)|, t+\tau'),$$

$$\gamma_{(2-i)i}(\|x_i\|_{[t,2t+\tau']}), \gamma_{2-i}^u(\|u_{2-i}\|_\infty)\}$$

$$\le \max\{\beta_{2-i}(x_{2-i}^*, t), \gamma_{(2-i)i}(\|x_i\|_{[t,8t]}), \gamma_{2-i}^u(\|u_{2-i}\|_\infty)\}. \quad \text{(D.5)}$$

By substituting (D.5) into (D.4), one has

$$\varlimsup_{t\to\infty} \|x_i\|_{[4t,8t]} = \varlimsup_{t\to\infty} \max\{\beta_i(x_i^*, 2t), \gamma_{i(2-i)} \circ \beta_{2-i}(x_{2-i}^*, t),$$

$$\gamma_{i(2-i)} \circ \gamma_{(2-i)i}(\|x_i\|_{[t,8t]}),$$

$$\gamma_{i(2-i)} \circ \gamma_{2-i}^u(\|u_{2-i}\|_\infty), \gamma_i^u(\|u_i\|_\infty)\}. \quad \text{(D.6)}$$

Note that

$$\varlimsup_{t\to\infty} \|x_i\|_{[t,8t]} = \varlimsup_{t\to\infty} \max\{\|x_i\|_{[t,2t]}, \|x_i\|_{[2t,4t]}, \|x_i\|_{[4t,8t]}\}$$

$$= \varlimsup_{t\to\infty} \|x_i\|_{[4t,8t]} \quad \text{(D.7)}$$

since $\varlimsup_{t\to\infty} \|x_i\|_{[t,2t]} = \varlimsup_{t\to\infty} \|x_i\|_{[2t,4t]} = \varlimsup_{t\to\infty} \|x_i\|_{[4t,8t]}$.

Then, from (D.6), with the small-gain condition (2.17) satisfied, it holds that

$$\varlimsup_{t\to\infty} \|x_i\|_{[t,8t]} \le \max\{\varlimsup_{t\to\infty} \beta_i(x_i^*, 2t),$$

$$\varlimsup_{t\to\infty} \gamma_{i(2-i)} \circ \beta_{2-i}(x_{2-i}^*, t),$$

$$\varlimsup_{t\to\infty} \gamma_{i(2-i)} \circ \gamma_{(2-i)i}(\|x_i\|_{[t,8t]}),$$

$$\gamma_{i(2-i)} \circ \gamma_{2-i}^u(\|u_{2-i}\|_\infty), \gamma_i^u(\|u_i\|_\infty)\}$$

$$\le \max\{\varlimsup_{t\to\infty} \beta_i(x_i^*, 2t),$$

$$\varlimsup_{t\to\infty} \gamma_{i(2-i)} \circ \beta_{2-i}(x_{2-i}^*, t),$$

$$\gamma_{i(2-i)} \circ \gamma_{2-i}^u(\|u_{2-i}\|_\infty), \gamma_i^u(\|u_i\|_\infty)\}$$

$$= \max\{\gamma_{i(2-i)} \circ \gamma_{2-i}^u(\|u_{2-i}\|_\infty), \gamma_i^u(\|u_i\|_\infty)\}. \quad \text{(D.8)}$$

The AG property is proved as

$$
\varlimsup_{t\to\infty} |x_i(t)| \leq \varlimsup_{t\to\infty} \|x_i\|_{[t,8t]}
$$
$$
\leq \max\{\gamma_{i(2-i)} \circ \gamma_{2-i}^u(\|u_{2-i}\|_\infty), \gamma_i^u(\|u_i\|_\infty)\}. \tag{D.9}
$$

This ends the proof of Theorem 2.1.

D.3   SKETCH OF PROOF OF THEOREM 3.2

Inspired by [130], the cyclic-small-gain theorem for large-scale dynamic networks composed of IOS subsystems can be proved in two steps:

1. Forward completeness of the system and boundedness of solutions for all $t \in [0, \infty)$;
2. IOS of the large-scale dynamic network.

For large-scale dynamic networks, the results are proved by induction. This is motivated by the proof of the cyclic-small-gain theorem for output-Lagrange input-to-output stable (OLIOS) systems in [134].

D.3.1   FORWARD COMPLETENESS OF THE SYSTEM AND BOUNDEDNESS OF SOLUTIONS

Pick any initial state $x(0)$ and any measurable and locally essentially bounded u. Suppose that $x(t)$ is right maximally defined on $[0, T)$ with T possibly infinite.

By (3.71), for $i = 1, \ldots, n$, it holds that

$$
|y_i(t)| \leq \max_{j=1,\ldots,n; j\neq i} \left\{ \sigma_i(|x_i(0)|), \gamma_{ij}\left(\|y_j\|_{[0,T)}\right), \gamma_i^u\left(\|u_j\|_{[0,T)}\right) \right\} \tag{D.10}
$$

for $t \in [0, T)$, where $\sigma_i(s) = \beta_i(s, 0)$ for $s \in \mathbb{R}_+$. Clearly, $\sigma_i \in \mathcal{K}$.

We first consider the case of $n = 2$. In this case, $\gamma_{i(3-i)} \circ \gamma_{(3-i)i} < \text{Id}$ for $i = 1, 2$.

By (D.10), one has for $i = 1, 2$,

$$
|y_i(t)| \leq \max\left\{ \sigma_i(|x_i(0)|), \gamma_{i(3-i)}\left(\|y_{3-i}\|_{[0,T)}\right), \gamma_i^u\left(\|u_j\|_{[0,T)}\right) \right\} \tag{D.11}
$$

for $t \in [0, T)$. By taking the supremum of $|y_i|$ over $[0, T)$ and defining

$$
\bar{\sigma}_i(s) = \max\{\sigma_i(s), \gamma_{i(3-i)} \circ \sigma_{3-i}(s)\}, \tag{D.12}
$$

one has

$$
\|y_i\|_{[0,T)} \leq \max\Big\{ \sigma_i(|x(0)|), \gamma_i^u\left(\|u_i\|_{[0,T)}\right), \gamma_{i(3-i)} \circ \gamma_{(3-i)i}\left(\|y_i\|_{[0,T)}\right),
$$
$$
\gamma_{i(3-i)} \circ \gamma_{3-i}^u\left(\|u_{3-i}\|_{[0,T)}\right) \Big\}. \tag{D.13}
$$

Since $\gamma_{i(3-i)} \circ \gamma_{(3-i)i} < \mathrm{Id}$, one achieves

$$\|y_i\|_{[0,T)} \leq \max\left\{\bar{\sigma}_i(|x(0)|), \gamma_i^u\left(\|u_i\|_{[0,T)}\right), \gamma_{i(3-i)} \circ \gamma_{3-i}^u\left(\|u_{3-i}\|_{[0,T)}\right)\right\}. \tag{D.14}$$

Hence, for any initial state $x(0)$, and any measurable and locally essentially bounded u, y is bounded over the interval $[0, T)$. By using the UO property (3.70), the state x is bounded over $[0, T)$. This means that the maximum interval for the definition of x is $[0, \infty)$. It again follows (D.14) that there exists $\bar{\sigma}, \bar{\gamma} \in \mathcal{K}$ such that for any initial state $x(0)$, and any measurable and locally essentially bounded u,

$$|y(t)| \leq \max\{\bar{\sigma}(|x(0)|), \bar{\gamma}(\|u\|_\infty)\} \tag{D.15}$$

for all $t \geq 0$.

Suppose that for any dynamic network with $n = n^*$, the existence and boundedness of the solutions on $[0, \infty)$ can be proved and property (D.15) holds for all $t \geq 0$, if the subsystems for $i = 1, \ldots, n^*$ are UO in the sense of (3.70), have the property (D.10), and satisfy the cyclic-small-gain condition.

We consider a dynamic network with $n = n^* + 1$, with the subsystems being UO in the sense of (3.70), having property (D.10) and satisfying the cyclic-small-gain condition.

By (D.10), one has for $i = 1, \ldots, n^* + 1$,

$$|y_i(t)| \leq \max_{j=1,\ldots,n^*+1; j \neq i}\left\{\sigma_i(|x_i(0)|), \gamma_{ij}\left(\|y_j\|_{[0,T)}\right), \gamma_i^u\left(\|u_i\|_{[0,T)}\right),\right.$$
$$\left.\gamma_{i(n^*+1)}\left(\|y_{n^*+1}\|_{[0,T)}\right)\right\}, \tag{D.16}$$

and thus, for $i = 1, \ldots, n^*$,

$$|y_i(t)| \leq \max_{j=1,\ldots,n^*; j \neq i; l=1,\ldots,n^*}\left\{\sigma_i(|x_i(0)|), \gamma_{ij}\left(\|y_j\|_{[0,T)}\right), \gamma_i^u\left(\|u_i\|_{[0,T)}\right),\right.$$
$$\gamma_{i(n^*+1)} \circ \sigma_{n^*+1}(|x_{n^*+1}(0)|), \gamma_{i(n^*+1)} \circ \gamma_{(n^*+1)l}\left(\|y_l\|_{[0,T)}\right),$$
$$\left.\gamma_{i(n^*+1)} \circ \gamma_{n^*+1}^u\left(\|u_{n^*+1}\|_{[0,T)}\right)\right\}$$
$$\leq \max_{j=1,\ldots,n^*; j \neq i}\left\{\breve{\sigma}_i(|x(0)|), \breve{\gamma}_{ij}\left(\|y_j\|_{[0,T)}\right), \breve{\gamma}_i^u\left(\|u\|_{[0,T)}\right)\right\}, \tag{D.17}$$

where

$$\breve{\sigma}_i(s) = \max\left\{\sigma_i(s), \gamma_{i(n^*+1)} \circ \sigma_{n^*+1}(s)\right\}$$
$$\breve{\gamma}_{ij}(s) = \max_{j=1,\ldots,n^*; j \neq i}\left\{\gamma_{ij}(s), \gamma_{i(n^*+1)} \circ \gamma_{(n^*+1)j}(s)\right\}$$
$$\breve{\gamma}_i^u(s) = \max\left\{\gamma_i^u(s), \gamma_{i(n^*+1)} \circ \gamma_{n^*+1}^u(s)\right\} \tag{D.18}$$

for $s \in \mathbb{R}_+$. We used the cyclic-small-gain condition $\gamma_{i(n^*+1)} \circ \gamma_{(n^*+1)i} < \mathrm{Id}$ to get the last inequality of (D.17).

Consider the dynamic network with the n^* subsystems being UO with zero offset and having property (D.17). Note that the new IOS gains $\bar{\gamma}_{ij}$ still satisfy the cyclic-small-gain condition according to [134, Lemma 5.3].

According to the hypothesis, for any initial state $x(0)$ and any measurable and locally essentially bounded u, the solution $(x_1(t), \ldots, x_{n^*}(t))$ exists and is bounded for $t \geq 0$, which implies the existence and boundedness of $x_{n^*+1}(t)$ for $t \geq 0$ by using the UO and IOS properties of the (n^*+1)-th subsystem. Then, for the dynamic network with $n = n^* + 1$, property (D.15) holds for $t \geq 0$.

D.3.2 INPUT-TO-OUTPUT STABILITY

The UO of the subsystems in the sense of (3.70) implies the UO of the dynamic network, i.e., there exist $\alpha^O \in \mathcal{K}_\infty$ such that

$$|x(t)| \leq \alpha^O \left(|x(0)| + \|u\|_\infty + \|y\|_\infty \right) \tag{D.19}$$

for $t \geq 0$.

For any specific initial state $x(0)$ and any measurable and locally essentially bounded u, with (D.15) proved, we define

$$\bar{c} = \alpha^O \left(|x(0)| + \|u\|_\infty + \max \left\{ \bar{\sigma}(|x(0)|), \bar{\gamma} \left(\|u\|_\infty \right) \right\} \right). \tag{D.20}$$

Then,

$$x(t) \leq \bar{c} \tag{D.21}$$

for $t \geq 0$.

By using the time-invariance property, property (3.71) implies

$$|y_i(t)| \leq \max_{j \neq i} \{ \beta_i(|x_i(t_0)|, t - t_0), \gamma_{ij} \left(\|y_j\|_{[t_0, \infty)]} \right), \gamma_i^u \left(\|u_i\|_\infty \right) \} \tag{D.22}$$

for all $0 \leq t_0 \leq t$. By choosing $t_0 = \mu t$ with $\mu \leq 0.5$ and using (D.21), one has

$$|y_i(t)| \leq \max_{j \neq i} \{ \beta_i(\bar{c}, (1 - \mu)t), \gamma_{ij} \left(\|y_j\|_{[\mu t, \infty)]} \right), \gamma_i^u \left(\|u_i\|_\infty \right) \}$$
$$\leq \max_{j \neq i} \{ \beta_i(\bar{c}, \mu t), \gamma_{ij} \left(\|y_j\|_{[\mu t, \infty)]} \right), \gamma_i^u \left(\|u_i\|_\infty \right) \} \tag{D.23}$$

for $t \geq 0$.

With the satisfaction of the cyclic-small-gain condition, by proving the existence of $\bar{\beta}_i \in \mathcal{KL}$, $\bar{\chi}_i \in \mathcal{K}$ satisfying $\bar{\chi}_i < \mathrm{Id}$, $\bar{\gamma}_i^u \in \mathcal{K}$, and $0 < \mu_i \leq 1$ such that

$$|y_i(t)| \leq \max \left\{ \bar{\beta}_i(\bar{c}, t), \bar{\chi}_i \left(\|y_i\|_{[\mu_i t, \infty)} \right), \bar{\gamma}_i^u \left(\|u\|_\infty \right) \right\} \tag{D.24}$$

for $i = 1, \ldots, n$, we can use Lemma D.1 to prove the IOS of the dynamic network.

Consider the case of $n = 2$. In this case, for $i = 1, 2$,

$$|y_i(t)| \leq \max\left\{\beta_i(\bar{c}, \mu t), \gamma_{i(3-i)}\left(\|y_{3-i}\|_{[\mu t, \infty)}\right), \gamma_i^u\left(\|u_i\|\right)\right\}, \qquad (D.25)$$

and

$$\|y_i\|_{[\mu t, \infty)} \leq \max_{\mu t \leq \tau < \infty}\left\{\beta_i(\bar{c}, \mu\tau), \gamma_{i(3-i)}\left(\|y_{3-i}\|_{[\mu\tau, \infty)}\right), \gamma_i^u\left(\|u_i\|\right)\right\}$$
$$\leq \max\left\{\beta_i(\bar{c}, \mu^2 t), \gamma_{i(3-i)}\left(\|y_{3-i}\|_{[\mu^2 t, \infty)}\right), \gamma_i^u\left(\|u_i\|\right)\right\}. \qquad (D.26)$$

By substituting (D.26) with i replaced by $3 - i$ into (D.25), one has

$$|y_i(t)| \leq \max\Big\{\beta_i(\bar{c}, \mu t), \gamma_{i(3-i)} \circ \beta_{3-i}(\bar{c}, \mu^2 t), \gamma_{i(3-i)} \circ \gamma_{(3-i)i}\left(\|y_i\|_{[\mu^2 t, \infty)}\right),$$
$$\gamma_{i(3-i)} \circ \gamma_{3-i}^u\left(\|u_{3-i}\|_\infty\right), \gamma_i^u\left(\|u_i\|_\infty\right)\Big\}. \qquad (D.27)$$

Thus, property (D.24) is proved for $n = 2$.

Suppose that property (D.24) can be proved for $n = n^*$. Consider the case of $n = n^* + 1$. In this case, by still using the time-invariance property, one has for $i = 1, \ldots, n^*$, Then,

$$|y_i(t)| \leq \max_{j=1,\ldots,n^*, j \neq i}\Big\{\beta_i(\bar{c}, \mu t), \gamma_{ij}\left(\|y_j\|_{[\mu t, \infty)}\right), \gamma_{i(n^*+1)}\left(\|y_{n^*+1}\|_{[\mu t, \infty)}\right),$$
$$\gamma_i^u\left(\|u_i\|_\infty\right)\Big\} \qquad (D.28)$$

for $i = 1, \ldots, n^*$, and

$$\|y_{n^*+1}\|_{[\mu t, \infty)} = \max_{\mu t \leq \tau < \infty} |y_{n^*+1}(\tau)|$$
$$\leq \max_{j=1,\ldots,n^*}\Big\{\beta_{n^*+1}(\bar{c}, \mu^2 t), \gamma_{(n^*+1)j}\left(\|y_j\|_{[\mu^2 t, \infty)}\right),$$
$$\gamma_{n^*+1}^u\left(\|u_{n^*+1}\|_\infty\right)\Big\}. \qquad (D.29)$$

By substituting (D.29) into (D.28), one has

$$|y_i(t)| \leq \max_{j=1,\ldots,n^*, j \neq i}\Big\{\hat{\beta}_i(\bar{c}, t), \hat{\gamma}_{ij}\left(\|y_j\|_{[\mu^2 t, \infty)}\right), \hat{\gamma}_i^u\left(\|u\|_\infty\right)\Big\} \qquad (D.30)$$

for $i = 1, \ldots, n^*$, where

$$\hat{\beta}_i(s, t) = \max\left\{\beta_i(s, \mu t), \gamma_{i(n^*+1)} \circ \beta_{n^*+1}(s, \mu^2 t)\right\} \qquad (D.31)$$
$$\hat{\gamma}_{ij}(s) = \max\left\{\gamma_{ij}(s), \gamma_{i(n^*+1)} \circ \gamma_{(n^*+1)j}(s)\right\} \qquad (D.32)$$
$$\hat{\gamma}_i^u(s) = \max\left\{\gamma_i^u(s), \gamma_{i(n^*+1)} \circ \gamma_{n^*+1}^u(s)\right\} \qquad (D.33)$$

for $s, t \geq 0$. Note that the new IOS gains $\hat{\gamma}_{ij}$ still satisfy the cyclic-small-gain condition according to [134, Lemma 5.3]. For the dynamic network composed of n^* subsystems satisfying (D.30), there exist $\bar{\beta}_i \in \mathcal{KL}$, $\bar{\chi}_i \in \mathcal{K}$ satisfying

$\bar{\chi}_i < \text{Id}$, $\bar{\gamma}_i^u \in \mathcal{K}$, and $0 < \mu_i \leq 1$ so that (D.24) holds for $i = 1, \ldots, n^*$. Note that one can pick the n^* subsystems arbitrarily from the $n^* + 1$ subsystems. Property (D.24) can be proved for $i = 1, \ldots, n^* + 1$.

By induction, property (D.24) is proved for each i-th subsystem with $i = 1, \ldots, n$ of the dynamic network under the cyclic-small-gain condition.

Then, with Lemma D.1, for $i = 1, \ldots, n$, there exist $\tilde{\beta}_i \in \mathcal{KL}$ and $\tilde{\gamma}_i^u \in \mathcal{K}$ such that

$$|y_i(t)| \leq \max \left\{ \tilde{\beta}_i(\bar{c}, t), \tilde{\gamma}_i^u \left(\|u\|_\infty \right) \right\} \tag{D.34}$$

for $t \geq 0$, and thus, there exist $\tilde{\beta} \in \mathcal{KL}$ and $\tilde{\gamma}^u \in \mathcal{K}$ such that

$$|y(t)| \leq \max \left\{ \tilde{\beta}(\bar{c}, t), \tilde{\gamma}^u \left(\|u\|_\infty \right) \right\} \tag{D.35}$$

for $t \geq 0$.

Recall the definition of $\bar{c}$ in (D.20). One can find $\beta \in \mathcal{KL}$ and $\gamma \in \mathcal{K}$ so that for any initial state $x(0)$ and any measurable and locally essentially bounded u,

$$|y(t)| \leq \max \left\{ \beta(|x(0)|, t), \gamma^u \left(\|u\|_\infty \right) \right\} \tag{D.36}$$

for $t \geq 0$. This ends the proof of Theorem 3.2.

D.4 PROOF OF THEOREM 3.6

For convenience of notation, we denote $V_{CH}(x) = \max_{i \in \mathcal{N}_{CH}} \{V_i(x_i)\}$ with $\mathcal{N}_{CH} = \mathcal{N}_C \cup \mathcal{N}_H$ and $V_D(x) = \max_{i \in \mathcal{N}_D} \{V_i(x_i)\}$. Then, $V(x) = \max \{V_{CH}(x), V_D(x)\}$. Denote $f(x, u) = [f_1^T(x, u_1), \ldots, f_N^T(x, u_N)]^T$.

Proof of Property 1

Under the conditions of Property 1, consider $t \notin \pi$.

Simply denote $x(t, t_0, \xi, u)$ as $x(t)$. We study the decreasing property of $V(x(t))$ at time t in the case of $V(x(t)) \geq \bar{u}(t)$ when $t \notin \pi$.

At time t, define

$$A = \{j \in \mathcal{N} : V_j(x_j(t)) = V(x(t))\}. \tag{D.37}$$

Recall that $V(x(t)) = \max \mathbb{V}(x(t)) = \max_{i \in \mathcal{N}} \{V_i(x_i(t))\}$. Since all the γ_{ij}'s $(i, j \in \mathcal{N}, i \neq j)$ are less than Id, it holds that

$$V_j(x_j(t)) \geq \max_{l \neq j} \{V_l(x_l(t))\} > \max_{l \neq j} \{\gamma_{jl}(V_l(x_l(t)))\} \tag{D.38}$$

for all $j \in A$. Furthermore, in the case of $V(x(t) \geq \bar{u}(t)$, with the definition of $\bar{u}$ in (3.161), it holds that $V_j(x_j(t)) = V(x(t)) \geq \bar{u}(t) \geq \gamma_{u_j}(|u_j(t)|)$ for all $j \in A$. It follows that

$$V_j(x_j(t)) \geq \max_{l \neq j} \{\gamma_{jl}(V_l(x_l(t))), \gamma_{u_j}(|u_j(t)|)\}. \tag{D.39}$$

If ∇V is defined at $x(t)$, then

$$\nabla V(x(t))f(x(t),u(t)) = \frac{d}{dr}\bigg|_{r=0} V(\phi(r)), \qquad (D.40)$$

where $\phi(r) = [\phi_1(r),\dots,\phi_N(r)]^T$ is the continuous solution of the initial-value problem

$$\dot{\phi}(r) = f(\phi(r),u(t)), \quad \phi(0) = x(t). \qquad (D.41)$$

In the case of $\mathcal{N}_{CH}\cap A \neq \emptyset$, for $j\in\mathcal{N}_{CH}\cap A$, if ∇V_j is well defined at $x_j(t)$, then from (3.125) and (D.39), it holds that

$$\nabla V_j(x_j(t))f_j(x(t),u_j(t)) \leq -\alpha_j(V_j(x_j(t))). \qquad (D.42)$$

Considering the continuity of f_j and the continuity of ∇V_j at $x_j(t)$, there exists a neighborhood $\mathcal{X} = \mathcal{X}_1\times\cdots\times\mathcal{X}_N$ of $x(t)$ such that for $j\in\mathcal{N}_{CH}\cap A$,

$$\nabla V_j(\zeta_j)f_j(\zeta,u_j(t)) \leq -\frac{1}{2}\alpha_j(V_j(x_j(t))) \qquad (D.43)$$

holds for all $\zeta = [\zeta_1^T,\dots,\zeta_N^T]^T \in \mathcal{X}$.

Because of the continuity of $\phi(r)$, there exists a $\bar{\delta} > 0$ such that $\phi(r)\in\mathcal{X}$ and

$$\max_{j\in A}\{V_j(\phi_j(r))\} = \max_{j\in\mathcal{N}}\{V_j(\phi_j(r))\} = V(\phi(r)) \qquad (D.44)$$

for $r\in[0,\bar{\delta})$.

For $j\in\mathcal{N}_{CH}\cap A$, from (D.43),

$$\frac{V_j(\phi_j(r)) - V_j(x_j(t))}{r} \leq -\frac{1}{2}\alpha_j(V_j(x_j(t))) \qquad (D.45)$$

holds for $r\in(0,\bar{\delta})$.

For $j\in\mathcal{N}_D$, since $f_j\equiv 0$,

$$\frac{V_j(\phi_j(r)) - V_j(x_j(t))}{r} = 0 \qquad (D.46)$$

holds for $r\in(0,\bar{\delta})$.

From (D.45) and (D.46), we have

$$\frac{V(\phi(r)) - V(x(t))}{r} = \frac{\max_{j\in A}\{V_j(\phi_j(r))\} - \max_{j\in A}\{V_j(x_j(t))\}}{r}$$
$$\leq 0 \qquad (D.47)$$

for $r\in(0,\bar{\delta})$. Hence, if ∇V is well defined at $x(t)$, then

$$V(x(t)) \geq \bar{u}(t) \Rightarrow \nabla V(x(t))f(x(t),u(t)) \leq 0. \qquad (D.48)$$

Note that $V(x)$ is smooth almost everywhere. By using the results of [86, Lemma 1] or [264, Section 2]), from (D.48), we conclude $V(x(t))$ is differentiable almost everywhere on the timeline and we achieve for any ξ, u, $t_0 \geq 0$,

$$V(x(t, t_0, \xi, u)) \geq \|\bar{u}\|_{[t_0, t]} \Rightarrow \dot{V}(x(t, t_0, \xi, u)) \leq 0 \qquad (D.49)$$

holds for almost all $t \in [t_0, \infty) \backslash \pi$. Property 1 in Theorem 3.6 is proved.

Proof of Property 2

Under the conditions of Property 2, consider $t \in \pi$.

For any ξ, u, and $t_0 \geq 0$, if $t > t_0$ and $t \notin \pi_i$, then $x_i(t, t_0, \xi, u)$ is continuous at time t and

$$V_i(x_i(t, t_0, \xi, u)) = V_i(x_i(t^-, t_0, \xi, u)) \leq V(x(t^-, t_0, \xi, u)); \qquad (D.50)$$

else if $t > t_0$ and $t \in \pi_i$ (obviously, $i \in \mathcal{N}_D \cup \mathcal{N}_H$), then

$$V_i(x_i(t, t_0, \xi, u))$$
$$= V_i(g_i((x_i(t^-, t_0, \xi, u), u_i(t^-))))$$
$$\leq (\mathrm{Id} - \rho_i)(\max_{l \neq i}\{\gamma_{il}(V_l(x_l(t^-, t_0, \xi, u))), V_i(x_i(t^-, t_0, \xi, u)), \gamma_{u_i}(|u_i(t^-)|)\})$$
$$\leq \max\{V(x(t^-, t_0, \xi, u)), \bar{u}(t^-)\}. \qquad (D.51)$$

From (D.50) and (D.51), when $t > t_0$ and $t \in \pi$, we obtain

$$V(x(t, t_0, \xi, u)) = \max_{i \in \mathcal{N}}\{V_i(x_i(t, t_0, \xi, u))\}$$
$$\leq \max\{V(x(t^-, t_0, \xi, u)), \bar{u}(t^-)\}. \qquad (D.52)$$

Property 2 in Theorem 3.6 is proved.

Uniformly Bounded-Input Bounded-State (UBIBS) Property

Note that $\bar{u}(t^-) \leq \|\bar{u}\|_{[t_0, t]}$ and $\bar{u}(t) \leq \|\bar{u}\|_{[t_0, t]}$ hold for any $t > t_0 \geq 0$. With Lemma C.7, by considering $V(x(t, t_0, \xi, u))$ as $\mu(t)$ and $\|\bar{u}\|_{[t_0, t]}$ as ω, from the properties (D.49) and (D.52), for any ξ, u, and $t_0 \geq 0$, we have

$$V(x(t, t_0, \xi, u)) \leq \max\{V(\xi), \|\bar{u}\|_{[t_0, t]}\} \qquad (D.53)$$

for all $t > t_0$. Note that $V(x(t_0, t_0, \xi, u)) = V(\xi)$. Thus, for any ξ, u, and $t_0 \geq 0$, property (D.53) holds for all $t \geq t_0$.

Proof of Property 3

Define

$$t_{D0} = \min\{t_{i0} : i \in \mathcal{N}_D\} - \delta \qquad (D.54)$$
$$t_{D(w+1)} - t_{Dw} = \overline{\delta t} + \delta, \quad w \in \mathbb{Z}_+ \backslash \{0\} \qquad (D.55)$$
$$\bar{\pi} = \{t_{D(2w)} : w \in \mathbb{Z}_+\}, \qquad (D.56)$$

where $0 < \delta < \min\{t_{i0} : i \in \mathcal{N}_D\}$ can be arbitrarily close to zero. The definition of t_{Dw} means that for each $i \in \mathcal{N}_D$, $(t_{D(2w)}, t_{D(2w+1)}] \cap \pi_i \neq \emptyset$ and of course $(t_{D(2w)}, t_{D(2w+2)}] \cap \pi_i \neq \emptyset$.

Denote $\Delta = [t_{D(2w)}, t_{D(2w+2)}]$. Property (D.53) implies that for any ξ and u,

$$V(x(t, t_{D(2w)}, \xi, u)) \leq \max\{V(\xi), \|\bar{u}\|_\Delta\} \tag{D.57}$$

for $t \in \Delta$.

If $V(\xi) \leq \|\bar{u}\|_\Delta$, then

$$V(x(t, t_{D(2w)}, \xi, u)) \leq \|\bar{u}\|_\Delta. \tag{D.58}$$

For each $i \in \mathcal{N}_D$, in the case of $V(\xi) > \|\bar{u}\|_\Delta$, we have

$$
\begin{aligned}
&V_i(x_i(t, t_{D(2w)}, \xi, u)) \\
=&V_i(g_i(x(t^-, t_{D(2w)}, \xi, u), u_i(t^-))) \\
\leq&(\mathrm{Id} - \rho_i)(\max\{V(x(t^-, t_{D(2w)}, \xi, u)), \gamma_{u_i}(|u_i(t^-)|)\}) \\
\leq&(\mathrm{Id} - \rho_i)(\max\{V(\xi), \|\bar{u}\|_\Delta\}) \\
=&(\mathrm{Id} - \rho_i)(V(\xi))
\end{aligned} \tag{D.59}
$$

for all $t \in (t_{D(2w)}, t_{D(2w+2)}) \cap \pi_i$.

Note that the state of each x_i-subsystem ($i \in \mathcal{N}_D$) keeps constant when $t \notin \pi_i$. For $i \in \mathcal{N}_D$, because $(t_{D(2w)}, t_{D(2w+1)}] \cap \pi_i \neq \emptyset$, from (D.59), we get

$$V_i(x_i(t, t_{D(2w)}, \xi, u)) \leq \max\{(\mathrm{Id} - \rho_i)(V(\xi)), \|\bar{u}\|_\Delta\}, \quad i \in \mathcal{N}_D \tag{D.60}$$

for $t \in [t_{D(2w+1)}, t_{D(2w+2)}]$.

Define $\rho(s) = \min_{i \in \mathcal{N}_D}\{\rho_i(s)\}$ for $s \geq 0$. Then,

$$V_D(x(t, t_{D(2w)}, \xi, u)) \leq \max\{(\mathrm{Id} - \rho)(V(\xi)), \|\bar{u}\|_\Delta\} \tag{D.61}$$

for $t \in [t_{D(2w+1)}, t_{D(2w+2)}]$. Clearly, ρ is continuous and positive definite, and $(\mathrm{Id} - \rho) \in \mathcal{K}_\infty$.

For $i \in \mathcal{N}_{CH}$, when $t \in [t_{D(2w+1)}, t_{D(2w+2)}] \backslash (\bigcup_{i \in \mathcal{N}_H}\{\pi_i\})$, only the continuous-time dynamics work. Consider $V_{CH}(x)$ as the ISS-Lyapunov function and V_D and u as the inputs of the interconnection of the continuous-time subsystems ($i \in \mathcal{N}_C$) and the hybrid subsystems ($i \in \mathcal{N}_H$). Using the cyclic-small-gain theorem for continuous-time dynamic networks in Section 3.1 and Property (D.61), we can find a continuous and positive definite function α_{CH} such that if ∇V_{CH} is defined at $x(t, t_{D(2w)}, \xi, u)$, then

$$
\begin{aligned}
&V_{CH}(x(t, t_{D(2w)}, \xi, u)) \geq \max\{V_D(x(t, t_{D(2w)}, \xi, u)), \bar{u}(t)\} \\
\Rightarrow\quad &\nabla V_{CH}(x(t, t_{D(2w)}, \xi, u)) f_{CH}(x(t, t_{D(2w)}, \xi, u), u(t)) \\
&\leq -\alpha_{CH}(V_{CH}(x(t, t_{D(2w)}, \xi, u)))
\end{aligned} \tag{D.62}
$$

holds for $t \in [t_{D(2w+1)}, t_{D(2w+2)}]\backslash(\bigcup_{i \in \mathcal{N}_H}\{\pi_i\})$, where f_{CH} is the vector of the continuous-time dynamics of the continuous-time subsystems and the hybrid subsystems.

Using (D.61) and $\bar{u}(t) \le \|\bar{u}\|_\Delta$, we have

$$V_{CH}(x(t, t_{D(2w)}, \xi, u)) \ge \max\{(\mathrm{Id} - \rho)(V(\xi)), \|\bar{u}\|_\Delta\}$$
$$\Rightarrow \quad \nabla V_{CH}(x(t, t_{D(2w)}, \xi, u))f_{CH}(x(t, t_{D(2w)}, \xi, u), u(t))$$
$$\le -\alpha_{CH}(V_{CH}(x(t, t_{D(2w)}, \xi, u))) \tag{D.63}$$

for $t \in [t_{D(2w+1)}, t_{D(2w+2)}]\backslash(\bigcup_{i \in \mathcal{N}_H}\{\pi_i\})$.

By using [86, Lemma 1] or [264, Section 2], we have $V_{CH}(x(t, t_{D(2w)}, \xi, u))$ is differentiable almost everywhere on the timeline and

$$V_{CH}(x(t, t_{D(2w)}, \xi, u)) \ge \max\{(\mathrm{Id} - \rho)(V(\xi)), \|\bar{u}\|_\Delta\}$$
$$\Rightarrow \dot{V}_{CH}(x(t, t_{D(2w)}, \xi, u)) \le -\alpha_{CH}(V_{CH}(x(t, t_{D(2w)}, \xi, u))) \tag{D.64}$$

for almost all $t \in [t_{D(2w+1)}, t_{D(2w+2)}]\backslash(\bigcup_{i \in \mathcal{N}_H}\{\pi_i\})$.

For $t \in [t_{D(2w+1)}, t_{D(2w+2)}] \cap (\bigcup_{i \in \mathcal{N}_H}\{\pi_i\})$, from (D.52) and (D.61), we have

$$V_{CH}(x(t, t_{D(2w)}, \xi, u))$$
$$\le V(x(t, t_{D(2w)}, \xi, u))$$
$$\le \max\{V(x(t^-, t_{D(2w)}, \xi, u)), \bar{u}(t^-)\}$$
$$= \max\{V_{CH}(x(t^-, t_{D(2w)}, \xi, u)), V_D(x(t^-, t_{D(2w)}, \xi, u)), \bar{u}(t^-)\}$$
$$\le \max\{(\mathrm{Id} - \rho)(V(\xi)), V_{CH}(x(t^-, t_{D(2w)}, \xi, u)), \bar{u}(t^-), \|\bar{u}\|_\Delta\}$$
$$\le \max\{(\mathrm{Id} - \rho)(V(\xi)), V_{CH}(x(t^-, t_{D(2w)}, \xi, u)), \|\bar{u}\|_\Delta\}. \tag{D.65}$$

For the last inequality above, We used the fact that $\bar{u}(t^-) \le \|\bar{u}\|_\Delta$ for $t \in [t_{D(2w+1)}, t_{D(2w+2)}]$.

By considering $V_{CH}(x(t, t_{D(2w)}, \xi, u))$ as $\mu(t)$, α_{CH} as φ and $\max\{(\mathrm{Id} - \rho)(V(\xi)), \|\bar{u}\|_\Delta\}$ as ω, with Lemma C.7, from (D.64) and (D.65), it can be proved that, for all $t \in [t_{D(2w+1)}, t_{D(2w+2)}]$,

$$V_{CH}(x(t, t_{D(2w)}, \xi, u) \le \max\{\nu(t), (\mathrm{Id} - \rho)(V(\xi)), \|\bar{u}\|_\Delta\}, \tag{D.66}$$

where $\nu(t)$ is the solution of $\dot{\nu} = -\alpha_{CH}(\nu)$ with $\nu(t_{D(2w+1)}) = V(\xi) \ge V_{CH}(x(t, t_{D(2w+1)}, \xi, u))$.

From Proposition 2.5 and Theorem 2.8 in [170], the uniform asymptotic stability of $\dot{\nu} = -\alpha_{CH}(\nu)$ implies the existence of $\beta_{CH} \in \mathcal{KL}$ satisfying $\beta_{CH}(s, 0) = s$ for all $s \ge 0$ such that $\nu(t) \le \beta_{CH}(V(\xi), t - t_{D(2w+1)})$ for all $t \in [t_{D(2w+1)}, t_{D(2w+2)}]$, and one can find a continuous and positive definite function ρ' satisfying $(\mathrm{Id} - \rho') \in \mathcal{K}_\infty$ such that

$$\nu(t_{D(2w+2)}) \le \beta_{CH}(V(\xi), t_{D(2w+2)} - t_{D(2w+1)})$$
$$\le \beta_{CH}(V(\xi), \max_{i \in \mathcal{N}_D} \overline{\delta t_i})$$
$$\le (\mathrm{Id} - \rho')(V(\xi)). \tag{D.67}$$

Define $\rho^*(s) = \min\{\rho(s), \rho'(s)\}$ for $s \in \mathbb{R}_+$. Then, from (D.66) and (D.67),

$$V_{CH}(x(t_{D(2w+2)}, t_{D(2w)}, \xi, u)) \leq \max\{(\mathrm{Id} - \rho^*)(V(\xi)), \|\bar{u}\|_\Delta\}, \qquad \text{(D.68)}$$

which together with (D.61) implies

$$V(x(t_{D(2w+2)}, t_{D(2w)}, \xi, u)) \leq \max\{(\mathrm{Id} - \rho^*)(V(\xi)), \|\bar{u}\|_\Delta\}. \qquad \text{(D.69)}$$

Define $\overline{\delta t}_D = 2\max_{w \in \mathbb{Z}_+}\{t_{D(2w+2)} - t_{D(2w)}\}$. Then, for any pair of nonnegative numbers (t, t_0) satisfying $t - t_0 \geq \overline{\delta t}_D$, there exists some $w \in \mathbb{Z}_+$ such that $[t_{D(2w)}, t_{D(2w+2)}] \in [t_0, t]$, and thus

$$\begin{aligned}
&V(x(t, t_0, \xi, u)) \\
&\leq \max\{V(x(t_{D(2w+2)}, t_0, \xi, u)), \|\bar{u}\|_{[t_{D(2w+2)}, t]}\} \\
&\leq \max\{(\mathrm{Id} - \rho^*)(V(x(t_{D(2w)}, t_0, \xi, u))), \|\bar{u}\|_{[t_{D(2w)}, t]}\} \\
&\leq \max\{(\mathrm{Id} - \rho^*)(V(\xi)), \|\bar{u}\|_{[t_0, t]}\}.
\end{aligned} \qquad \text{(D.70)}$$

The ISS of the dynamic network can be proved based on (D.53) and (D.70) following a similar approach as in the proof of [86, Theorem 1].

E Proofs of Technical Lemmas in Chapter 4

E.1 PROOF OF LEMMA 4.2

For simplicity, we use S_k instead of $S_k(\bar{x}_k)$ for $k = 1, \ldots, i-1$. We only consider the case of $e_i > 0$. The proof for the case of $e_i < 0$ is similar.

Consider the recursive definition of S_k's in (4.55). For $k = 1, \ldots, i-1$, the strictly decreasing property of the κ_k's implies

$$\max S_k = \kappa_k(x_k - \max S_{k-1} - \bar{w}_k), \tag{E.1}$$

$$\min S_k = \kappa_k(x_k - \min S_{k-1} + \bar{w}_k). \tag{E.2}$$

The continuous differentiability of the κ_k's implies the continuous differentiability of $\max S_k$ with respect to x_k and $\max S_{k-1}$ for $k = 1, \ldots, i-1$. Using the property of composition of continuously differentiable functions, we can see $\max S_{i-1}$ is continuously differentiable with respect to $\bar{x}_{i-1}$ and thus $\nabla \max S_{i-1}$ is continuous with respect to $\bar{x}_{i-1}$.

In the case of $e_i > 0$, the dynamics of e_i can be rewritten as

$$\begin{aligned}
\dot{e}_i &= \dot{x}_i - \nabla \max S_{i-1} \dot{\bar{x}}_{i-1} \\
&= x_{i+1} + \Delta_i(\bar{x}_i, d) - \nabla \max S_{i-1} \dot{\bar{x}}_{i-1} \\
&:= x_{i+1} + \phi_i^*(\bar{x}_i, d). \tag{E.3}
\end{aligned}$$

Note that $\dot{\bar{x}}_{i-1} = [x_2 + \Delta_1(\bar{x}_1, d), \ldots, x_i + \Delta_{i-1}(\bar{x}_{i-1}, d)]^T$. With Assumption 4.1, one can find a $\psi_{\dot{\bar{x}}_{i-1}} \in \mathcal{K}_\infty$ such that $|\dot{\bar{x}}_{i-1}| \leq \psi_{\dot{\bar{x}}_{i-1}}(|[\bar{x}_i^T, d^T]^T|)$. Since $\nabla \max S_{i-1}$ is continuous with respect to $\bar{x}_{i-1}$, one can find a $\psi_{\phi_i^*}^0 \in \mathcal{K}_\infty$ such that

$$|\phi_i^*(\bar{x}_i, d)| \leq \psi_{\phi_i^*}^0(|[\bar{x}_i^T, d^T]^T|). \tag{E.4}$$

To prove (4.57), for each $k = 1, \ldots, i-1$, we look for a $\psi_{x_{k+1}} \in \mathcal{K}_\infty$ such that $|x_{k+1}| \leq \psi_{x_{k+1}}(|[\bar{e}_{k+1}^T, W_k^T]^T|)$.

For $k = 1, \ldots, i-1$, from the definitions of e_{k+1} in (4.50), we can observe $\min S_k \leq x_{k+1} - e_{k+1} \leq \max S_k$ and thus

$$|x_{k+1}| \leq \max\{|\max S_k|, |\min S_k|\} + |e_{k+1}|. \tag{E.5}$$

For each $k = 1, \ldots, i-1$, define $\kappa_k^0(s) = |\kappa_k(s)|$ for $s \in \mathbb{R}_+$. Because κ_k is odd, strictly decreasing, and radially unbounded, we have $\kappa_k^0 \in \mathcal{K}_\infty$. From

(E.1), we have

$$|\max S_k| \le \kappa_k^0(|x_k - \max S_{k-1} - \bar{w}_k|)$$
$$\le \kappa_k^0(|x_k - \max S_{k-1}| + |\bar{w}_k|)$$
$$\le \kappa_k^0(|\max S_{k-1}| + |\min S_{k-1}| + |e_k| + \bar{w}_k). \qquad (E.6)$$

We used $|x_k - \max S_{k-1}| \le |\max S_{k-1}| + |\min S_{k-1}| + |e_k|$, which was derived from $\min S_k \le x_{k+1} - e_{k+1} \le \max S_k$. Similarly, we can also get

$$|\min S_k| \le \kappa_k^0(|\max S_{k-1}| + |\min S_{k-1}| + |e_k| + \bar{w}_k). \qquad (E.7)$$

For each x_{k+1} $(k = 1, \ldots, i-1)$, using (E.5) and repeatedly using (E.6) and (E.7), one can find a $\psi_{x_{k+1}} \in \mathcal{K}_\infty$ such that $|x_{k+1}| \le \psi_{x_{k+1}}(|[\bar{e}_{k+1}^T, W_k^T]^T|)$. This, together with (E.4), leads to the satisfaction of (4.57).

If the ψ_{Δ_k}'s for $k = 1, \ldots, i$ are Lipschitz on compact sets, then all the class $\mathcal{K}_\infty$ functions determining $\psi_{\phi_i^*}$ are Lipschitz on compact sets, and one can find a $\psi_{\phi_i^*} \in \mathcal{K}_\infty$ which is Lipschitz on compact sets.

E.2   PROOF OF LEMMA 4.3

By convention, $S_{i1}(y_i, \bar{\xi}_{i1}) := S_{i1}(y_i)$. We simply use S_{ik} instead of $S_{ik}(y_i, \bar{\xi}_{ik})$ for $k = 1, \ldots, j - 1$. We only consider the case of $e_{ij} > 0$. The proof for the case of $e_{ij} < 0$ is similar.

Consider the definition of S_{i1} in (4.171) and the iteration-type definitions of S_{ik}'s in (4.178). The strictly decreasing properties of the $\kappa_{(.)}$'s imply

$$\max S_{i1} = \kappa_{i1}(y_i - \bar{d}_i^m), \qquad (E.8)$$
$$\max S_{ik} = \kappa_{ik}(\xi_{ik} - \max S_{i(k-1)}), \quad k = 2, \ldots, j - 1, \qquad (E.9)$$
$$\min S_{i1} = \kappa_{i1}(y_i + \bar{d}_i^m), \qquad (E.10)$$
$$\min S_{ik} = \kappa_{ik}(\xi_{ik} - \min S_{i(k-1)}), \quad k = 2, \ldots, j - 1. \qquad (E.11)$$

The continuous differentiability of the $\kappa_{(.)}$'s implies the continuous differentiability of $\max S_{i1}$ with respect to y_i and the continuous differentiability of $\max S_{ik}$ with respect to ξ_{ik} and $\max S_{i(k-1)}$ for $k = 2, \ldots, j - 1$. Using the property of the composition of continuously differentiable functions, we can see $\max S_{i(j-1)}$ is continuously differentiable with respect to $[y_i, \bar{\xi}_{i(j-1)}^T]^T$ and thus $\nabla \max S_{i(j-1)}$ in (E.12) is continuous with respect to $[y_i, \bar{\xi}_{i(j-1)}^T]^T$.

In the case of $e_{ij} > 0$, the dynamics of e_{ij} can be derived as

$$\dot{e}_{ij} = \dot{\xi}_{ij} - \nabla \max S_{i(j-1)}[\dot{y}_i, \dot{\bar{\xi}}_{i(j-1)}^T]^T$$
$$= \xi_{i(j+1)} + \phi_{ij}(y_i, \xi_{i2}, d_i) - \nabla \max S_{i(j-1)}[\dot{y}_i, \dot{\bar{\xi}}_{i(j-1)}^T]^T$$
$$:= \xi_{i(j+1)} + \phi_{ij}'(z_i, y_i, \bar{\xi}_{ij}, w_i, d_i). \qquad (E.12)$$

Specifically, $\xi_{i(n_i+1)} = u_i$. We used (4.160), (4.161), and (4.162) to get the last equality. Also from (4.160), (4.161), and (4.162), we can see $|\phi_{ij}(y_i, \xi_{i2}, d_i)|$ is bounded by a $\mathcal{K}_\infty$ function of $|[y_i, \xi_{i2}, d_i]^T|$, and $|[\dot{y}_i, \dot{\xi}^T_{i(j-1)}]^T|$ is bounded by a $\mathcal{K}_\infty$ function of $|[z_i^T, y_i, \bar{\xi}_{ij}^T, w_i^T, d_i]^T|$. Thus, we can conclude that $|\phi'_{ij}(y_i, \bar{\xi}_{ij}, w_i, d_i)|$ is bounded by a $\mathcal{K}_\infty$ function of $|[z_i^T, y_i, \bar{\xi}_{ij}^T, w_i^T, d_i]^T|$. Note that $e_{i0} = [\zeta_i^T, z_i^T]^T$. To prove (4.181), we show that for each $k = 1, \ldots, j-1$, $|\xi_{i(k+1)}|$ is bounded by a $\mathcal{K}_\infty$ function of $|[\bar{e}^T_{i(k+1)}, d_i]^T|$.

From the definitions of e_{i2} in (4.172) and $e_{i(k+1)}$ $(k = 2, \ldots, j-1)$ in (4.179), we have $\min S_{ik} \leq \xi_{i(k+1)} - e_{i(k+1)} \leq \max S_{ik}$ and thus

$$|\xi_{i(k+1)}| \leq \max\{|\max S_{ik}|, |\min S_{ik}|\} + |e_{i(k+1)}|. \tag{E.13}$$

Define $\kappa_{ik}^o(s) = |\kappa_{ik}(s)|$ for $s \in \mathbb{R}_+$. Since κ_{ik} is odd, strictly decreasing, and radially unbounded, $\kappa_{ik}^o \in \mathcal{K}_\infty$. From (E.8) and (E.9), we have

$$|\max S_{i1}| \leq \kappa_{i1}^o(|y_i| + \bar{d}_i^m) \tag{E.14}$$

$$|\max S_{ik}| = \kappa_{ik}^o(|\xi_{ik} - e_{ik} + e_{ik} - \max S_{i(k-1)}|)$$
$$\leq \kappa_{ik}^o(|\max S_{i(k-1)}| + |\min S_{i(k-1)}| + |e_{i(k-1)}|),$$
$$k = 2, \ldots, j-1. \tag{E.15}$$

Similarly, we can also obtain

$$|\min S_{i1}| \leq \kappa_{i1}^o(|y_i| + \bar{d}_i^m) \tag{E.16}$$

$$|\min S_{ik}| \leq \kappa_{ik}^o(|\max S_{i(k-1)}| + |\min S_{i(k-1)}| + |e_{i(k-1)}|),$$
$$k = 2, \ldots, j-1. \tag{E.17}$$

Note that $y_i = e_{i1}$. Using (E.13) along with a repeated application of (E.14)–(E.17), we can prove that for each $k = 1, \ldots, j-1$, $|\xi_{i(k+1)}|$ is bounded by a $\mathcal{K}_\infty$ function of $|[\bar{e}_{i(k+1)}, d_i]^T|$.

E.3   PROOF OF LEMMA 4.5

We simply use S_k to denote $S_k(X_{1k}, X_{2k})$ for $k = 1, \ldots, j-1$. We only consider the case of $e_j > 0$. The case of $e_j < 0$ is similar.

Consider the iterative definitions of S_k in (4.311). For $k = 1, \ldots, j-1$, the positive and nondecreasing properties of the μ_j's and the strictly decreasing properties of the θ_j's imply

$$\max S_k = \max\{\mu_k(|X_{1k} + \delta_{1k}|, |X_{2k} + \delta_{2k}|)\theta_k(z_k - \max S_{k-1} - \bar{d}_k):$$
$$- \bar{D}_{1k} \leq \delta_{1k} \leq \bar{D}_{1k}, -\bar{D}_{2k} \leq \delta_{2k} \leq \bar{D}_{2k}\}. \tag{E.18}$$

By iteratively using (E.18), we can see that $\max S_k$ is continuously differentiable almost everywhere with respect to $[X_{1k}^T, X_{2k}^T]^T$.

Denote $X_k = [X_{1k}^T, X_{2k}^T]^T$. Define

$$\partial \max S_{j-1} = \bigcap_{\epsilon > 0} \bigcap_{\mu(\tilde{\mathcal{M}})=0} \overline{\mathrm{co}} \nabla \max S_{j-1}(\mathcal{B}_\epsilon(X_{j-1}) \backslash \tilde{\mathcal{M}}), \qquad (\text{E.19})$$

where $\mathcal{B}_\epsilon(X_{j-1})$ is a ball of radius ϵ around X_{j-1}. Then, $\partial \max S_{j-1}$ is convex, compact, and upper semi-continuous.

In the case of $e_j > 0$, the e_j-subsystem can be represented with a differential inclusion as

$$\dot{e}_j \in \left\{ z_{j+1} + \phi_j(X_{1j}, X_{2j}) - \phi_j^0 : \phi_j^0 \in \partial \max S_{j-1} \dot{X}_{j-1} \right\}$$
$$:= \left\{ z_{j+1} + \phi_j^* : \phi_j^* \in \Phi_j^*(X_{1j}, X_{2j}) \right\}, \qquad (\text{E.20})$$

where

$$\Phi_j^*(X_{1j}, X_{2j}) = \left\{ \phi_j(X_{1j}, X_{2j}) - \phi_j^0 : \phi_j^0 \in \partial \max S_{j-1} \dot{X}_{j-1} \right\}. \qquad (\text{E.21})$$

Because $\phi_j(X_{1j}, X_{2j})$ and $\dot{X}_{j-1}$ are locally Lipschitz, and $\partial \max S_{j-1}$ is convex, compact, and upper semi-continuous, Φ_j^* is convex, compact, and upper semi-continuous.

$\max S_k$ can be considered as a discontinuous function of X_{1k}, X_{2k}, z_k, $\max S_{k-1}$, and $\bar{d}_k$. From the definitions of μ_k and θ_k, there exist $\varphi_{\overline{S}_k}^0$ positive and nondecreasing, and $\psi_{\overline{S}_k}^0 \in \mathcal{K}_\infty$ such that

$$|\max S_k| \leq \varphi_{\overline{S}_k}^0(|X_{1k}|, |X_{2k}|)\psi_{\overline{S}_k}^0(|[z_k, \max S_{k-1}, \bar{d}_k]^T|). \qquad (\text{E.22})$$

We can also calculate

$$\nabla \max S_k \dot{X}_k = \frac{\partial \mu_k}{\partial [X_{1k}^T, X_{2k}^T]^T} [\dot{X}_{1k}^T, \dot{X}_{2k}^T]^T \theta_k(z_k - \max S_{k-1} - \bar{d}_k)$$
$$+ \frac{\partial \theta_k}{\partial [z_k, \max S_{k-1}]^T} [\dot{z}_k, \nabla \max S_{k-1} \dot{X}_{k-1}]^T, \qquad (\text{E.23})$$

where $(\delta_{1k}, \delta_{2k})$ may take the value of (D_{1k}, D_{2k}) or $(-D_{1k}, -D_{2k})$, depending on the sign of $\theta_k(z_k - \max S_{k-1} - \bar{d}_k)$.

From the definition of $\dot{z}_k$, there exist $\varphi_{\nabla \overline{S}_k}^0$ positive and nondecreasing, and $\psi_{\nabla \overline{S}_k}^0 \in \mathcal{K}_\infty$ such that

$$|\nabla \max S_k \dot{X}_k| \leq \varphi_{\nabla \overline{S}_k}^0(|X_{1(k+1)}|, |X_{2(k+1)}|, \nabla \max S_{k-1}) \times$$
$$\psi_{\nabla \overline{S}_k}^0(|[z_k, z_{k+1}, \max S_{k-1}, \nabla \max S_{k-1}, \bar{d}_k]^T|). \qquad (\text{E.24})$$

By recursively using (E.22) and (E.24), one can find $\varphi_{\nabla \overline{S}_{j-1}}^1$ positive and nondecreasing and $\psi_{\nabla \overline{S}_{j-1}}^1 \in \mathcal{K}_\infty$ such that

$$|\nabla \max S_{j-1} \dot{X}_{j-1}| \leq \varphi_{\nabla \overline{S}_{j-1}}^1(|X_{1j}|, |X_{2j}|)\psi_{\nabla \overline{S}_{j-1}}^1(|[Z_j^T, \bar{D}_{j-1}^T]^T|). \qquad (\text{E.25})$$

As for (E.22), there exist $\varphi^0_{\underline{S}_k}$ positive and nondecreasing, and $\psi^0_{\underline{S}_k} \in \mathcal{K}_\infty$ such that

$$|\min S_k| \leq \varphi^0_{\underline{S}_k}(|X_{1k}|, |X_{2k}|)\psi^0_{\underline{S}_k}(\|[z_k, \max S_{k-1}, \bar{d}_k]^T\|). \tag{E.26}$$

The definitions of e_{k+1} $(k = 1, \ldots, j-1)$ in (4.313) implies

$$|z_{k+1}| \leq \max\{|\max S_k|, |\min S_k|\} + |e_{k+1}|. \tag{E.27}$$

With (E.27), and iteratively using (E.22) and (E.26), for each $k = 1, \ldots, j-1$, one can find φ_{z_k} positive and nondecreasing and $\psi_{z_k} \in \mathcal{K}_\infty$ such that

$$|z_{k+1}| \leq \varphi_{z_k}(|X_{1k}|, |X_{2k}|)\psi_{z_k}(\|[E^T_{k+1}, \bar{D}^T_k]^T\|). \tag{E.28}$$

Thus, from (E.25) and (E.28), we can find that $\varphi^2_{\nabla \overline{S}_{j-1}}$ positive and nondecreasing and $\psi^2_{\nabla \overline{S}_{j-1}} \in \mathcal{K}_\infty$ such that

$$|\nabla \max S_{j-1}\dot{X}_{j-1}| \leq \varphi^2_{\nabla \overline{S}_{j-1}}(|X_{1j}|, |X_{2j}|)\psi^2_{\nabla \overline{S}_{j-1}}(\|[E^T_j, \bar{D}^T_{j-1}]^T\|). \tag{E.29}$$

From the definition of Φ^*_j in (E.21) and the definition of $\partial \max S_{j-1}\dot{X}_{j-1}$ in (E.19), we can find $\varphi^*_j : \mathbb{R}_+ \times \mathbb{R}_+ \to \mathbb{R}_+$ is positive and nondecreasing with respect to the two variables, and $\psi^*_j \in \mathcal{K}_\infty$, such that for any $\phi^*_j \in \Phi^*_j(X_{1j}, X_{2j})$, (4.305) holds.

E.4 PROOF OF LEMMA 4.6

With (4.305) satisfied, one can find $\psi^{e_k}_{\Phi^*_j}$ $(k = 1, \ldots, j+1)$ and $\psi^{d_k}_{\Phi^*_j} \in \mathcal{K}_\infty$ $(k = 1, \ldots, j-1)$, such that for any $\phi^*_j \in \Phi^*_j(X_{1j}, X_{2j})$, it holds that

$$|\phi^*_j| \leq \varphi^*_j(|X_{1j}|, |X_{2j}|)\left(\sum_{k=1}^{j} \psi^{e_k}_{\Phi^*_j}(|e_k|) + \sum_{k=1}^{j-1} \psi^{d_k}_{\Phi^*_j}(\bar{d}_k)\right). \tag{E.30}$$

By convenience, define $\gamma^{e_j}_{e_j} = \mathrm{Id}$. Define

$$\Pi_j(s) = \sum_{k=1}^{j} \psi^{e_k}_{\Phi^*_j} \circ \alpha_V^{-1} \circ \left(\gamma^{e_k}_{e_j}\right)^{-1} \circ \alpha_V(s)$$

$$+ \sum_{k=1}^{j-1} \psi^{d_k}_{\Phi^*_j} \circ \left(\gamma^{d_k}_{e_j}\right)^{-1} \circ \alpha_V(s) \tag{E.31}$$

for $s \in \mathbb{R}_+$. Then, $\Pi_j \in \mathcal{K}_\infty$.

For any $0 < c_j < 1$, $\epsilon_j > 0$, one can find a $\nu_j : \mathbb{R}_+ \to \mathbb{R}_+$ positive, nondecreasing and continuously differentiable on $(0, \infty)$ and satisfying

$$(1 - c_j)\nu_j((1 - c_j)s)s \geq \Pi_j(s) + \frac{\ell_j}{2}s + \alpha_V^{-1} \circ \left(\gamma^{e_{j+1}}_{e_j}\right)^{-1} \circ \alpha_V(s) \tag{E.32}$$

for $s \in \mathbb{R}_+$.

One can also find a $\mu_j : \mathbb{R}_+ \times \mathbb{R}_+ \to \mathbb{R}_+$, which is continuously differentiable on $(0, \infty) \times (0, \infty)$, positive, and nondecreasing, such that

$$\min\{\mu_j(|X_{1j} + \delta_{1j}|, |X_{2j} + \delta_{2j}|) : -\bar{D}_{1j} \le \delta_{1j} \le \bar{D}_{1j}, -\bar{D}_{2j} \le \delta_{2j} \le \bar{D}_{2j}\}$$
$$\ge \max\{\varphi_j^*(|X_{1j}|, |X_{2j}|), 1\} \tag{E.33}$$

for all $X_{1j}, X_{2j} \in \mathbb{R}^j$.

Define $\kappa_j(a_1, a_2, a_3) = \mu_j(|a_1|, |a_2|)\theta_j(a_3)$ with $\theta_j(a_3) = -\nu_j(|a_3|)a_3$ for $a_1, a_2, a_3 \in \mathbb{R}$.

Recall that $V_k(e_k) = \alpha_V(|e_k|) = \frac{1}{2}e_k^2$ for $k = 1, \ldots, n+1$. We use V_k instead of $V_k(e_k)$ for convenience. Consider the case of

$$V_j \ge \max_{k=1,\ldots,j-1}\left\{\gamma_{e_j}^{e_k}(V_k), \gamma_{e_j}^{e_{j+1}}(V_{j+1}), \gamma_{e_j}^{d_k}(\bar{d}_k), \gamma_{e_j}^{d_j}(\bar{d}_j), \epsilon_j\right\} \tag{E.34}$$

with $\gamma_{e_j}^{d_j}(s) = \alpha_V\left(\frac{s}{c_j}\right)$ for $s \in \mathbb{R}_+$.

In this case, for any $\phi_j^* \in \Phi_j^*(X_{1j}, X_{2j})$, it holds that

$$|\phi_j^* + e_{j+1}| \le \mu_j(|X_{1j}, X_{2j}|)\left(\Pi_j(|e_j|) + \alpha_V^{-1} \circ \left(\gamma_{e_j}^{e_{j+1}}\right)^{-1} \circ \alpha_V(|e_j|)\right), \tag{E.35}$$

we can also get

$$\bar{d}_j \le c_j|e_j| \tag{E.36}$$

$$\bar{d}_k \le \left(\gamma_{e_j}^{d_k}\right)^{-1} \circ \alpha_V(|e_j|), \quad k = 1, \ldots, j-1 \tag{E.37}$$

$$|e_j| \ge \sqrt{2\epsilon_j}. \tag{E.38}$$

With $0 < c_j < 1$, for $p_{j-1} \in S_{j-1}$ and $|\delta| \le \bar{d}_j$, we have

$$|z_j - p_{j-1} + \delta_j| \ge (1 - c_j)|e_j| \tag{E.39}$$

$$\mathrm{sgn}(z_j - p_{j-1} + \delta_j) = \mathrm{sgn}(e_j) \tag{E.40}$$

and thus

$$\nu_j(|z_j - p_{j-1} + \delta_j|)|z_j - p_{j-1} + \delta_j|$$
$$\ge (1 - c_j)\nu_j((1 - c_j)|e_j|)|e_j|. \tag{E.41}$$

In the case of (E.34), for $p_{j-1} \in S_{j-1}$ and $|\delta_j| \le \bar{d}_j$, for any $\phi_j^* \in$

$\Phi_j^*(X_{1j}, X_{2j})$, we have

$$\nabla V_j(z_{j+1} - e_{j+1} + \phi_j^* + e_{j+1})$$

$$= e_j\Big(-\mu_j(|X_{1j}|, |X_{2j}|)\nu_j(|z_j - p_{j-1} + \delta_j|)(z_j - p_{j-1} + \delta_j) + \phi_j^* + e_{j+1}\Big)$$

$$\leq -\mu_j(|X_{1j}|, |X_{2j}|)\nu_j(|z_j - p_{j-1} + \delta_j|)|z_j - p_{j-1} + \delta_j||e_j|$$
$$\quad + |\phi_j^* + e_{j+1}||e_j|$$

$$\leq -\mu_j(|X_{1j}|, |X_{2j}|)\nu_j((1 - c_j)|e_j|)(1 - c_j)|e_j|^2$$

$$\quad + \mu_j(|X_{1j}|, |X_{2j}|)\left(\Pi_j(|e_j|) + \alpha_V^{-1} \circ \left(\gamma_{e_j}^{e_{j+1}}\right)^{-1} \circ \alpha_V(|e_j|)\right)|e_j|$$

$$\leq -\frac{\ell_j}{2}\mu_j(|X_{1j}|, |X_{2j}|)|e_j|^2$$

$$\leq -\ell_j V_j. \tag{E.42}$$

F Proofs of Technical Lemmas in Chapter 5

F.1 PROOF OF LEMMA 5.1

We simply use $\breve{S}_k$ and S_k to denote $\breve{S}_k(\bar{x}_k)$ and $S_k(\bar{x}_k)$ for $1 \leq k \leq i-1$. We only consider the case of $e_i > 0$. The proof for the case of $e_i < 0$ is similar.

Consider the recursive definitions of $\breve{S}_k$ and S_k in (5.25)–(5.26). With condition 5.4 satisfied, we have $0 \leq b_k < 1$ and $a_k \geq 0$ for $1 \leq k \leq n$. For $1 \leq k \leq i-1$, the strictly decreasing properties of the κ_k's imply

$$\max S_k = \max\left\{ d_{k2} \max \breve{S}_k : \frac{1}{1+b_{k+1}} \leq d_{k2} \leq \frac{1}{1-b_{k+1}} \right\} \tag{F.1}$$

$$\max \breve{S}_k = \kappa_k(x_k - \max \breve{S}_{k-1} - b_k|x_k| - (1-b_k)a_k). \tag{F.2}$$

By iteratively using (F.2), we can see that $\max \breve{S}_{i-1}$ is continuously differentiable almost everywhere with respect to $\bar{x}_{i-1}$. From (F.1), $\max S_{i-1}$ is continuously differentiable almost everywhere with respect to $\max \breve{S}_{i-1}$. Thus, $\max S_{i-1}$ is continuously differentiable almost everywhere with respect to $\bar{x}_{i-1}$.

Considering the definition of e_i in (5.27) with $k = i-1$, when $e_i > 0$, we can represent the e_i-subsystem with a differential equation

$$\dot{e}_i = x_{i+1} + \Delta_i(\bar{x}_i) - \nabla \max S_{i-1} \dot{\bar{x}}_{i-1} \tag{F.3}$$

wherever $\max S_{i-1}$ is continuously differentiable, or equivalently, wherever $\nabla \max S_{i-1}$ exists. Because $\max S_{i-1}$ is continuously differentiable almost everywhere, $\nabla \max S_{i-1}$ is discontinuous and thus the e_i-subsystem is a discontinuous system. We represent the e_i-subsystem with a differential inclusion by embedding the discontinuous $\nabla \max S_{i-1}$ into a set-valued map

$$\partial \max S_{i-1} = \bigcap_{\epsilon > 0} \bigcap_{\mu(\tilde{\mathcal{M}})=0} \overline{\mathrm{co}} \nabla \max S_{i-1}(\mathcal{B}_\epsilon(\bar{x}_{i-1}) \backslash \tilde{\mathcal{M}}), \tag{F.4}$$

where $\mathcal{B}_\epsilon(\bar{x}_{i-1})$ is a ball of radius ϵ around $\bar{x}_{i-1}$ and $\tilde{\mathcal{M}}$ represents all sets of zero measure (i.e., $\mu(\tilde{\mathcal{M}}) = 0$). Then, $\partial \max S_{i-1}$ is convex, compact, and upper semi-continuous (see [84] for recent results on such properties for discontinuous systems).

Then, in the case of $e_i > 0$, the e_i-subsystem can be represented with a differential inclusion as

$$\begin{aligned}
\dot{e}_i &\in \{x_{i+1} + \Delta_i(\bar{x}_i) - \phi_i : \phi_i \in \partial \max S_{i-1} \dot{\bar{x}}_{i-1}\} \\
&:= \{x_{i+1} - e_{i+1} + \phi_i^* : \phi_i^* \in \Phi_i^*(\bar{x}_i, e_{i+1})\},
\end{aligned} \tag{F.5}$$

where

$$\Phi_i^*(\bar{x}_i, e_{i+1}) = \{e_{i+1} + \Delta_i(\bar{x}_i) - \phi_i : \phi_i \in \partial \max S_{i-1}\dot{\bar{x}}_{i-1}\}. \qquad (F.6)$$

Since $\Delta_i(\bar{x}_i)$ and $\dot{\bar{x}}_{i-1}$ are locally Lipschitz, and $\partial \max S_{i-1}$ is convex, compact, and upper semi-continuous, $\Phi_i^*(\bar{x}_i, e_{i+1})$ is convex, compact, and upper semi-continuous.

For system (5.1)–(5.2), with condition (5.6), $|\Delta_i(\bar{x}_i)|$ is bounded by a $\mathcal{K}_\infty$ function of $|\bar{x}_i|$ and $|\dot{\bar{x}}_{i-1}| = |[\dot{x}_1, \ldots, \dot{x}_i]^T|$ is bounded by a $\mathcal{K}_\infty$ function of $|\bar{x}_i|$. Hence, there exists a $\psi_{\Phi_i^*1} \in \mathcal{K}_\infty$ such that for any $\phi_i^* \in \Phi_i^*(\bar{x}_i, e_{i+1})$, it holds that

$$|\phi_i^*| \le \psi_{\Phi_i^*1}(|[\bar{x}_i^T, e_{i+1}]^T|). \qquad (F.7)$$

We show that $|\bar{x}_i|$ is bounded by a $\mathcal{K}_\infty$ function of $|\bar{e}_i|$ and $\bar{a}_{i-1}$. The definitions of e_{k+1} $(1 \le k \le i - 1)$ in (5.27) imply

$$|x_{k+1}| \le \max\{|\max S_k|, |\min S_k|\} + |e_{k+1}|. \qquad (F.8)$$

Define $\kappa_k^o(s) = |\kappa_k(s)|$ for $s \in \mathbb{R}_+$. Then, $\kappa_k^o \in \mathcal{K}_\infty$. From (F.1) and (F.2), for $1 \le k \le i - 1$,

$$|\max S_k| \le \left|\frac{1}{1 - b_{k+1}}\right| |\max \breve{S}_k| \qquad (F.9)$$

$$\begin{aligned}
|\max \breve{S}_k| &\le \kappa_k^o((1 + b_k)|x_k| + |\max \breve{S}_{k-1}| + (1 - b_k)a_k)) \\
&\le \kappa_k^o((1 + b_k)(|x_k - e_k| + |e_k|) + |\max \breve{S}_{k-1}| + (1 - b_k)a_k) \\
&\le \kappa_k^o\Big((1 + b_k)(|\max \breve{S}_{k-1}| + |\min \breve{S}_{k-1}| + |e_k|) \\
&\qquad + |\max \breve{S}_{k-1}| + (1 - b_k)a_k\Big). \qquad (F.10)
\end{aligned}$$

In the same way, for $1 \le k \le i - 1$, we can also get

$$|\min S_k| \le \left|\frac{1}{1 - b_{k+1}}\right| |\min \breve{S}_k| \qquad (F.11)$$

$$\begin{aligned}
|\min \breve{S}_k| &\le \kappa_k^o\Big((1 + b_k)(|\max \breve{S}_{k-1}| + |\min \breve{S}_{k-1}| + |e_k|) \\
&\qquad + |\min \breve{S}_{k-1}| + (1 - b_k)a_k\Big). \qquad (F.12)
\end{aligned}$$

Note that $x_1 = e_1$. With (F.8), and iteratively using (F.9)–(F.12), we can prove that for each $1 \le k \le i - 1$, $|x_{k+1}|$ is bounded by a $\mathcal{K}_\infty$ function of $|\bar{e}_{k+1}|$ and $\bar{a}_k$. This, together with (F.7), implies that there exists a $\Psi_{\Phi_i^*} \in \mathcal{K}_\infty$ such that for any $\phi_i^* \in \Phi_i^*(\bar{x}_i, e_{i+1})$, it holds that

$$|\phi_i^*| \le \psi_{\Phi_i^*}(|[\bar{e}_{i+1}^T, \bar{a}_{i-1}^T]^T|). \qquad (F.13)$$

This ends the proof.

F.2   PROOF OF LEMMA 5.3

We simply use S_k instead of $S_k(\bar{x}_k, \bar{\mu}_{k1}, \bar{\mu}_{k2})$ for $k = 1, \ldots, i-1$. We only consider the case of $e_i > 0$.

Consider the recursive definition of S_k's in (5.104). For $k = 1, \ldots, i-1$, the strictly decreasing property of the κ_k's implies

$$\max S_k = \kappa_k(x_k - \max S_{k-1} - \max\{c_{k1}|e_k|, \mu_{k1}\}) + \mu_{k2}, \qquad (F.14)$$

$$\min S_k = \kappa_k(x_k - \min S_{k-1} + \max\{c_{k1}|e_k|, \mu_{k1}\}) - \mu_{k2}. \qquad (F.15)$$

From the iteration type definition of e_k's for $k = 1, \ldots, i-1$, e_{i-1} is continuous and differentiable almost everywhere with respect to $\bar{x}_{i-1}, \bar{\mu}_{(i-2)1}, \bar{\mu}_{(i-2)2}$.

Since κ_k's are continuously differentiable for $k = 1, \ldots, i-1$, using (F.14), we can see $\max S_{i-1}$ is continuously differentiable almost everywhere with respect to $\bar{x}_{i-1}, \bar{\mu}_{(i-1)1}, \bar{\mu}_{(i-1)2}$.

Considering the definition of e_i in (5.105) with $k = i-1$, when $e_i > 0$, we can represent the e_i-subsystem with a differential equation

$$\dot{e}_i = x_{i+1} + \Delta_i(\bar{x}_i, z) - \nabla \max S_{i-1}[\dot{\bar{x}}_{i-1}, 0_{(i-1)}, 0_{(i-1)}]^T \qquad (F.16)$$

with $0_{(i-1)}$ being the vector composed of $i-1$ zero elements, wherever $\max S_{i-1}$ is continuously differentiable, or equivalently, $\nabla \max S_{i-1}$ exists. Because $\max S_{i-1}$ is continuously differentiable almost everywhere, $\nabla \max S_{i-1}$ is discontinuous and thus the e_i-subsystem is a discontinuous system. We represent the e_i-subsystem with a differential inclusion by embedding the discontinuous $\nabla \max S_{i-1}$ into a set-valued map

$$\partial \max S_{i-1} = \bigcap_{\varepsilon > 0} \bigcap_{\tau(\tilde{\mathcal{M}})=0} \overline{\text{co}} \nabla \max S_{i-1}(\mathcal{B}_\varepsilon(\zeta_{i-1}) \backslash \tilde{\mathcal{M}}), \qquad (F.17)$$

where $\mathcal{B}_\varepsilon(\zeta_{i-1})$ is an open ball of radius ε around $\zeta_{i-1} := [\bar{x}_{i-1}^T, \bar{\mu}_{(i-1)1}^T, \bar{\mu}_{(i-1)2}^T]^T$, and $\tilde{\mathcal{M}}$ represents all sets of zero measure (i.e., $\tau(\tilde{\mathcal{M}}) = 0$).

Then, in the case of $e_i > 0$, the e_i-subsystem can be represented with a differential inclusion as

$$\dot{e}_i \in \{x_{i+1} + \Delta_i(\bar{x}_i, z) - \varphi_i : \varphi_i \in \partial \max S_{i-1}[\dot{\bar{x}}_{i-1}^T, 0_{(i-1)}, 0_{(i-1)}]^T\}$$

$$:= \{x_{i+1} + \phi_i : \phi_i \in \Phi_i(\bar{x}_i, \bar{\mu}_{(i-1)1}, \bar{\mu}_{(i-1)2}, z)\}, \qquad (F.18)$$

where

$$\Phi_i(\bar{x}_i, \bar{\mu}_{(i-1)1}, \bar{\mu}_{(i-1)2}, z)$$

$$= \{\Delta_i(\bar{x}_i, z) - \varphi_i : \varphi_i \in \partial \max S_{i-1}[\dot{\bar{x}}_{i-1}^T, 0_{(i-1)}, 0_{(i-1)}]^T\}. \qquad (F.19)$$

Because Δ_i and $\dot{\bar{x}}_i$ are locally Lipschitz and $\partial \max S_{i-1}$ is convex, compact, and upper semi-continuous, Φ_i is convex, compact, and upper semi-continuous. Considering the definition of $\partial \max S_{i-1}$, one can find a continuous function $\bar{s}_{i-1}$ such that for all $\bar{x}_{i-1}, \bar{\mu}_{(i-1)1}, \bar{\mu}_{(i-1)2}$, any $s_{i-1} \in \partial \max S_{i-1}$ satisfies $|s_{i-1}| \leq \bar{s}_{i-1}(\bar{x}_{i-1}, \bar{\mu}_{(i-1)1}, \bar{\mu}_{(i-1)2})$. Thus, for all $\bar{x}_{i-1}, \bar{\mu}_{(i-1)1}, \bar{\mu}_{(i-1)2}, z$, any $\phi_i \in \Phi_i(\bar{x}_i, \bar{\mu}_{(i-1)1}, \bar{\mu}_{(i-1)2}, z)$ satisfies

$$|\phi_i| \leq |\Delta_i(\bar{x}_i, z)| + \bar{s}_{i-1}(\bar{x}_{i-1}, \bar{\mu}_{(i-1)1}, \bar{\mu}_{(i-1)2})|\dot{\bar{x}}_{i-1}|. \tag{F.20}$$

From (5.74)–(5.75) and Assumption 5.4, $\Delta_i(\bar{x}_i, z)$ is bounded by a $\mathcal{K}_\infty$ function of $(\bar{x}_i, z)$ and $\dot{\bar{x}}_{i-1}$ is bounded by a $\mathcal{K}_\infty$ function of $(\bar{x}_i, z)$. Thus, there exists a $\lambda_{\Phi_i}^0 \in \mathcal{K}_\infty$ such that for any $\phi_i \in \Phi_i(\bar{x}_i, \bar{\mu}_{(i-1)1}, \bar{\mu}_{(i-1)2}, z)$, it holds that

$$|\phi_i| \leq \lambda_{\Phi_i}^0(|(\bar{x}_i, \bar{\mu}_{(i-1)1}, \bar{\mu}_{(i-1)2}, z)|). \tag{F.21}$$

For the purpose of (5.107), for each $k = 1, \ldots, i-1$, we find a $\lambda_{x_{k+1}} \in \mathcal{K}_\infty$ such that $|x_{k+1}| \leq \lambda_{x_{k+1}}(|(\bar{e}_{k+1}, \bar{\mu}_{k1}, \bar{\mu}_{k2})|)$. For $k = 1, \ldots, i-1$, from the definitions of x_{k+1} in (5.105), we have $\min S_k \leq x_{k+1} - e_{k+1} \leq \max S_k$ and thus

$$|x_{k+1}| \leq \max\{|\max S_k|, |\min S_k|\} + |e_{k+1}|. \tag{F.22}$$

For each $k = 1, \ldots, i-1$, define $\kappa_k^o(s) = \kappa_k(|s|)$ for $s \in \mathbb{R}_+$. Because κ_k is odd, strictly decreasing, and radially unbounded, $\kappa_k^o \in \mathcal{K}_\infty$. From (F.14), we have

$$\begin{aligned}
|\max S_k| &\leq \kappa_k^o(|x_k - \max S_{k-1} - \max\{c_{k1}|e_k|, \mu_{k1}\}|) + \mu_{k2} \\
&\leq \kappa_k^o(|x_k - \max S_{k-1}| + \max\{c_{k1}|e_k|, \mu_{k1}\}) + \mu_{k2} \\
&\leq \kappa_k^o(|\max S_{k-1}| + |\min S_{k-1}| + |e_k| + \max\{c_{k1}|e_k|, \mu_{k1}\}) \\
&\quad + \mu_{k2}.
\end{aligned} \tag{F.23}$$

In (F.23), we used the fact that $\min S_{k-1} \leq x_k - e_k \leq \max S_{k-1}$ and thus $\min S_{k-1} - \max S_{k-1} + e_k \leq x_k - \max S_{k-1} \leq e_k$, to arrive at $|x_k - \max S_{k-1}| \leq |\max S_{k-1}| + |\min S_{k-1}| + |e_k|$. Similarly, we obtain

$$\begin{aligned}
|\min S_k| &\leq \kappa_k^o(|\max S_{k-1}| + |\min S_{k-1}| + |e_k| + \max\{c_{k1}|e_k|, \mu_{k1}\}) \\
&\quad + \mu_{k2}.
\end{aligned} \tag{F.24}$$

For each x_{k+1} $(k = 1, \ldots, i-1)$, using (F.22), (F.23), and (F.24), one can find a $\lambda_{x_{k+1}} \in \mathcal{K}_\infty$ such that $|x_{k+1}| \leq \lambda_{x_{k+1}}(|(\bar{e}_{k+1}, \bar{\mu}_{k1}, \bar{\mu}_{k2})|)$. This, together with (F.21), guarantees that there exists a $\lambda_{\Phi_i} \in \mathcal{K}_\infty$ such that for all $(\bar{x}_i, \bar{\mu}_{(i-1)1}, \bar{\mu}_{(i-1)2}, z)$, any $\phi_i \in \Phi_i(\bar{x}_i, \bar{\mu}_{(i-1)1}, \bar{\mu}_{(i-1)2}, z)$ satisfies

$$|\phi_i| \leq \lambda_{\Phi_i}(|(\bar{e}_i, z, \bar{\mu}_{(i-1)1}, \bar{\mu}_{(i-1)2})|), \tag{F.25}$$

where $\bar{e}_i := [e_1, \ldots, e_i]^T$.

Define

$$\begin{aligned}
&\Phi_i^*(e_{i+1}, \bar{x}_i, \bar{\mu}_{(i-1)1}, \bar{\mu}_{(i-1)2}, z) \\
&= \{\phi_i + e_{i+1} : \phi_i \in \Phi_i(\bar{x}_i, \bar{\mu}_{(i-1)1}, \bar{\mu}_{(i-1)2}, z)\}. \tag{F.26}
\end{aligned}$$

From (5.105), $x_{i+1} - e_{i+1} \in S_i(\bar{x}_i, \bar{\mu}_{i1}, \bar{\mu}_{i2})$. Then, equation (F.18) can be rewritten as (5.106).

From (F.25), we can find a $\lambda_{\Phi_i^*} \in \mathcal{K}_\infty$ such that any $\phi_i^* \in \Phi_i^*(e_{i+1}, \bar{x}_i, \bar{\mu}_{(i-1)1}, \bar{\mu}_{(i-1)2}, z)$ satisfies (5.107) for all $(e_{i+1}, \bar{x}_i, \bar{\mu}_{(i-1)1}, \bar{\mu}_{(i-1)2}, z)$.

By also considering the cases of $e_i = 1$ and $e_i < 0$, Lemma 5.3 can be proved.

F.3   PROOF OF LEMMA 5.4

Note that $e_0 = z$. With (5.107) satisfied, one can find $\lambda_{\Phi_i^*}^{e_k} \in \mathcal{K}_\infty$ for $k = 0, \ldots, i+1$ and $\lambda_{\Phi_i^*}^{\mu_{k1}}, \lambda_{\Phi_i^*}^{\mu_{k2}} \in \mathcal{K}_\infty$ for $k = 1, \ldots, i-1$ such that for any $\phi_i^* \in \Phi_i^*(e_{i+1}, \bar{x}_i, \bar{\mu}_{(i-1)1}, \bar{\mu}_{(i-1)2}, z)$, it holds that

$$|\phi_i^*| \leq \Sigma_{k=1}^{i+1} \lambda_{\Phi_i^*}^{e_k}(|e_k|) + \Sigma_{k=1}^{i-1} \left(\lambda_{\Phi_i^*}^{\mu_{k1}}(\mu_{k1}) + \lambda_{\Phi_i^*}^{\mu_{k2}}(\mu_{k2}) \right). \qquad (F.27)$$

By convenience, let $\gamma_{e_i}^{e_i} = \mathrm{Id}$. Define

$$\Pi_i(s) = \lambda_{\Phi_i^*}^{e_0} \circ \underline{\alpha}_0^{-1} \circ \left(\gamma_{e_i}^{e_0} \right)^{-1} \circ \alpha_V(s) + \Sigma_{k=1}^{i+1} \lambda_{\Phi_i^*}^{e_k} \circ \alpha_V^{-1} \circ \left(\gamma_{e_i}^{e_k} \right)^{-1} \circ \alpha_V(s)$$
$$+ \Sigma_{k=1}^{i-1} \lambda_{\Phi_i^*}^{\mu_{k1}} \circ \left(\gamma_{e_i}^{\mu_{k1}} \right)^{-1} \circ \alpha_V(s) + \Sigma_{k=1}^{i-1} \lambda_{\Phi_i^*}^{\mu_{k2}} \circ \left(\gamma_{e_i}^{\mu_{k2}} \right)^{-1} \circ \alpha_V(s)$$
$$+ \frac{\iota_i}{2} s \qquad (F.28)$$

for $s \in \mathbb{R}_+$. Then, $\Pi_i \in \mathcal{K}_\infty$.

From Lemma 1 in [123], for any $0 < c_{i1}, c_{i2} < 1$, $\epsilon_i > 0$, one can find a $\nu_i : \mathbb{R}_+ \to \mathbb{R}_+$ that is positive, nondecreasing and continuously differentiable on $(0, \infty)$, and satisfies

$$(1 - c_{i2})(1 - c_{i1})\nu_i \left((1 - c_{i1})s \right) s \geq \Pi_i(s) \qquad (F.29)$$

for $s \geq \sqrt{2\epsilon_i}$. With the ν_i satisfying (F.29), define $\kappa_i(r) = -\nu_i(|r|)r$ for $r \in \mathbb{R}$. Noticing that $\lim_{t \to 0^+} \frac{d\kappa_i(r)}{dr} = \lim_{t \to 0^-} \frac{d\kappa_i(r)}{dr}$, κ_i is continuously differentiable, odd, strictly decreasing, and radially unbounded.

Recall that $V_k(e_k) = \alpha_V(|e_k|) = \frac{1}{2}e_k^2$ for $k = 1, \ldots, n$. We use V_k instead of $V_k(e_k)$ for $k = 1, \ldots, n$. Consider the case of

$$V_i \geq \max_{k=1,\ldots,i-1} \left\{ \begin{array}{l} \gamma_{e_i}^{e_0}(V_0), \gamma_{e_i}^{e_k}(V_k), \gamma_{e_i}^{e_{i+1}}(V_{i+1}), \\ \gamma_{e_i}^{\mu_{k1}}(\mu_{k1}), \gamma_{e_i}^{\mu_{k2}}(\mu_{k2}), \gamma_{e_i}^{\mu_{i1}}(\mu_{i1}), \gamma_{e_i}^{\mu_{i2}}(\mu_{i2}), \epsilon_i \end{array} \right\}. \qquad (F.30)$$

In this case, we have

$$\Pi_i(|e_i|) - \frac{\iota_i}{2}|e_i| \geq \phi_i^* \qquad (F.31)$$

for all $\phi_i^* \in \Phi_i^*(e_{i+1}, \bar{x}_i, \bar{\mu}_{(i-1)1}, \bar{\mu}_{(i-1)2}, z)$. And it also holds that

$$\mu_{i1} \leq c_{i1}|e_i| \qquad (F.32)$$
$$\mu_{i2} \leq c_{i2}\bar{\kappa}_i \left((1 - c_{i1})|e_i| \right) |e_i| \qquad (F.33)$$
$$|e_i| \geq \sqrt{2\epsilon_i}. \qquad (F.34)$$

When $e_i \neq 0$, with $0 < c_{i1} < 1$, for $\varsigma_{i-1} \in S_{i-1}$ and $|b_{i1}| \leq \max\{c_{i1}|e_i|, \mu_{i1}\} = c_{i1}|e_i|$, we have

$$|x_i - \varsigma_{i-1} + b_{i1}| \geq (1 - c_{i1})|e_i| \qquad (F.35)$$

$$\mathrm{sgn}(x_i - \varsigma_{i-1} + b_{i1}) = \mathrm{sgn}(e_i) \qquad (F.36)$$

and thus

$$\nu_i(|x_i - \varsigma_{i-1} + b_{i1}|)|x_i - \varsigma_{i-1} + b_{i1}| \geq (1 - c_{i1})\nu_i((1 - c_{i1})|e_i|)|e_i|. \qquad (F.37)$$

In the case of (F.30), for any $\phi_i^* \in \Phi_i^*(e_{i+1}, \bar{x}_i, \bar{\mu}_{(i-1)1}, \bar{\mu}_{(i-1)2}, z)$, with $\varsigma_{i-1} \in S_{i-1}$, $|b_{i1}| \leq \max\{c_{i1}|e_i|, \mu_{i1}\}$ and $|b_{i2}| \leq \mu_{i2}$, using (F.31)–(F.37), we have

$$
\begin{aligned}
&\nabla V_i \left(\kappa_i(x_i - \varsigma_{i-1} + b_{i1}) + b_{i2} + \phi_i^* \right) \\
&= e_i \left(-\nu_i(|x_i - \varsigma_{i-1} + b_{i1}|)(x_i - \varsigma_{i-1} + b_{i1}) + b_{i2} + \phi_i^* \right) \\
&\leq -\nu_i(|x_i - \varsigma_{i-1} + b_{i1}|)|x_i - \varsigma_{i-1} + b_{i1}||e_i| + |b_{i2}||e_i| + |\phi_i^*||e_i| \\
&\leq -(1 - c_{i2})(1 - c_{i1})\nu_i((1 - c_{i1})|e_i|)|e_i|^2 + \Pi_i(|e_i|)|e_i| - \frac{\iota_i}{2}|e_i|^2 \\
&\leq -\frac{\iota_i}{2}|e_i|^2 = -\iota_i \alpha_V(|e_i|), \qquad (F.38)
\end{aligned}
$$

which implies (5.111).

F.4   PROOF OF LEMMA 5.5

For convenience of notation, define $v_n = u$. Note that $S_1(\bar{x}_1, \bar{\mu}_{11}, \bar{\mu}_{12})$ defined in (5.101) is in the form of (5.104) with $S_0(\bar{x}_0, \bar{\mu}_{01}, \bar{\mu}_{02}) := \{0\}$. Then, $v_0 \in S_0(\bar{x}_0, \bar{\mu}_{01}, \bar{\mu}_{02})$. Suppose that $v_{i-1} \in S_{i-1}(\bar{x}_{i-1}, \bar{\mu}_{(i-1)1}, \bar{\mu}_{(i-1)2})$. We will find ς_{i-1}, b_{i1} and b_{i2} satisfying $\varsigma_{i-1} \in S_{i-1}(\bar{x}_{i-1}, \bar{\mu}_{(i-1)1}, \bar{\mu}_{(i-1)2})$, $|b_{i1}| \leq \max\{c_{i1}|e_i|, \mu_{i1}\}$, and $|b_{i2}| \leq \mu_{i2}$, respectively, such that

$$v_i = \kappa_i(x_i - \varsigma_{i-1} + b_{i1}) + b_{i2} \in S_i(\bar{x}_i, \bar{\mu}_{i1}, \bar{\mu}_{i2}). \qquad (F.39)$$

By applying this reasoning repeatedly, property (5.118) can be proved. We consider only the case of $e_i \geq 0$. The proof for the case of $e_i < 0$ is similar. We study the following cases (A) and (B).

(A) $|\kappa_i(q_{i1}(x_i - v_{i-1}, \mu_{i1}))| \leq M_{i2}\mu_{i2}$.
       With Assumption 5.6 satisfied, one can find a $|b_{i2}| \leq \mu_{i2}$ such that

$$q_{i2}(\kappa_i(q_{i1}(x_i - v_{i-1}, \mu_{i1})), \mu_{i2}) = \kappa_i(q_{i1}(x_i - v_{i-1}, \mu_{i1})) + b_{i2}. \qquad (F.40)$$

(A1) $|x_i - v_{i-1}| \leq M_{i1}\mu_{i1}$.

In this case, Assumption 5.6 implies that there exists a $|b_{i1}| \leq \mu_{i1}$ such that

$$q_{i1}(x_i - v_{i-1}, \mu_{i1}) = x_i - v_{i-1} + b_{i1}. \tag{F.41}$$

Choose $\varsigma_{i-1} = v_{i-1}$. Then, $\varsigma_{i-1} \in S_{i-1}(\bar{x}_{i-1}, \bar{\mu}_{(i-1)1}, \bar{\mu}_{(i-1)2})$ and

$$q_{i1}(x_i - v_{i-1}, \mu_{i1}) = x_i - \varsigma_{i-1} + b_{i1}. \tag{F.42}$$

(A2) $|x_i - v_{i-1}| > M_{i1}\mu_{i1}$.

In this case, Assumption 5.6 implies that there exists a $|b_{i1}| \leq \mu_{i1}$ such that

$$q_{i1}(x_i - v_{i-1}, \mu_{i1}) = \operatorname{sgn}(x_i - v_{i-1})M_{i1}\mu_{i1} + b_{i1}. \tag{F.43}$$

We study the following two cases:

- $e_i > 0$. Recall (5.105) and (5.116). In this case, we have $x_i > v_{i-1}$ and

$$x_i - v_{i-1} > M_{i1}\mu_{i1} \geq e_i$$
$$= x_i - \max S_{i-1}(\bar{x}_{i-1}, \bar{\mu}_{(i-1)1}, \bar{\mu}_{(i-1)2}). \tag{F.44}$$

One can find a $\varsigma_{i-1} \in [v_{i-1}, \max S_{i-1}(\bar{x}_{i-1}, \bar{\mu}_{(i-1)1}, \bar{\mu}_{(i-1)2})]$ such that $x_i - \varsigma_{i-1} = M_{i1}\mu_{i1}$ and thus

$$q_{i1}(x_i - v_{i-1}, \mu_{i1}) = x_i - \varsigma_{i-1} + b_{i1}. \tag{F.45}$$

- $e_i = 0$. In this case, by using (5.105), we have $x_i \in S_{i-1}(\bar{x}_{i-1}, \bar{\mu}_{(i-1)1}, \bar{\mu}_{(i-1)2})$ and can directly find a $\varsigma_{i-1} \in S_{i-1}(\bar{x}_{i-1}, \bar{\mu}_{(i-1)1}, \bar{\mu}_{(i-1)2})$, which is closer to x_i than v_{i-1} such that $x_i - \varsigma_{i-1} = \operatorname{sgn}(x_i - v_{i-1})M_{i1}\mu_{i1}$ and thus

$$q_{i1}(x_i - v_{i-1}, \mu_{i1}) = x_i - \varsigma_{i-1} + b_{i1}. \tag{F.46}$$

From (F.42) and (F.46), in the case of $|\kappa_i(q_{i1}(x_i - v_{i-1}))| \leq M_{i2}\mu_{i2}$, we can find $\varsigma_{i-1} \in S_{i-1}(\bar{x}_{i-1})$, $|b_{i1}| \leq \mu_{i1}$ and $|b_{i2}| \leq \mu_{i2}$ such that

$$v_i = q_{i2}(\kappa_i(q_{i1}(x_i - v_{i-1}))) = \kappa_i(x_i - \varsigma_{i-1} + b_{i1}) + b_{i2}. \tag{F.47}$$

(B) $|\kappa_i(q_{i1}(x_i - v_{i-1}, \mu_{i1}))| > M_{i2}\mu_{i2}$.

Before the discussions, we give the following lemma.

Lemma F.1 *Under the conditions of Lemma 5.5, if $|\kappa_i(q_{i1}(x_i - v_{i-1}, \mu_{i1}))| > M_{i2}\mu_{i2}$, then*

$$\operatorname{sgn}(x_i - v_{i-1}) = \operatorname{sgn}(q_{i1}(x_i - v_{i-1}, \mu_{i1})). \tag{F.48}$$

The proof of Lemma F.1 is in Appendix F.4.1

Note that κ_i is an odd and strictly decreasing function. Then, we have

$$
\begin{aligned}
\operatorname{sgn}(\kappa_i(q_{i1}(x_i - v_{i-1}, \mu_{i1}))) &= -\operatorname{sgn}(q_{i1}(x_i - v_{i-1}, \mu_{i1})) \\
&= -\operatorname{sgn}(x_i - v_{i-1}).
\end{aligned}
\tag{F.49}
$$

Under Assumption 5.6, using Lemma F.1, one can find a $|b_{i2}| \le \mu_{i2}$ such that

$$
\begin{aligned}
&q_{i2}(\kappa_i(q_{i1}(x_i - v_{i-1}, \mu_{i1})), \mu_{i2}) \\
&= \operatorname{sgn}(\kappa_i(q_{i1}(x_i - v_{i-1}, \mu_{i1}))) M_{i2}\mu_{i2} + b_{i2} \\
&= -\operatorname{sgn}(x_i - v_{i-1}) M_{i2}\mu_{i2} + b_{i2}.
\end{aligned}
\tag{F.50}
$$

(B1) $e_i > 0$.

In this case, using (5.105), we have $x_i > v_{i-1}$ and thus

$$
\kappa_i(q_{i1}(x_i - v_{i-1}, \mu_{i1})) < 0,
\tag{F.51}
$$
$$
q_{i2}(\kappa_i(q_{i1}(x_i - v_{i-1}, \mu_{i1}))) = -M_{i2}\mu_{i2} + b_{i2}.
\tag{F.52}
$$

With $|\kappa_i(q_{i1}(x_i - v_{i-1}, \mu_{i1}))| > M_{i2}\mu_{i2}$, property (F.51) implies

$$
\kappa_i(q_{i1}(x_i - v_{i-1}, \mu_{i1})) < -M_{i2}\mu_{i2}.
\tag{F.53}
$$

Consider the following two cases:

− $x_i - v_{i-1} \le M_{i1}\mu_{i1}$. In this case, under Assumption 5.6, one can find a $|b'_{i1}| \le \mu_{i1}$ such that

$$
\begin{aligned}
\kappa_i(x_i - v_{i-1} + b'_{i1}) &= \kappa_i(q_{i1}(x_i - v_{i-1}, \mu_{i1})) \\
&< -M_{i2}\mu_{i2}.
\end{aligned}
\tag{F.54}
$$

− $x_1 - v_{i-1} > M_{i1}\mu_{i1}$. In this case, under Assumption 5.6, one can find a $|b'_{i1}| \le \mu_{i1}$ such that

$$
\begin{aligned}
\kappa_i(M_{i1}\mu_{i1} + b'_{i1}) &= \kappa_i(q_{i1}(x_i - v_{i-1}, \mu_{i1})) \\
&< -M_{i2}\mu_{i2}.
\end{aligned}
\tag{F.55}
$$

By using the strictly decreasing property of κ_i, we have

$$
\kappa_i(x_i - v_{i-1} + b'_{i1}) < \kappa_i(M_{i1}\mu_{i1} + b'_{i1}) < -M_{i2}\mu_{i2}.
\tag{F.56}
$$

Thus, in both the cases above, one can find a $|b'_{i1}| \le \mu_{i1}$ such that

$$
\kappa_i(x_i - v_{i-1} + b'_{i1}) < \kappa_i(M_{i1}\mu_{i1} + b'_{i1}) < -M_{i2}\mu_{i2}.
\tag{F.57}
$$

From (5.117), we have

$$
\bar{\kappa}_i((1 - c_{i1})|e_i|) < M_{i2}\mu_{i2}.
\tag{F.58}
$$

From the definition of e_i, using $e_i > 0$, we have

$$\kappa_i(x_i - \max S_{i-1}(\bar{x}_{i-1}, \bar{\mu}_{(i-1)1}, \bar{\mu}_{(i-1)2}) - c_{i1}e_i)$$
$$> -M_{i2}\mu_{i2}. \tag{F.59}$$

From (F.57) and (F.59), by using the continuity of κ_i, one can find a $\varsigma_{i-1} \in [v_{i-1}, \max S_{i-1}(\bar{x}_{i-1}, \bar{\mu}_{(i-1)1}, \bar{\mu}_{(i-1)2})]$ and a $b_{i1} \in [-c_{i1}e_i, b'_{i1}]$ such that

$$\kappa_i(x_i - \varsigma_{i-1} + b_{i1}) = -M_{i2}\mu_{i2}. \tag{F.60}$$

Recall (F.52). We have

$$v_i = q_{i2}(\kappa_i(q_{i1}(x_i - v_{i-1}, \mu_{i1})))$$
$$= \kappa_i(x_i - \varsigma_{i-1} + b_{i1}) + b_{i2}. \tag{F.61}$$

(B2) $e_i = 0$.

In this case, $x_i \in S_{i-1}(\bar{x}_{i-1}, \bar{\mu}_{(i-1)1}, \bar{\mu}_{(i-1)2})$. From Lemma F.1, we have $x_i - v_{i-1} \neq 0$. Consider the following two cases:

- $|x_i - v_{i-1}| \leq M_{i1}\mu_{i1}$. In this case, define $\varsigma'_{i-1} = v_{i-1}$. With Assumption 5.6, one can find a $|b'_{i1}| \leq \mu_{i1}$ such that

$$\kappa_i(x_i - \varsigma'_{i-1} + b'_{i1}) = \kappa_i(q_{i1}(x_i - v_{i-1}, \mu_{i1}))$$
$$\begin{cases} > M_{i2}\mu_{i2}, & \text{if } x_i < \varsigma'_{i-1} \\ < -M_{i2}\mu_{i2}, & \text{if } x_i > \varsigma'_{i-1}. \end{cases} \tag{F.62}$$

We used $|\kappa_i(q_{i1}(x_i - v_{i-1}, \mu_{i1}))| > M_{i2}\mu_{i2}$ and (F.49) for the last part of (F.62).

- $|x_i - v_{i-1}| > M_{i1}\mu_{i1}$. In this case, under Assumption 5.6, one can find a $|b'_{i1}| \leq \mu_{i1}$ such that

$$\kappa_i(\text{sgn}(x_i - v_{i-1})M_{i1}\mu_{i1} + b'_{i1})$$
$$= \kappa_i(q_{i1}(x_i - v_{i-1}, \mu_{i1})) \begin{cases} > M_{i2}\mu_{i2}, & \text{if } x_i < v_{i-1} \\ < -M_{i2}\mu_{i2}, & \text{if } x_i > v_{i-1}. \end{cases} \tag{F.63}$$

In the case of $|x_i - v_{i-1}| > M_{i1}\mu_{i1}$, one can find a $\varsigma'_{i-1} \in [x_i, v_{i-1}]$ satisfying $\text{sgn}(x_i - \varsigma'_{i-1}) = \text{sgn}(x_i - v_{i-1})$ and $\text{sgn}(x_i - v_{i-1})M_{i1}\mu_{i1} = x_i - \varsigma'_{i-1}$. In this way, we achieve

$$\kappa_i(x_i - \varsigma'_{i-1} + b'_{i1}) \begin{cases} > M_{i2}\mu_{i2}, & \text{if } x_i < \varsigma'_{i-1} \\ < -M_{i2}\mu_{i2}, & \text{if } x_i > \varsigma'_{i-1}. \end{cases} \tag{F.64}$$

Note that $\kappa_i(x_i - x_i + 0) = \kappa_i(0) = 0$. By using the continuity of κ_i, one can find a $\varsigma_{i-1} \in [x_i, \varsigma'_{i-1}]$ and a $b_{i1} \in [0, b'_{i1}]$ such that

$$\text{sgn}(x_i - \varsigma_{i-1}) = \text{sgn}(x_i - \varsigma'_{i-1}) = \text{sgn}(x_i - v_{i-1}) \qquad (F.65)$$

$$\kappa_i(x_i - \varsigma_{i-1} + b_{i1}) = -\text{sgn}(x_i - v_{i-1})M_{i2}\mu_{i2}. \qquad (F.66)$$

Clearly, $\varsigma_{i-1} \in S_{i-1}(\bar{x}_{i-1}, \bar{\mu}_{(i-1)1}, \bar{\mu}_{(i-1)2})$ and $|b_{i1}| \leq \mu_{i1}$. Recall (F.50). We have

$$v_i = q_{i2}(\kappa_i(q_{i1}(x_i - v_{i-1}, \mu_{i1})))$$
$$= \kappa_i(x_i - \varsigma_{i-1} + b_{i1}) + b_{i2}. \qquad (F.67)$$

Considering both cases (A) and (B), the proof of Lemma 5.5 is concluded.

F.4.1   PROOF OF LEMMA F.1

Consider the following two cases:

- $|x_i - v_{i-1}| > M_{i1}\mu_{i1}$. In this case, under Assumption 5.6, one can find a $|b_{i1}| \leq \mu_{i1}$ such that

$$q_{i1}(x_i - v_{i-1}, \mu_{i1}) = \text{sgn}(x_i - v_{i-1})M_{i1}\mu_{i1} + b_{i1} \qquad (F.68)$$

Note that $M_{i1} > 2$. Thus,

$$\text{sgn}(x_i - v_{i-1}) = \text{sgn}(q_{i1}(x_i - v_{i-1}, \mu_{i1})). \qquad (F.69)$$

- $|x_i - v_{i-1}| \leq M_{i1}\mu_{i1}$. In this case, under Assumption 5.6, one can find a $|b_{i1}| \leq \mu_{i1}$ such that

$$q_{i1}(x_i - v_{i-1}, \mu_{i1}) = x_i - v_{i-1} + b_{i1}. \qquad (F.70)$$

Condition $|\kappa_i(q_{i1}(x_i - v_{i-1}, \mu_{i1}))| > M_{i2}\mu_{i2}$ implies $q_{i1}(x_i - v_{i-1}, \mu_{i1}) \neq 0$.

If $\text{sgn}(x_i - v_{i-1}) \neq \text{sgn}(q_{i1}(x_i - v_{i-1}, \mu_{i1}))$, then $\text{sgn}(b_{i1}) = \text{sgn}(q_{i1}(x_i - v_{i-1}, \mu_{i1}))$ and $|b_{i1}| > |x_i - v_{i-1}|$. Thus, $|x_i - v_{i-1} + b_{i1}| \leq |b_{i1}| \leq \mu_{i1}$. Note that $\frac{1}{M_{i1}} < c_{i1} < 0.5$. Then, we can derive

$$|\kappa_i(q_{i1}(x_i - v_{i-1}, \mu_{i1}))| \leq \bar{\kappa}_i(\mu_{i1})$$
$$\leq \bar{\kappa}_i\left(\frac{1 - c_{i1}}{c_{i1}}\mu_{i1}\right)$$
$$< \bar{\kappa}_i((1 - c_{i1})M_{i1}\mu_{i1})$$
$$= M_{i2}\mu_{i2}, \qquad (F.71)$$

which leads to a contradiction with $|\kappa_i(q_{i1}(x_i - v_{i-1}))| > M_{i2}\mu_{i2}$. We used (5.128) for the last equality in (F.71).

F.5   PROOF OF LEMMA 5.8

Recall that if $\chi_1, \chi_2 \in \mathcal{K}_\infty$ satisfies $\chi_1(s) > \chi_2(s)$ for $s \in \mathbb{R}_+$, then $(\mathrm{Id} - \tilde{\chi}) \circ \chi_1(s) \geq \chi_2(s)$ for $s \in \mathbb{R}_+$ with $\tilde{\chi} := \mathrm{Id} - \chi_2 \circ \chi_1^{-1}$ being continuous and positive definite. For each $i = 1, \ldots, n$, with (5.115) satisfied, we can find a continuous and positive definite ρ_i such that

$$\sigma_i \circ \alpha_V \left(\frac{1}{c_{i1}} s \right) \leq (\mathrm{Id} - \rho_i) \circ \sigma_i \circ \alpha_V (M_{i1} s) \tag{F.72}$$

$$\sigma_i \circ \alpha_V \left(\frac{1}{1 - c_{i1}} \bar{\kappa}_i^{-1} \left(\frac{1}{c_{i2}} s \right) \right) \leq (\mathrm{Id} - \rho_i) \circ \sigma_i \circ \alpha_V \left(\frac{1}{1 - c_{i1}} \bar{\kappa}_i^{-1} (M_{i2} s) \right) \tag{F.73}$$

for all $s \in \mathbb{R}_+$. Define $\rho(s) = \min_{i=1,\ldots,n}\{\rho_i(s)\}$ for $s \in \mathbb{R}_+$. Then, ρ is continuous and positive definite. Using (5.126), (5.127), (5.132), and (5.138), we have

$$B_2(\bar{\mu}_{n1}(t), \bar{\mu}_{n2}(t)) \leq (\mathrm{Id} - \rho)(\Theta(t)) \tag{F.74}$$

for any $t \in \mathbb{R}_+$.

Note that the zooming variables $\mu_{n1}(t)$ and $\mu_{n2}(t)$ are constant on $[t_k, t_{k+1})$, that is, $\mu_{n1}(t) = \mu_{n1}(t_k)$ and $\mu_{n2}(t) = \mu_{n2}(t_k)$ for $t \in [t_k, t_{k+1})$. Suppose (5.148) holds. We study the following two cases:

(a) $V(e(X(t_{k+1}), \bar{\mu}_{n1}(t_k), \bar{\mu}_{n2}(t_k))) < \max\{(\mathrm{Id} - \rho)(\Theta(t_k)), \theta_0\}$.
(b) $V(e(X(t_{k+1}), \bar{\mu}_{n1}(t_k), \bar{\mu}_{n2}(t_k))) \geq \max\{(\mathrm{Id} - \rho)(\Theta(t_k)), \theta_0\}$. In this case, from (5.131), (5.148), and (F.74), it follows that $V(e(X(t), \bar{\mu}_{n1}(t), \bar{\mu}_{n2}(t)))$ is strictly decreasing for $t \in [t_k, t_{k+1})$ and

$$\max\{(\mathrm{Id} - \rho)(\Theta(t_k)), \theta_0\} \leq V(e(X(t), \bar{\mu}_{n1}(t), \bar{\mu}_{n2}(t)))$$
$$\leq \Theta(t_k) \tag{F.75}$$

for all $t \in [t_k, t_{k+1})$. By using (5.131), we have

$$V(e(X(t_{k+1}), \bar{\mu}_{n1}(t_k), \bar{\mu}_{n2}(t_k)))$$
$$\leq V(e(X(t_k), \bar{\mu}_{n1}(t_k), \bar{\mu}_{n2}(t_k)))$$
$$\quad - \int_{t_k}^{t_{k+1}} \alpha(V(e(X(\tau), \bar{\mu}_{n1}(\tau), \bar{\mu}_{n2}(\tau)))) d\tau$$
$$\leq \Theta(t_k) - t_d \cdot \min_{\max\{(\mathrm{Id}-\rho)(\Theta(t_k)), \theta_0\} \leq v \leq \Theta(t_k)} \alpha(v)$$
$$\leq \Theta(t_k) - t_d \cdot \min_{(\mathrm{Id}-\rho)(\Theta(t_k)) \leq v \leq \Theta(t_k)} \alpha(v), \tag{F.76}$$

where $t_d = t_{k+1} - t_k$. Define $\rho'(s) = t_d \cdot \min_{(\mathrm{Id}-\rho)(s) \leq v \leq s} \alpha(v)$ for $s \in \mathbb{R}_+$. Then, it can be directly verified that ρ' is continuous and positive definite and that

$$V(e(X(t_{k+1}), \bar{\mu}_{n1}(t_k), \bar{\mu}_{n2}(t_k))) \leq (\mathrm{Id} - \rho')(\Theta(t_k)). \tag{F.77}$$

Lemma 5.8 is proved by finding a continuous and positive definite function $\bar{\rho}$ such that $(\mathrm{Id} - \bar{\rho}) \in \mathcal{K}_\infty$ and $(\mathrm{Id} - \bar{\rho})(s) \geq \max\{(\mathrm{Id} - \rho)(s), (\mathrm{Id} - \rho')(s)\}$ for $s \in \mathbb{R}_+$.

References

1. D. Aeyels and J. Peuteman. A new asymptotic stability criterion for nonlinear time-variant differential equations. *IEEE Transactions on Automatic Control*, 43:968–971, 1998.
2. A. P. Aguiar and A. M. Pascoal. Coordinated path-following control for nonlinear systems with logic-based communication. In *Proceedings of the 46th Conference on Decision and Control*, pages 1473–1479, 2007.
3. M. A. Aizerman and F. R. Gantmacher. *Absolute Stability of Regulator Systems*. Holden Day, San Francisco, 1963.
4. B. D. O. Anderson. The small gain theorem, the passivity theorem and their equivalence. *Journal of Franklin Institute*, 293:105–115, 1972.
5. B. D. O. Anderson and S. Vongpanitlerd. *Network Analysis and Synthesis: A Modern Systems Approach*. Prentice-Hall, New Jersey, 1973.
6. D. Angeli and E. D. Sontag. Forward completeness, unbounded observability and Their Lyapunov characterizations. *Systems & Control Letters*, 38:209–217, 1999.
7. D. Angeli, E. D. Sontag, and Y. Wang. A characterization of integral input-to-state stability. *IEEE Transactions on Automatic Control*, 45:1082–1097, 2000.
8. M. Arcak. Passivity as a design tool for group coordination. *IEEE Transactions on Automatic Control*, 52:1380–1390, 2007.
9. M. Arcak. Passivity approach to network stability analysis and distributed control synthesis. In W. Levine, editor, *The Control Handbook*, pages 1–18. CRC Press, 2010.
10. M. Arcak and P. V. Kokotović. Nonlinear observers: A circle criterion design and robustness analysis. *Automatica*, 37:1923–1930, 2001.
11. A. Astolfi, D. Karagiannis, and R. Ortega. *Nonlinear and Adaptive Control with Applications*. Springer, 2008.
12. K. J. Åström and B. Wittenmark. *Adaptive Control*. Pearson Education, second edition, 1995.
13. J.-P. Aubin and H. Frankowska. *Set-Valued Analysis*. Birkhäuser, Boston, 1990.
14. A. Bacciotti. *Local Stabilizability of Nonlinear Control Systems*. World Scientific, 1992.
15. A. Bacciotti and L. Rosier. Lyapunov stability and Lagrange stability: Inverse theorems for discontinuous systems. *Mathematics of Control, Signals and Systems*, 11:101–125, 1998.
16. A. Bacciotti and L. Rosier. *Liapunov Functions and Stability in Control Theory*. Springer-Verlag, Berlin, second edition, 2005.
17. H. Bai, M. Arcak, and J. Wen. *Cooperative Control Design: A Systematic, Passivity-Based Approach*. Springer, New York, 2011.
18. D. P. Bertsekas and J. N. Tsitsiklis. *Parallel and Distributed Computation: Numerical Methods*. Athena Scientific, 1997.
19. R. W. Brockett and D. Liberzon. Quantized feedback stabilization of linear

systems. *IEEE Transactions on Automatic Control*, 45:1279–1289, 2000.

20. B. Brogliato, R. Lozano, B. Maschke, and O. Egeland. *Dissipative Systems Analysis and Control: Theory and Applications*. Springer-Verlag, London, second edition, 2007.

21. C. I. Byrnes, F. Delli Priscoli, and A. Isidori. *Output Regulation of Uncertain Nonlinear Systems*. Birkhäuser, 1997.

22. C. Cai and A. R. Teel. Input-output-to-state stability for discrete-time systems. *Automatica*, 44:326–336, 2008.

23. C. Cai and A. R. Teel. Characterizations of input-to-state stability for hybrid systems. *Systems & Control Letters*, 58:47–53, 2009.

24. Y. Cao and W. Ren. Distributed coordinated tracking with reduced interaction via a variable structure approach. *IEEE Transactions on Automatic Control*, 57:33–48, 2012.

25. D. Carnevale, A. R. Teel, and D. Nešić. A Lyapunov proof of an improved maximum allowable transfer interval for networked control systems. *IEEE Transactions on Automatic Control*, 52:892–897, 2007.

26. F. Ceragioli and C. De Persis. Discontinuous stabilization of nonlinear systems: Quantized and switching controls. *Systems & Control Letters*, 56:461–473, 2007.

27. J. W. Chapman, M. D. Ilic, C. A. King, L. Eng, and H. Kaufman. Stabilizing a multimachine power system via decentralized feedback linearizing excitation control. *IEEE Transactions on Power Systems*, 8:830–839, 1993.

28. C. T. Chen. *Linear System Theory and Design*. Oxford University Press, 1999.

29. P.-C. Chen, R. Salcedo, Q. Zhu, F. de Leon, D. Czarkowski, Z. P. Jiang, V. Spitsa, Z. Zabar, and R. E. Uosef. Analysis of voltage profile problems due to the penetration of distributed generation in low-voltage secondary distribution networks. *IEEE transactions on Power Delivery*, 27:2020–2028, 2012.

30. T. Chen and J. Huang. A small-gain approach to global stabilization of nonlinear feedforward systems with input unmodeled dynamics. *Automatica*, 46:1028–1034, 2010.

31. Z. Chen and J. Huang. A general formulation and solvability of the global robust output regulation problem. *IEEE Transactions on Automatic Control*, 50:448–462, 2005.

32. Z. Chen and J. Huang. Global robust output regulation problem for output feedback systems. *IEEE Transactions on Automatic Control*, 50:117–121, 2005.

33. D. Cheng, X. Hu, and T. Shen. *Analysis and Design of Nonlinear Control Systems*. Springer, 2011.

34. H.-D. Chiang. *Direct Methods for Stability Analysis of Electric Power Systems: Theoretical Foundation, BCU Methodologies, and Applications*. Wiley, 2011.

35. F. H. Clarke, Yu. S. Ledyaev, R. J. Stern, and P. R. Wolenski. Asymptotic stability and smooth Lyapunov functions. *Journal of Differential Equations*, 149:69–114, 1998.

36. F. H. Clarke, Yu. S. Ledyaev, R. J. Stern, and P. R. Wolenski. *Nonsmooth Analysis and Control Theory*. Springer, New York, 1998.

37. J.-M. Coron. *Control and Nonlinearity.* American Mathematical Society, 2007.

38. J. Cortés. Discontinuous dynamical systems: A tutorial on solutions, non-smooth analysis, and stability. *IEEE Control Systems Magazine*, 28:36–73, 2008.

39. J. Cortés, S. Martjnez, and F. Bullo. Robust rendezvous for mobile autonomous agents via proximity graphs in arbitrary dimensions. *IEEE Transactions on Automatic Control*, 51:1289–1298, 2006.

40. B. d'Andréa-Novel, G. Bastin, and G. Campion. Dynamic feedback linearization of nonholonomic wheeled mobile robots. In *Proceedings of the 1992 IEEE International Conference on Robotics and Automation*, pages 2527–2532, 1992.

41. S. Dashkovskiy, H. Ito, and F. Wirth. On a small gain theorem for ISS networks in dissipative Lyapunov form. *European Journal of Control*, 17:357–365, 2011.

42. S. Dashkovskiy and M. Kosmykov. Stability of networks of hybrid ISS systems. In *Proceedings of Joint 48th IEEE Conference on Decision and Control and 28th Chinese Control Conference*, pages 3870–3875, 2009.

43. S. Dashkovskiy, B. S. Rüffer, and F. R. Wirth. An ISS small-gain theorem for general networks. *Mathematics of Control, Signals and Systems*, 19:93–122, 2007.

44. S. Dashkovskiy, B. S. Rüffer, and F. R. Wirth. Small gain theorems for large scale systems and construction of ISS Lyapunov functions. *SIAM Journal on Control and Optimization*, 48:4089–4118, 2010.

45. P. De Leenheer, D. Angeli, and E. D. Sontag. On predator-prey systems and small-gain theorems. *Mathematical Biosciences and Engineering*, 2:25–42, 2005.

46. C. De Persis. n-bit stabilization of n-dimensional nonlinear systems in feed-forward form. *IEEE Transactions on Automatic Control*, 50:299–311, 2005.

47. J. P. Desai, J. P. Ostrowski, and V. Kumar. Modeling and control of formations of nonholonomic mobile robots. *IEEE Transactions on Robotics and Automation*, 17:905–908, 2001.

48. C. A. Desoer and M. Vidyasagar. *Feedback Systems: Input-Output Properties.* Academic Press, New York, 1975.

49. W. E. Dixon, D. M. Dawson, E. Zergeroglu, and A. Behal. *Nonlinear Control of Wheeled Mobile Robots.* Springer, London, 2001.

50. K. D. Do. Formation tracking control of unicycle-type mobile robots with limited sensing ranges. *IEEE Transactions on Control Systems Technology*, 16:527–538, 2008.

51. W. Dong and J. A. Farrell. Decentralized cooperative control of multiple nonholonomic dynamic systems with uncertainty. *Automatica*, 45:706–710, 2009.

52. T. Donkers and M. Heemels. Output-based event-triggered control with guaranteed $\mathcal{L}_\infty$-gain and improved and decentralized event-triggering. *IEEE Transactions on Automatic Control*, 57:1362–1376, 2012.

53. P. M. Dower and M. R. James. Dissipativity and nonlinear systems with finite power gain. *International Journal of Robust and Nonlinear Control*, 8:699–724, 1998.

54. J. Doyle, B. Francis, and A. Tannenbaum. *Feedback Control Systems.* MacMillan Publishing Co., 1992.

55. N. Elia and S. K. Mitter. Stabilization of linear systems with limited information. *IEEE Transactions on Automatic Control,* 46:1384–1400, 2001.

56. G. A. Enciso and E. D. Sontag. Global attractivity, I/O monotone small-gain theorems, and biological delay systems. *Discrete and Continuous Dynamical Systems,* 14:549–578, 2006.

57. H. R. Everett. *Sensors for Mobile Robots: Theory and Application.* A. K. Peters, 1995.

58. J. A. Fax and R. M. Murray. Information flow and cooperative control of vehicle formation. *IEEE Transactions on Automatic Control,* 49:1465–1476, 2004.

59. H. Federer. *Geometric Measure Theory.* Springer-Verlag, New York, 1969.

60. A. F. Filippov. *Differential Equations with Discontinuous Righthand Sides.* Kluwer Academic Publishers, 1988.

61. M. Fliess, J. L. Lévine, P. Martin, and P. Rouchon. Flatness and defect of non-linear systems: Introductory theory and examples. *International Journal of Control,* 61:1327–1361, 1995.

62. F. Forni, R. Sepulchre, and A. van der Schaft. On differential passivity of physical systems. *CoRR,* abs/1309.2558, 2013.

63. F. Forni, R. Sepulchre, and A. van der Schaft. On differentially dissipative dynamical systems. In *Proceedings of the 9th IFAC Symposium on Nonlinear Control Systems,* pages 15–20, 2013.

64. R. A. Freeman. Global internal stabilization does not imply global external stabilizability for small sensor disturbances. *IEEE Transactions on Automatic Control,* 40:2119–2122, 1995.

65. R. A. Freeman and P. V. Kokotović. Global robustness of nonlinear systems to state measurement disturbances. In *Proceedings of the 32nd IEEE Conference on Decision and Control,* pages 1507–1512, 1993.

66. R. A. Freeman and P. V. Kokotović. *Robust Nonlinear Control Design: State-space and Lyapunov Techniques.* Birkhäuser, Boston, 1996.

67. M. Fu and L. Xie. The sector bound approach to quantized feedback control. *IEEE Transactions on Automatic Control,* 50:1698–1711, 2005.

68. R. Ghabcheloo, A. P. Aguiar, A. Pascoal, C. Silvestre, I. Kaminer, and J. Hespanha. Coordinated path-following in the presence of communication losses and time delays. *SIAM Journal on Control and Optimization,* 48:234–265, 2009.

69. R. Goebel. Set-valued Lyapunov functions for difference inclusions. *Automatica,* 47:127–132, 2011.

70. R. Goebel, R. G. Sanfelice, and A. R. Teel. *Hybrid Dynamical Systems: Modeling, Stability, and Robustness.* Princeton University Press, 2012.

71. R. Goebel and A. R. Teel. Solution to hybrid inclusions via set and graphical convergence with stability theory applications. *Automatica,* 42:573–587, 2006.

72. W. M. Griggs, B. D. O. Anderson, A. Lanzon, and M. Rotkowitz. A stability result for interconnections of nonlinear systems with "mixed" small gain and passivity properties. In *Proceedings of the 46th IEEE Conference on Decision and Control,* pages 4489–4494, 2011.

73. L. Grüne. *Asymptotic Behavior of Dynamical and Control Systems under*

Perturbation and Discretization. Springer, Berlin, 2002.

74. L. Grüne. Input-to-state dynamical stability and its Lyapunov function characterization. *IEEE Transactions on Automatic Control*, 47:1499–1504, 2002.

75. Y. Guo, D. J. Hill, and Y. Wang. Nonlinear decentralized control of large-scale power systems. *Automatica*, 36:1275–1289, 2000.

76. W. M. Haddad and V. Chellaboina. *Nonlinear Dynamical Systems and Control: A Lyapunov-Based Approach.* Princeton University Press, 2008.

77. W. M. Haddad and S. G. Nersesov. *Stability and Control of Large-Scale Dynamical Systems: A Vector Dissipative Systems Approach.* Princeton University Press, 2011.

78. W. Hahn. *Stability of Motion.* Springer-Verlag, Berlin, 1967.

79. E. J. Hancock and A. Papachristodoulou. Structured sum of squares for networked systems analysis. In *Proceedings of the 50th IEEE Conference on Decision and Control and European Control Conference*, pages 7236–7241, 2011.

80. J. Harmard, A. R. Rapaport, and F. Mazenc. Output tracking of continuous bioreactors through recirculation and by-pass. *Automatica*, 42:1025–1032, 2006.

81. T. Hayakawa, H. Ishii, and K. Tsumura. Adaptive quantized control for linear uncertain discrete-time systems. *Automatica*, 45:692–700, 2009.

82. T. Hayakawa, H. Ishii, and K. Tsumura. Adaptive quantized control for nonlinear uncertain systems. *Systems & Control Letters*, 58:625–632, 2009.

83. W. P. M. H. Heemels, K. H. Johansson, and P. Tabuada. An introduction to event-triggered and self-triggered control. In *Proceedings of the 51st IEEE Conference on Decision and Control*, pages 3270–3285, 2012.

84. W. P. M. H. Heemels and S. Weiland. Input-to-state stability and interconnections of discontinuous dynamical systems. *Automatica*, 44:3079–3086, 2008.

85. J. P. Hespanha, D. Liberzon, D. Angeli, and E. D. Sontag. Nonlinear norm-observability notions and stability of switched systems. *IEEE Transactions on Automatic Control*, 50:154–168, 2005.

86. J. P. Hespanha, D. Liberzon, and A. R. Teel. Lyapunov conditions for input-to-state stability of impulsive systems. *Automatica*, 44:2735–2744, 2008.

87. D. J. Hill. Dissipativeness, stability theory and some remaining problems. In C. I. Byrnes, C. F. Martin, and R. E. Saeks, editors, *Analysis and Control of Nonlinear Systems*, Amsterdam, 1988. North-Holland.

88. D. J. Hill. A generalization of the small-gain theorem for nonlinear feedback systems. *Automatica*, 27:1043–1045, 1991.

89. D. J. Hill and P. J. Moylan. The stability of nonlinear dissipative systems. *IEEE Transactions on Automatic Control*, 21:708–711, 1976.

90. D. J. Hill and P. J. Moylan. Stability results for nonlinear feedback systems. *Automatica*, 13:377–382, 1977.

91. D. J. Hill and P. J. Moylan. Connections between finite-gain and asymptotic stability. *IEEE Transactions on Automatic Control*, 25:931–936, 1980.

92. D. J. Hill and P. J. Moylan. Dissipative dynamical systems: Basic input-output and state properties. *Journal of Franklin Institute*, 5:327–357, 1980.

93. D. J. Hill and P. J. Moylan. General instability results for interconnected systems. *SIAM Journal on Control and Optimization*, 21:256–279, 1983.

94. D. J. Hill, C. Wen, and G. C. Goodwin. Stability analysis of decentralised robust adaptive control. *Systems & Control Letters*, 11:277–284, 1988.

95. S. Hirche, T. Matiakis, and M. Buss. A distributed controller approach for delay-independent stability of networked control systems. *Automatica*, 45:1828–1836, 2009.

96. J. Hirschhorn. *Kinematics and Dynamics of Plane Mechanisms*. McGraw-Hill Book Company, 1962.

97. Y. Hong, G. Chen, and L. Bushnell. Distributed observers design for leader-following control of multi-agent networks. *Automatica*, 44:846–850, 2008.

98. Y. Hong, L. Gao, D. Cheng, and J. Hu. Lyapunov-based approach to multia-gent systems with switching jointly connected interconnection. *IEEE Transactions on Automatic Control*, 52:943–948, 2007.

99. J. Huang. *Nonlinear Output Regulation: Theory and Applications*. SIAM, Philadelphia, 2004.

100. J. Huang and Z. Chen. A general framework for tackling the output regulation problem. *IEEE Transactions on Automatic Control*, 49:2203–2218, 2004.

101. S. Huang, M. R. James, D. Nešić, and P. M. Dower. Analysis of input-to-state stability for discrete time nonlinear systems via dynamic programming. *Automatica*, 41:2055–2065, 2005.

102. S. Huang, M. R. James, D. Nešić, and P. M. Dower. A unified approach to controller design for achieving ISS and related properties. *IEEE Transactions on Automatic Control*, 50:1681–1697, 2005.

103. I.-A. F. Ihle, M. Arcak, and T. I. Fossen. Passivity-based designs for synchronized path-following. *Automatica*, 43:1508–1518, 2007.

104. B. Ingalls, E. D. Sontag, and Y. Wang. Generalizations of asymptotic gain characterizations of ISS to input-to-output stability. In *Proceedings of the 2001 American Control Conference*, pages 704–708, 2001.

105. A. Isidori. *Nonlinear Control Systems*. Springer, London, third edition, 1995.

106. A. Isidori. *Nonlinear Control Systems II*. Springer-Verlag, London, 1999.

107. H. Ito. State-dependent scaling problems and stability of interconnected iISS and ISS systems. *IEEE Transactions on Automatic Control*, 51:1626–1643, 2006.

108. H. Ito and Z. P. Jiang. Necessary and sufficient small gain conditions for integral input-to-state stable systems: A Lyapunov perspective. *IEEE Transactions on Automatic Control*, 54:2389–2404, 2009.

109. H. Ito, Z. P. Jiang, and P. Pepe. Construction of Lyapunov-Krasovskii functionals for networks of iISS retarded systems in small-gain formulation. *Automatica*, 49:3246–3257, 2013.

110. H. Ito, P. Pepe, and Z. P. Jiang. A small-gain condition for iISS of interconnected retarded systems based on Lyapunov-Krasovskii functionals. *Automatica*, 46:1646–1656, 2010.

111. A. Jadbabaie, J. Lin, and A. Morse. Coordination of groups of mobile autonomous agents using nearest neighbor rules. *IEEE Transactions on Automatic Control*, 48:988–1001, 2003.

112. M. R. James. A partial differential inequality for dissipative nonlinear systems. *Automatica*, 21:315–320, 1993.

113. Y. Jiang and Z. P. Jiang. Robust adaptive dynamic programming for large-scale systems with an application to multimachine power systems. *IEEE*

Transactions on Circuits and Systems–II: Express Briefs, 59:693–697, 2012.

114. Y. Jiang and Z. P. Jiang. Robust adaptive dynamic programming with an application to power systems. *IEEE Transactions on Neural Networks and Learning Systems*, 24:1150–1156, 2013.

115. Z. P. Jiang. A combined backstepping and small-gain approach to adaptive output-feedback control. *Automatica*, 35:1131–1139, 1999.

116. Z. P. Jiang. Decentralized and adaptive nonlinear tracking of large-scale systems via output feedback. *IEEE Transactions on Automatic Control*, 45:2122–2128, 2000.

117. Z. P. Jiang. Decentralized disturbance attenuating output-feedback trackers for large-scale nonlinear systems. *Automatica*, 38:1407–1415, 2002.

118. Z. P. Jiang. Decentralized control for large-scale nonlinear systems: A review of recent results. *Dynamics of Continuous, Discrete and Impulsive Systems Series B: Algorithms and Applications*, 11:537–552, 2004.

119. Z. P. Jiang and M. Arcak. Robust global stabilization with ignored input dynamics: An input-to-state stability (ISS) small-gain approach. *IEEE Transactions on Automatic Control*, 46:1411–1415, 2001.

120. Z. P. Jiang and Y. Jiang. Robust adaptive dynamic programming for linear and nonlinear systems: An overview. *European Journal of Control*, 19:417–425, 2013.

121. Z. P. Jiang, Y. Lin, and Y. Wang. Nonlinear small-gain theorems for discrete-time large-scale systems. In *Proceedings of the 27th Chinese Control Conference*, pages 704–708, 2008.

122. Z. P. Jiang and T. Liu. Quantized nonlinear control–A survey. *Acta Automatica Sinica*, 2013. In press.

123. Z. P. Jiang and I. M. Y. Mareels. A small-gain control method for non-linear cascade systems with dynamic uncertainties. *IEEE Transactions on Automatic Control*, 42:292–308, 1997.

124. Z. P. Jiang and I. M. Y. Mareels. Robust nonlinear integral control. *IEEE Transactions on Automatic Control*, 46:1336–1342, 2001.

125. Z. P. Jiang, I. M. Y. Mareels, and D. J. Hill. Robust control of uncertain non-linear systems via measurement feedback. *IEEE Transactions on Automatic Control*, 44:807–812, 1999.

126. Z. P. Jiang, I. M. Y. Mareels, and Y. Wang. A Lyapunov formulation of the nonlinear small-gain theorem for interconnected systems. *Automatica*, 32:1211–1215, 1996.

127. Z. P. Jiang and H. Nijmeijer. Tracking control of mobile robots: A case study in backstepping. *Automatica*, 33:1393–1399, 1997.

128. Z. P. Jiang and L. Praly. Preliminary results about robust Lagrange stability in adaptive nonlinear regulation. *International Journal of Adaptive Control and Signal Processing*, 6:285–307, 1992.

129. Z. P. Jiang, D. W. Repperger, and D. J. Hill. Decentralized nonlinear output-feedback stabilization with disturbance attenuation. *IEEE Transactions on Automatic Control*, 46:1623–1629, 2001.

130. Z. P. Jiang, A. R. Teel, and L. Praly. Small-gain theorem for ISS systems and applications. *Mathematics of Control, Signals and Systems*, 7:95–120, 1994.

131. Z. P. Jiang and Y. Wang. Input-to-state stability for discrete-time nonlinear systems. *Automatica*, 37:857–869, 2001.

132. Z. P. Jiang and Y. Wang. A converse Lyapunov theorem for discrete-time systems with disturbances. *Systems & Control Letters*, 45:49–58, 2002.

133. Z. P. Jiang and Y. Wang. Small-gain theorems on input-to-output stability. In *Proceedings of the 3rd International DCDIS Conference*, pages 220–224, 2003.

134. Z. P. Jiang and Y. Wang. A generalization of the nonlinear small-gain theorem for large-scale complex systems. In *Proceedings of the 7th World Congress on Intelligent Control and Automation*, pages 1188–1193, 2008.

135. I. Karafyllis and Z. P. Jiang. A small-gain theorem for a wide class of feedback systems with control applications. *SIAM Journal on Control and Optimization*, 46:1483–1517, 2007.

136. I. Karafyllis and Z. P. Jiang. Necessary and sufficient Lyapunov-like conditions for robust nonlinear stabilization. *ESAIM: Control, Optimisation and Calculus of Variations*, 16:887–928, 2009.

137. I. Karafyllis and Z. P. Jiang. *Stability and Stabilization of Nonlinear Systems.* Springer, London, 2011.

138. I. Karafyllis and Z. P. Jiang. A vector small-gain theorem for general nonlinear control systems. *IMA Journal of Mathematical Control and Information*, 28:309–344, 2011.

139. I. Karafyllis and Z. P. Jiang. Global stabilization of nonlinear systems based on vector control Lyapunov functions. *IEEE Transactions on Automatic Control*, 58:2550–2562, 2013.

140. I. Karafyllis, C. Kravaris, L. Syrou, and G. Lyberatos. A vector Lyapunov function characterization of input-to-state stability with application to robust global stabilization of the chemostat. *European Journal of Control*, 14:47–61, 2008.

141. I. Karafyllis and J. Tsinias. Non-uniform in time input-to-state stability and the small-gain theorem. *IEEE Transactions on Automatic Control*, 49:196–216, 2004.

142. C. Kellett and A. R. Teel. Smooth Lyapunov functions and robustness of stability for difference inclusions. *Systems & Control Letters*, 52:395–405, 2004.

143. C. Kellett and A. R. Teel. On the robustness of $\mathcal{KL}$-stability for difference inclusions: Smooth discrete-time Lyapunov functions. *SIAM Journal on Control and Optimization*, 44:777–800, 2005.

144. H. K. Khalil. *Nonlinear Systems.* Prentice-Hall, New Jersey, third edition, 2002.

145. H. Koeppl, D. Densmore, G. Setti, and M. di Bernardo, editors. *Design and Analysis of Biomolecular Circuits.* Springer, 2011.

146. N. N. Krasovskii. *Stability of Motion.* Stanford University Press, 1963.

147. P. Krishnamurthy and F. Khorrami. Decentralized control and disturbance attenuation for large-scale nonlinear systems in generalized output-feedback canonical form. *Automatica*, 39:1923–1933, 2003.

148. P. Krishnamurthy, F. Khorrami, and Z. P. Jiang. Global output feedback tracking for nonlinear systems in generalized output-feedback canonical form. *IEEE Transactions on Automatic Control*, 47:814–819, 2002.

149. M. Krstić. *Delay Compensation for Nonlinear, Adaptive, and PDE Systems.* Birkhäuser, Boston, 2009.

150. M. Krstić. Input delay compensation for forward complete and feedforward nonlinear systems. *IEEE Transactions on Automatic Control*, 55:287–303, 2010.

151. M. Krstić and H. Deng. *Stabilization of Nonlinear Uncertain Systems*. Springer-Verlag, London, 1998.

152. M. Krstić, D. Fontaine, P. V. Kokotović, and J. Paduano. Useful nonlinearities and global bifurcation control of jet engine stall and surge. *IEEE Transactions on Automatic Control*, 43:1739–1745, 1998.

153. M. Krstić, I. Kanellakopoulos, and P. V. Kokotović. *Nonlinear and Adaptive Control Design*. John Wiley & Sons, New York, 1995.

154. D. S. Laila and D. Nešić. Lyapunov based small-gain theorem for parameterized discrete-time interconnected ISS systems. *IEEE Transactions on Automatic Control*, 48:1783–1788, 2003.

155. V. Lakshmikantham and S. Leela. *Differential and Integral Inequalities: Theory and Applications. Volume I–Ordinary Differential Equations*. Academic Press, New York, 1969.

156. V. Lakshmikantham, S. Leela, and A. A. Martyuk. *Practical Stability of Nonlinear Systems*. World Scientific, New Jersey, 1990.

157. Y. Lan, G. Yan, and Z. Lin. Synthesis of distributed control of coordinated path following. *IEEE Transactions on Automatic Control*, 56:1170–1175, 2011.

158. J. LaSalle and S. Lefschetz. *Stability by Liapunov's Direct Method with Applications*. Academic Press, New York, 1961.

159. Y. S. Ledyaev and E. D. Sontag. A Lyapunov characterization of robust stabilization. *Nonlinear Analysis*, 37:813–840, 1999.

160. M. D. Lemmon. Event-triggered feedback in control, estimation, and optimization. In A. Bemporad, M. Heemels, and M. Johansson, editors, *Networked Control Systems*, Lecture Notes in Control and Information Sciences, pages 293–358, Berlin, 2010. Springer-Verlag.

161. Q. Li and Z. P. Jiang. Flocking control of multi-agent systems with application to nonholonomic multi-robots. *Kybernetika*, 45:84–100, 2009.

162. T. Li, M. Fu, L. Xie, and J. F. Zhang. Distributed consensus with limited communication data rate. *IEEE Transactions on Automatic Control*, 56:279–292, 2011.

163. D. Liberzon. Hybrid feedback stabilization of systems with quantized signals. *Automatica*, 39:1543–1554, 2003.

164. D. Liberzon. *Switching in Systems and Control*. Birkhäuser, Boston, 2003.

165. D. Liberzon. Quantization, time delays, and nonlinear stabilization. *IEEE Transactions on Automatic Control*, 51:1190–1195, 2006.

166. D. Liberzon. Observer-based quantized output feedback control of nonlinear systems. In *Proceedings of the 17th IFAC World Congress*, pages 8039–8043, 2008.

167. D. Liberzon and D. Nešić. Stability analysis of hybrid systems via small-gain theorems. In J. P. Hespanha and A. Tiwari, editors, *Proceedings of the 9th International Workshop on Hybrid Systems: Computation and Control*, Lecture Notes in Computer Science, pages 421–435, Berlin, 2006. Springer.

168. D. Liberzon and D. Nešić. Input-to-state stabilization of linear systems with quantized state measurements. *IEEE Transactions on Automatic Control*,

52:767–781, 2007.

169. D. Liberzon, D. Nešić, and A. R. Teel. Lyapunov-based small-gain theorems for hybrid systems. Submitted to *IEEE Transactions on Automatic Control*.

170. Y. Lin, E. D. Sontag, and Y. Wang. A smooth converse Lyapunov theorem for robust stability. *SIAM Journal on Control and Optimization*, 34:124–160, 1996.

171. Z. Lin, B. Francis, and M. Maggiore. Necessary and sufficient graphical conditions for formation control of unicycles. *IEEE Transactions on Automatic Control*, 50:121–127, 2005.

172. Z. Lin, B. Francis, and M. Maggiore. State agreement for continuous-time coupled nonlinear systems. *SIAM Journal on Control and Optimization*, 46:288–307, 2007.

173. B. Liu and D. J. Hill. Stability via hybrid-event-time Lyapunov function and impulsive stabilization for discrete-time delayed switched systems. *SIAM Journal on Control and Optimization*, 2013. In press.

174. B. Liu, D. J. Hill, and K. L. Teo. Input-to-state stability for a class of hybrid dynamical systems via hybrid time approach. In *Proceedings of the 48th IEEE Conference on Decision and Control and 28th Chinese Control Conference*, pages 3926–3931, 2009.

175. J. Liu and N. Elia. Quantized feedback stabilization of non-linear affine systems. *International Journal of Control*, 77:239–249, 2004.

176. S. Liu, L. Xie, and H. Zhang. Distributed consensus for multi-agent systems with delays and noises in transmission channels. *Automatica*, 47:920–934, 2011.

177. T. Liu. *Input-to-State Stability Methods for Nonlinear and Network Systems*. PhD thesis, The Australian National University, 2011.

178. T. Liu, D. J. Hill, and Z. P. Jiang. Lyapunov formulation of ISS cyclic-small-gain in continuous-time dynamical networks. *Automatica*, 47:2088–2093, 2011.

179. T. Liu, D. J. Hill, and Z. P. Jiang. Lyapunov formulation of the large-scale, ISS cyclic-small-gain theorem: The discrete-time case. *Systems & Control Letters*, 61:266–272, 2012.

180. T. Liu and Z. P. Jiang. Distributed nonlinear control of mobile autonomous multi-agents. Submitted to *Automatica*.

181. T. Liu and Z. P. Jiang. Event-triggered control of nonlinear systems with partial state feedback. Submitted to *Automatica*.

182. T. Liu and Z. P. Jiang. A small-gain approach to robust event-triggered control of nonlinear systems. Submitted to *IEEE Transactions on Automatic Control*.

183. T. Liu and Z. P. Jiang. Distributed control of nonlinear uncertain systems: A cyclic-small-gain approach. *Acta Automatica Sinica*, 2013. In press.

184. T. Liu and Z. P. Jiang. Distributed formation control of nonholonomic mobile robots without global position measurements. *Automatica*, 49:592–600, 2013.

185. T. Liu and Z. P. Jiang. Distributed output-feedback control of nonlinear multi-agent systems. *IEEE Transactions on Automatic Control*, 58:2912–2917, 2013.

186. T. Liu, Z. P. Jiang, and D. J. Hill. Robust control of nonlinear strict-feedback systems with measurement errors. In *Proceedings of the 50th IEEE Conference on Decision and Control and European Control Conference*, pages 2034–2039,

2011.

187. T. Liu, Z. P. Jiang, and D. J. Hill. Decentralized output-feedback control of large-scale nonlinear systems with sensor noise. *Automatica*, 48:2560–2568, 2012.

188. T. Liu, Z. P. Jiang, and D. J. Hill. Lyapunov formulation of the ISS cyclic-small-gain theorem for hybrid dynamical networks. *Nonlinear Analysis: Hybrid Systems*, 6:988–1001, 2012.

189. T. Liu, Z. P. Jiang, and D. J. Hill. Quantized stabilization of strict-feedback nonlinear systems based on ISS cyclic-small-gain theorem. *Mathematics of Control, Signals, and Systems*, 24:75–110, 2012.

190. T. Liu, Z. P. Jiang, and D. J. Hill. A sector bound approach to feedback control of nonlinear systems with state quantization. *Automatica*, 48:145–152, 2012.

191. T. Liu, Z. P. Jiang, and D. J. Hill. Small-gain based output-feedback controller design for a class of nonlinear systems with actuator dynamic quantization. *IEEE Transactions on Automatic Control*, 57:1326–1332, 2012.

192. Z. Liu and L. Guo. Synchronization of multi-agent systems without connectivity assumptions. *Automatica*, 45:2744–2753, 2009.

193. W. Lohmiller and J.-J. E. Slotine. On contraction analysis for non-linear systems. *Automatica*, 34:683–696, 1998.

194. Q. Lu, Y. Sun, and S. Mei. *Nonlinear Control Systems and Power System Dynamics*. Kluwer Academic, 2001.

195. A. M. Lyapunov. *The General Problem of the Stability of Motion*. Taylor & Francis, London, 1992.

196. M. Maggiore and K. Passino. Output feedback control for stabilizable and incompletely observable nonlinear systems: Jet engine stall and surge control. In *Proceedings of the 2000 American Control Conference*, pages 3626–3630, 2000.

197. M. Malisoff and F. Mazenc. *Constructions of Strict Lyapunov Functions*. Springer-Verlag, London, 2009.

198. I. M. Y. Mareels and R. R. Bitmead. Non-linear dynamics in adaptive control: Chaotic and periodic stabilization. *Automatica*, 22:641–655, 1986.

199. I. M. Y. Mareels and D. J. Hill. Monotone stability of nonlinear feedback systems. *Journal of Mathematical Systems, Estimation, and Control*, 2:275–291, 1992.

200. S. J. Mason. Feedback theory–Some properties of signal flow graphs. *Proceedings of the IRE*, 41:1144–1156, 1953.

201. S. J. Mason. Feedback theory–Further properties of signal flow graphs. *Proceedings of the IRE*, 44:920–926, 1956.

202. F. Mazenc and Z. P. Jiang. Global output feedback stabilization of a chemostat with an arbitrary number of species. *IEEE Transactions on Automatic Control*, 55:2570–2575, 2010.

203. F. Mazenc, M. Malisoff, and J. Harmand. Further results on stabilization of periodic trajectories for a chemostat with two species. *IEEE Transactions on Automatic Control*, 53:66–74, 2008.

204. M. Mesbahi and M. Egerstedt. *Graph Theoretic Methods in Multiagent Networks*. Princeton University Press, New Jersey, 2010.

205. A. N. Michel and R. K. Miller. *Qualitative Analysis of Large Scale Dynamical*

Systems. Academic Press, New York, 1977.

206. F. K. Moore and E. M. Greitzer. A theory of post-stall transients in axial compression systems-Part I: Development of equations. *Journal of Engineering for Gas Turbines and Power*, 108:68–76, 1986.

207. P. J. Moylan and D. J. Hill. Stability criteria for large-scale systems. *IEEE Transactions on Automatic Control*, 23:143–149, 1978.

208. D. Nešić and D. Liberzon. A small-gain approach to stability analysis of hybrid systems. In *Proceedings of the 44th IEEE Conference on Decision and Control*, pages 5409–5414, 2005.

209. D. Nešić and D. Liberzon. A unified framework for design and analysis of networked and quantized control systems. *IEEE Transactions on Automatic Control*, 54:732–747, 2009.

210. D. Nešić and A. R. Teel. Changing supply functions in input to state stable systems: The discrete-time case. *IEEE Transactions on Automatic Control*, 46:960–962, 2001.

211. D. Nešić and A. R. Teel. A Lyapunov-based small-gain theorem for hybrid ISS systems. In *Proceedings of the 47th IEEE Conference on Decision and Control*, pages 3380–3385, 2008.

212. P. Ögren, M. Egerstedt, and X. Hu. A control Lyapunov function approach to multiagent coordination. *IEEE Transactions on Robotics and Automation*, 18:847–851, 2002.

213. P. Ögren, E. Fiorelli, and N. Leonard. Cooperative control of mobile sensor networks: Adaptive gradient climbing in a distributed network. *IEEE Transactions on Automatic Control*, 49:1292–1302, 2004.

214. R. Olfati-Saber. Flocking for multi-agent dynamic systems: Algorithms and theory. *IEEE Transactions on Automatic Control*, 51:401–420, 2006.

215. R. Olfati-Saber and R. M. Murray. Consensus problems in networks of agents with switching topology and time-delays. *IEEE Transactions on Automatic Control*, 49:1520–1533, 2004.

216. R. Pates and G. Vinnicombe. Stability certificates for networks of heterogeneous linear systems. In *Proceedings of the 51st IEEE Conference on Decision and Control*, pages 6915–6920, 2012.

217. P. Pepe. On the asymptotic stability of coupled delay differential and continuous time difference equations. *Automatica*, 41:107–112, 2005.

218. P. Pepe and H. Ito. On saturation, discontinuities, and delays, in iISS and ISS feedback control redesign. *IEEE Transactions on Automatic Control*, 57:1125–1140, 2012.

219. P. Pepe, I. Karafyllis, and Z. P. Jiang. On the Liapunov-Krasovskii methodology for the ISS of systems described by coupled delay differential and difference equations. *Automatica*, 44:2266–2273, 2008.

220. R. Postoyan and D. Nešić. Trajectory based small gain theorems for parameterized systems. In *Proceedings of 2010 American Control Conference*, pages 184–189, 2010.

221. H. G. Potrykus, F. Allgöwer, and S. J. Qin. The character of an idempotent-analytic nonlinear small gain theorem. In L. Benvenuti, A. De Santis, and L. Farina, editors, *Positive Systems*, volume 294 of *Lecture Notes in Control and Information Science*, pages 361–368, Berlin, 2003. Springer.

222. L. Praly and Z. P. Jiang. Stabilization by output feedback for systems with

ISS inverse dynamics. *Systems & Control Letters*, 21:19–33, 1993.

223. L. Praly and Y. Wang. Stabilization in spite of matched unmodeled dynamics and an equivalent definition of input-to-state stability. *Mathematics of Control, Signals and Systems*, 9:1–33, 1996.

224. Z. Qu. *Cooperative Control of Dynamical Systems: Applications to Autonomous Vehicles*. Springer-Verlag, 2009.

225. Z. Qu, J. Wang, and R. A. Hull. Cooperative control of dynamical systems with application to autonomous vehicles. *IEEE Transactions on Automatic Control*, 53:894–911, 2008.

226. A. Rantzer. Distributed performance analysis of heterogeneous systems. In *Proceedings of the 49th IEEE Conference on Decision and Control*, pages 2682–2685, 2010.

227. W. Ren. On consensus algorithms for double-integrator dynamics. *IEEE Transactions on Automatic Control*, 53:1503–1509, 2008.

228. W. Ren and R. Beard. *Distributed Consensus in Multi-vehicle Cooperative Control*. Springer-Verlag, London, 2008.

229. W. Ren, R. Beard, and E. Atkins. Information consensus in multivehicle cooperative control. *IEEE Control Systems Magazine*, 27:71–82, 2007.

230. R. Rifford. Existence of Lipschitz and semiconcave control-Lyapunov functions. *SIAM Journal on Control and Optimization*, 39:1043–1064, 2000.

231. B. S. Rüffer. *Monotone Systems, Graphs, and Stability of Large-Scale Interconnected Systems*. PhD thesis, University of Bremen, Germany, 2007.

232. I. W. Sandberg. On the $\mathcal{L}_2$-boundedness of solutions of nonlinear functional equations. *Bell System Technical Journal*, 43:1581–1599, 1964.

233. I. W. Sandberg. An observation concerning the application of the contraction mapping fixed-point theorem and a result concerning the norm-boundedness of solutions of nonlinear functional equations. *Bell System Technical Journal*, 44:1809–1812, 1965.

234. A. V. Savkin and R. J. Evans. *Hybrid Dynamical Systems: Controller and Sensor Switching Problems*. Birkhäuser, Boston, 2002.

235. R. Sepulchre, M. Jankovic, and P. V. Kokotović. *Constructive Nonlinear Control*. Springer-Verlag, New York, 1997.

236. G. Shi and Y. Hong. Global target aggregation and state agreement of nonlinear multi-agent systems with switching topologies. *Automatica*, 45:1165–1175, 2009.

237. L. Shi and S. K. Singh. Decentralized control for interconnected uncertain systems: Extensions to high-order uncertainties. *International Journal of Control*, 57:1453–1468, 1993.

238. H. Shim and A. R. Teel. Asymptotic controllability and observability imply semiglobal practical asymptotic stabilizability by sampled-data output feedback. *Automatica*, 39:441–454, 2003.

239. D. D. Šiljak. *Decentralized Control of Complex Systems*. Academic Press, Boston, 1991.

240. D. D. Šiljak. Dynamic graphs. *Nonlinear Analysis: Hybrid Systems*, 2:544–567, 2008.

241. E. D. Sontag. Smooth stabilization implies coprime factorization. *IEEE Transactions on Automatic Control*, 34:435–443, 1989.

242. E. D. Sontag. Further facts about input to state stabilization. *IEEE Trans-*

actions on Automatic Control, 35:473–476, 1990.

243. E. D. Sontag. Comments on integral variants of ISS. *Systems & Control Letters*, 34:93–100, 1998.

244. E. D. Sontag. *Mathematical Control Theory: Deterministic Finite-Dimensional Systems*. Springer, New York, second edition, 1998.

245. E. D. Sontag. Clocks and insensitivity to small measurement errors. *ESAIM: Control, Optimisation and Calculus of Variations*, 4:537–557, 1999.

246. E. D. Sontag. Asymptotic amplitudes and Cauchy gains: A small-gain principle and an application to inhibitory biological feedback. *Systems & Control Letters*, 47:167–179, 2002.

247. E. D. Sontag. Input to state stability: Basic concepts and results. In P. Nistri and G. Stefani, editors, *Nonlinear and Optimal Control Theory*, pages 163–220, Berlin, 2007. Springer-Verlag.

248. E. D. Sontag and A. R. Teel. Changing supply functions in input/state stable systems. *IEEE Transactions on Automatic Control*, 40:1476–1478, 1995.

249. E. D. Sontag and Y. Wang. On characterizations of the input-to-state stability property. *Systems & Control Letters*, 24:351–359, 1995.

250. E. D. Sontag and Y. Wang. New characterizations of input-to-state stability. *IEEE Transactions on Automatic Control*, 41:1283–1294, 1996.

251. E. D. Sontag and Y. Wang. Notions of input to output stability. *Systems & Control Letters*, 38:351–359, 1999.

252. E. D. Sontag and Y. Wang. Lyapunov characterization of input to output stability. *SIAM Journal on Control and Optimization*, 39:226–249, 2001.

253. H. Su, X. Wang, and Z. Lin. Flocking of multi-agents with a virtual leader. *IEEE Transactions on Automatic Control*, 54:293–307, 2009.

254. Y. Su and J. Huang. Cooperative output regulation of linear multi-agent systems. *IEEE Transactions on Automatic Control*, 57:1062–1066, 2012.

255. P. Tabuada. Event-triggered real-time scheduling of stabilizing control tasks. *IEEE Transactions on Automatic Control*, 52:1680–1685, 2007.

256. H. Tanner, A. Jadbabaie, and G. Pappas. Stable flocking of mobile agents, Part I: Fixed topology. In *Proceedings of the 42nd IEEE Conference on Decision and Control*, pages 2010–2015, 2003.

257. H. Tanner, A. Jadbabaie, and G. Pappas. Stable flocking of mobile agents, Part II: Dynamic topology. In *Proceedings of the 42nd IEEE Conference on Decision and Control*, pages 2016–2021, 2003.

258. H. G. Tanner, G. J. Pappas, and V. Kummar. Leader-to-formation stability. *IEEE Transactions on Robotics and Automation*, 20:443–455, 2004.

259. A. R. Teel. Input-to-state stability and the nonlinear small-gain theorem. Private communications.

260. A. R. Teel. A nonlinear small gain theorem for the analysis of control systems with saturation. *IEEE Transactions on Automatic Control*, 41:1256–1270, 1996.

261. A. R. Teel. Connections between Razumikhin-type theorems and the ISS nonlinear small gain theorem. *IEEE Transactions on Automatic Control*, 43:960–964, 1998.

262. A. R. Teel and L. Praly. Tools for semiglobal stabilization by partial state and output feedback. *SIAM Journal on Control and Optimization*, 33:1443–1488, 1995.

263. A. R. Teel and L. Praly. On assigning the derivative of a disturbance attenuation control Lyapunov function. *Mathematics of Control, Signals, and Systems*, 13:95–124, 2000.

264. A. R. Teel and L. Praly. A smooth Lyapunov function from a class-$\mathcal{KL}$ estimate involving two positive functions. *ESAIM Control Optimisation and Calculus of Variations*, 5:313–367, 2000.

265. S. Tiwari, Y. Wang, and Z. P. Jiang. Nonlinear small-gain theorems for large-scale time-delay systems. *Dynamics of Continuous, Discrete and Impulsive Systems Series A: Mathematical Analysis*, 19:27–63, 2012.

266. J. Tsinias. Sufficient Lyapunov-like conditions for stabilization. *Mathematics of Control, Signals, and Systems*, 2:343–357, 1989.

267. J. Tsinias and N. Kalouptsidis. Output feedback stabilization. *IEEE Transactions on Automatic Control*, 35:951–954, 1990.

268. A. J. van der Schaft. *$\mathcal{L}_2$-Gain and Passivity Techniques in Nonlinear Control*. Lecture Notes in Control and Information Sciences. Springer-Verlag, London, 1996.

269. V. I. Vorotnikov. *Partial Stability and Control*. Birkhäuser, Boston, 1998.

270. C. Wang, S. S. Ge, D. J. Hill, and G. Chen. An ISS-modular approach for adaptive neural control of pure-feedback systems. *Automatica*, 42:723–731, 2006.

271. X. Wang, Y. Hong, J. Huang, and Z. P. Jiang. A distributed control approach to a robust output regulation problem for multi-agent systems. *IEEE Transactions on Automatic Control*, 55:2891–2895, 2010.

272. X. Wang and M. D. Lemmon. On event design in event-triggered feedback systems. *Automatica*, 47:2319–2322, 2011.

273. C. Wen and J. Zhou. Decentralized adaptive stabilization in the presence of unknown backlash-like hysteresis. *Automatica*, 43:426–440, 2007.

274. C. Wen, J. Zhou, and W. Wang. Decentralized adaptive backstepping stabilization of interconnected systems with dynamic input and output interaction. *Automatica*, 45:55–67, 2009.

275. P. Wieland, R. Sepulchre, and F. Allgöwer. An internal model principle is necessary and sufficient for linear output synchronization. *Automatica*, 47:1068–1074, 2011.

276. J. C. Willems. The generation of Lyapunov functions for input-output stable systems. *SIAM Journal on Control*, 9:105–134, 1971.

277. J. C. Willems. Dissipative dynamical systems, Part I: General theory; Part II: Linear systems with quadratic supply rates. *Archive for Rationale Mechanics Analysis*, 45:321–393, 1972.

278. J. C. Willems. The behavioral approach to open and interconnected systems. *IEEE Control Systems Magazine*, 27:46–99, 2007.

279. S. Xie and L. Xie. Decentralized global robust stabilization of a class of large-scale interconnected minimum-phase nonlinear systems. In *Proceedings of the 37th IEEE Conference on Decision and Control*, pages 1482–1487, 1998.

280. H. Xin, Z. Qu, J. Seuss, and A. Maknouninejad. A self organizing strategy for power flow control of photovoltaic generators in a distribution network. *IEEE Transactions on Power Systems*, 26:1462–1473, 2011.

281. V. A. Yakubovich, G. A. Leonov, and A. Kh. Gelig. *Stability of Stationary Sets in Control Systems with Discontinuous Nonlinearities*. World Scientific

Press, 2004.

282. C. Yu, B. D. O. Anderson, S. Dasgupta, and B. Fidan. Control of minimally persistent formations in the plane. *SIAM Journal on Control and Optimization*, 48:206–233, 2009.

283. G. Zames. On the input-output stability of time-varying nonlinear feedback systems–Part I: Conditions derived using concepts of loop gain, conicity, and positivity. *IEEE Transactions on Automatic Control*, 11:228–238, 1966.

284. G. Zames. On the input-output stability of time-varying nonlinear feedback systems–Part II: Conditions involving circles in the frequency plane and sector nonlinearities. *IEEE Transactions on Automatic Control*, 11:465–476, 1966.

285. A. Zečević and D. D. Šiljak. Control design with arbitrary information structure constraints. *Automatica*, 44:2642–2647, 2008.

286. A. Zečević and D. D. Šiljak. *Control of Complex Systems: Structural Constraints and Uncertainty*. Springer, 2010.

287. J. Zhao, D. J. Hill, and T. Liu. Synchronization of dynamical networks with nonidentical nodes: Criteria and control. *IEEE Transactions on Circuits and Systems–I: Regular Papers*, 58:584–594, 2011.

288. K. Zhou, J. C. Doyle, and K. Glover. *Robust and Optimal Control*. Prentice Hall, 1995.

Index